NONLINEAR ANALYSIS AND VARIATIONAL PROBLEMS

Springer Optimization and Its Applications

VOLUME 35

Aims and Scope

Optimization has been expanding in all directions at an astonishing rate during the last few decades. New algorithmic and theoretical techniques have been developed, the diffusion into other disciplines has proceeded at a rapid pace, and our knowledge of all aspects of the field has grown even more profound. At the same time, one of the most striking trends in optimization is the constantly increasing emphasis on the interdisciplinary nature of the field. Optimization has been a basic tool in all areas of applied mathematics, engineering, medicine, economics and other sciences.

The series *Optimization and Its Applications* publishes undergraduate and graduate textbooks, monographs and state-of-the-art expository works that focus on algorithms for solving optimization problems and also study applications involving such problems. Some of the topics covered include nonlinear optimization (convex and nonconvex), network flow problems, stochastic optimization, optimal control, discrete optimization, multiobjective programming, description of software packages, approximation techniques and heuristic approaches.

NONLINEAR ANALYSIS AND VARIATIONAL PROBLEMS

In Honor of George Isac

Edited By

PANOS M. PARDALOS
Department of Industrial and Systems Engineering,
University of Florida,
Gainesville, Florida

THEMISTOCLES M. RASSIAS
Department of Mathematics,
National Technical University of Athens,
Athens, Greece

AKHTAR A. KHAN
School of Mathematical Sciences,
Rochester Institute of Technology,
Rochester, New York

Springer

Editors

Panos M. Pardalos
Department of Industrial & Systems Engineering
University of Florida
303 Weil Hall
Gainesville FL 32611-6595
USA
pardalos@ise.ufl.edu

Akhtar A. Khan
School of Mathematical Sciences
Rochester Institute of Technology
85 Lomb Memorial Drive
Rochester NY 14623-5602
USA
aaksma@rit.edu

Themistocles M. Rassias
Department of Mathematics
National Technical University of Athens
Zagoras Street 4
151 25 Athens
Paradissos, Amaroussion
Greece

ISSN 1931-6828
ISBN 978-1-4614-2481-9 e-ISBN 978-1-4419-0158-3
DOI 10.1007/978-1-4419-0158-3
Springer New York Dordrecht Heidelberg London

Mathematics Subject Classification (2000): 47H10, 90C33, 90C29

Cover illustration: Photo entitled "Back to Creation" taken by Elias Tyligadas

Printed on acid-free paper

Springer is part of Springer Science+Business Media (www.springer.com)

With our deepest appreciation, we dedicate this volume to the memory of our dearest friend and eminent mathematician, George Isac

Preface

The papers published in this volume focus on some of the most recent developments in complementarity theory, variational principles, stability theory of functional equations, nonsmooth optimization, and various other important topics of nonlinear analysis and optimization.

This volume was initially planned to celebrate Professor George Isac's 70th birthday by bringing together research scientists from mathematical domains which have long benefited from Isac's active research passion. Unfortunately, George Isac passed away in February 2009 at the age of 69.

George Isac received his Ph.D. in 1973 from the Institute of Mathematics of the Romanian Academy of Sciences. He made outstanding contributions in several branches of pure and applied mathematics, including complementarity theory, variational inequalities, fixed point theory, scalar and vector optimization, theory of cones, eigenvalue problems, convex analysis, variational principles and regularization methods, as well as a number of other topics. In his long and outstanding career, he wrote more than 200 papers and 13 books. Professor Isac was an avid traveler who visited more than 70 universities around the globe and delivered approximately 180 research presentations. He also authored seven books on poetry. During his scientific career he collaborated with numerous mathematicians. His research papers contain very deep, original and beautiful results. Through his significant contributions, he earned a distinguished position and became an internationally renowned leading scholar in his research fields. Professor Isac's prolific career was supported by the love and affection of his wife, Viorica. In fact her dedication, was so strong that she typed most of Isac's manuscripts for his papers and books. We offer our sincerest sympathies to Viorica Isac on her monumental loss. Her husband was not only a wonderful mathematician but also a outstanding human being who will be greatly missed.

The submitted works of eminent research scientists from the international mathematical community are dedicated to the memory of this leading mathematician and very special colleague and friend, George Isac.

The contributions are organized into two parts. Part I focuses on selected topics in nonlinear analysis, in particular, stability issues for functional equations, and fixed point theorems.

In Chapter 1, Agratini and Andrica present a survey focusing on linear positive operators having the degree of exactness null and fixing the monomial of the second degree.

In their contribution, Amyari and Sadeghi present a Mazur–Ulam type theorem in non-Archimedean strictly convex 2-normed spaces and give some properties of mappings on non-Archimedean strictly 2-convex 2-normed spaces.

The emphasis of Cădariu and Radu is on extending some results of Isac and Rassias on ψ-additive mappings by giving a stability theorem for functions defined on generalized α-normed spaces and taking values in β-normed spaces.

The objective of Constantinescu's contribution is on some investigates of W*-tensor products of W*-algebras.

In his contribution, Dragomir introduces a perturbed version of the median principle and presents its applications for various Riemann–Stieltjes integral and Lebesgue integral inequalities.

M. Eshaghi-Gordji et al. undertake the issues related to the stability of a mixed type additive, quadratic, cubic and quartic functional equation.

A short survey about the Hyers–Ulam stability of ψ-additive mappings is given by Găvruţa and Găvruţa in Chapter 7.

In their contribution, Jun and Kim investigate the generalized Hyers–Ulam stability problem for quadratic functional equations in several variables and obtain an asymptotic behavior of quadratic mappings on restricted domains.

The focus of Jung and Rassias is to apply the fixed point method for proving the Hyers–Ulam–Rassias stability of a logarithmic functional equation.

In their work, Park and Cui use the fixed point method to prove the generalized Hyers–Ulam stability of homomorphisms in C^*-algebras and Lie C^*-algebras and of derivations of C^*-algebras and Lie C^*-algebras for the 3-variable Cauchy functional equation.

In their paper, Park and Rassias use the fixed point method to prove the generalized Hyers–Ulam stability of certain functional equations in real Banach spaces.

The focus of Precup is on presenting new compression and expansion type critical point theorems in a conical shell of a Hilbert space identified with its dual.

The aim of Rus's contribution is to give some Hyers–Ulam–Rassias stability results for Volterra and Fredholm integral equations by using some Gronwall lemmas.

In his contribution, Turinici presents a detailed study of Brezis–Browder's principle. He shows that the version of Brezis–Browder's principle for general separable sets is a logical equivalent of the Zorn–Bourbaki maximality result. In addition, several other interesting connections are established.

The second part of this volume discusses several important aspects of vector optimization and non-smooth optimization, as well as variational problems.

In Chapter 15, Balaj and O'Regan make use of the Kakutani–Fan–Glicksberg fixed point theorem to give an existence theorem for a generalized vector quasi-equilibrium problem.

In their contribution, Cojocaru et al. give a new method of tracking the dynamics of an equilibrium problem using an evolutionary variational inequalities and hybrid dynamical systems approach. They apply their approach to describe the time evolution of a differentiated product market model under incentive policies with a finite life span.

In Chapter 17, Daniele et al. give an overview of recent developments in the theory of generalized projections both in non-pivot Hilbert spaces and strictly convex and smooth Banach spaces. They also study the equivalence between solutions of variational inequalities and critical points of projected dynamical systems.

Eichfelder and Jahn's aim is to present various foundations of a new field of research in optimization unifying semidefinite and copositive programming, called set-semidefinite optimization.

Giannessi and Khan extend the notion of image of a variational inequality by introducing the notion of an envelope for a variational inequality.

In Chapter 20, Goeleven develops a new approach to study a class of nonlinear generalized ordered complementarity problems.

Ha's chapter presents a unified framework for the study of strong efficient solutions, weak efficient solutions, positive proper efficient solutions, Henig global proper efficient solutions, Henig proper efficient solutions, super-efficient solutions, Benson proper efficient solutions, Hartley proper efficient solutions, Hurwicz proper efficient solutions and Borwein proper efficient solutions of a set-valued optimization problem with or without constraints.

The contribution of Isac and Németh presents some mean value theorems for the scalar derivatives which are then used to develop a new method applicable to the study of the existence of nontrivial solutions of complementarity problems.

The chapter by Isac and Tammer presents new necessary conditions for approximate solutions of vector-valued optimization problems in general spaces by introducing an axiomatic approach for a scalarization scheme.

Lukkassen et al. undertake homogenization of sequences of integral functionals with natural growth conditions. Some means are analyzed and used to discuss some fairly new bounds for the homogenized integrand corresponding to integrands which are periodic in the spatial variable. Several applications are given.

In their contribution, Moldovan and Gowda employ duality and complementarity ideas and Z-transformations as well as discuss equivalent ways of describing the existence of common linear/quadratic Lyapunov functions for switched linear systems.

Motreanu's focus is on the necessary conditions of optimality for general mathematical programming problems on a product space. Interesting applications to an optimal control problem governed by an elliptic differential inclusion are given.

In his contribution, Pascali's focus is on studying variational inequalities with S-mappings.

In Chapter 28, a new completely generalized co-complementarity problem for fuzzy mappings is introduced. By using the definitions of p-relaxed accretive and p-strongly accretive mappings, the authors propose an iterative algorithm for computing the approximate solutions, and establish its convergence.

The contribution of Wolkowicz is aimed to illustrate how optimization can be used to derive known and new theoretical results about perturbations of matrices and sensitivity of eigenvalues.

It is our immense pleasure to express our utmost and deepest gratitude to all of the scientists who, by their works, participated in this tribute to honor Professor George Isac. We are grateful to the referees of the enclosed contributions. One of the editors (AAK) expresses his sincere gratitude to Prof. Sophia Maggelakis and Prof. Patricia Clark of RIT and Prof. G. Jailan Zalmai of NMU for their kindness and support.

Panos M. Pardalos
Themistocles M.Rassias
Akhtar A. Khan
May, 2009

Biographical Sketch of George Isac

George Isac (1940–2009)

George Isac was born on April 1, 1940, in Filipesti, Romania, a village in the district of Braila. His father was a village schoolteacher who fought in the First World War where he lost an arm. His mother was a housewife. George was the youngest of three children—he had an older sister and an older brother.

George enjoyed a happy childhood as evoked in his nostalgia-filled poems. His poetry has had much success in Romanian communities all over the world. It was said that on the first day of class, on the way to school, his father told him: I would like you to be the best student in your class. Young George took his father's wish as an order and he strived with all his power to maintain this status during all of his student years. He remained deeply attached not only to his birth village but also to the village school. After the fall of the communist regime in Romania, he visited the school several times and funded a scholarship to be granted every year to the best student.

In 1955, George was admitted to the very prestigious college Nicolaie Balcescu in Braila. The impressive number of great Romanian intellectuals who have graduated from that college is certainly due to the fact that the college had exceptionally good teachers. One of them was the Romanian literature teacher who believed that George was a real artistic gem and who was certain that George would follow a higher education in the arts. However, George realized that under the communist regime, it was too difficult for people to succeed in such a career without making major moral concessions and he was not willing to make such concessions.

So it was, facing the prospect of producing a big disappointment to his favorite teacher, he made the dramatic decision to enroll in the mathematics program at the University of Bucharest. Still, he returned every year, full of emotion, to his beloved college in Braila to long walks in Braila's parks and along the banks of the Danube River. It was there that he met Viorica Georgescu, who on May 8, 1965, became his beloved wife and his inspiration for countless love poems that were filled with enormous gratitude to her. Viorica gave him the most precious gift one could receive: two wonderful children: Catalin and Roxana—the pride of the Isac family.

At the University of Bucharest, George was a remarkable student and upon graduation he was offered a position in the Department of Analysis, whose chairman at that time was George Marinescu. There was an immediate and deep chemistry between the two Georges and they soon began to work together on a pioneering book on analysis on ultra-metric fields (published in 1976). George Isac had wonderful memories regarding his mentor George Marinescu and started to write a biography about him that unfortunately remains unfinished.

George and a friend, Ion Ichim, were working part-time at the Institute of Mathematics, where I was a senior researcher nominated to direct doctoral studies. They both came to my office one day and asked me to accept them as students in a Ph.D. program. Knowing their value, I joyfully accepted them, as well as their subjects of research: the area of functional analysis for George and the area of potential theory for Ion.

In the meantime, in February 1972 I left Romania. At that time, the work on their Ph.D. thesis was advanced but unfinished. Officially, Professor Marinescu was nominated to take charge of directing their thesis, but due to his health problems, my friend Aurel Cornea performed the real work. It was a pleasant job for him and he offered them not only his help but also his friendship after graduation.

George Isac was offered a contract with University of Kinshasa in Zaire and decided to accept it in order to escape to the West. Before his departure he paid a visit to Aurel Cornea and told him about his intentions, demonstrating great confidence in him, given the fact that the country was studded with secret police informers. He mentioned that his intention was to go alone at first given the tough conditions there and then bring his family after a while.

"Are you crazy?" asked Aurel. "Don't you know what a source of blackmail a family left behind is for the secret police?" George insisted that his plan was sound, since he didn't want to expose his family to hardship. Aurel walked thoughtfully through the room, then suddenly stopped and said: "This is what I have to say: Uncle George (a slightly ironical, yet kind address), take your family and go there. If it is too harsh, then put it on my account."

George obeyed Aurel's advice and fortunately had no regrets on his decision. He had an amazingly successful career in Canada, a country in which his family enjoyed every moment, and he was forever grateful to his friend Aurel for the advice to not leave them behind. Except for a short period of time at the University of Sherbrook, he always worked for the Royal Military College, first in Saint-Jean-sur-Richelieu, Quebec, and later in Kingston. He was also associated with Queen's University, where he directed graduate student work.

While he was still in Romania, he started to pay attention to applied mathematics, a field in which he was able to use his functional analysis knowledge. He taught some courses in that field. In Canada, George evolved in his field and he achieved many accomplishments in applied mathematics, solving problems of complementarity, fixed point theorems with applications to decision theory, game theory, optimal control, Pareto problems, and nonlinear analysis, etc. Some basic concepts of these domains such as nuclear cones have been introduced by him.

George's record of lifelong publications numbers over 200 papers and authorship or coauthorship of 13 books. There are around 640 quotations of his work in collaboration with 258 mathematicians. The world mathematics community sanctioned his mathematics contribution and he appeared as an invited speaker at countless international conferences and congresses. He was a member of the editorial committees for many mathematics periodicals. He received awards of excellence in mathematics, such as the "Spiru Haret" prize of the Romanian Academy of Science in 2003. He was also nominated for the title of Doctor Honoris Causa of the University Babeş-Bolyai, Cluj-Napoca, Romania.

One's cannot talk about George Isac without saying something related to his poems, which were an important component of his personality. He started to write poetry only later in his life, probably because of the stress he experienced or because of an excessively busy mathematics schedule that did not allow him the peace of mind necessary for such an activity. However, he carried with him all of his life a kind of poetic archive, which overflowed tempestuously when the right time came, producing an impressive seven volumes of poems in just one decade, the eighth volume waiting for posthumous publication.

George's poems are dominated by nostalgic memories of early childhood and adolescence. His birth village appears as the sacred place where forefathers' traditions are still alive with colors and scents specific to every season, with flowers, birds, bugs, with rivers, cemeteries, wheat fields, vineyards, forest hills, and the usual childhood preoccupations such as flower picking in spring or tobogganing in winter. A lot of poems are dedicated to his parents' house featuring an impressive garden and a very attentive mother. However, there are also dominant philosophical problems, such as those related to life and death, treated mainly through the sieve of ancient oriental philosophy that he studied thoroughly.

As the religious man he was, George tackled the problem of life after death, trying to use poetic metaphors in order to revive in us the shiver of the absolute truth. There is also advice to not give too much importance to the superficial aspects of life, but rather concentrate on those that are deep and essential. He makes an acid indictment of our modern society, which shows signs of moral decadence.

George's dreadful sickness surprised him while he was in full stride, which made it even worse. He had all kinds of mathematics projects in mind or under way: a new book, the biography of George Marinescu, and a book of personal recollections related to the communist era in Romania, including a recollection of his father's tragic experience. It was not meant to be, and I regret that these marvelous projects are now forever lost.

Through his mathematics research, through his poems and through his teaching, George Isac brought a lot of light into this world. He lived life fully, offering us a rich harvest, similar to that of his birthplace fields described in his poems. Now, a new name must be added to the long list of science or arts personalities on the crown of Nicolae Balcescu College: that of George Isac, renowned personality in science and arts.

Benglen, April 2009 *Corneliu Constantinescu*

Contents

List of Contributors

Octavian Agratini
Babeş-Bolyai University, Faculty of Mathematics and Computer Science, 400084 Cluj-Napoca, Romania, e-mail: agratini@math.ubbcluj.ro

Maryam Amyari
Department of Mathematics, Faculty of Science, Islamic Azad University-Mashhad Branch, Mashhad 91735, Iran, e-mail: amyari@mshdiau.ac.ir and maryam_amyari@yahoo.com

Tudor Andrica
Babeş-Bolyai University, Faculty of Mathematics and Computer Science, 400084 Cluj-Napoca, Romania, e-mail: tudor_an@yahoo.com

Mircea Balaj
Department of Mathematics, University of Oradea, 410087 Oradea, Romania, email: mbalaj@uoradea.ro and mbalaj@uoradea.ro

M. Cojocaru
Department of Mathematics and Statistics, University of Guelph, Guelph, ON, Canada, e-mail: mcojocar@uoguelph.ca

Liviu Cădariu
"Politehnica" University of Timişoara, Department of Mathematics, Piaţa Victoriei 2, 300006, Timişoara, Romania, e-mail: liviu.cadariu@mat.upt.ro and lcadariu@yahoo.com

Corneliu Constantinescu
Bodenacherstr. 53, CH 8121 Benglen, Switzerland, e-mail: constant@math.ethz.ch

Jianlian Cui
Department of Mathematical Sciences, Tsinghua University, Beijing 100084, P.R. China, e-mail: jcui@math.tsinghua.edu.cn

Patrizia Daniele
Department of Mathematics and Computer Science, University of Catania, Catania, Italy, e-mail: daniele@dmi.unict.it

S.S. Dragomir
Research Group in Mathematical Inequalities and Applications, School of Engineering and Science, Victoria University, P.O. Box 14428, Melbourne City, VIC, Australia 8001, e-mail: sever.dragomir@vu.edu.au

Gabriele Eichfelder
Department Mathematik, Universität Erlangen-Nürnberg, Martensstr. 3, D-91058 Erlangen, Germany, e-mail: Gabriele.Eichfelder@am.uni-erlangen.de

M. Eshaghi-Gordji
Department of Mathematics, Semnan University, P.O. Box 35195-363, Semnan, Iran, e-mail: madjid.eshaghi@gmail.com

L. Găvruţa
Universitatea Politehnica din Timişoara, Departamentul de Matematică, Piaţa Victoriei no.2, 300006 Timişoara, Romania, e-mail: gavruta_laura@yahoo.com

P. Găvruţa
Universitatea Politehnica din Timişoara, Departamentul de Matematică, Piaţa Victoriei no. 2, 300006 Timişoara, Romania, e-mail: pgavruta@yahoo.com

F. Giannessi
Department of Mathematics, Faculty of Natural Sciences, University of Pisa, Via F. Buonarroti, 56127, Pisa, Italy,

Sofia Giuffré
D.I.M.E.T., Mediterranean University, Reggio Calabria, Italy, e-mail: sofia.giuffre@unirc.it

D. Goeleven
IREMIA, University of La Reunion, 97400 Saint-Denis, France, e-mail: goeleven@univ-reunion.fr

M. Seetharama Gowda
Department of Mathematics and Statistics, University of Maryland, Baltimore County, Baltimore, MD 21250, USA, e-mail: gowda@math.umbc.edu

T.X.D. Ha
Researcher, Hanoi Institute of Mathematics, Hanoi, Vietnam

S. Hawkins
Department of Mathematics and Statistics, University of Guelph, Guelph, ON, Canada, e-mail: mcojocar@uoguelph.ca

Syed Shakaib Irfan
College of Engineering, Qassim University, P.O. Box 6677, Buraidah 51452, Al-Qassim, Kingdom of Saudi Arabia, e-mail: shakaib11@rediffmail.com

G. Isac[†]
Department of Mathematics and Computer Science, Royal Military College of Canada, P.O. Box 17000, STN Forces Kingston, Ontario K7K 7B4, Canada

Johannes Jahn
Department Mathematik, Universität Erlangen-Nürnberg, Martensstr. 3, D-91058 Erlangen, Germany, e-mail: jahn@am.uni-erlangen.de

Kil-Woung Jun
Department of Mathematics, Chungnam National University, 220 Yuseong-Gu, Daejeon, 305-764, Korea, e-mail: kwjun@cnu.ac.kr

Soon-Mo Jung
Mathematics Section, College of Science and Technology, Hongik University, 339-701 Jochiwon, Republic of Korea, e-mail: smjung@hongik.ac.kr

S. Kaboli-Gharetapeh
Department of Mathematics, Payame Noor University of Mashhad, Mashhad, Iran, e-mail: simin.kaboli@gmail.com

A.A. Khan
School of Mathematical Sciences, Rochester Institute of Technology, 85 Lomb Memorial Drive, Rochester, NY, USA, e-mail: aaksma@rit.edu

Hark-Mahn Kim
Department of Mathematics, Chungnam National University, 220 Yuseong-Gu, Daejeon, 305-764, Korea, e-mail: hmkim@cnu.ac.kr

Dag Lukkassen
Narvik University College and Norut Narvik, P.O.B. 385 N-8505 Narvik, Norway

Antonino Maugeri
Department of Mathematics and Computer Science, University of Catania, Catania, Italy, e-mail: maugeri@dmi.unict.it

Annette Meidell
Narvik University College, P.O.B. 385 N-8505 Narvik, Norway

Melania M. Moldovan
Department of Mathematics and Statistics, University of Maryland, Baltimore County, Baltimore, MD 21250, USA, e-mail: melania1@umbc.edu

M.S. Moslehian
Department of Pure Mathematics and Center of Excellence in Analysis on Algebraic Structures (CEAAS), Ferdowsi University of Mashhad, P.O. Box 1159, Mashhad 91775, Iran, e-mail: moslehian@ferdowsi.um.ac.ir and moslehian@ams.org

Dumitru Motreanu
University of Perpignan, Department de Mathématiques, 66860 Perpignan, France, e-mail: motreanu@univ-perp.fr

S.Z. Németh
The University of Birmingham, School of Mathematics, The Watson Building, Edgbaston, B15 2TT Birmingham, UK, e-mail: nemeths@for.mat.bham.ac.uk

Donal O'Regan
Department of Mathematics, National University of Ireland, Galway, Ireland

Choonkil Park
Department of Mathematics, Hanyang University, Seoul 133-791, South Korea, e-mail: baak@hanyang.ac.kr

Dan Pascali
Courant Institute of Mathematical Sciences, New York University, New York, NY, USA, e-mail: dp39@nyu.edu

Lars-Erik Persson
Department of Mathematics, Lulea University, S-97187 Lulea, Sweden

Stephane Pia
Department of Mathematics and Computer Science, University of Catania, Catania, Italy, e-mail: pia@dmi.unict.it

Radu Precup
Department of Applied Mathematics, Babeş-Bolyai University, 400084 Cluj, Romania, e-mail: r.precup@math.ubbcluj.ro

Viorel Radu
West University of Timişoara, Faculty of Mathematics and Computer Science, Department of Mathematics, Vasile Pârvan 4, 300223, Timişoara, Romania, e-mail: radu@math.uvt.ro

Themistocles M. Rassias
Department of Mathematics, National Technical University of Athens, Zografou Campus, 15780 Athens, Greece, e-mail: trassias@math.ntua.gr

Ioan A. Rus
Babeş-Bolyai University, Department of Applied Mathematics, Kogălniceanu Nr. 1, 400084 Cluj-Napoca, Romania, e-mail: iarus@math.ubbcluj.ro

Ghadir Sadeghi
Department of Pure Mathematics, Ferdowsi University of Mashhad, P.O. Box 1159, Mashhad 91775, Iran; and Banach Mathematical Research Group (BMRG), Mashhad, Iran, e-mail: ghadir54@yahoo.com and gh.sadeghi@math.um.ac.ir

Abul Hasan Siddiqi
B.M.A.S. Engineering College, Agra 282007, U.P., India, e-mail: siddiqi.abulhasan@gmail.com

C. Tammer
Institute of Mathematics, Martin-Luther University of Halle-Wittenberg, 06099 Halle, Germany

H. Thille
Department of Economics, University of Guelph, Guelph, ON, Canada,
e-mail: hthille@uoguelph.ca

E. Thommes
Department of Physics, University of Guelph, Guelph, ON, Canada,
e-mail: ethommes@uoguelph.ca

Mihai Turinici
"A. Myller" Mathematical Seminar; "A. I. Cuza" University, 11, Copou Boulevard,
700506 Iaşi, Romania, e-mail: mturi@uaic.ro

Henry Wolkowicz
Department of Combinatorics and Optimization, University of Waterloo, Waterloo,
Ontario N2L 3G1, Canada, e-mail: hwolkowi@uwater100.ca

S. Zolfaghari
Department of Mathematics, Semnan University, P.O. Box 35195-363, Semnan,
Iran, e-mail: zolfaghgrys@yahoo.com

Part I
Nonlinear Analysis

Chapter 1
Discrete Approximation Processes of King's Type

Octavian Agratini and Tudor Andrica

Dedicated to the memory of Professor George Isac

Abstract This survey paper is focused on linear positive operators having the degree of exactness null and fixing the monomial of the second degree. The starting point is represented by J.P. King's paper appearing in 2003. Our first aim is to sum up results obtained in the past five years. The second aim is to present a general class of discretizations following the features of the operators introduced by King.

1.1 Introduction

Let $C([a,b])$ be the Banach space of all real-valued and continuous functions defined on $[a,b]$, equipped with the norm $\|\cdot\|$ of the uniform convergence. Let e_n be the monomial of n degree, $n \in \mathbb{N}_0 := \{0\} \cup \mathbb{N}$. Bohman–Korovkin's theorem states: if $(L_n)_{n\geq 1}$ is a sequence of positive linear operators mapping $C([a,b])$ into itself such that $\lim_n \|L_n e_i - e_i\| = 0$ for $i = 0,1,2$, then one has $\lim_n \|L_n f - f\| = 0$ for every $f \in C([a,b])$.

Many classical linear positive operators have the degree of exactness one, this meaning they preserve the monomials e_0 and e_1. On the other hand, it is well known that if a linear positive operator reproduces all three test functions of the Bohman–Korovkin criterion, then it is the identity operator of the space. A question arises: What is known about operators which fix the monomials e_0 and e_2?

Octavian Agratini
Babeş-Bolyai University, Faculty of Mathematics and Computer Science, 400084 Cluj-Napoca, Romania, e-mail: agratini@math.ubbcluj.ro

Tudor Andrica
Babeş-Bolyai University, Faculty of Mathematics and Computer Science, 400084 Cluj-Napoca, Romania, e-mail: tudor_an@yahoo.com

P.M. Pardalos et al. (eds.), *Nonlinear Analysis and Variational Problems*,
Springer Optimization and Its Applications 35, DOI 10.1007/978-1-4419-0158-3_1,

In 2003, J.P. King [10] was the first to present an example of linear positive operators enjoying this property. These operators are of Bernstein type and are given as follows. $V_n : C([0,1]) \to C([0,1])$,

$$(V_n f)(x) = \sum_{x=0}^{n} \binom{n}{k} (r_n^*(x))^k (1 - r_n^*(x))^{n-k} f\left(\frac{k}{n}\right), \quad x \in [0,1], \tag{1.1}$$

where $r_n^* : [0,1] \to [0,1]$,

$$r_n^*(x) = \begin{cases} x^2, \ n = 1, \\ -\dfrac{1}{2(n-1)} + \sqrt{\dfrac{n}{n-1}x^2 + \dfrac{1}{4(n-1)^2}}, \ n = 2,3,\ldots \end{cases} \tag{1.2}$$

From approximation theory point of view, this sequence is useful. In spite of the fact that the operators have the degree of exactness null, the order of approximation is at least as good as the order of Bernstein operators.

King's construction quickly gained popularity and, during the past five years, several authors studied classes of linear positive operators which preserve the test functions e_0 and e_2. There are new papers on this topic constantly coming out and wide generalizations being studied.

The current paper aims to summarize the main results obtained in this area. Also, this survey should stimulate further research.

1.2 Further Results on V_n Type Operators

King also established quantitative estimates for V_n in terms of the classical first-order modulus $\omega_1(f;\cdot)$. New quantitative estimates involving the second modulus of continuity $\omega_2(f;\cdot)$ have been obtained by H. Gonska and P. Piţul. They proved [9; Theorem 2.1] the following result.

Theorem 1.1. *Let V_n, $n \in \mathbb{N}$, be defined by (1.1). For each $f \in C[0,1]$ and $x \in [0,1]$ one has*

$$|(V_n f)(x) - f(x)| \le \sqrt{x - r_n^*(x)}\,\omega_1\left(f; \sqrt{x - r_n^*(x)}\right) + (1+x)\,\omega_2\left(f; \sqrt{x - r_n^*(x)}\right).$$

Consequently, if $f \in C^1([0,1])$, one obtains the approximation order $\mathcal{O}\left(\sqrt{x - r_n^*(x)}\right)$, $n \to \infty$. For $f \in C^2([0,1])$ the approximation order is $\mathcal{O}(x - r_n^*(x))$, $n \to \infty$.

Setting the powers of an operator L by $L^0 = I_X$, $L^1 = L$, $L^{m+1} = L \circ L^m$, $m \in \mathbb{N}$, where I_X indicates the identity operator on the space X, the iterates of V_n have been obtained [9; Theorem 3.2].

Theorem 1.2. *Let V_n, $n \in \mathbb{N}$, be defined by (1.1). If $n \in \mathbb{N}$ is fixed, then for all $f \in C([0,1])$ and $x \in [0,1]$, one has*

$$\lim_{m\to\infty}(V_n^m f)(x) = f(0) + (f(1) - f(0))x^2 = (V_1 f)(x).$$

As regards the approximation process $(V_n)_n$, the transition from uniform convergence to A-statistical convergence was done by O. Duman and C. Orhan [6]. At first, we briefly recall elements of this type of convergence. Let $A = (a_{j,n})_{j\geq 1, n\geq 1}$ be an infinite summability matrix. For a given sequence $x := (x_n)_{n\geq 1}$ the A-transform of x, denoted by $Ax := ((Ax)_j)$, is defined by $(Ax)_j = \sum_{n=1}^{\infty} a_{j,n}x_n$ provided the series converges for each j. A is regular if $\lim_n x_n = L$ implies $\lim_j (Ax)_j = L$. Further on, we assume that A is a non-negative regular summability matrix and K is a subset of $\mathbb{N}$. The A-density of K, denoted by $\delta_A(K)$, is defined by $\delta_A(K) := \lim_j \sum_{n\in K} a_{j,n}$ provided the limit exists. A sequence $x := (x_n)_{n\geq 1}$ is said to be A-statistical convergent to the real number L if, for every $\varepsilon > 0$,

$$\delta_A(\{n \in \mathbb{N} : |x_n - L| \geq \varepsilon\}) = 0$$

takes place. This limit is denoted by $st_A - \lim x = L$.

From this moment, assume that A is a non-negative regular summability matrix such that $\lim_j \max_n \{a_{j,n}\} = 0$ holds. On the basis of [11], we can choose an infinite subset K of $\mathbb{N} \setminus \{1\}$ such that $\delta_A(K) = 0$. Define the functions p_n, $n \in \mathbb{N}$, by

$$p_n(x) = \begin{cases} x^2, & \text{if } n = 1, \\ r_n^*(x), & \text{if } n \notin K \cup \{1\}, \\ 0, & \text{otherwise}, \end{cases} \tag{1.3}$$

where r_n^* is defined by (1.2). Each function p_n is continuous on $[0,1]$ and $p_n([0,1]) \subset [0,1]$. One has $st_A - \lim_n p_n(x) = x$, $x \in [0,1]$. An analog of King's result proved in [6; Theorem 2.3] will be read as follows.

Theorem 1.3. *Let V_n be defined by (1.1) such that r_n^* is replaced by p_n described at (1.3). Then, for all $f \in C([0,1])$ and all $x \in [0,1]$,*

$$st_A - \lim_n |(V_n f)(x) - f(x)| = 0$$

holds.

Another recent paper [4] centers around a family of sequences of linear Bernstein-type operators depending on a real parameter $\alpha \geq 0$ and preserving e_0 and the polynomial $e_2 + \alpha e_1$. Let $\alpha \geq 0$ be fixed. For each $n \geq 2$, the authors consider the function $r_{n,\alpha} : [0,1] \to \mathbb{R}$ defined by

$$r_{n,\alpha}(x) := -\frac{n\alpha + 1}{2(n-1)} + \sqrt{\frac{(n\alpha+1)^2}{4(n-1)^2} + \frac{n(\alpha x + x^2)}{n-1}} \tag{1.4}$$

and the operator $B_{n,\alpha} : C([0,1]) \to C([0,1])$ given by

$$(B_{n,\alpha}f)(x) := \sum_{k=0}^{n} p_{n,k,\alpha}(x) f\left(\frac{k}{n}\right), \tag{1.5}$$

$$p_{n,k,\alpha}(x) := \binom{n}{k} r_{n,\alpha}^k(x)(1 - r_{n,\alpha}(x))^{n-k}.$$

If we choose $\alpha = 0$, $B_{n,0}$ becomes the V_n operator introduced by King. Also, the relations (1.4), (1.5) guarantee $B_{n,\alpha}e_0 = e_0$, $B_{n,\alpha}e_1 = r_{n,\alpha}$ and $B_{n,\alpha}(e_2 + \alpha e_1) = e_2 + \alpha e_1$.

Besides qualitative and quantitative results regarding the sequence $(B_{n,\alpha})_{n\geq 2}$, the authors obtain the following asymptotic formula of Voronovskaja type.

Theorem 1.4. *Let $B_{n,\alpha}$ be defined by (1.5). For $x \in (0,1)$ one has*

$$\lim_n 2n((B_{n,\alpha}f)(x) - f(x)) = x(1-x)\left(f''(x) - \frac{2}{2x+\alpha} f'(x)\right)$$

provided f has the required regularity conditions at the point x.

We conclude this section with q-Bernstein operators. To formulate the results we need the following definitions.

Let $q > 0$. For any $n \in \mathbb{N}_0$, the q-integer $[n]_q$ is defined by

$$[n]_q := 1 + q + \cdots + q^{n-1} \quad (n \in \mathbb{N}), \quad [0]_q := 0$$

and the q-factorial $[n]_q!$ by

$$[n]_q! := [1]_q[2]_q \ldots [n]_q \quad (n \in \mathbb{N}), \quad [0]_q! := 1.$$

For each integer $k \in \{0, 1, \ldots, n\}$, the q-binomial coefficient is defined by

$$\begin{bmatrix} n \\ k \end{bmatrix}_q := \frac{[n]_q!}{[k]_q![n-k]_q!}.$$

Clearly, for $q = 1$, one gets $[n]_1 = n$, $[n]_1! = n!$, $\begin{bmatrix} n \\ k \end{bmatrix}_1 = \binom{n}{k}$.

The q-Bernstein polynomials of $f : [0,1] \to \mathbb{C}$ introduced by G.M. Phillips [13] are defined as follows

$$(B_n f)(q;x) := \sum_{k=0}^{n} f\left(\frac{[k]_q}{[n]_q}\right) \begin{bmatrix} n \\ k \end{bmatrix}_q x^k \prod_{\nu=0}^{n-1-k} (1 - q^\nu x), \quad x \in [0,1],\ n \in \mathbb{N}. \tag{1.6}$$

We mention that an empty product is taken to be equal to 1.

For $q = 1$, the polynomials $(B_n f)(1;\cdot)$ are classical Bernstein polynomials. In what follows we consider $0 < q < 1$; in this case q-Bernstein polynomials are positive linear operators on $C([0,1])$. These operators satisfy the following properties

$$(B_n e_0)(q;x) = 1, \quad (B_n e_1)(q;x) = x, \quad (B_n e_2)(q;x) = x^2 + \frac{x - x^2}{[n]_q}. \tag{1.7}$$

The Phillips results are the basis of many research papers, and the comprehensive survey due to S. Ostrovska [12] gives a good perspective of these achievements.

Our aim is to modify the q-Bernstein operators $(B_n f)(q;\cdot)$, $n \geq 2$, into the King variant. Setting

$$r_{n,q}(x) := -\frac{1}{2([n]_q - 1)} + \sqrt{\frac{[n]_q}{[n]_q - 1}x^2 + \frac{1}{4([n]_q - 1)^2}}, \quad x \in [0,1], \tag{1.8}$$

for each $n \geq 2$, we consider the operator

$$(B_n^* f)(q;x) := (B_n f)(q; r_{n,q}(x)), \quad f \in \mathbb{R}^{[0,1]}, \tag{1.9}$$

where $(B_n f)(q;\cdot)$ is given by (1.5).

For the particular case $q = 1$, $(B_n^* f)(1;\cdot)$ turns into $V_n f$, King's example.

Since $(B_n^* e_1)(q;\cdot) = r_{n,q}$ and $\lim\limits_n r_{n,q}(x) = \left(\sqrt{4q^2x^2 + (1-q)^2} - 1 + q\right)/2q$, $q \in (0,1)$, based on the Bohman–Korovkin theorem, it is obvious that our sequence does not form an approximation process on the space $C([0,1])$. In order to satisfy this property, for each $n \geq 2$, the constant $q \in (0,1)$ will be replaced by a number $q_n \in (0,1)$.

Theorem 1.5. *Let $(q_n)_{n\geq 2}$, $0 < q_n < 1$, be a sequence such that $\lim\limits_n q_n = 1$ and $\lim\limits_n q_n^n$ exists. Let $(B_n^* f)(q_n;\cdot)$ be defined as in (1.9). For any $f \in C([0,1])$ one has*

$$\lim_n (B_n^* f)(q_n;x) = f(x), \text{ uniformly in } x \in [0,1].$$

Proof. The assumptions made upon the sequence $(q_n)_{n\geq 2}$ guarantee that $\lim\limits_n [n]_{q_n} = \infty$. Examining (1.8), this implies

$$\lim_n r_{n,q_n}(x) = x, \text{ uniformly in } x \in [0,1],$$

and, consequently, $\lim\limits_n (B_n^* e_1)(q_n;\cdot) = e_1$. Also, relations (1.9) and (1.7) ensure $\lim\limits_n (B_n^* e_j)(q_n;\cdot) = e_j$, $j \in \{0,2\}$. Since the requirements of the Bohman–Korovkin theorem are satisfied, the conclusion follows. □

1.3 A General Class in Study

The object of this section is to present a class of discrete operators reproducing the third test function of the Bohman–Korovkin theorem. This class is defined on certain subspaces of $C(J)$, $J \subset \mathbb{R}$. We take into account two kinds of intervals: $J = [0,1]$

and $J=\mathbb{R}_+ := [0,\infty)$, respectively. Let $I_n \subset \mathbb{N}$ be a set of indices. Following [1], we consider a sequence $(L_n)_{n\ge1}$ of linear positive operators acting on $C(J)$ and defined by

$$(L_n f)(x) = \sum_{k\in I_n} u_{n,k}(x) f(x_{n,k}), \quad x\in J,\ f\in\mathscr{F}(J), \tag{1.10}$$

where $u_{n,k}\in C(J)$ is a positive real-valued function for each $(n,k)\in\mathbb{N}\times I_n$, $(x_{n,k})_{k\in I_n}$ is a mesh of nodes on J, and

$$\mathscr{F}(J) := \{f\in C(J):\ \text{the series in (1.10) is convergent}\}.$$

We note that the right-hand side of (1.10) could be a finite sum. In this case, $\mathscr{F}(J)$ is just $C(J)$. For each $n\in\mathbb{N}$, we assume that the following identities

$$(L_n e_0)(x) = 1, \quad (L_n e_1)(x) = x, \quad (L_n e_2)(x) = a_n x^2 + b_n x, \quad x\in J, \tag{1.11}$$

are fulfilled, where $a_n>0$, $b_n\ge 0$. Knowing that $u_{n,k}\ge 0$, $k\in I_n$, the first identity from (1.11) implies that each $u_{n,k}$ belongs to $C_B(J)$, the space of all real-valued continuous and bounded functions on J. In regard to the sequences of real numbers $(a_n)_{n\ge1}$ and $(b_n)_{n\ge1}$ we assume

$$\lim_{n\to\infty} a_n = 1, \quad \lim_{n\to\infty} b_n = 0. \tag{1.12}$$

Relations (1.11) and (1.12) guarantee that $(L_n)_{n\ge1}$ is a strong approximation process on any compact $\mathscr{K}\subset J$, this meaning $\lim_n (L_n f)(x) = f(x)$ uniformly for every $x\in\mathscr{K}$ and $f\in\mathscr{F}(J)$.

For each $n\in\mathbb{N}$, we define the continuous function $v_n : J\to\mathbb{R}_+$,

$$v_n(x) = \frac{1}{2a_n}\left(\sqrt{b_n^2 + 4a_n x^2} - b_n\right), \quad x\in J. \tag{1.13}$$

Starting with (1.10), we introduce the operators

$$(L_n^* f)(x) = \sum_{k\in I_n} u_{n,k}(v_n(x)) f(x_{n,k}), \quad x\in J,\ f\in\mathscr{F}(J). \tag{1.14}$$

On the basis of (1.11), the following identities [1]

$$L_n^* e_0 = e_0, \quad L_n^* e_1 = v_n, \quad L_n^* e_2 = e_2 \tag{1.15}$$

hold. Consequently, one has $\lim_n L_n^* f = f$ uniformly on compact subintervals of J for every $f\in\mathscr{F}(J)$.

Considering $\varphi_x : J\to\mathbb{R}$, $\varphi_x(t) = t-x$, $x\in J$, the second central moment of L_n^* has the form

$$(L_n^*\varphi_x)^2(x) = 2x(x - v_n(x)), \quad x\in J.$$

By using (1.13) and the fact that L_n^* is a positive operator, one gets

$$0 \le v_n(x) \le x, \quad x \in J.$$

Our aim is to explore the rate of convergence. For $J = [0,1]$ we use the modulus of continuity. For $J = \mathbb{R}_+$ we use a weighted modulus associated to the Banach space $(E_\alpha, \|\cdot\|_\alpha)$, $\alpha \ge 2$, where

$$E_\alpha := \{f \in C(\mathbb{R}_+) : \ w_\alpha(x)f(x) \text{ is convergent as } x \to \infty\},$$

and $\|f\|_\alpha := \sup\limits_{x\ge 0} w_\alpha(x)|f(x)|$. The weight w_α is given by $w_\alpha(x) = (1+x^\alpha)^{-1}$. Since $\alpha \ge 2$, the test functions e_j, $j \in \{0,1,2\}$, belong to E_α.

Theorem 1.6. *Let L_n^*, $n \in \mathbb{N}$, be defined by (1.14), where $J = [0,1]$. For every $f \in C(J)$, one has*

$$|(L_n^* f)(x) - f(x)| \le \left(1 + \frac{1}{\delta}\widetilde{v}_n(x)\right)\omega(f;\delta), \quad x \in J,\ \delta > 0, \tag{1.16}$$

where

$$\widetilde{v}_n(x) = \sqrt{2x(x - v_n(x))}, \quad x \in J. \tag{1.17}$$

The proof can be found in [1, Theorem 3.1.(ii)].

Since $x \in J = [0,1]$, the following upper bound for $e_1 - v_n$ can be easily established

$$x - v_n(x) = x + \frac{b_n}{2a_n} - \sqrt{\frac{x^2}{a_n} + \frac{b_n^2}{4a_n^2}} \le \left(1 - \frac{1}{\sqrt{a_n}}\right)x + \frac{b_n}{2a_n} \le \frac{|a_n - 1|}{\sqrt{a_n}} + \frac{b_n}{2a_n}.$$

Consequently, $\widetilde{v}_n(x) \le \left(\frac{2|a_n-1|}{\sqrt{a_n}} + \frac{b_n}{a_n}\right)^{1/2}$ which tends to zero for n tending to infinity. Choosing in (1.16) δ to be equal with this quantity, we obtain

$$|(L_n^* f)(x) - f(x)| \le 2\omega\left(f; \left(\frac{2|a_n-1|}{\sqrt{a_n}} + \frac{b_n}{a_n}\right)^{1/2}\right), \quad x \in [0,1].$$

For $f \in E_\alpha$ we consider the following weighted modulus

$$\Omega_{w_\alpha}(f;\delta) := \sup_{\substack{x\ge 0 \\ 0<h\le\delta}} w_\alpha(x+h)|f(x+h) - f(x)|, \quad \delta > 0.$$

Theorem 1.7. *Let L_n^*, $n \in \mathbb{N}$, be defined by (1.14), where $J = \mathbb{R}_+$. For every $f \in \mathscr{F}(J) \cap E_\alpha$, one has*

$$|(L_n^* f)(x) - f(x)| \le \sqrt{(L_n^*\mu_x^2)(x)}\left(1 + \frac{\widetilde{v}_n(x)}{\delta}\right)\Omega_{w_\alpha}(f;\delta), \quad x \ge 0,\ \delta > 0,$$

where $\widetilde{v}_n$ is given at (1.17) and $\mu_x(t) := 1 + (x + |t-x|)^\alpha$, $t \ge 0$.

The proof can be found in [1; Theorem 3.2].

Let $n \geq 2$. Choosing $I_n = \{0,1,\ldots,n\}$, $J=[0,1]$, $x_{n,k} = k/n$,

$$u_{n,k}(x) = \binom{n}{k} x^k (1-x)^{n-k},\ 0 \leq k \leq n,$$

L_n becomes a Bernstein operator and we get

$$a_n = 1 - \frac{1}{n}, \quad b_n = \frac{1}{n}, \quad v_n(x) = r_n^*(x), \quad x \in [0,1],$$

where r_n^* is defined by (1.2). Consequently, L_n^* turns into King operator V_n.

In what follows, starting from classical Baskakov operators we present the modified variant L_n^*.

Example 1.8. In (1.10) we choose $J=[0,\infty)$, $I_n = \mathbb{N}$, $x_{n,k} = k/n$ and

$$u_{n,k}(x) = \binom{n+k-1}{k} x^k (1+x)^{-n-k},\ x \geq 0.$$

The requirements (1.11) are fulfilled. It is known that $a_n = 1+1/n$ and $b_n = 1/n$, $n \in \mathbb{N}$, consequently (1.12) holds. We obtain

$$v_n(x) = \frac{\sqrt{1+4n(n+1)x^2} - 1}{2(n+1)}, \quad x \geq 0,\ n \in \mathbb{N},$$

and, following (1.14), the modified Baskakov operators are defined by

$$(L_n^* f)(x) = \sum_{k=0}^{\infty} \binom{n+k-1}{k} \frac{v_n^k(x)}{(1+v_n(x))^{n+k}} f\left(\frac{k}{n}\right), \quad x \geq 0,\ n \in \mathbb{N},$$

where $f \in E_2$.

Remark 1.9. For $J = \mathbb{R}_+$, our operator L_n^* maps $C_B(\mathbb{R}_+)$ in $C_B(\mathbb{R}_+)$ because of the first identity of relation (1.15). Here $C_B(\mathbb{R}_+)$ denotes the space of all real-valued continuous and bounded functions defined on $\mathbb{R}_+$. On the basis of [8; Theorem 1], relations (1.15) lead us to the following result.

If $st - \lim\limits_n \|v_n - e_1\|_K = 0$, then $st - \lim\limits_n \|L_n^* f - f\|_K = 0$, for any function f belonging to $C_B(\mathbb{R}_+)$, where $K \subset \mathbb{R}_+$ is a compact and $\|\cdot\|_K$ is the norm of the uniform convergence on K.

Remark 1.10. For general operators of King's type, a generalization to the m-dimensional case will be read as follows. Let K_m be a compact and convex subset of the Euclidean space $\mathbb{R}^m$. It was shown by Volkov [15] that the following $m+2$ functions: $\mathbf{1}, pr_1, \ldots, pr_n, \sum\limits_{j=1}^{m} pr_j^2$, are test functions for $C(K_m)$. Here $\mathbf{1}$ stands for the constant function on K_m of value 1 and pr_j, $1 \leq j \leq m$, represent the canonical projections on K_m, i.e., $pr_j(x) = x_j$ for every $x = (x_i)_{1 \leq i \leq m} \in K_m$. Considering

$\widetilde{L}_n^*: C(K_m) \to C(K_m)$ such that $\widetilde{L}_n^* \mathbf{1} = \mathbf{1}$ and $\widetilde{L}_n^* \left(\sum_{j=1}^{m} pr_j^2 \right) = \sum_{j=1}^{m} pr_j^2$, we get

$$\widetilde{L}_n^*(\|\cdot - x\|^2; x) = \widetilde{L}_n^* \left(\sum_{j=1}^{m} (\cdot - x_j)^2; x \right) = 2\|x\|^2 - 2\sum_{j=1}^{m} x_j E_{n,j}(x),$$

where $\|\cdot\|$ stands for the Euclidean norm in $\mathbb{R}^m$. Here $E_{n,j}(x) := \widetilde{L}_n^*(pr_j; x)$, $x \in K_m$. Let $\mu_n := 2\|x\|^2 - 2\sum_{j=1}^{m} x_j E_{n,j}(x)$.

Taking into account a result established by Censor [5; Eq. (5)], we obtain

$$\|\widetilde{L}_n^* f - f\|_{K_m} \le 2\omega(f; \sqrt{\mu_n}),$$

where $\omega(f;\cdot)$ is the modulus of continuity defined for $f \in C(K_m)$ as follows

$$\omega(f;\delta) = \max_{\substack{x,y \in K_m \\ d(x,y) \le \delta}} |f(x) - f(y)|,$$

$d(\cdot,\cdot)$ being the Euclidean distance.

At the end of this section we emphasize other achievements as regards the sequences of operators of King's type. In [3], for a compact interval J and $I_n = \{0, 1, \ldots, n\}$, the limit of iterates of L_n^* operators defined as in (1.14) has been established. The tensor product extension of L_n^* to the bidimensional case was investigated in [2]. As a particular case, a modified variant of the bivariate Bernstein–Chlodovsky operators was presented. Recently, O. Duman and M.A. Özarslan [7] studied the modified Szász–Mirakjan operators S_n^* which preserve e_0 and e_2. They also proved that the order of approximation of a function f by $S_n^* f$ is at least as good as the order of approximation of f by $S_n f$.

Important results have been obtained by L. Rempulska and K. Tomczak [14] who entered upon certain sequences of linear positive operators acting on $C_p(I)$, $p \in \mathbb{N}_0$, the space of all functions $f: I \to \mathbb{R}$ with the property that $f w_p$ is bounded and uniformly continuous on I, where $w_0(x) = 1$ and $w_p(x) = (1 + x^p)^{-1}$, for $p \in \mathbb{N}$.

References

1. O. Agratini, *Linear operators that preserve some test functions*, International Journal of Mathematics and Mathematical Sciences, Vol. 2006, Article ID 94136, pp. 11, DOI 10.1155/IJMMS.
2. O. Agratini, *On a class of linear positive bivariate operators of King type*, Studia Univ. "Babeş-Bolyai", Mathematica, **51**(2006), f. 4, 13–22.
3. O. Agratini, *On the iterates of a class of summation-type linear positive operators*, Computers Mathematics with Applications, **55**(2008), 1178–1180.

4. D. Cárdenas-Morales, P. Garrancho, F.J. Muñoz-Delgado, *Shape preserving approximation by Bernstein-type operators which fix polynomials*, Applied Mathematics and Computation, **182**(2006), 1615–1622.
5. E. Censor, *Quantitative results for positive linear approximation operators*, J. Approx. Theory, **4**(1971), 442–450.
6. O. Duman, C. Orhan, *An abstract version of the Korovkin approximation theorem*, Publ. Math. Debrecen, **69**(2006), f. 1-2, 33–46.
7. O. Duman, M.A. Özarslan, *Szász-Mirakjan type operators providing a better error estimation*, Applied Math. Letters, **20**(2007), 1184–1188.
8. A.D. Gadjiev, C. Orhan, *Some approximation theorems via statistical convergence*, Rocky Mountain J. Math., **32**(2002), 129–138.
9. H. Gonska, P. Piţul, *Remarks on an article of J.P. King*, Schriftenreihe des Fachbereichs Mathematik, SM-DU-**596**, 2005, Universität Duisburg-Essen, 1-8.
10. J.P. King, *Positive linear operators which preserve x^2*, Acta Math. Hungar., **99**(2003), f. 3, 203–208.
11. E. Kolk, *Matrix summability of statistical convergent sequences*, Analysis, **13**(1993), 77–83.
12. S. Ostrovska, *The first decade of the q-Bernstein polynomials: results and perspectives*, Journal of Mathematical Analysis and Approximation Theory, **2**(2007), Number 1, 35–51.
13. G.M. Phillips, *Bernstein polynomials based on the q-integers*, Ann. Numer. Math., **4**(1997), 511–518.
14. L. Rempulska, K. Tomczak, *Approximation by certain linear operators preserving x^2*, Turk. J. Math., **32**(2008), 1–11.
15. V.I. Volkov, *On the convergence of sequences of linear positive operators in the space of continuous functions of two variables* (in Russian), Dokl. Akad. Nauk SSSR (N.S.), **115**(1957), 17–19.

Chapter 2
Isometrics in Non-Archimedean Strictly Convex and Strictly 2-Convex 2-Normed Spaces

Maryam Amyari and Ghadir Sadeghi

Dedicated to the memory of Professor George Isac

Abstract In this paper, we present a Mazur–Ulam type theorem in non-Archimedean strictly convex 2-normed spaces and present some properties of mappings on non-Archimedean strictly 2-convex 2-normed spaces.

2.1 Introduction and Preliminaries

A non-Archimedean field is a field $\mathscr{K}$ equipped with a function (valuation) $|\cdot|$: $\mathscr{K} \to [0,\infty)$ such that for all $r,s \in \mathscr{K}$:
(i) $|r| = 0$ if and only if $r = 0$,
(ii) $|rs| = |r|\,|s|$,
(iii) $|r+s| \leq \max\{|r|,|s|\}$.
Clearly $|1| = |-1| = 1$ and $|n| \leq 1$ for all $n \in \mathbf{N}$. An example of a non-Archimedean valuation is the mapping

$$|x| = \begin{cases} 1 & x \neq 0 \\ 0 & x = 0 \end{cases}$$

This valuation is called trivial.
In 1897, Hensel [4] discovered the p-adic numbers. Fix a prime number p. For any nonzero rational number x, there exists a unique integer $n_x \in \mathbf{Z}$ such that $x = \frac{a}{b}p^{n_x}$,

Maryam Amyari
Department of Mathematics, Faculty of Science, Islamic Azad University-Mashhad Branch, Mashhad 91735, Iran, e-mail: amyari@mshdiau.ac.ir and maryam_amyari@yahoo.com

Ghadir Sadeghi
Department of Pure Mathematics, Ferdowsi University of Mashhad, P.O. Box 1159, Mashhad 91775, Iran, and Banach Mathematical Research Group (BMRG), Mashhad Iran, e-mail: ghadir54@yahoo.com and gh.sadeghi@math.um.ac.ir

P.M. Pardalos et al. (eds.), *Nonlinear Analysis and Variational Problems*, Springer Optimization and Its Applications 35, DOI 10.1007/978-1-4419-0158-3_2,

where neither of the integers a and b is divisible by p. Then $|x|_p := p^{-n_x}$ defines a non-Archimedean norm on $\mathbf{Q}$. The completion of $\mathbf{Q}$ with respect to the metric $d(x,y) = |x-y|_p$ is denoted by $\mathbf{Q}_p$ which is called the p-adic number field; cf. [14].

Let $\mathscr{X}$ be a vector space over a field $\mathscr{K}$ with a non-Archimedean valuation $|\cdot|$. A function $\|\cdot\| : \mathscr{X} \to [0,\infty)$ is said to be a non-Archimedean norm if it satisfies the following conditions for all $x,y \in \mathscr{X}$ and $r \in \mathscr{K}$:

(i) $\|x\| = 0$ if and only if $x = 0$,

(ii) $\|rx\| = |r|\|x\|$,

(iii) $\|x+y\| \leq \max\{\|x\|,\|y\|\}$.

Then $(\mathscr{X},\|\cdot\|)$ is called a non-Archimedean normed space. A non-Archimedean normed space is called strictly convex if $\|x+y\| = \max\{\|x\|,\|y\|\}$ and $\|x\| = \|y\|$ imply $x = y$. Theory of non-Archimedean normed spaces is not trivial, for instance there may not be any unit vector; see [9] and references therein.

Definition 2.1. Let $\mathscr{X}$ be a vector space of dimension greater than 1 over a field $\mathscr{K}$ with a non-Archimedean valuation $|\cdot|$. A function $\|\cdot,\cdot\| : \mathscr{X} \times \mathscr{X} \to \mathbf{R}$ is said to be a non-Archimedean 2-norm if it satisfies the following conditions:

(i) $\|x,y\| = 0$ if and only if x,y are linearly dependent,

(ii) $\|x,y\| = \|y,x\|$,

(iii) $\|rx,y\| = |r|\|x,y\| \quad (r \in \mathscr{K}, x,y \in \mathscr{X})$,

(iv) the strong triangle inequality

$$\|x,y+z\| \leq \max\{\|x,y\|,\|x,z\|\} \quad (x,y,z \in X).$$

Then $(\mathscr{X},\|\cdot,\cdot\|)$ is called a non-Archimedean 2-normed space.

Let $\|x,y\| \neq \|x,z\|$ in inequality (iv). Without loss of generality we may assume that $\|x,y\| > \|x,z\|$, then we have

$$\|x,y+z\| \leq \|x,y\| \tag{2.1}$$

and

$$\|x,y\| \leq \max\{\|x,y+z\|,\|x,z\|\} = \|x,y+z\|. \tag{2.2}$$

Inequalities (2.1) and (2.2) imply that

$$\|x,y+z\| = \max\{\|x,y\|,\|x,z\|\},$$

which is interesting in its own right.

Example 2.2. Let $(\mathscr{X},\|\cdot\|)$ be a non-Archimedean normed space over a valued field $\mathscr{K}$. Then $\|\cdot,\cdot\|$ on $\mathscr{X}^2$ defined by

$$\|x,y\| = \begin{cases} \|x\|\,\|y\| & x \text{ and } y \text{ are linearly independent} \\ 0 & \text{otherwise} \end{cases}$$

is a 2-norm and $(\mathscr{X},\|\cdot,\cdot\|)$ is a non-Archimedean 2-normed space.

Definition 2.3. Let $\mathscr{X}$ and $\mathscr{Y}$ be non-Archimedean 2-normed spaces and $f : \mathscr{X} \to \mathscr{Y}$ be a mapping. Then f is called a 2-isometry if

$$\|x-z, y-z\| = \|f(x)-f(z), f(y)-f(z)\|$$

for all x, y and z in $\mathscr{X}$; cf. [2].

For non-zero vectors x, y in $\mathscr{X}$, let $\mathscr{V}(x,y)$ denote the subspace of $\mathscr{X}$ generated by x and y.

Definition 2.4. Suppose $\mathscr{X}$ is a non-Archimedean 2-normed space over a valued field $\mathscr{K}$ with $|2| = 1$, then $\mathscr{X}$ is called strictly convex if $\|x+y, z\| = \max\{\|x,z\|, \|y,z\|\}$, $\|x,z\| = \|y,z\| \neq 0$ and $z \notin \mathscr{V}(x,y)$ imply that $x = y$.

The theory of isometric mappings had its beginning in the classical paper [7] by S. Mazur and S. Ulam, who proved that every isometry of a normed real vector space onto another normed real vector space is a linear mapping up to translation. A number of mathematicians have had deal with Mazur–Ulam theorem; see [5, 10, 11, 12, 13, 15, 16] and references therein. Mazur–Ulam theorem is not valid in the contents of non-Archimedean normed spaces, in general. As a counterexample take $\mathbf{R}$ with the trivial non-Archimedean valuation and define $f : \mathbf{R} \to \mathbf{R}$ by $f(x) = x^3$. Then f is clearly a surjective isometry and $f(0) = 0$, but f is not linear; cf. [8]. H.Y. Chu [2] studied the notation of 2-isometry and proved the Mazur–Ulam problem in 2-normed spaces.

Example 2.5. Suppose $\mathbf{R}^2$ is the vector space over field $\mathbf{R}$ with non-Archimedean trivial valuation $|.|$. Then the function $\|.,.\| : \mathbf{R}^2 \times \mathbf{R}^2 \to \mathbf{R}$ is defined by

$$\|(x_1,x_2),(y_1,y_2)\| = |x_1,x_2||y_1,y_2|,$$

where

$$|x,y| = \begin{cases} 1 & x \neq 0 \text{ and } y \neq 0 \\ 0 & \text{otherwise} \end{cases}$$

is defined as a non-Archimedean 2-norm. Define $f : \mathbf{R}^2 \to \mathbf{R}^2$ by $f(x,y) = (x^3, y^3)$. Then f is clearly a 2-isometry and $f(0,0) = 0$, but f is not additive.

In this paper, by using the terminology and some ideas of [1], [2], [3], [6], and [8], we establish a Mazur–Ulam type theorem in the framework of non-Archimedean 2-normed spaces.

2.2 Non-Archimedean Strictly Convex 2-Normed Spaces

We begin this section with the following useful lemma.

Lemma 2.6. *Let $(\mathscr{X}, \|\cdot,\cdot\|)$ be a non-Archimedean 2-normed space over a valued field $\mathscr{K}$. Then $\|x,y\| = \|x, y+rx\|$ for all $x, y \in \mathscr{X}$ and all $r \in \mathscr{K}$.*

Proof. Let $x,y \in \mathscr{X}$, then

$$\|x,y+rx\| \leq \max\{\|x,y\|,\|x,rx\|\} = \|x,y\| \tag{2.3}$$

and

$$\begin{aligned}\|x,y\| = \|x,y+rx-rx\| &\leq \max\{\|x,y+rx\|,\|x,-rx\|\} \\ &= \|x,y+rx\|.\end{aligned} \tag{2.4}$$

It follows from inequalities (2.3) and (2.4) that $\|x,y+rx\| = \|x,y\|$. □

Definition 2.7. Let $\mathscr{X}$ be a non-Archimedean linear space over a valued field $\mathscr{K}$ and x,y,z be mutually disjoint of $\mathscr{X}$. Then x,y,z are said to be collinear if $x-y = r(x-z)$ for some $r \in \mathscr{K}$.

Lemma 2.8. *Let $(\mathscr{X},\|.,.\|)$ be a non-Archimedean 2-normed space over a valued field $\mathscr{K}$ which is strictly convex and let $x,y \in \mathscr{X}$. Then $\frac{x+y}{2}$ is the unique member t of $\mathscr{X}$, collinear with x,y satisfying*

$$\|x-z,x-t\| = \|y-t,y-z\| = \|x-z,y-z\|$$

for all $z \in \mathscr{X}$ with $\|x-z,y-z\| \neq 0$.

Proof. Set $t = \frac{x+y}{2}$. Then t,x,y are collinear. By Lemma 2.6 we have

$$\begin{aligned}\|x-z,x-t\| &= \|x-z,x-\frac{x+y}{2}\| = \|x-z,\frac{x-y}{2}\| \\ &= \frac{1}{|2|}\|x-z,x-y\| = \|x-z,z-y\| = \|x-z,y-z\|;\end{aligned}$$

$$\begin{aligned}\|y-z,y-t\| &= \|y-z,y-\frac{x+y}{2}\| = \|y-z,\frac{y-x}{2}\| \\ &= \frac{1}{|2|}\|y-z,y-x\| = \|y-z,z-x\| = \|x-z,y-z\|.\end{aligned}$$

Assume that another $s \in \mathscr{X}$, collinear with x,y satisfies,

$$\|x-z,x-s\| = \|y-s,y-z\| = \|x-z,y-z\|.$$

Then

$$\begin{aligned}\|x-z,x-\frac{t+s}{2}\| &\leq \max\left\{\|x-z,\frac{x-t}{2}\|,\|x-z,\frac{x-s}{2}\|\right\} \\ &= \|x-z,y-z\|.\end{aligned} \tag{2.5}$$

Similarly

$$\|y-z,y-\frac{t+s}{2}\| \leq \|x-z,y-z\|. \tag{2.6}$$

we have

$$\begin{aligned}\|x-z,y-z\| &= \|x-z,y-x\| \\ &\leq \max\left\{\|x-z,x-\frac{t+s}{2}\|,\|x-z,y-\frac{t+s}{2}\|\right\}.\end{aligned} \tag{2.7}$$

It follows from (ii) and (iv) that

$$\begin{aligned}\|x-z,y-\frac{t+s}{2}\| &= \|y-z+x-y,y-\frac{t+s}{2}\| \\ &\leq \max\left\{\|y-z,y-\frac{t+s}{2}\|,\|x-y,y-\frac{t+s}{2}\|\right\}.\end{aligned} \tag{2.8}$$

Since t,x,y and s,x,y are collinear we have

$$\|x-y,y-\frac{t+s}{2}\| \leq \max\left\{\|x-y,\frac{y-t}{2}\|,\|x-y,\frac{y-s}{2}\|\right\} = 0. \tag{2.9}$$

It follows from (2.8) and (2.9) that

$$\|x-z,y-\frac{t+s}{2}\| \leq \|y-z,y-\frac{t+s}{2}\|. \tag{2.10}$$

If both inequalities (2.5) and (2.6) were strict, then by inequalities (2.7) and (2.10) we would have

$$\|x-z,y-z\| \leq \max\left\{\|x-z,x-\frac{t+s}{2}\|,\|y-\frac{t+s}{2},y-z\|\right\} < \|x-z,y-z\|,$$

a contradiction. So at least one of the equalities holds in (2.5) and (2.6). Without lose of generality assume that equality holds in (2.5). Then

$$\|x-z,\frac{x-t}{2}+\frac{x-s}{2}\| = \max\left\{\|x-z,\frac{x-t}{2}\|,\|x-z,\frac{x-s}{2}\|\right\}.$$

By the strict convexity we obtain $\frac{x-t}{2}=\frac{x-s}{2}$, that is $t=s$. □

Theorem 2.9. *Suppose that $\mathscr{X}$ and $\mathscr{Y}$ are non-Archimedean 2-normed spaces over a valued field $\mathscr{K}$ such that $\mathscr{Y}$ is strictly convex. Assume that $f(x), f(y)$, and $f(z)$ are collinear when x,y, and z are collinear. If $f:\mathscr{X}\to\mathscr{Y}$ is a 2-isometry, then $f-f(0)$ is an additive mapping.*

Proof. Let $g(x)=f(x)-f(0)$. Then g is an 2-isometry and $g(0)=0$. Let $x,y\in\mathscr{X}$ with $x\neq y$. Since $\dim\mathscr{X}>1$, there exists an element $z\in\mathscr{X}$ such that $\|x-z,y-z\|\neq 0$. Since g is a 2-isometry, we have

$$\|g(x)-g(z),g(x)-g\left(\frac{x+y}{2}\right)\| = \|x-z,x-\frac{x+y}{2}\|$$

$$= \|x-z, \frac{x-y}{2}\|$$
$$= \|x-z, x-y\| = \|x-z, y-z\|$$
$$= \|g(x)-g(z), g(y)-g(z)\|$$

and similarly we obtain

$$\begin{aligned}\|g(y)-g\left(\frac{x+y}{2}\right), g(y)-g(z)\| &= \|y-\frac{x+y}{2}, y-z\| \\ &= \|y-x, y-z\| = \|x-z, y-z\| \\ &= \|g(x)-g(z), g(y)-g(z)\|.\end{aligned}$$

Since $\frac{x+y}{2}, x$ and y collinear, $g\left(\frac{x+y}{2}\right), g(x)$ and $g(y)$ are also collinear. It follows from Lemma 2.8 that

$$g\left(\frac{x+y}{2}\right) = \frac{g(x)+g(y)}{2},$$

for all $x, y \in \mathscr{X}$. Hence $g = f - f(0)$ is additive since $g(0) = 0$. □

Example 2.10. Consider $\mathbf{Q}_2 \times \mathbf{Q}_2$ with norm $|(\alpha, \beta)| = \max\{|\alpha|_2, |\beta|_2\}$. Then $\mathbf{Q}_2 \times \mathbf{Q}_2$ is a non-Archimedean normed space such that is not strictly convex. Define 2-norm $|\cdot, \cdot|$ on $\mathbf{Q}_2 \times \mathbf{Q}_2$ by

$$|u, v| = \begin{cases} |u||v| & u \text{ and } v \text{ are linearly independent} \\ 0 & \text{otherwise} \end{cases}$$

Now define the mapping $f : \mathbf{Q}_2 \times \mathbf{Q}_2 \to \mathbf{Q}_2 \times \mathbf{Q}_2$ by

$$f(\alpha, \beta) = \begin{cases} (\alpha^2, \beta^2) & \alpha = \frac{a}{b}2^0 \text{ and } \beta = \frac{c}{d}2^0 \\ (\alpha, \beta) & \text{otherwise.} \end{cases}$$

Then f is a 2-isometry and $f(0,0) = 0$ but f is not additive. Therefore the assumption that $\mathscr{Y}$ is strictly convex cannot be omitted in Theorem 2.9.

2.3 Non-Archimedean Strictly 2-Convex 2-Normed Spaces

Definition 2.11. Suppose $\mathscr{X}$ is a non-Archimedean 2-normed space over a valued field $\mathscr{K}$ with $|3| = 1$, then $\mathscr{X}$ is called strictly 2-convex if for all $x, y, z \in \mathscr{X}$ which $\|x+z, y+z\| = \max\{\|x, y\|, \|x, z\|, \|y, z\|\}$ and $\|x, y\| = \|x, z\| = \|y, z\| \neq 0$, we have $z = x + y$.

Definition 2.12. Suppose $\mathscr{X}$ is a non-Archimedean 2-normed space over a valued field $\mathscr{K}$ and a, b, c are three non-collinear points in $\mathscr{X}$, then

$$\begin{aligned}T(a,b,c) = \{x \in \mathscr{X} : &\|a-b, b-c\| \\ &= \max\{\|a-x, b-x\|, \|a-c, x-c\|, \|x-c, b-c\|\}\}\end{aligned}$$

is called the triangle with vertices a, b, and c.

Definition 2.13. A point p of a 2-normed space $(\mathscr{X}, \|.,.\|)$ is called 2-normed midpoint of three non-collinear points a, b, and c of $\mathscr{X}$ ($\|a-c, b-c\| \neq 0$) if

$$\|a-p, b-p\| = \|a-c, p-c\| = \|p-c, b-c\| = \|a-b, b-c\|.$$

If p is a 2-normed midpoint of a, b, and c, then p is called a center of $T(a,b,c)$

Lemma 2.14. *Let $\mathscr{X}$ be a non-Archimedean 2-normed space over a valued field $\mathscr{K}$ with $|3| = 1$ and $x, y, z \in \mathscr{X}$, are non-collinear. Then $u = \frac{x+y+z}{3}$ is the element of $\mathscr{X}$, collinear with x, y, z satisfying*

$$\|x-z, y-z\| = \|x-u, y-u\| = \|x-z, u-z\| = \|u-z, y-z\|.$$

Proof.

$$\begin{aligned}\|x-u, y-u\| &= \|x-\frac{x+y+z}{3}, y-\frac{x+y+z}{3}\| = \|2x-y-z, 2y-x-z\| \\ &= \|2x-y-z, 3y-3x\| = \|2x-y-z, y-x\| \\ &= \|y-x, x-z\| = \|x-z, y-z\|\end{aligned}$$

and

$$\|x-z, u-z\| = \|x-z, \frac{x+y+z}{3}-z\| = \|x-z, x+y-2z\| = \|x-z, y-z\|$$

$$\|u-z, y-z\| = \|\frac{x+y+z}{3}-z, y-z\| = \|x+y-2z, y-z\| = \|x-z, y-z\|.$$

□

Theorem 2.15. *Let $\mathscr{X}$ be a non-Archimedean 2-normed space over a valued field $\mathscr{K}$ with $|3| = 1$. Then the following statements are equivalent for all $a, b, c \in \mathscr{X}$,*
(i) $(\mathscr{X}, \|.,.\|)$ is strictly 2-convex.
(ii) $T(a,b,c)$ has a unique center.

Proof. Let $(\mathscr{X}, \|.,.\|)$ be strictly 2-convex. By Lemma 2.14 it is a clear that $\frac{a+b+c}{3}$ is a center of $T(a,b,c)$. If x is a center of $T(a,b,c)$, then we have

$$\|a-x, b-x\| = \|a-c, x-c\| = \|x-c, b-c\| = \|a-c, b-c\|$$

or

$$\|a-x, b-x\| = \|a-x, x-c\| = \|x-c, b-x\| = \|a-c, b-c\|.$$

Put $X = a-x, Y = b-x$ and $Z = x-c$. Then

$$\|X+Z, Y+Z\| = \|X, Y\| = \|X, Z\| = \|Y, Z\| \neq 0.$$

$\mathscr{X}$ is a strictly 2-convex, then $Z = X+Y$, namely, $x = \frac{a+b+c}{3}$.

Suppose that $T(a,b,c)$ has a unique center, and also $(\mathscr{X}, \|.,.\|)$ is not strictly 2-convex, then there exist x, y, and z in $\mathscr{X}$ such that

$$\|x+y,y+z\| = \max\{\|x,y\|,\|x,z\|,\|y,z\|\}$$

and $\|x,y\| = \|x,z\| = \|y,z\| \neq 0$ such that $x+y \neq z$. Thus $x \neq z-y$ and $y \neq z-x$. Put $a = z-y, b = z-x$ and $c = z$ in $T(a,b,c)$. Then $\|a-c,b-c\| \neq 0$. We will show that $t = z-(x+y) \in T(a,b,c)$.

$$\|a-c,b-c\| = \|z-y-z,z-x-z\| = \|x,y\|$$

$$\|a-t,b-t\| = \|z-y-z+x+y,z-x-z+x+y\| = \|x,y\|$$

$$\|a-c,t-c\| = \|z-y-z,z-x-y-z\| = \|y,x+y\| = \|x,y\|$$

and

$$\|t-c,b-c\| = \|z-x-y-z,z-x-z\| = \|x+y,y\| = \|x,y\|.$$

On the other hand, t is a center of $T(a,b,c)$. Hence $t = \frac{a+b+c}{3}$. Therefore $2(x+y) = 0$ or $\|x,y\| = 0$, which is a contradiction. □

Theorem 2.16. *Let $\mathscr{X}$ and $\mathscr{Y}$ be non-Archimedean 2-normed spaces over a valued field $\mathscr{K}$ and $\mathscr{Y}$ is strictly 2-convex. Assume that $f(x), f(y)$, and $f(z)$ are collinear when x,y, and z are collinear .If $f : \mathscr{X} \to \mathscr{Y}$ is an 2-isometry, then for all x,y, and z in $\mathscr{X}$ we have*

$$f\left(\frac{x+y+z}{3}\right) = \frac{f(x)+f(y)+f(z)}{3}.$$

Proof. Let $g(x) = f(x)-f(0)$. Then g is an 2-isometry and $g(0) = 0$. Let $x,y \in \mathscr{X}$ with $x \neq y$. Since $\dim \mathscr{X} > 1$, there exists an element $z \in \mathscr{X}$ such that $\|x-z,y-z\| \neq 0$. Since g is a 2-isometry, we have

$$\begin{aligned}
&\|g(x)-g\left(\frac{x+y+z}{3}\right), \ g(y)-g\left(\frac{x+y+z}{3}\right)\| \\
&\qquad = \|x-\frac{x+y+z}{3},y-\frac{x+y+z}{3}\| \quad \text{(by Lemma 2.14)} \\
&\qquad = \|x-z,y-z\| = \|g(x)-g(z),g(y)-g(z)\|
\end{aligned}$$

and

$$\begin{aligned}
\|g(x)-g(z),g\left(\frac{x+y+z}{3}\right)-g(z)\| &= \|x-z,\frac{x+y+z}{3}-z\| \quad \text{(by Lemma 2.14)} \\
&= \|x-z,y-z\| = \|g(x)-g(z),g(y)-g(z)\|
\end{aligned}$$

$$\begin{aligned}
\|g\left(\frac{x+y+z}{3}\right)-g(z),g(y)-g(z)\| &= \|\frac{x+y+z}{3}-z,y-z\| \quad \text{(by Lemma 2.14)} \\
&= \|x-z,y-z\| = \|g(x)-g(z),g(y)-g(z)\|.
\end{aligned}$$

Therefore $g(\frac{x+y+z}{3})$ is a center of $T(g(x),g(y),g(z))$. Since $\mathscr{Y}$ is strictly 2-convex by Theorem 2.15 we have

$$g\left(\frac{x+y+z}{3}\right)=\frac{g(x)+g(y)+g(z)}{3}.$$

□

Example 2.17. Consider $\mathbf{Q}_3\times\mathbf{Q}_3$ with norm $|(\alpha,\beta)|=\max\{|\alpha|_3,|\beta|_3\}$. Then $\mathbf{Q}_3\times\mathbf{Q}_3$ is a non-Archimedean normed space such that it is not strictly 2-convex. Define the 2-norm $|\cdot,\cdot|$ on $\mathbf{Q}_3\times\mathbf{Q}_3$ by

$$|u,v|=\begin{cases}|u||v| & u \text{ and } v \text{ are linearly independent}\\ 0 & \text{otherwise.}\end{cases}$$

Now define the mapping $f:\mathbf{Q}_3\times\mathbf{Q}_3\to\mathbf{Q}_3\times\mathbf{Q}_3$ by

$$f(\alpha,\beta)=\begin{cases}(2\alpha,2\beta) & \alpha=\frac{a}{b}3^0\text{and}\beta=\frac{c}{d}3^0\\ (\alpha,\beta) & \text{otherwise.}\end{cases}$$

Then f is a 2-isometry and $f(0,0)=0$, but if we put $x=(1,1), y=(3,3)$, and, $z=(0,0)$, then

$$f\left(\frac{x+y+z}{3}\right)\neq\frac{f(x)+f(y)+f(z)}{3}.$$

Therefore the assumption that $\mathscr{Y}$ is strictly 2- convex cannot be omitted in Theorem 2.16.

References

1. J.A. Baker, *Isometries in normed spaces*, Amer. Math. Monthly **78** (1971), 655–658.
2. H.Y. Chu, *On the Mazur–Ulam problem in linear 2-normed spaces*, J. Math. Anal. Appl. **327** (2007), 1041–1045.
3. R.W. Freese, Y.J. Cho and S.S. Kim, *Strictly 2-convex linear 2-normed spaces*, J.Korean Math. Soc. **29** (1992), 391–400.
4. K. Hensel, *Über eine neue Begründung der Theorie der algebraischen Zahlen*, Jahresber. Deutsch. Math. Verein **6** (1897), 83–88.
5. D.H Hyers, G. Isac and Th.M. Rassias, *On the Hyers-Ulam stability of ψ-additive mappings*, J. Approximation Theory. **72** (1993), no. 2, 131–137.
6. G. Isac and Th.M. Rassias, *Stability of Functional Equations in Several Variables*, Progress in Nonlinear Differential Equations and their Applications, 34. Birkhäuser Boston, Inc., Boston, MA, 1998.
7. S. Mazur and S. Ulam, *Sur les transformation isometriques d'espaces vectoriels normes*, C. R. Acad. Sci. Paris **194** (1932), 946–948.
8. M.S. Moslehian and Gh. Sadeghi, *A Mazur–Ulam theorem in non-Archimedean normed spaces*, Nonlinear Anal. *69* (2008), no. 10, 3405–3408.
9. Th.M. Rassias and M.S. Moslehian, *Stability of functional equations in non-Archimedean spaces*, Appl. Anal. Discrete Math. **1** (2007), no. 2, 325–334.
10. Th.M. Rassias and P. Semrl, *On the Mazur–Ulam theorem and the Aleksandrov problem for unit distance preserving mapping*, Proc. Amer. Math. Soc. **114** (1992), 989–993.
11. Th.M. Rassias and P. Wagner, *Volume preserving mappings in the spirit of the Mazur-Ulam theorem*, Aequationes Math. **66** (2003), no. 1-2, 85–89.

12. Th.M. Rassias and S. Xiang, *On Mazur-Ulam theorem and mappings which preserve distances*, Nonlinear Funct. Anal. Appl. 5 (2000), no. 2, 61–66.
13. Th.M. Rassias and S. Xiang, *On mappings with conservative distances and the Mazur-Ulam theorem*, Univ. Beograd. Publ. Elektrotehn. Fak. Ser. Mat. **11** (2000), 1–8.
14. A.C.M. van Rooij, *Non-Archimedean Functional Analysis*, Monographs and Textbooks in Pure and Applied Math., 51. Marcel Dekker, New York, 1978.
15. J. Wang, *On the generalizations of the Mazur-Ulam isometric theorem*, J. Math. Anal. Appl. **263** (2001), no. 2, 510–521.
16. S. Xiang, *On the Mazur-Ulam theorem and the solution of two problems of Rassias*, Nonlinear Funct. Anal. Appl. **12** (2007), no. 1, 99–105.

Chapter 3
Fixed Points and Generalized Stability for ψ-Additive Mappings of Isac–Rassias Type

Liviu Cădariu and Viorel Radu

Dedicated to the memory of Professor George Isac

Abstract Some results of G. Isac and Th. M. Rassias on ψ-additive mappings will be slightly extended by proving a stability theorem for functions defined on generalized α-normed spaces and taking values in β-normed spaces. We will show that some well-known theorems concerning the stability of Cauchy's functional equation can be obtained as consequences of our results.

3.1 Introduction

Starting from a question of S. M. Ulam concerning the stability of group homomorphisms, D. H. Hyers gave a purely constructive solution in the case of Cauchy functional equation in Banach spaces (see, e.g., [14], [17]). For the sake of convenience we note here his result:

Proposition 3.1. *Suppose that $(E, ||\cdot||_E)$ is a real normed space, $(F, ||\cdot||_F)$ is a real Banach space and $f : E \to F$ is a given function such that the following condition holds*

$$||f(x+y) - f(x) - f(y)||_F \leq \delta, \textit{ for all } x, y \in E,$$

for some $\delta > 0$. Then there exists a unique additive function $a : E \to F$ such that

Liviu Cădariu
"Politehnica" University of Timişoara, Department of Mathematics, Piaţa Victoriei 2, 300006, Timişoara, Romania, e-mail: liviu.cadariu@mat.upt.ro, lcadariu@yahoo.com

Viorel Radu
West University of Timişoara, Faculty of Mathematics and Computer Science, Department of Mathematics, Vasile Pârvan 4, 300223, Timişoara, Romania, e-mail: radu@math.uvt.ro

P.M. Pardalos et al. (eds.), *Nonlinear Analysis and Variational Problems*,
Springer Optimization and Its Applications 35, DOI 10.1007/978-1-4419-0158-3_3,

$$\|f(x) - a(x)\|_F \leq \delta, \textit{ for all } x \in E. \tag{3.1}$$

This phenomenon, called *Hyers–Ulam stability*, has been extensively investigated for different functional equations or inequalities (see the expository papers [15, 9, 29] and the books [17, 22, 7]).

Subsequently, the result of Hyers has been generalized by considering unbounded Cauchy differences ([1], [2], [28], [8]).

Although different approaches (for example, the invariant mean technique [33, 34], or based on the sandwich theorems [25]), have been used to obtain stability results for functional equations, nevertheless, almost all proofs used the *direct method*, imagined by D. H. Hyers: starting from the given function f, the additive function $a : E \to F$ verifying (3.1) is explicitly constructed by the following formula:

$$a(x) = \lim_{n\to\infty} \frac{1}{2^n} f(2^n x).$$

As it was observed in [26], the *existence* of the solution a and the *estimation* (3.1) can be easily obtained from the *fixed point alternative* for a strictly contractive operator defined on a suitable generalized metric function space:

$$Jh(x) := \frac{1}{q} h(qx),$$

so that the stability properties are related to some fixed points. Also in [26], the same method is used to prove a result of G. Isac and Th. M. Rassias concerning the stability of ψ-additive mappings (see Proposition 3.16 below).

It is worth noticing that G. Isac and Th. M. Rassias (see [18, 19]) introduced the concept of ψ-additive mappings and proved the following necessary and sufficient stability property:

Proposition 3.2. *Let E_1 be a real normed space and E_2 be a real Banach space. Let us consider a function $\psi : [0;\infty) \to [0;\infty)$, which satisfies the following conditions*

(I) $\lim_{t\to\infty} \frac{\psi(t)}{t} = 0$;

(II) $\psi(ts) \leq \psi(t)\psi(s)$, *for all* $t,s \in [0;\infty)$;

(III) $\psi(t) < t$, *for all* $t > 1$.

(IV) ψ *is monotone increasing on* $\mathbb{R}_+$.

(V) $\psi(t+s) \leq \psi(t) + \psi(s)$, *for all* $t,s \in [0;\infty)$.

Then the mapping $f : E_1 \to E_2$ satisfies the inequality

$$|||f(x+y) - f(x) - f(y)||| \leq \varepsilon(\psi(\|x\|) + \psi(\|y\|)$$

for some $\varepsilon > 0$ and for all $x,y \in E_1$, iff there exists a constant δ and a unique additive function $a : E_1 \to E_2$, such that

$$|||f(x) - a(x)||| \leq \delta\psi(\|x\|), \forall x \in E_1.$$

In the current note we will slightly extend this result by proving a stability theorem for mappings defined on generalized α-normed spaces and taking values in β-normed spaces. In fact, we prove a stability theorem for Cauchy equations in β-normed spaces (cf. [3] for the direct method and [6] for a particular case). We also show that some well-known theorems of Th. M. Rassias, Z. Gajda, J. M. Rassias, G. Isac and Th. M Rassias, K. W. Kim and H. M. Jun, can be obtained as applications of our results.

3.2 Stability Properties for Cauchy Equation in β-Normed Spaces

For the sake of convenience, we recall some necessary notions and results, used in the sequel.

Let E be a vector space over the real or the complex field $\mathbb{K}, \alpha \in \mathbb{R}_+$, and $\beta \in (0,1]$.

Definition 3.3. A mapping $||.||_\alpha : E \to \mathbb{R}_+$ is called an *h-functional of order* α iff it has the property

$$(h_\alpha) : ||\lambda \cdot x||_\alpha \leq |\lambda|^\alpha \cdot ||x||_\alpha, \textit{ for all } \lambda \in \mathbb{K}, \text{for all } x \in E.$$

As usual, E is identified with $E \times \{0\}$ in $E \times E$, so that $||x||_\alpha = ||(x,0)||_\alpha$ for all $x \in E$ and each h-functional of order α on $E \times E$.

Definition 3.4. A mapping $||\cdot||_\beta : E \to \mathbb{R}_+$ is called a β*-norm* iff it has the following properties:

$n^I_\beta : ||x||_\beta = 0 \Longleftrightarrow x = 0;$

$n^{II}_\beta : ||\lambda \cdot x||_\beta = |\lambda|^\beta \cdot ||x||_\beta, \text{for all } x \in E;$

$n^{III}_\beta : ||x+y||_\beta \leq ||x||_\beta + ||y||_\beta, \text{for all } x, y \in E.$

For more details see, e.g., [32].

In the sequel, we shall also make use of the following alternative of fixed point ([24], see also [31] or [35]). For an extensive treatment of fixed point theory with applications to a variety of problems in nonlinear analysis the reader is referred to the book of G. Isac, D. H. Hyers and Th. M. Rassias [16].

Proposition 3.5. *Suppose we are given a complete generalized metric space* (X,d)*, i.e., one for which d may assume infinite values, and a strictly contractive mapping* $A : X \to X$*, with the Lipschitz constant* $L < 1$*. Then, for each given element* $x \in X$*,* ***either***

(A_1) $d(A^n x, A^{n+1} x) = +\infty$*, for all* $n \geq 0$*,*

or

(A_2) *There exists k such that* $d(A^n x, A^{n+1} x) < +\infty$*, for all* $n \geq k$*.*

Actually, if (A_2) *holds, then*

(A_{21}) *The sequence* $(A^n x)$ *is convergent to a fixed point* y^* *of* A;
(A_{22}) y^* *is the unique fixed point of* A *in* $Y := \{y \in X,\ d(A^k x, y) < +\infty\}$;
(A_{23}) $d(y, y^*) \le \frac{1}{1-L} d(y, Ay)$, *for all* $y \in Y$.

Remark 3.6. The fixed point y^*, if it exists, is not necessarily unique *in the whole space* X; it may depend on the starting approximation. It is worth noting that, in the case (A_2), the pair (Y,d) is a complete *metric* space and $A(Y) \subset Y$. Therefore the properties $(A_{21})-(A_{23})$ follow from Banach's Contraction Principle (cf. [13], [23], [31] or [35]).

In 1996, G. Isac and Th. M. Rassias [20] were the first to prove new fixed point theorems by applying the Hyers–Ulam stability approach.

The following theorem, extending a stability result in [6], is proven by using the fixed point method.

Theorem 3.7. *Let* E_1 *be a linear space over a (real or complex) field,* E_2 *a complete* β*-normed space, and* $q_i = \begin{cases} m, & \text{if } i=0 \\ \frac{1}{m}, & \text{if } i=1 \end{cases}$, *with* $m \ge 2$ *a fixed integer. Assume that the mapping* $f : E_1 \to E_2$ *satisfies the condition* $f(0) = 0$ *and an inequality of the form*

$$(\mathbf{C}_\varphi) \qquad \|f(x+y) - f(x) - f(y)\|_\beta \le \varphi(x,y), \text{ for all } x,y \in E_1,$$

where $\varphi : E_1 \times E_1 \to \mathbb{R}_+$.

If there exists a positive constant $L < 1$ *such that the mapping*

$$x \to \theta(x) = \sum_{j=2}^{m} \varphi\left(\frac{x}{m}, \frac{j-1}{m}x\right)$$

has the property

$$(\mathbf{H_i}) \qquad \theta(x) \le L \cdot q_i^\beta \cdot \theta\left(\frac{x}{q_i}\right), \text{ for all } x \in E_1,$$

and the mapping φ *satisfies the condition*

$$(\mathbf{H_i^*}) \qquad \lim_{n\to\infty} \frac{\varphi(q_i^n x, q_i^n y)}{q_i^{n\beta}} = 0,$$

then there exists a unique additive mapping $a : E_1 \to E_2$ *such that*

$$(\mathbf{Est_i}) \qquad \|f(x) - a(x)\|_\beta \le \frac{L^{1-i}}{1-L}\theta(x), \text{for all } x \in E_1.$$

Proof. Consider the set

$$X := \{h : E_1 \to E_2, h(0) = 0\}$$

and introduce the *generalized metric* on X:

$$(GM) \quad d(g,h) = d_{\theta\beta}(g,h) = \inf\left\{C \in \overline{\mathbb{R}}_+, \|g(x) - h(x)\|_\beta \le C\theta(x), \forall x \in E_1\right\}$$

It is easy to see that (X,d) *is complete*.

Now we will consider the mapping $J : X \to X$, defined by

$$(OP_i) \qquad Jh(x) := \frac{1}{q_i} h(q_i x).$$

Remember that $q_i = m^{1-2i}, i \in \{0,1\}$.

Step 1. By using the hypothesis $(\mathbf{H}_i)$, we show that J is strictly contractive on X. Namely, we have, for any $g,h \in X$:

$$d(g,h) < C \Longrightarrow \|g(x) - h(x)\|_\beta \le C\theta(x), \text{ for all } x \in E_1 \Longrightarrow$$

$$\left\| \frac{1}{q_i} g(q_i x) - \frac{1}{q_i} h(q_i x) \right\|_\beta \le LC\theta(x), \forall x \in E_1 \Longrightarrow d(Jg, Jh) \le LC.$$

Therefore we see that

$$(CC) \qquad d(Jg,Jh) \le Ld(g,h), \text{ for all } g,h \in X,$$

that is, J is a *strictly contractive* self-mapping of X, with the Lipschitz constant L.

Step 2. We show that $d(f,Jf) < \infty$.

If we set $y = jx$ in the condition $(\mathbf{C}_\varphi)$, then we see that

$$\|f((j+1)x) - f(x) - f(jx)\|_\beta \le \varphi(x, jx), \textit{ for all } x \in E_1,$$

hence

$$\|f(mx) - mf(x)\|_\beta \le \sum_{j=1}^{m-1} \|f((j+1)x) - f(x) - f(jx)\|_\beta \le \sum_{j=2}^{m} \varphi(x,(j-1)x), \tag{3.2}$$

for all $x \in E_1$.

If $(\mathbf{H_0})$ holds, then

$$\left\| \frac{f(mx)}{m} - f(x) \right\|_\beta \le \frac{1}{m^\beta} \sum_{j=2}^{m} \varphi(x,(j-1)x) = \frac{\theta(mx)}{m^\beta} \le L\theta(x),$$

hence

$$\|Jf(x) - f(x)\|_\beta \le L \cdot \theta(x), \text{ for all } x \in E_1.$$

On the other side, replacing x by $\frac{x}{m}$ in (3.2), we obtain

$$\left\|f(x)-mf\left(\frac{x}{m}\right)\right\|_\beta \le \sum_{j=2}^{m}\varphi\left(\frac{x}{m},\frac{j-1}{m}x\right)=\theta(x).$$

If $(\mathbf{H_1})$ is satisfied, then

$$\|Jf(x)-f(x)\|_\beta \le 1\cdot\theta(x),\ \text{for all } x\in E_1,$$

that is, $d(f,Jf)\le L^{1-i}<\infty, i\in\{0,1\}$.

Step 3. Using the fixed point alternative (see Proposition 3.5), we obtain the existence of a mapping $a: X\to X$ with the following three properties:

- The mapping a is a fixed point of J, that is

$$a(mx)=m\,a(x)\,,\text{for all } x\in E_1 \tag{3.3}$$

and a is the unique fixed point of J in the set

$$Y=\{g\in X, d(f,g)<+\infty\}.$$

This says that a is the unique mapping with *both* the properties (3.3)–(3.4), where

$$\exists C\in(0,\infty)\ \text{such that}\ \|a(x)-f(x)\|_\beta\le C\theta(x),\ \text{for all } x\in E_1. \tag{3.4}$$

- $d(J^n f,a)\to 0$, which implies the equality

$$\lim_{n\to\infty}\frac{f(q_i{}^n x)}{q_i{}^n}=a(x),\ \text{for all } x\in E_1. \tag{3.5}$$

- $d(f,a)\le \frac{1}{1-L}d(f,Jf)$, which implies the inequality

$$d(f,a)\le\frac{L^{1-i}}{1-L},$$

whence $(\mathbf{Est_i})$ is seen to be true.

Step 4. The additivity of a immediately follows from $(\mathbf{C}_\varphi)$, $(\mathbf{H_i^*})$ and (3.5): If in $(\mathbf{C}_\varphi)$ we replace x by $q_i^n x$ and y by $q_i^n y$, then we obtain

$$\left\|\frac{f(q_i{}^n(x+y))}{q_i{}^n}-\frac{f(q_i{}^n x)}{q_i{}^n}-\frac{f(q_i{}^n y)}{q_i{}^n}\right\|_\beta \le \frac{\varphi(q_i{}^n x, q_i{}^n y)}{q_i{}^{n\beta}},\ \text{for all } x,y\in E_1,$$

and letting $n\to\infty$ we get

$$a(x+y)=a(x)+a(y),\ \text{for all } x,y\in E_1,$$

which ends the proof. □

As an immediate consequence of Theorem 3.7, we have the following

Corollary 3.8. *Let E_1, E_2 be two linear spaces over the same (real or complex) field. Suppose that we are given an h-functional of order α on $E_1\times E_1$ and E_2 is a com-*

plete β-normed space, with $\alpha \neq \beta$. In these conditions we have the following ***stability property****:*
For each $\varepsilon > 0$, there exists $\delta(\varepsilon) > 0$ such that, for every mapping $f : E_1 \to E_2$ which satisfies

$$(\mathbf{C}_{\alpha\beta}) \qquad ||f(x)+f(y)-f(x+y)||_\beta \leq \delta(\varepsilon)\cdot||(x,y)||_\alpha, \textit{ for all } x,y \in E_1,$$

there exists a unique mapping $a : E_1 \to E_2$, with the properties

$$(\mathbf{Add}) \qquad a(x+y) = a(x)+a(y), \textit{ for all } x,y \in E_1$$

and

$$(\mathbf{Est}_{\alpha\beta}) \qquad ||f(x)-a(x)||_\beta \leq \varepsilon\cdot\sum_{j=2}^{m}||x,(j-1)x||_\alpha, \textit{ for all } x \in E_1.$$

Proof. Having in mind Theorem 3.7, we take

$$\varphi(x,y) := \delta(\varepsilon)\cdot\|(x,y)\|_\alpha, \text{ for all } x,y \in E_1$$

(appearing in the hypothesis $(\mathbf{C}_{\alpha\beta})$). Since

$$\frac{\varphi(q_i{}^n x, q_i{}^n y)}{q_i{}^{n\beta}} \leq \delta(\varepsilon)\cdot q_i{}^{n(\alpha-\beta)}\cdot\|(x,y)\|_\alpha \to 0,$$

then $(\mathbf{H_i^*})$ is true.

For $q_0 = m$ and $\alpha - \beta < 0$, we have

$$\begin{aligned}\theta(mx) &= \delta(\varepsilon)\cdot\sum_{j=2}^{m}||x,(j-1)x||_\alpha \leq \delta(\varepsilon)\cdot m^\alpha\sum_{j=2}^{m}\left|\left|\frac{x}{m},\frac{(j-1)x}{m}\right|\right|_\alpha \\ &= m^{\alpha-\beta}\cdot m^\beta\cdot\theta(x) = L\cdot m^\beta\cdot\theta(x).\end{aligned}$$

For $q_1 = \frac{1}{m}$ and $\alpha - \beta > 0$, we have

$$\begin{aligned}\theta(x) &= \delta(\varepsilon)\cdot\sum_{j=2}^{m}\left|\left|\frac{x}{m},\frac{(j-1)x}{m}\right|\right|_\alpha \leq \delta(\varepsilon)\cdot m^{-\alpha}\sum_{j=2}^{m}||x,(j-1)x||_\alpha \\ &= m^{\beta-\alpha}\cdot m^{-\beta}\cdot\theta(mx) = L\cdot m^{-\beta}\cdot\theta(mx).\end{aligned}$$

Hence either $(\mathbf{H}_0)$ holds with $L = m^{\alpha-\beta} < 1$ or $(\mathbf{H}_1)$ holds with $L = m^{\beta-\alpha} < 1$. Then there exists a *unique additive* mapping $a : E_1 \to E_2$ such that either

$$(\mathbf{Est_0}) \qquad \|f(x)-a(x)\|_\beta \leq \frac{L}{1-L}\theta(x), \text{ for all } x \in E_1$$

holds, with $L = m^{\alpha-\beta}$, or

$$(\mathbf{Est_1}) \qquad \|f(x)-a(x)\|_\beta \le \frac{1}{1-L}\theta(x), \text{ for all } x \in E_1$$

holds, with $L = m^{\beta-\alpha}$.

Thus, the inequality $(\mathbf{Est}_{\alpha\beta})$ holds true for $\delta(\varepsilon) = \varepsilon \cdot (m^\beta - m^\alpha)$, respectively $\delta(\varepsilon) = \varepsilon \cdot (m^\alpha - m^\beta)$. □

For $m = 2$ in Theorem 3.7, we obtain our result proved in [6]:

Theorem 3.9. *Let E_1 be a linear space over a (real or complex) field, E_2 a complete β-normed space, and $q_i = \begin{cases} 2, & \text{if } i = 0 \\ \frac{1}{2}, & \text{if } i = 1 \end{cases}$. Assume that the mapping $f : E_1 \to E_2$ satisfies the condition $f(0) = 0$ and an inequality of the form*

$$(\mathbf{C}_\varphi) \qquad \|f(x+y) - f(x) - f(y)\|_\beta \le \varphi(x,y), \text{ for all } x,y \in E_1$$

where $\varphi : E_1 \times E_1 \to \mathbb{R}_+$.

If, for some positive constant $L < 1$, the mapping

$$x \to \theta(x) = \varphi\left(\frac{x}{2}, \frac{x}{2}\right)$$

verifies

$$(\mathbf{H}_i) \qquad \theta(x) \le L \cdot q_i^\beta \cdot \theta\left(\frac{x}{q_i}\right), \text{ for all } x \in E_1,$$

and the mapping φ satisfies

$$(\mathbf{H_i^*}) \qquad \lim_{n\to\infty} \frac{\varphi(q_i^n x, q_i^n y)}{q_i^{n\beta}} = 0,$$

then there exists a unique additive mapping $a : E_1 \to E_2$ such that

$$(\mathbf{Est_i}) \qquad \|f(x)-a(x)\|_\beta \le \frac{L^{1-i}}{1-L}\theta(x), \text{ for all } x \in E_1.$$

Another application is obtained by taking

$$\varphi(x,y) := \delta(\varepsilon) \cdot \|(x,y)\|_\alpha, \text{ for all } x,y \in E_1$$

in Theorem 3.9 (or by using Corollary 3.8 for $m = 2$):

Corollary 3.10. *Let E_1, E_2 be two linear spaces over the same (real or complex) field. Suppose that we are given an h-functional of order α on $E_1 \times E_1$ and E_2 is a complete β-normed space, with $\alpha \neq \beta$. In these conditions we have the following* ***stability property****:*
For each $\varepsilon > 0$, there exists $\delta(\varepsilon) > 0$ such that, for every mapping $f : E_1 \to E_2$ which satisfies

$(\mathbf{C}_{\alpha\beta})$ $\qquad ||f(x)+f(y)-f(x+y)||_\beta \leq \delta(\varepsilon)\cdot||(x,y)||_\alpha,$ *for all* $x,y\in E_1,$

there exists a unique mapping $a: E_1 \to E_2$, *with the properties*

$(\mathbf{Add})$ $\qquad a(x+y)=a(x)+a(y),$ *for all* $x,y\in E_1$

and

$(\mathbf{Est}_{\alpha\beta})$ $\qquad ||f(x)-a(x)||_\beta \leq \varepsilon\cdot||(x,x)||_\alpha,$ *for all* $x\in E_1$.

Proof. Indeed, since

$$\frac{\varphi\left(q_i{}^n x, q_i{}^n y\right)}{q_i{}^{n\beta}} \leq \delta(\varepsilon)\cdot q_i{}^{n(\alpha-\beta)}\cdot\|(x,y)\|_\alpha \to 0,$$

then $(\mathbf{H}_\mathbf{i}^*)$ is true.

Moreover, it is easy to see that either $(\mathbf{H}_0)$ holds with $L=2^{\alpha-\beta}<1$ or $(\mathbf{H}_1)$ holds with $L=2^{\beta-\alpha}<1$. Therefore there exists a *unique additive* mapping $a: E_1 \to E_2$ such that either

$(\mathbf{Est_0})$ $$\|f(x)-a(x)\|_\beta \leq \frac{L}{1-L}\theta(x), \text{ for all } x\in E_1$$

holds, with $L=2^{\alpha-\beta}$, or

$(\mathbf{Est_1})$ $$\|f(x)-a(x)\|_\beta \leq \frac{1}{1-L}\theta(x), \text{ for all } x\in E_1$$

holds, with $L=2^{\beta-\alpha}$.

Thus, the inequality $(\mathbf{Est}_{\alpha\beta})$ holds true for $\delta(\varepsilon)=\varepsilon\cdot(2^\beta-2^\alpha)$, respectively $\delta(\varepsilon)=\varepsilon\cdot(2^\alpha-2^\beta)$. □

3.3 Other Examples and Applications

Using Theorems 3.7, 3.9 or Corollary 3.10, we can obtain some known stability theorems. It should be noted that in particular cases we only have to recognize, as we have made in the proof of Corollaries 3.8 and 3.10, that the conditions of type (H_i) and (H_i^*) do really hold.

Let E_1 be a normed space and E_2 a Banach space, with the norms $||\cdot||$, respectively $|||\cdot|||$.

Example 3.11. Let us consider $\varphi(x,y)=||(x,y)||_p := \delta(\varepsilon)\cdot(||x||^p+||y||^p)$, with $\delta(\varepsilon)=\frac{\varepsilon}{2}|2-2^p|$, and $||z||_\beta := |||z|||$, which is a 1-*norm* on E_2. For all $p\neq 1$ the estimation relation will have the form

$$|||f(x)-a(x)||| \leq \varepsilon\cdot||x||^p, \text{ for all } x\in E_1.$$

For $p \in (0,1)$ we obtain a partial result of Th. M. Rassias [28] (see also T. Aoki [1]) and particular cases of the results in [8]. Notice that in the year 1990, during the 27th International Symposium on Functional Equations, Th. M. Rassias asked the question whether his theorem (see [28]) can also be proved for real values of $p > 1$. In the year 1991, Z. Gajda [10], following the same approach as in Th. M. Rassias [28], by replacing the value of n with $-n$, gave an affirmative solution to this question for real values of p greater than 1.

Example 3.12. Let us set $\varphi(x,y) = ||(x,y)||_{p+q} := \delta(\varepsilon)\cdot(||x||^p \cdot ||y||^q)$, with $\delta(\varepsilon) = \frac{\varepsilon}{2}(2-2^{p+q})$ and $||z||_\beta := |||z|||$, which is a 1- *norm* on E_2. For $0 \leq p+q < 1$ we obtain the stability result of J. M. Rassias [27], with the estimation

$$|||f(x)-a(x)||| \leq \varepsilon \cdot ||x||^{p+q}, \text{ for all } x \in E_1.$$

Example 3.13. Th. M. Rassias and P. Šemrl [30] considered a control function $H(||x||,||y||)$, where $H:\mathbb{R}_+^2 \to \mathbb{R}_+$ is homogenous mapping of order p, with $p > 0, p \neq 1$. They proved that $||f(x)-T(x)||$ is bounded by $\frac{H(1,1)}{|2-2^p|}\cdot||x||^p$.

K. W. Jun and H. M. Kim [21] further extended the above result in [30]:

Proposition 3.14. *Let us consider E_1 a real normed space, E_2 a Banach space, $p > 0$, with $p \neq 1$ and $H:\mathbb{R}_+^2 \to \mathbb{R}_+$, a function such that $H(tx,ty) \leq t^p H(x,y)$, for all $t,x,y \in \mathbb{R}_+$. Assume that the mapping $f: E_1 \to E_2$ satisfies an inequality of the form*

$$|||f(x+y)-f(x)-f(y)||| \leq H(||x||,||y||), \forall x,y \in E_1.$$

Then there exists a unique additive mapping $a: E_1 \to E_2$, such that:

$$|||f(x)-T(x)||| \leq \frac{1}{|2-2^p|}\cdot H(||x||,||x||) \leq \frac{H(1,1)}{|2-2^p|}\cdot||x||^p, \forall x,y \in E_1.$$

This is an immediate consequence of our Corollary 3.10. Indeed let us consider $p \in (0,1)\cup(1,\infty)$ and a mapping $H:\mathbb{R}_+\times\mathbb{R}_+ \longrightarrow \mathbb{R}_+$, with the properties

(a) $H(0,0) = 0$;

(b) $H(tx,ty) \leq t^p \cdot H(x,y)$, for all $x,y,t \in \mathbb{R}_+$.

Now if we take $||(x,y)||_p := \frac{1}{\delta(\varepsilon)}\cdot H(||x||,||y||)$, which is an h-functional of order p on $E_1 \times E_1$, and $||z||_1 := |||z|||$, then we obtain Proposition 3.14.

Example 3.15. In 1993, G. Isac and Th. M. Rassias [18] proved the following stability result for the additive Cauchy functional equation by introducing ψ-additive mappings. They also indicated the connection of Hyers–Ulam stability of mappings with the concept of asymptotic stability leading to new open problems.

Proposition 3.16. *Let E_1 be a real normed space and E_2 be a real Banach space. Let us consider a function $\psi:[0;\infty)\to[0;\infty)$, which satisfies the following three conditions*

(*I*) $\lim\limits_{t\longrightarrow\infty}\frac{\psi(t)}{t} = 0$;

(II) $\psi(ts) \le \psi(t)\psi(s)$, *for all* $t, s \in [0;\infty)$;
(III) $\psi(t) < t$, *for all* $t > 1$.
If a mapping $f : E_1 \to E_2$ *satisfies the inequality*

$$|||f(x+y) - f(x) - f(y)||| \le \varepsilon(\psi(\|x\|) + \psi(\|y\|)$$

for some $\varepsilon > 0$ *and for all* $x, y \in E_1$*, then there exists a unique additive function* $a : E_1 \to E_2$*, such that*

$$|||f(x) - a(x)||| \le \frac{2\varepsilon}{2 - \psi(2)}\psi(\|x\|), \forall x \in E_1.$$

The *proof* can be obtained by using Corollary 3.10:

Consider $\|z\|_\beta = ||z||_1 := |||z|||$, which is a 1-*norm* on E_2 and $||(x,y)||_1 := \frac{1}{\delta(\varepsilon)} \cdot (\psi(||x||) + \psi(||y||)$, with $\delta(\varepsilon) = \varepsilon(2 - \psi(2))$, which is an h-functional of order 1 on $E_1 \times E_1$, because

$$||(\lambda x, \lambda y)||_1 = \frac{\psi(||\lambda x||) + \psi(||\lambda y||)}{\delta(\varepsilon)} \le \frac{\psi(|\lambda|) \cdot (\psi(||x||) + \psi(||y||))}{\delta(\varepsilon)} \le |\lambda| \cdot ||(x,y)||_1.$$

Example 3.17. We give a slight generalization of Proposition 3.2 by considering mappings defined on α-normed spaces and taking values in β-normed spaces. The proof of our result is given by using Theorem 3.7 (see also [12]).

Theorem 3.18. *Let* E_1 *be an* α*-real normed space and* E_2 *be a complete* β*-normed space. Let us consider a function* $\psi : [0;\infty) \to [0;\infty)$*, which satisfies the following conditions*

(II) $\psi(ts) \le \psi(t)\psi(s)$, *for all* $t, s \in [0;\infty)$;
(III′) there exists an integer $m \ge 2$ *such that* $\psi(m)^\alpha < m^\beta$.
(IV) ψ *is monotone increasing on* $\mathbb{R}_+$.
(V) $\psi(t+s) \le \psi(t) + \psi(s)$, *for all* $t, s \in [0;\infty)$.
The mapping $f : E_1 \to E_2$ *satisfies the inequality*

$$||f(x+y) - f(x) - f(y)||_\beta \le \varepsilon(\psi(\|x\|_\alpha) + \psi(\|y\|_\alpha)$$

for some $\varepsilon > 0$ *and for all* $x, y \in E_1$*, iff there exists a constant* $\delta = \delta(\varepsilon, m)$ *and a unique additive function* $a : E_1 \to E_2$*, such that*

$$(Est_{IR}) \qquad ||f(x) - a(x)||_\beta \le \delta(\varepsilon, m) \cdot \psi(\|x\|_\alpha), \forall x \in E_1.$$

Proof. For the direct implication, we consider in Theorem 3.7

$$\varphi(x,y) := \varepsilon \cdot (\psi(||x||_\alpha) + \psi(||y||_\alpha)), \forall x, y \in E_1.$$

It is clear that

$$\theta(x) := \sum_{j=2}^{m} \varphi\left(\frac{x}{m}, \frac{j-1}{m}x\right) = \varepsilon\left((m-1)\psi\left(\left|\left|\frac{x}{m}\right|\right|_\alpha\right) + \sum_{j=2}^{m} \psi\left(\left|\left|\frac{j-1}{m}x\right|\right|_\alpha\right)\right)$$

We will show that the conditions $(\mathbf{H_0})$ and $(\mathbf{H_0^*})$ are satisfied.

If $q_0 = m$, from (II), it follows

$$\begin{aligned}\theta(mx) &= \varepsilon\left((m-1)\psi(||x||_\alpha) + \sum_{j=2}^{m}\psi((j-1)||x||_\alpha)\right)\\ &\le \varepsilon\cdot\psi(m)^\alpha\left((m-1)\psi\left(\frac{||x||_\alpha}{m^\alpha}\right) + \sum_{j=2}^{m}\psi\left(\frac{j-1}{m^\alpha}||x||_\alpha\right)\right)\\ &= m^\beta\cdot\frac{\psi(m)^\alpha}{m^\beta}\cdot\theta(x) = m^\beta\cdot L\cdot\theta(x), \forall x\in E_1,\end{aligned}$$

therefore $(\mathbf{H_0})$ holds, with $L := \frac{\psi(m)^\alpha}{m^\beta} < 1$.

Since

$$\frac{\varphi(m^n x, m^n y)}{m^{\beta n}} \le \varepsilon\cdot\left(\frac{\psi(m)^\alpha}{m^\beta}\right)^n\cdot(\psi(||x||_\alpha)+\psi(||x||_\alpha)) \overset{n\to\infty}{\longrightarrow} 0,$$

then $(\mathbf{H_0^*})$ is true.

Then there exists a *unique additive* mapping $a: E_1 \to E_2$ such that

$$||f(x) - a(x)||_\beta \le \frac{L}{1-L}\theta(x), \text{ for all } x\in E_1$$

holds, with $L = \frac{\psi(m)^\alpha}{m^\beta}$. But

$$\theta(x) \le \varepsilon\cdot\psi(||x||_\alpha)\cdot\left((m-1)\psi\left(\frac{1}{m}\right) + \sum_{j=2}^{m}\psi\left(\frac{j-1}{m}\right)\right), \forall x\in E_1.$$

Thus, the inequality (Est_{IR}) holds true for

$$\delta(\varepsilon, m) = \varepsilon\cdot\frac{\psi(m)^\alpha}{m^\beta - \psi(m)^\alpha}\left((m-1)\psi\left(\frac{1}{m}\right) + \sum_{j=2}^{m}\psi\left(\frac{j-1}{m}\right)\right).$$

Conversely, we have, by using (IV) and (V), that

$$\begin{aligned}&||f(x+y) - f(x) - f(y)||_\beta\\ &\le ||f(x+y) - a(x+y)||_\beta + ||f(x) - a(x)||_\beta + ||f(y) - a(y)||_\beta\\ &\le \delta(\varepsilon,m)\cdot(\psi(||x+y||_\alpha) + \psi(||x||_\alpha) + \psi(||y||_\alpha))\\ &\le \delta(\varepsilon,m)\cdot(\psi(||x||_\alpha + ||y||_\alpha) + \psi(||x||_\alpha) + \psi(||y||_\alpha))\\ &\le 2\delta(\varepsilon,m)\cdot(\psi(\|x\|_\alpha) + \psi(\|y\|_\alpha)).\end{aligned}$$

□

Remark 3.19. For $\alpha = \beta = 1$ in Theorem 3.18 we obtain Proposition 3.2.

References

1. Aoki, T., *On the stability of the linear transformation in Banach spaces,* J. Math. Soc. Japan *2* (1950), 64–66.
2. Bourgin, D. G., *Classes of transformations and bordering transformations,* Bull. Amer. Math. Soc. *57* (1951), 223–237.
3. Cădariu, L., *A general theorem of stability for the Cauchy's equation*, Bull. Şt. Univ. "Politehnica" Timişoara, Seria Matematică -Fizică, Tom *47(61)*, no. 2(2002), 14–28.
4. Cădariu, L. & Radu, V., *Fixed points and the stability of Jensen's functional equation, J. Inequal. Pure and Appl. Math.*, *4(1)* (2003), Art.4 (http://jipam.vu.edu.au).
5. Cădariu L. & Radu V., *Fixed points and the stability of quadratic functional equations*, Analele Universităţii de Vest din Timişoara, *41*(1) (2003), 25–48.
6. Cădariu L. & Radu V., *On the stability of the Cauchy functional equation: a fixed points approach*, Iteration theory (ECIT '02), (J. Sousa Ramos, D. Gronau, C. Mira, L. Reich, A. N. Sharkovsky - Eds.), Grazer Math. Ber., Bericht Nr. *346* (2004), 43–52.
7. Czerwik S., *Functional Equations and Inequalities in Several Variables*, World Scientific Publishing Company, Singapore, 2002.
8. Forti, G. L., *An existence and stability theorem for a class of functional equations*, Stochastica. *4* (1980), 23–30.
9. Forti, G. L., *Hyers-Ulam stability of functional equations in several variables*, Aequationes Math. *50* (1995), 143–190.
10. Gajda, Z., *On stability of additive mappings*, Int. J. Math. Math. Sci., *14* (1991), 431–434.
11. Găvruţa, P., *A generalization of the Hyers Ulam Rassias stability of approximately additive mappings,* J. Math. Anal. Appl. *184* (1994) , 431–436.
12. Găvruţa, P., *On a problem of G. Isac and Th. M. Rassias concerning the stability of mappings*, J. Math. Anal. Appl. *261*, No. 2 (2001), 543–553.
13. Goebel, K. & Kirk, W. A., *Topics in Metric Fixed Point Theory*, Cambridge Studies in Advanced Mathematics, *28*, Cambridge, CUP, viii+244 pp, 1990.
14. Hyers, D. H., *On the stability of the linear functional equation*, Proc. Natl. Acad. Sci. USA *27* (1941), 222–224.
15. Hyers D.H. & Rassias Th. M., *Approximate homomorphisms*, Aequationes Mathematicae *44* (1992), 125–153.
16. Hyers, D. H., Isac G., & Rassias, Th. M., *Topics in Nonlinear Analysis and Applications*, World Scientific Publishing Co., Singapore, 1997.
17. Hyers, D. H., Isac G., & Rassias, Th. M., *Stability of Functional Equations in Several Variables*, Birkhäuser, Basel, 1998.
18. Isac G. & Rassias, Th. M., *On the Hyers Ulam stability of ψ-additive mappings*, J. Approx. Theory, *72* (1993), 131–137.
19. Isac G. & Rassias, Th. M., *Functional inequalities for approximately additive mappings*, in "Stability of mappings of Hyers-Ulam type" (Th. M. Rassias and J. Tabor, Eds.) pp. 117–125, Hadronic Press, Palm Harbor, Florida, 1994.
20. Isac G. & Rassias, Th. M., *Stability of $\psi-$additive mappings: Applications to nonlinear analysis*, Int. J. Math. Math. Sci. *19* (2) (1996), 219–228.
21. Jun, K. W. & Kim, H. M., *Remarks on the stability of additive functional equation*, Bull. Korean Math. Soc., *38* (2001), No. 4, 679–687.
22. Jung, S. M., *Hyers-Ulam-Rassias Stability of Functional Equations in Mathematical Analysis*, Hadronic Press, Palm Harbor, Florida, 2002.
23. Kirk, W. A. & Sims, B. (Eds.), *Handbook of Metric Fixed Point Theory*, Kluwer Academic Publishers, Dordrecht, xiii+703 pp, 2001.
24. Margolis, B. & Diaz, J. B., *A fixed point theorem of the alternative for contractions on a generalized complete metric space*, Bull. Amer. Math. Soc., *74* (1968), 305–309.
25. Páles, Zs., *Generalized stability of the Cauchy functional equation.* Aequationes Math., *56(3)* (1998), 222–232.

26. Radu, V., *The fixed point alternative and the stability of functional equations*, Fixed Point Theory *4*, (2003), No. 1, 91–96.
27. Rassias, J. M., *On a new approximation of approximately linear mappings by linear mappings,* Discuss. Math. *7* (1985), 193–196.
28. Rassias, Th. M., *On the stability of the linear mapping in Banach spaces*, Proc. Amer. Math. Soc., *72* (1978), 297–300.
29. Rassias, Th. M., *On the stability of functional equations and a problem of Ulam*, Acta Appl. Math., *62* (2000), 23–130.
30. Rassias, Th. M. & Šemrl, P., *On the Hyers-Ulam stability of linear mappings*, J. Math. Anal. Appl. *173* (1993), 325–338.
31. Rus, I. A., *Principles and Applications of Fixed Point Theory*, Ed. Dacia, Cluj-Napoca, 1979 (in Romanian).
32. Simons, S., *Boundedness in linear topological spaces*, Trans. Amer. Math. Soc. *113* (1964), 169–180.
33. Székelyhidi, L., *On a stability theorem*, C. R. Math. Rep. Acad. Sci. Canada, *3(5)* (1981), 253–255.
34. Székelyhidi, L., *The stability of linear functional equations,* C. R. Math. Rep. Acad. Sci. Canada, *3(2)* (1981), 63–67.
35. Zeidler, E., *Nonlinear Functional Analysis and Its Applications. I. Fixed-Point Theorems,* Springer-Verlag, New York, 1986.

Chapter 4
A Remark on W*-Tensor Products of W*-Algebras

Corneliu Constantinescu

Dedicated to the memory of Professor George Isac

Abstract Let E be a W*-algebra, T a hyperstonian compact space, $\mathscr{C}(T)$ the W*-algebra of continuous scalar valued functions on T, and $\mathscr{F}(T,E)$ the set of bounded maps $x : T \longrightarrow E$ such that for every element a of the predual of E the function

$$T \longrightarrow \mathbb{K}, \quad t \longmapsto \langle x_t \, , a \rangle$$

is continuous. We define for every $x \in \mathscr{F}(T,E)$ an element $\tilde{x} \in \mathscr{C}(T)\bar{\otimes}E$ such that the map

$$\mathscr{F}(T,E) \longrightarrow \mathscr{C}(T)\bar{\otimes}E, \quad x \longmapsto \tilde{x}$$

is a bijective isometry of ordered involutive Banach spaces (where this structure on $\mathscr{F}(T,E)$ is defined pointwise). In general $\mathscr{F}(T,E)$ is not an algebra for the pointwise multiplication, but for $x, y, z \in \mathscr{F}(T,E)$ we characterize the case when $\tilde{x}\tilde{y} = \tilde{z}$.

4.1 Introduction

Let T be a hyperstonian compact space, $\mathscr{C}(T)$ the W*-algebra of continuous scalar valued functions on T, E a W*-algebra, and $l^\infty(T,E)$ the W*-algebra of bounded maps of T in E (isomorphic to the C*-direct product of the family $(E)_{t\in T}$). As it is known, $\mathscr{C}(T) \otimes E$ is isomorphic to the C*-subalgebra $\mathscr{C}(T,E)$ of $l^\infty(T,E)$ of continuous maps $x : T \longrightarrow E$. We try to give a similar description for $\mathscr{C}(T)\bar{\otimes}E$. For this we denote by $\mathscr{F}(T,E)$ the set of elements x of $l^\infty(T,E)$ such that for every element a of the predual of E, the map

Corneliu Constantinescu
Bodenacherstr. 53, CH 8121 Benglen, Switzerland, e-mail: constant@math.ethz.ch

P.M. Pardalos et al. (eds.), *Nonlinear Analysis and Variational Problems*, Springer Optimization and Its Applications 35, DOI 10.1007/978-1-4419-0158-3_4,

$$T \longrightarrow \mathbb{K}, \quad t \longmapsto \langle x_t \, , a \rangle$$

is continuous and show that for every $x \in \mathscr{F}(T,E)$ there is exactly one $\tilde{x} \in \mathscr{C}(T)\bar{\otimes}E$ such that

$$\tilde{x}(\mu \otimes a) = \int \langle x_t \, , a \rangle \, \mathrm{d}\mu(t)$$

for all a belonging to the predual of E and for all elements μ belonging to the predual of $\mathscr{C}(T)$ (these are exactly the normal measures on T). Moreover $\mathscr{F}(T,E)$ is an ordered involutive Banach subspace of $l^\infty(T,E)$ and the map

$$\mathscr{F}(T,E) \longrightarrow \mathscr{C}(T)\bar{\otimes}E, \quad x \longmapsto \tilde{x}$$

is a bijective isometry of ordered involutive Banach spaces (Theorem 4.3). In general $\mathscr{F}(T,E)$ is not a subalgebra of $l^\infty(T,E)$ but for $x,y,z \in \mathscr{F}(T,E)$ we give a characterization of the case when $\tilde{x}\tilde{y} = \tilde{z}$ (Theorem 4.12).

We use the notation and terminology of [1]. For W*-tensor products of W*-algebras we use [2]. In the sequel we give a list of some notations used in this paper.

1. $\mathbb{K}$ denotes the field of real numbers $\mathbb{R}$ or the field of complex numbers. The whole theory is developed in parallel for the real and complex case (but the proofs coincide). $\mathbb{N}$ denotes the set of natural numbers ($0 \notin \mathbb{N}$).
2. For all i, j we denote by δ_{ij} the Kronecker's symbol:

$$\delta_{ij} := \begin{cases} 1 & \text{if} \quad i = j \\ 0 & \text{if} \quad i \neq j \end{cases} .$$

3. If X is a set, then we denote for every subset A of X by $e_A := e_A^X$ the characteristic function of A in X, i.e., the function on X equal to 1 on A and equal to 0 on $X \setminus A$.
4. If F is an ordered vector space, then F_+ denotes the convex cone of positive elements of F.
5. If T is a hyperstonian space, then a subset of T is called nowhere dense if its closure contains no interior points. A countable union of nowhere dense sets of T is nowhere dense. If $P(t)$ is a statement about a point $t \in T$, then we say that $P(t)$ holds for almost all $t \in T$ if the set of $t \in T$ for which $P(t)$ does not hold is nowhere dense.
6. For every normed space E we denote by $E^\#$ its unit ball, i.e.,

$$E^\# := \{ \, x \in E \mid \| \, x \, \| \leq 1 \, \} \, .$$

7. Let E be a W*-algebra. We denote by $\ddot{E}$ the predual of E and by $E_{\ddot{E}}$ the vector space E endowed with the locally convex topology of pointwise convergence on $\ddot{E}$ (i.e., with the weak*-topology). In particular $E^\#_{\ddot{E}}$ is compact.
8. If I is a set and E is a W*-algebra, then $l^\infty(I,E)$ denotes the W*-algebra of bounded maps $x : I \longrightarrow E$ (i.e., the C*-direct product of the family $(E)_{i \in I}$).
9. If T is a hyperstonian space and E is a W*-algebra, then $\mathscr{C}(T,E)$ denotes the C*-subalgebra of $l^\infty(T,E)$ of continuous maps, $\mathscr{C}(T)$ the W*-algebra $\mathscr{C}(T,\mathbb{K})$,

$\mathscr{C}(T)^{\pi}$ the predual of $\mathscr{C}(T)$ (i.e., $\mathscr{C}(T)^{\pi} := \overset{\cdot\cdot}{\widehat{\mathscr{C}(T)}}$); its elements are exactly the normal measures on T. For every $\mu \in \mathscr{C}(T)^{\pi}_{+}$, $Supp\,\mu$ denotes the support of the positive normal measure μ on T.

10. If $(H_i)_{i\in I}$ is a family of Hilbert spaces, then $\{i \in I\}\, H_i$ denotes the Hilbert sum of this family.
11. Let H be a Hilbert space. $\mathscr{L}(H)$ denotes the W*-algebra of operators on H. For $\xi, \eta \in H$ we put

$$\langle\cdot\,|\,\eta\,\rangle\xi : H \longrightarrow H, \quad \zeta \longmapsto \langle\zeta\,|\,\eta\,\rangle\xi ,$$

$$\widetilde{\widehat{\langle\cdot\,|\,\eta\,\rangle\xi}} : \mathscr{L}(H) \longrightarrow \mathbb{K}, \quad u \longmapsto \langle u\xi\,|\,\eta\,\rangle ;$$

then $\langle\cdot\,|\,\eta\,\rangle\xi \in \mathscr{L}(H)$ and $\widetilde{\widehat{\langle\cdot\,|\,\eta\,\rangle\xi}} \in \overset{\cdot\cdot}{\widehat{\mathscr{L}(H)}}$.

12. If F, G are Banach spaces, then $F \odot G$ denotes the algebraic tensor product of F and G, γ the greatest subcross norm on $F \odot G$, $F \otimes_{\gamma} G$ the completion of $F \odot G$ with respect to γ, and γ' the dual norm of γ on $F' \odot G'$, where F' and G' denote the duals of F and G, respectively ([3] T3).

Throughout this paper T is a hyperstonian space, E a von Neumann algebra on a Hilbert space H, and $\mathscr{F}(T,E) := \{\, x \in l^{\infty}(T,E) \mid T \longrightarrow E_{\ddot{E}}, \quad t \longmapsto x_t$ is continuous$\}$

4.2 The Ordered Involutive Banach Space

Proposition 4.1. *Let U be an open dense set of T and $x \in l^{\infty}(U,E)$ such that*

$$U \longrightarrow \mathbb{K}, \quad t \longmapsto \langle x_t\xi\,|\,\eta\,\rangle$$

is continuous for every $\xi, \eta \in H$. Then there is a unique $y \in \mathscr{F}(T,E)$ such that $x_t = y_t$ for every $t \in U$.

The uniqueness is obvious. Let $a \in \ddot{E}$. By [1] Proposition 6.3.4.2 f), there is a sequence $(\xi_n, \eta_n)_{n\in\mathbb{N}}$ in $H \times H$ such that

$$\sum_{n\in\mathbb{N}} \|\, \xi_n \,\|\|\, \eta_n \,\| = \|\, a \,\|$$

and

$$\langle z, a\rangle = \sum_{n\in\mathbb{N}} \langle z\xi_n\,|\,\eta_n\,\rangle$$

for every $z \in E$. Then

$$U \longrightarrow \mathbb{K}, \quad t \longmapsto \sum_{n\in\mathbb{N}} \langle x_t\xi_n \,|\, \eta_n \rangle = \langle x_t \,, a \rangle$$

is continuous. Thus the map

$$U \longrightarrow E_{\ddot{E}}, \quad t \longmapsto x_t$$

is continuous. Since $E^{\#}_{\ddot{E}}$ is compact, this map may be extended to a continuous map

$$y : T \longrightarrow E_{\ddot{E}}$$

which possesses the required properties. □

Lemma 4.2. *Put*

$$\mathscr{M} := \mathscr{C}(T)^{\pi}_{+}, \qquad M := (\{\mu \in \mathscr{M}\} L^2(\mu)) \otimes H$$

and denote by N the subset of M of the vectors of the form

$$\sum_{i\in I} f_i \otimes \xi_i \,,$$

where $(f_i)_{i\in I}$ is a finite family in an $L^2(\mu)$ for a $\mu \in \mathscr{M}$ such that

$$\{\, t\in T \mid f_i(t) \neq 0\} \cap \{\, t \in T \mid f_j(t) \neq 0\} = \emptyset$$

for all distinct $i, j \in I$ and $\xi_i \in H$ for every $i \in I$. Then N is dense in M.

For $\mu \in \mathscr{M}$, A, B clopen sets of $Supp\,\mu$, and $\xi, \eta \in H$,

$$e_A \otimes \xi + e_B \otimes \eta = e_{A\setminus B} \otimes \xi + e_{B\setminus A} \otimes \eta + e_{A\cap B} \otimes (\xi + \eta) \in N.$$

By complete induction

$$\sum_{i\in I} e_{A_i} \otimes \xi_i \in N$$

for every finite family $(e_{A_i})_{i\in I}$ of clopen sets of $Supp\,\mu$ and $\xi_i \in H$ for every $i \in I$. The continuous functions on $Supp\,\mu$ can be approximated uniformly by step functions and $\mathscr{C}(Supp\,\mu)$ is dense in $L^2(\mu)$. Moreover $\mathscr{M}$ is closed with respect to finite sums. Thus the set of functions of the above form

$$\sum_{i\in I} e_{A_i} \otimes \xi_i$$

is dense in M. By the above, N is dense in M. □

Theorem 4.3. *For every $x \in \mathscr{F}(T,E)$ there is a unique $\tilde{x} \in \mathscr{C}(T)\bar{\otimes}E$ such that*

$$\langle \tilde{x}(f \otimes \xi) \,|\, g \otimes \eta \rangle = \int \langle x_t \xi \,|\, \eta \rangle\, f(t)\overline{g(t)}\, \mathrm{d}\mu(t)$$

for all $\mu \in \mathscr{C}(T)^{\pi}_{+}$, $f, g \in L^2(\mu)$, and $\xi, \eta \in H$ and the map

$$\mathscr{F}(T,E) \longrightarrow \mathscr{C}(T)\bar{\otimes}E, \quad x \longmapsto \tilde{x}$$

is a bijective isometry of ordered involutive Banach spaces.

Let us consider the linear map

$$\tilde{x} : \mathscr{C}(T)^{\pi} \odot \ddot{E} \longrightarrow \mathrm{I\!K}, \quad \mu \otimes a \longmapsto \int \langle x_t \,, a \rangle \, \mathrm{d}\mu(t) \,.$$

If $(\mu_i, a_i)_{i\in I}$ is a finite family in $\mathscr{C}(T)^{\pi} \times \ddot{E}$, then

$$\left| \tilde{x}\left(\sum_{i\in I} \mu_i \otimes a_i \right) \right| = \left| \sum_{i\in I} \int \langle x_t \,, a_i \rangle \, \mathrm{d}\mu_i(t) \right| \le \sum_{i\in I} \| x \| \, \| a_i \| \, \| \mu_i \|$$

so

$$\left| \tilde{x}\left(\sum_{i\in I} \mu_i \otimes a_i \right) \right| \le \| x \| \, \gamma\left(\sum_{i\in I} \mu_i \otimes a_i \right),$$

$$\gamma'(\tilde{x}) \le \| x \| \,,$$

where γ denotes the greatest subcross norm on $\mathscr{C}(T)^{\pi} \odot \ddot{E}$ and γ' its dual norm ([3] T.3). In particular $\tilde{x}$ may be extended continuously to $\mathscr{C}(T)^{\pi} \otimes_{\gamma} \ddot{E}$ and $\gamma'(\tilde{x}) \le \| x \|$.

Let $t \in T$, $a \in \ddot{E}$, $\| a \| \le 1$, and $\varepsilon > 0$. There is an open neighborhood U of t such that

$$| \langle x_s \,, a \rangle - \langle x_t \,, a \rangle | < \varepsilon$$

for all $s \in U$. Let $\mu \in \mathscr{C}(T)^{\pi}_{+}$ with

$$Supp\, \mu \subset U \,, \qquad \mu(U) = 1 \,.$$

Then

$$\gamma(\mu \otimes a) = \| \mu \| \| a \| \le 1$$

so

$$\gamma'(\tilde{x}) \ge | \langle \tilde{x}, \mu \otimes a \rangle | = \left| \int \langle x_s \,, a \rangle \, \mathrm{d}\mu(s) \right|$$

$$\ge \left| \int \langle x_t \,, a \rangle \, \mathrm{d}\mu(s) \right| - \left| \int \langle x_s - x_t \,, a \rangle \, \mathrm{d}\mu(s) \right| \ge | \langle x_t \,, a \rangle | - \varepsilon \,.$$

Since a and ε are arbitrary

$$\gamma'(\tilde{x}) \ge \| x_t \| \,, \qquad \gamma'(\tilde{x}) \ge \| x \| \,, \qquad \gamma'(\tilde{x}) = \| x \| \;.$$

Since $\mathscr{C}(T)\bar{\otimes}E$ is isomorphic to the dual of $\mathscr{C}(T)^{\pi} \otimes_{\gamma} E$ we may identify $\tilde{x}$ with an element of $\mathscr{C}(T)\bar{\otimes}E$ and $\| \tilde{x} \| = \| x \|$. Thus the map

$$\mathscr{F}(T,E) \longrightarrow \mathscr{C}(T)\bar{\otimes}E, \quad x \longmapsto \tilde{x}$$

is a linear map preserving the norms and this map is obviously involutive.

Let $u \in \mathscr{C}(T)\bar{\otimes}E$. For $a \in \ddot{E}$ the map

$$\mathscr{C}(T)^{\pi} \longrightarrow \mathrm{I\!K}, \quad \mu \longmapsto \langle u, \mu \otimes a \rangle$$

is continuous with norm at most $\| u \| \| a \|$ so there is an $\tilde{a} \in \mathscr{C}(T)$ with

$$\langle u, \mu \otimes a \rangle = \langle \tilde{a}, \mu \rangle , \qquad \| \tilde{a} \| \leq \| u \| \| a \| \ .$$

For $t \in T$, the map

$$\ddot{E} \longrightarrow \mathrm{I\!K}, \quad a \longmapsto \tilde{a}(t)$$

is continuous with norm at most $\| u \|$ so there is an $x_t \in E$ with

$$\langle x_t , a \rangle = \tilde{a}(t)$$

for every $a \in \ddot{E}$. Then $\| x \| \leq \| u \|$ and $x \in \mathscr{F}(T,E)$. For $(\mu,a) \in \mathscr{C}(T)^{\pi} \times \ddot{E}$,

$$\langle \tilde{x}, \mu \otimes a \rangle = \int \langle x_t , a \rangle \, \mathrm{d}\mu(t) = \int \tilde{a}(t) \, \mathrm{d}\mu(t) = \langle \tilde{a}, \mu \rangle = u(\mu \otimes a),$$

i.e., $\tilde{x} = u$ and the map is surjective.

Let $x \in \mathscr{F}(T,E)$, $\mu \in \mathscr{C}(T)^{\pi}$, $f,g \in L^2(\mu)$, and $\xi, \eta \in H$. Then

$$\langle \tilde{x}(f \otimes \xi) \,|\, g \otimes \eta \rangle = \left\langle \tilde{x}, \widetilde{\langle \cdot \,|\, g \otimes \eta \rangle (f \otimes \xi)} \right\rangle$$

$$= \left\langle \tilde{x}, \widetilde{\langle \cdot \,|\, g \rangle f} \otimes \widetilde{\langle \cdot \,|\, \eta \rangle \xi} \right\rangle = \int \left\langle x_t , \widetilde{\langle \cdot \,|\, \eta \rangle \xi} \right\rangle f(t)\overline{g(t)} \, \mathrm{d}\mu(t)$$

$$= \int \langle x_t \xi \,|\, \eta \rangle f(t)\overline{g(t)} \, \mathrm{d}\mu(t) \, .$$

The uniqueness is obvious.

Let $x \in \mathscr{F}(T,E)$ and assume $\tilde{x} \geq 0$. For $\mu \in \mathscr{C}(T)^{\pi}_{+}$, $f \in L^2(\mu)$, and $\xi \in H$,

$$0 \leq \langle \tilde{x}(f \otimes \xi) \,|\, f \otimes \xi \rangle = \int \langle x_t \xi \,|\, \xi \rangle |f(t)|^2 \, \mathrm{d}\mu(t)$$

so

$$\langle x_t \xi \,|\, \xi \rangle \geq 0$$

for all $t \in \mathit{Supp}\,\mu$. By continuity

$$\langle x_t \xi \,|\, \xi \rangle \geq 0$$

for all $t \in T$. Since ξ is arbitrary, $x_t \in E_+$ for all $t \in T$, i.e., $x \in \mathscr{F}(T,E)_+$.

Let now $x \in \mathscr{F}(T,E)_+$. If we put $\mathscr{M} := \mathscr{C}(T)^{\pi}_{+}$, then $\mathscr{C}(T)\bar{\otimes}E$ is faithfully represented on the Hilbert space $(\{\mu \in \mathscr{M}\} L^2(\mu)) \otimes H$. Let $\mu \in \mathscr{M}$ and let $(f_i, \xi_i)_{i \in I}$ be a finite family in $L^2(\mu) \times H$ such that

$$\{\, t \in T \mid f_i(t) \neq 0 \} \cap \{\, t \in T \mid f_j(t) \neq 0 \} = \emptyset$$

for all distinct $i, j \in I$. Then

$$\left\langle \tilde{x}\left(\sum_{i\in I} f_i \otimes \xi_i\right) \,\middle|\, \sum_{i\in I} f_i \otimes \xi_i \right\rangle = \sum_{i,j\in I} \langle \tilde{x}(f_i \otimes \xi_i) \,|\, f_j \otimes \xi_j \rangle$$

$$= \sum_{i,j\in I} \int \langle x_t \xi_i \,|\, \xi_j \rangle f_i(t)\overline{f_j(t)}\,\mathrm{d}\mu(t) = \sum_{i\in I} \int \langle x_t \xi_i \,|\, \xi_i \rangle |f_i(t)|^2\,\mathrm{d}\mu(t) \geq 0\,.$$

By Lemma 4.2, the set of vectors of the above form $\sum_{i\in I} f_i \otimes \xi_i$ is dense in

$$(\{\mu \in \mathscr{M}\} L^2(\mu)) \otimes H\,,$$

so we have $\tilde{x} \geq 0$. □

Remark 4.4. In general $\mathscr{F}(T,E)$ is not a subalgebra of $l^\infty(T,E)$.

a) Let T be the *Stone–Čech* compactification of $\mathbb{N}$, l^2 the Hilbert space of square-summable sequences in $\mathbb{K}$ with the canonical orthonormal basis $(e_n)_{n\in\mathbb{N}}$, $E := \mathscr{L}(l^2)$, and $x \in \mathscr{F}(T,E)$ defined by

$$x_n = \langle \cdot \,|\, e_1 \rangle e_n$$

for every $n \in \mathbb{N}$ (Proposition 4.1). For $\xi, \eta \in l^2$ and $n \in \mathbb{N}$,

$$\left\langle x_n\,, \widetilde{\langle \cdot \,|\, \eta \rangle \xi} \right\rangle = \langle \xi \,|\, e_1 \rangle \langle e_n \,|\, \eta \rangle\,, \qquad \left\langle x_n^*\,, \widetilde{\langle \cdot \,|\, \eta \rangle \xi} \right\rangle = \langle \xi \,|\, e_n \rangle \langle e_1 \,|\, \eta \rangle$$

so $x = x^* = 0$ on $T \setminus \mathbb{N}$. For $n \in \mathbb{N}$,

$$x_n^* x_n = (\langle \cdot \,|\, e_n \rangle e_1) \circ (\langle \cdot \,|\, e_1 \rangle e_n) = \langle \cdot \,|\, e_1 \rangle e_1\,,$$

$$\left\langle x_n^* x_n\,, \widetilde{\langle \cdot \,|\, e_1 \rangle e_1} \right\rangle = \langle x_n^* x_n e_1 \,|\, e_1 \rangle = 1$$

so $x^* x \notin \mathscr{F}(T,E)$.

b) In this example E is even commutative. Let T be the *Stone–Čech* compactification of $\mathbb{N}$, λ the Lebesgue measure on the interval $[0,1]$, E the canonical von Neumann algebra $L^\infty(\lambda)$ on the Hilbert space $L^2(\lambda)$, $(f_n)_{n\in\mathbb{N}}$ the Rademacher sequence of functions, and $x \in \mathscr{F}(T,E)$ defined by $x_n := f_n$ for every $n \in \mathbb{N}$ (Proposition 4.1). For $\xi, \eta \in L^2(\lambda)$,

$$\langle x_n \xi \,|\, \eta \rangle = \int f_n(t)\xi(t)\overline{\eta(t)}\,\mathrm{d}\lambda(t)\,, \qquad \langle x_n^2 \xi \,|\, \eta \rangle = \langle \xi \,|\, \eta \rangle$$

so $x = 0$ on $T \setminus \mathbb{N}$ and $x^2 \notin \mathscr{F}(T,E)$.

Corollary 4.5. *The following are equivalent for every downward directed set $\mathscr{G}$ of $\mathscr{F}(T,E)$:*

(a) $\inf_{x\in\mathscr{G}}\tilde{x}=0$ *(where the infimum is considered in $\mathscr{C}(T)\bar{\otimes}E$).*
(b) For every $\xi\in H$,

$$\inf_{x\in\mathscr{G}}\langle x_t\xi\,|\,\xi\,\rangle=0$$

for almost all $t\in T$.

$a\Rightarrow b$. For $\mu\in\mathscr{C}(T)^{\pi}_{+}$, $f\in L^2(\mu)$, and $\xi\in H$, by Theorem 4.3 (and [1] Theorem 4.4.1.8 b)),

$$0=\left\langle\,(\inf_{x\in\mathscr{G}}\tilde{x})(f\otimes\xi)\,|\,f\otimes\xi\,\right\rangle=\inf_{x\in\mathscr{G}}\langle\tilde{x}(f\otimes\xi)\,|\,f\otimes\xi\,\rangle$$

$$=\inf_{x\in\mathscr{G}}\int\langle x_t\xi\,|\,\xi\,\rangle\,|f(t)|^2\,\mathrm{d}\mu(t)=\int\inf_{x\in\mathscr{G}}\langle x_t\xi\,|\,\xi\,\rangle\,|f(T)|^2\,\mathrm{d}\mu(t)$$

so

$$\inf_{x\in\mathscr{G}}\langle x_t\xi\,|\,\xi\,\rangle=0$$

for μ-almost all $t\in T$. Since μ is arbitrary

$$\inf_{x\in\mathscr{G}}\langle x_t\xi\,|\,\xi\,\rangle=0$$

for almost all $t\in T$.

$b\Rightarrow a$. Let $\mu\in\mathscr{C}(T)^{\pi}_{+}$ and $(f_i,\xi_i)_{i\in I}$ a finite family in $L^2(\mu)\times H$ such that

$$\{\,t\in T\mid f_i(t)\neq 0\}\cap\{\,t\in T\mid f_j(t)\neq 0\}=\emptyset$$

for all distinct $i,j\in I$. By Theorem 4.3 (and [1] Theorem 4.4.1.8 b))

$$\left\langle\,(\inf_{x\in\mathscr{G}}\tilde{x})\left(\sum_{i\in I}f_i\otimes\xi_i\right)\middle|\,\sum_{i\in I}f_i\otimes\xi_i\,\right\rangle$$

$$=\inf_{x\in\mathscr{G}}\left\langle\,\tilde{x}\left(\sum_{i\in I}f_i\otimes\xi_i\right)\middle|\,\sum_{i\in I}f_i\otimes\xi_i\,\right\rangle=\inf_{x\in\mathscr{G}}\sum_{i,j\in I}\langle\,\tilde{x}(f_i\otimes\xi_i)\,|\,f_j\otimes\xi_j\,\rangle$$

$$=\inf_{x\in\mathscr{G}}\sum_{i,j\in I}\int\langle\,x_t\xi_i\,|\,\xi_j\,\rangle\,f_i(t)\overline{f_j(t)}\,\mathrm{d}\mu(t)$$

$$=\inf_{x\in\mathscr{G}}\sum_{i\in I}\int\langle x_t\xi_i\,|\,\xi_i\,\rangle\,|f_i(t)|^2\,\mathrm{d}\mu(t)=0\,.$$

By Lemma 4.2

$$\inf_{x\in\mathscr{G}}\tilde{x}=0.$$

□

4.3 The Multiplication

Proposition 4.6. *If $x \in \mathscr{F}(T,E)$ and $y \in \mathscr{C}(T,E)$ then*

$$xy, yx \in \mathscr{F}(T,E), \qquad \widetilde{xy} = \tilde{x}\tilde{y}\,, \widetilde{yx} = \tilde{y}\tilde{x}\,.$$

The assertion is obvious if $y(T)$ is finite. The general case follows from the fact that

$$\{\, y \in \mathscr{C}(T,E) \mid y(T) \text{ is finite} \,\}$$

is dense in $\mathscr{C}(T,E)$. □

Proposition 4.7. *Let $x \in \mathscr{F}(T,E)$, $\mu \in \mathscr{C}(T)_+^\pi$, $\xi \in H$, and*

$$\alpha := \sup_{t \in T} \|\, x_t \xi \,\|_H \,.$$

Then for $f \in L^2 := L^2(\mu)$,

$$\|\, \tilde{x}(f \otimes \xi) \,\|_{L^2 \otimes H} \leq \alpha \,\|\, f \,\|_{L^2} \,.$$

Let $(g_i, \eta_i)_{i \in I}$ be a finite family in $L^2(\mu) \times H$ such that $\langle\, \eta_i \,|\, \eta_j \,\rangle = \delta_{ij}$ for all $i, j \in I$ and

$$y : T \longrightarrow H, \quad t \longmapsto \sum_{i \in I} g_i(t)\eta_i \,.$$

Then for $t \in T$,

$$\|\, y(t) \,\|_H^2 = \sum_{i \in I} |g_i(t)|^2$$

so $\|\, y \,\|_H \in L^2(\mu)$,

$$\|\, (\|\, y \,\|_H) \,\|_{L^2}^2 = \int \sum_{i \in I} |g_i(t)|^2 \,\mathrm{d}\mu(t) = \left\| \sum_{i \in I} g_i \otimes \eta_i \right\|_{L^2 \otimes H}^2 .$$

By Theorem 4.3,

$$\left| \left\langle \tilde{x}(f \otimes \xi) \,\middle|\, \sum_{i \in I} g_i \otimes \eta_i \right\rangle \right| = \left| \int \left\langle f(t) x_t \xi \,\middle|\, \sum_{i \in I} g_i(t)\eta_i \right\rangle \mathrm{d}\mu(t) \right|$$

$$\leq \int |f(t)| \;\|\, x_t \xi \,\|_H \,\|\, y(t) \,\|_H \;\mathrm{d}\mu(t) \leq \alpha \,\|\, f \,\|_{L^2} \,\|\, (\|\, y \,\|_H) \,\|_{L^2}$$

$$= \alpha \,\|\, f \,\|_{L^2} \left\| \sum_{i \in I} g_i \otimes \eta_i \right\|_{L^2 \otimes H}$$

so

$$\|\, \tilde{x}(f \otimes \xi) \,\|_{L^2 \otimes H} \leq \alpha \,\|\, f \,\|_{L^2} \,.$$

□

Proposition 4.8. *Let $x,y \in \mathscr{F}(T,E)$ such that*

$$T \longrightarrow H, \quad t \longmapsto y_t\xi$$

is continuous for every $\xi \in H$. Then $xy \in \mathscr{F}(T,E)$ and $\tilde{x}\tilde{y} = \widetilde{xy}$.

For $\xi, \eta \in H$ and $t, t_0 \in T$,

$$|\langle (x_t y_t - x_{t_0} y_{t_0})\xi \,|\, \eta \rangle| \leq |\langle x_t(y_t - y_{t_0})\xi \,|\, \eta \rangle| + |\langle (x_t - x_{t_0}) y_{t_0}\xi \,|\, \eta \rangle|$$

$$\leq \| x \| \, \| y_t\xi - y_{t_0}\xi \| \, \| \eta \| + |\langle (x_t - x_{t_0}) y_{t_0}\xi \,|\, \eta \rangle|$$

so

$$\lim_{t \to t_0} \langle x_t y_t \xi \,|\, \eta \rangle = \langle x_{t_0} y_{t_0} \xi \,|\, \eta \rangle \, .$$

By Proposition 4.1, $xy \in \mathscr{F}(T,E)$.

Let $\mu \in \mathscr{C}(T)^{\pi}_{+}$, $(f,\xi) \in L^2(\mu) \times H$, and $\varepsilon > 0$. There is a finite pairwise disjoint family $(U_i)_{i \in I}$ of clopen sets of T and a $(t_i)_{i \in I} \in \prod_{i \in I} U_i$ such that $T = \cup_{i \in I} U_i$ and

$$\| y_t\xi - y_{t_i}\xi \| < \varepsilon$$

for every $i \in I$ and $t \in U_i$. We denote by π the orthogonal projection of H on the vector subspace of H generated by ξ and put

$$z : T \longrightarrow \mathscr{L}(H), \quad t \longmapsto y_{t_i} \circ \pi \quad (t \in U_i),$$

$$g : T \longrightarrow \mathrm{I\!R}, \quad t \longmapsto \| (y_t - z_t)\xi \| \, .$$

Then $z \in \mathscr{C}(T, \mathscr{L}(H))$, $g \in \mathscr{C}(T)$, and $\| g \| < \varepsilon$. By Proposition 4.6, $xz \in \mathscr{F}(T, \mathscr{L}(H))$ and $\widetilde{xz} = \tilde{x}\tilde{z}$ so

$$\| \widetilde{xy}(f \otimes \xi) - \tilde{x}\tilde{y}(f \otimes \xi) \|_{L^2 \otimes H}$$

$$\leq \| \widetilde{\overbrace{xy - xz}}(f \otimes \xi) \|_{L^2 \otimes H} + \| \tilde{x}(\tilde{z} - \tilde{y})(f \otimes \xi) \|_{L^2 \otimes H} \, .$$

By Proposition 4.7,

$$\| \widetilde{\overbrace{xy - xz}}(f \otimes \xi) \|_{L^2 \otimes H} \leq \varepsilon \, \| x \| \, \| f \|_{L^2} \, ,$$

$$\| \tilde{x}(\tilde{z} - \tilde{y})(f \otimes \xi) \|_{L^2 \otimes H} \leq \varepsilon \, \| x \| \, \| f \|_{L^2}$$

so

$$\| (\widetilde{xy} - \tilde{x}\tilde{y})(f \otimes \xi) \|_{L^2 \otimes H} \leq 2\varepsilon \, \| x \| \, \| f \|_{L^2} \, ,$$

$$\widetilde{xy}(f \otimes \xi) = \tilde{x}\tilde{y}(f \otimes \xi)$$

and by Theorem 4.3, $\widetilde{xy} = \tilde{x}\tilde{y}$. □

Lemma 4.9. *Let A be a Borel set of T such that $T \setminus A$ is nowhere dense. Then the union of clopen sets of T contained in A is dense in T.*

Let U be the union of clopen sets of T contained in A and let $\mu \in \mathscr{C}(T)^{\pi}_{+}$ with

$$U \cap (Supp\,\mu) = \emptyset .$$

There is an increasing sequence $(K_n)_{n\in \mathbb{N}}$ of compact subsets of A with

$$\mu(A) = \sup_{n\in \mathbb{N}} \mu(K_n) .$$

For every $n \in \mathbb{N}$ let U_n be the interior of K_n. Then $U_n \subset U$ and

$$\mu(K_n) = \mu(U_n) = 0$$

for every $n \in \mathbb{N}$ so

$$\mu(T) = \mu(A) = 0$$

and U is dense in T. □

Proposition 4.10. *For every $x \in \mathscr{F}(T,E)$ and every separable Hilbert subspace K of H there is a dense open set U of T such that*

$$U \longrightarrow H, \quad t \longmapsto x_t \xi$$

is continuous for every $\xi \in K$.

Let $(\xi_n)_{n\in \mathbb{N}}$ be a dense sequence in K. For $n \in \mathbb{N}$ and $t \in T$,

$$\| x_t \xi_n \| = \sup_{\eta \in H^{\#}} \langle x_t \xi_n \,|\, \eta \rangle$$

so

$$T \longrightarrow \mathbb{R}, \quad t \longmapsto \| x_t \xi_n \|$$

is lower semicontinuous. By [1] Proposition 1.7.2.13 $a \Rightarrow c$, there is an $f_n \in \mathscr{C}(T)$ such that $\| x_t \xi_n \| \leq f_n(t)$ for every $t \in T$ and

$$\{ t \in T \mid \| x_t \xi_n \| \neq f_n(t) \}$$

is nowhere dense. The set

$$A := \bigcap_{n\in \mathbb{N}} \{ t \in T \mid \| x_t \xi_n \| = f_n(t) \}$$

is therefore a G_δ-set of T and $T \setminus A$ is nowhere dense. By Lemma 4.9, there is an open dense set U of T contained in A. For $t, t_0 \in U$ and $n \in \mathbb{N}$,

$$\| x_t \xi_n - x_{t_0} \xi_n \|^2 = \langle x_t \xi_n \,|\, x_t \xi_n \rangle - 2\,re \langle x_t \xi_n \,|\, x_{t_0} \xi_n \rangle + \langle x_{t_0} \xi_n \,|\, x_{t_0} \xi_n \rangle$$

$$= f_n(t)^2 - 2\,re \langle x_t \xi_n \,|\, x_{t_0} \xi_n \rangle + f_n(t_0)^2 ,$$

$$\lim_{t\to t_0} \| x_t \xi_n - x_{t_0} \xi_n \|^2 = f_n(t_0)^2 - 2 f_n(t_0)^2 + f_n(t_0)^2 = 0 .$$

The assertion follows now from the fact that $(\xi_n)_{n\in \mathbb{N}}$ is dense in K. □

Proposition 4.11. *Let* $x,y \in \mathscr{F}(T,E)$, $\xi \in H$, $\| \xi \| = 1$,

$$\pi : H \longrightarrow H, \quad \eta \longmapsto \langle \eta \,|\, \xi \rangle \xi ,$$

$$y' : T \longrightarrow \mathscr{L}(H), \quad t \longmapsto y_t \circ \pi .$$

(a) $y' \in \mathscr{F}(T,\mathscr{L}(H))$.
(b) There is a unique $u \in \mathscr{F}(T,\mathscr{L}(H))$ *such that* $u_t = x_t y'_t$ *for almost all* $t \in T$.
(c) $\tilde{u} = \tilde{x}\widetilde{y'}$ *and for* $\mu \in \mathscr{C}(T)^{\pi}_{+}$, $f \in L^2(\mu)$, *and* $\eta \in H$,

$$\widetilde{y'}(f \otimes \eta) = \langle \eta \,|\, \xi \rangle \tilde{y}(f \otimes \xi), \qquad \tilde{u}(f \otimes \xi) = \langle \eta \,|\, \xi \rangle \tilde{x}\tilde{y}(f \otimes \xi) .$$

a) For $\eta, \zeta \in H$ and $t \in T$,

$$\langle y'_t \eta \,|\, \zeta \rangle = \langle \eta \,|\, \xi \rangle \langle y_t \xi \,|\, \zeta \rangle$$

so

$$T \longrightarrow \mathrm{I\!K}, \quad t \longmapsto \langle y'_t \eta \,|\, \zeta \rangle$$

is continuous and the assertion follows from Proposition 4.1.

b) By Proposition 4.10, there is an open dense set U of T such that

$$T \longrightarrow \mathrm{I\!K}, \quad t \longmapsto y_t \xi$$

is continuous. Put

$$\mathscr{N} := \left\{ \nu \in \mathscr{C}(T)^{\pi}_{+} \;\middle|\; Supp\, \nu \subset U \right\},$$

$$V := \bigcup_{\nu \in \mathscr{N}} Supp\, \nu .$$

V is an open dense set of T.

Let $\nu \in \mathscr{N}$, $S := Supp\, \nu$, and

$$x_S : S \longrightarrow \mathscr{L}(H), \quad t \longmapsto x_t .$$

The map

$$y'_S : S \longrightarrow H, \quad t \longmapsto y'_t \eta$$

is continuous for all $\eta \in H$. By Proposition 4.8, $y'_S, x_S y'_S \in \mathscr{F}(S,\mathscr{L}(H))$, and $\widetilde{x_S y'_S} = \tilde{x}_S \widetilde{y'}_S$. In particular for $\eta, \zeta \in H$ the map

$$V \longrightarrow \mathrm{I\!K}, \quad t \longmapsto \langle x_t y'_t \eta \,|\, \zeta \rangle$$

is continuous so by Proposition 4.1, there is a $u \in \mathscr{F}(T,\mathscr{L}(H))$ with

$$u_t = x_t y'_t$$

for all $t \in V$. The uniqueness is obvious.

c) For $\nu \in \mathscr{N}$, $g,h \in L^2(\nu)$, and $\eta,\zeta \in H$, by Theorem 4.3,

$$\left\langle \widetilde{y'}(g\otimes\eta)\,|\,h\otimes\zeta \right\rangle = \int \langle y'_t\eta\,|\,\zeta\,\rangle g(t)\overline{h(t)}\,\mathrm{d}\nu(t)$$

$$= \langle \eta\,|\,\xi\,\rangle \int \langle y_t\xi\,|\,\zeta\,\rangle g(t)\overline{h(t)}\,\mathrm{d}\nu(t) = \langle \eta\,|\,\xi\,\rangle \langle \tilde{y}(g\otimes\xi)\,|\,h\otimes\zeta\,\rangle\,,$$

$$\widetilde{y'}(g\otimes\eta) = \langle \eta\,|\,\xi\,\rangle \tilde{y}(g\otimes\xi)\,,$$

$$\tilde{u}(g\otimes\eta) = \widetilde{xy'}(g\otimes\eta) = \tilde{x}\widetilde{y'}(g\otimes\eta) = \langle \eta\,|\,\xi\,\rangle \tilde{x}\tilde{y}(g\otimes\xi)\,.$$

There is a sequence $(U_n)_{n\in\mathbb{N}}$ of pairwise disjoint clopen subsets of

$$V\cap(Supp\,\mu)$$

such that

$$Supp\,\mu \setminus \left(\bigcup_{n\in\mathbb{N}} U_n\right)$$

is nowhere dense. We put for every $n \in \mathbb{N}$,

$$f_n : T \longrightarrow \mathbb{K}\,, \quad t \longmapsto \begin{cases} f(t) & \text{if} \quad t\in U_n \\ 0 & \text{if} \quad t\in T\setminus U_n \end{cases}.$$

Then $f_n \in L^2(\mu)$ for every $n \in \mathbb{N}$ and

$$f = \sum_{n\in\mathbb{N}} f_n \quad (\text{in } L^2(\mu))$$

so by the above,

$$\widetilde{y'}(f\otimes\eta) = \sum_{n\in\mathbb{N}} \widetilde{y'}(f_n\otimes\eta) = \langle \eta\,|\,\xi\,\rangle \sum_{n\in\mathbb{N}} \tilde{y}(f_n\otimes\xi) = \langle \eta\,|\,\xi\,\rangle \tilde{y}(f\otimes\xi)\,,$$

$$\tilde{u}(f\otimes\eta) = \sum_{n\in\mathbb{N}} \tilde{u}(f_n\otimes\eta) = \langle \eta\,|\,\xi\,\rangle \sum_{n\in\mathbb{N}} \tilde{x}\tilde{y}(f_n\otimes\xi) = \langle \eta\,|\,\xi\,\rangle \tilde{x}\tilde{y}(f\otimes\xi)\,.$$ □

Theorem 4.12. *The following are equivalent for all $x,y,z \in \mathscr{F}(T,E)$ and $\xi \in H$:*

(a) $x_ty_t\xi = z_t\xi$ *for almost all* $t \in T$.

(b) For all $\mu \in \mathscr{C}(T)^{\pi}_{+}$ *and* $f \in L^2(\mu)$,

$$\tilde{x}\tilde{y}(f\otimes\xi) = \tilde{z}(f\otimes\xi)\,.$$

We assume $\|\,\xi\,\| = 1$ and put

$$\pi : H \longrightarrow H\,, \quad \eta \longmapsto \langle \eta\,|\,\xi\,\rangle\xi\,,$$

$$y' : T \longrightarrow \mathscr{L}(H)\,, \quad t \longmapsto y_t\circ\pi\,, \qquad z' : T \longrightarrow \mathscr{L}(H)\,, \quad t \longmapsto z_t\circ\pi\,.$$

$a \Rightarrow b$. For $\eta \in H$,

$$x_t y'_t \eta = \langle \eta \,|\, \xi \,\rangle x_t y_t \xi \qquad z'_t \eta = \langle \eta \,|\, \xi \,\rangle z_t \xi$$

so $x_t y'_t = z'_t$ for almost all $t \in T$. By Proposition 4.11,

$$\tilde{x}\tilde{y}(f \otimes \xi) = \tilde{x}\widetilde{y'}(f \otimes \xi) = \widetilde{z'}(f \otimes \xi) = \tilde{z}(f \otimes \xi)\,.$$

$b \Rightarrow a$. By Proposition 4.11, there is a unique $u \in \mathscr{F}(T, \mathscr{L}(H))$ such that $u_t = x_t y'_t$ for almost all $t \in T$ and for $\mu \in \mathscr{C}(T)^{\pi}_{+}$, $f \in L^2(\mu)$, and $\eta \in H$,

$$\tilde{u}(f \otimes \eta) = \langle \eta \,|\, \xi \,\rangle \tilde{x}\tilde{y}(f \otimes \xi) = \langle \eta \,|\, \xi \,\rangle \tilde{z}(f \otimes \xi) = \widetilde{z'}(f \otimes \eta)$$

so $\tilde{u} = \widetilde{z'}$, $u = z'$ (Theorem 4.3), and

$$\{\, t \in T \mid x_t y_t \xi \neq z_t \xi \,\} = \{\, t \in T \mid x_t y_t \xi \neq z'_t \xi \,\} = \{\, t \in T \mid x_t y_t \xi \neq u_t \xi \,\}$$

is nowhere dense. □

Corollary 4.13. *If H is separable, then the following are equivalent for all $x, y, z \in \mathscr{F}(T, E)$:*

(a) $\tilde{x}\tilde{y} = \tilde{z}$.
(b) $x_t y_t = z_t$ *for almost all* $t \in T$.

Let $(\xi_n)_{n \in \mathbb{N}}$ be a dense sequence in H and for every $n \in \mathbb{N}$ put

$$A_n := \{\, t \in T \mid x_t y_t \xi_n \neq z_t \xi_n \,\}\,.$$

$a \Rightarrow b$. By Theorem 4.12 $b \Rightarrow a$, A_n is nowhere dense for every $n \in \mathbb{N}$ so

$$\{\, t \in T \mid x_t y_t \neq z_t \,\} = \bigcup_{n \in \mathbb{N}} A_n$$

is nowhere dense.

$b \Rightarrow a$. By Theorem 4.12 $a \Rightarrow b$, for all $\mu \in \mathscr{C}(T)^{\pi}_{+}$, $f \in L^2(\mu)$, and $\xi \in H$,

$$\tilde{x}\tilde{y}(f \otimes \xi) = \tilde{z}(f \otimes \xi)$$

so $\tilde{x}\tilde{y} = \tilde{z}$. □

Corollary 4.14. *Assume H separable and let $x \in \mathscr{F}(T, E)$.*

(a) $\tilde{x}$ *is unitary iff* x_t *is unitary for almost all* $t \in T$.
(b) $\tilde{x}$ *is an orthogonal projection iff* x_t *is an orthogonal projection for almost all* $t \in T$.

a) If $\tilde{x}$ is unitary, then

$$\widetilde{x^*}\tilde{x} = \tilde{x}\widetilde{x^*} = 1$$

so by Corollary 4.13 $a \Rightarrow b$,

$$x_t^* x_t = x_t x_t^* = 1$$

for almost all $t \in T$, i.e., x_t is unitary for almost all $t \in T$. If this condition is fulfilled, then by Corollary 4.13 $b \Rightarrow a$,

$$\tilde{x}^* \tilde{x} = \widetilde{x^* x} = 1\,, \qquad \tilde{x}\tilde{x}^* = \widetilde{x x^*} = 1\,,$$

i.e., $\tilde{x}$ is unitary.

b) If $\tilde{x}$ is an orthogonal projection, then $\tilde{x}\tilde{x} = \tilde{x}$ so by Corollary 4.13 $a \Rightarrow b$, $x_t x_t = x_t$ for almost all $t \in T$, i.e., x_t is an orthogonal projection for almost all $t \in T$. If this condition is fulfilled, then by Corollary 4.13 $b \Rightarrow a$, $\tilde{x}\tilde{x} = \tilde{x}$, i.e., $\tilde{x}$ is an orthogonal projection. □

Remark 4.15. The next example shows that we cannot drop the separability condition in the above Corollary. Let $\mathbb{T}$ be the one-dimensional torus, i.e., $\mathbb{T} := \mathbb{R}/\mathbb{Z}$ or

$$\mathbb{T} := \left\{\, (s,t) \in \mathbb{R}^2 \;\middle|\; s^2 + t^2 = 1 \,\right\},$$

λ the Lebesgue measure on $\mathbb{T}$, T the hyperstonian space with $\mathscr{C}(T) = L^\infty(\lambda)$, F the von Neumann algebra $L^\infty(\lambda)$ of multiplication operators on the Hilbert space $L^2(\lambda)$, and H the Hilbert space

$$H := \{\alpha \in \mathbb{T}\} L^2(\lambda)\,.$$

As E we take the von Neumann algebra on H obtained as the C*-direct product of the family of von Neumann algebras $(F)_{\alpha \in \mathbb{T}}$. Let

$$i : \mathscr{C}(\mathbb{T}) \longrightarrow L^\infty(\lambda) = \mathscr{C}(T)$$

be the inclusion map and $\varphi : T \longrightarrow \mathbb{T}$ the continuous surjective map such that

$$if = f \circ \varphi$$

for every $f \in \mathscr{C}(\mathbb{T})$ ([1] Proposition 4.1.2.15). For every $\alpha \in \mathbb{T}$ put

$$\psi_\alpha : \mathbb{T} \longrightarrow \mathbb{T}, \quad \beta \longmapsto \alpha + \beta$$

and for every $n \in \mathbb{N}$ put

$$A_n := \left[-\frac{1}{2^n}, -\frac{1}{2^{n+1}}\right[\cup \left]\frac{1}{2^{n+1}}, \frac{1}{2^n}\right] \subset \mathbb{T}\,.$$

Then for $\alpha \in \mathbb{T}$, $(\psi_\alpha(A_n))_{n \in \mathbb{N}}$ are pairwise disjoint and

$$\bigcup_{n \in \mathbb{N}} \psi_\alpha(A_n) = \mathbb{T} \setminus \{\alpha\}\,.$$

Let further $(f_n)_{n\in\mathbb{N}}$ be the sequence of Rademacher functions on $\mathbb{T}$. We define $x \in l^\infty(T,E)$ by

$$x_{t,\alpha} := \begin{cases} f_n \circ \varphi & \text{if} \quad \varphi(t) \in \psi_\alpha(A_n) \text{ for an } n \in \mathbb{N} \\ 0 & \text{if} \quad \varphi(t) = \alpha \end{cases} .$$

Then $x \in \mathscr{F}(T,E)$ and $\tilde{x}$ is a self-adjoint unitary element of $\mathscr{C}(T)\bar{\otimes}E$, but x_t is not invertible for every $t \in T$.

Corollary 4.16. *If H is separable, then the following are equivalent for all $x \in \mathscr{F}(T,E)$:*

(a) $\tilde{x}$ is invertible.
(b) There is a $y \in \mathscr{F}(T,E)$ such that x_t is invertible and $x_t^{-1} = y_t$ for almost all $t \in T$.

$a \Rightarrow b$. There is a $y \in \mathscr{F}(T,E)$ with $\tilde{y} = \tilde{x}^{-1}$. Then

$$\tilde{x}\tilde{y} = \tilde{y}\tilde{x} = 1$$

and by Corollary 4.13 $a \Rightarrow b$,

$$x_t y_t = y_t x_t = 1$$

for almost all $t \in T$. Thus x_t is invertible and $x_t^{-1} = y_t$ for almost all $t \in T$.

$b \Rightarrow a$. For almost all $t \in T$,

$$x_t y_t = y_t x_t = 1$$

so by Corollary 4.13 $b \Rightarrow a$, $\tilde{x}\tilde{y} = \tilde{y}\tilde{x} = 1$, i.e., $\tilde{x}$ is invertible. □

References

1. Constantinescu, C., *C*-algebras*, Elsevier, 2001.
2. Takesaki, M., *Theory of Operator Algebra I,* Springer, 2002.
3. Wegge-Olsen, N. E., *K-theory and C*-algebras,* Oxford University Press, 1993.

Chapter 5
The Perturbed Median Principle for Integral Inequalities with Applications

S.S. Dragomir

Dedicated to the memory of Professor George Isac

Abstract In this paper, a perturbed version of the median principle introduced by the author in [1] is developed. Applications for various Riemann–Stieltjes integral and Lebesgue integral inequalities are also provided.

5.1 Introduction

In analytic inequalities theory, there are many results involving the sup-norm of a function or of its derivative. To give only two examples, we recall the *Ostrowski inequality*:

$$\left| f(x) - \frac{1}{b-a} \int_a^b f(t)\,dt \right| \leq \left[\frac{1}{4} + \left(\frac{x - \frac{a+b}{2}}{b-a} \right)^2 \right] (b-a) \left\| f' \right\|_{[a,b],\infty} \tag{5.1}$$

for any $x \in [a,b]$, where f is absolutely continuous on $[a,b]$ and $f' \in L_\infty [a,b]$, i.e., $\|f'\|_{[a,b],\infty} := ess \sup_{t\in[a,b]} |f'(t)| < \infty$, and the *Čebyšev inequality*

S.S. Dragomir
Research Group in Mathematical Inequalities & Applications, School of Engineering and Science, Victoria University, P.O. Box 14428, Melbourne City, VIC, Australia 8001, e-mail: sever.dragomir@vu.edu.au

P.M. Pardalos et al. (eds.), *Nonlinear Analysis and Variational Problems*, Springer Optimization and Its Applications 35, DOI 10.1007/978-1-4419-0158-3_5,

$$\left|\frac{1}{b-a}\int_a^b f(t)g(t)dt - \frac{1}{b-a}\int_a^b f(t)dt \cdot \frac{1}{b-a}\int_a^b g(t)dt\right| \qquad (5.2)$$
$$\leq \frac{1}{12}(b-a)^2 \|f'\|_{[a,b],\infty} \|g'\|_{[a,b],\infty},$$

provided that f and g are absolutely continuous with $f', g' \in L_\infty[a,b]$.

Since, in order to estimate $\left\|f^{(r)}\right\|_{[a,b],\infty}$, in practice it is usually necessary to find the quantities

$$M_r := \sup_{t\in[a,b]} f^{(r)}(t) \quad \text{and} \quad m_r := \inf_{t\in[a,b]} f^{(r)}(t)$$

(as, obviously, $\left\|f^{(r)}\right\|_{[a,b],\infty} = \max\{|M_r|, |m_r|\}$), the knowledge of $\left\|f^{(r)}\right\|_{[a,b],\infty}$ may be as difficult as the knowledge of M_r and m_r.

As pointed out in [1], it is natural, therefore, to try to establish inequalities where instead of $\left\|f^{(r)}\right\|_{[a,b],\infty}$ one would have the positive quantity $M_r - m_r$. This can be also useful since for functions whose derivatives $f^{(r)}$ have a "modest variation" the quantity $M_r - m_r$ may be much smaller than $\left\|f^{(r)}\right\|_{[a,b],\infty}$.

In order to address this problem, the author has stated in [1] the "median principle" that can be formalized as follows:

Theorem 5.1 (Median Principle). *Let $\mathscr{P}_n^\circ$ be the class of polynomials*

$$\{P_n | P_n(x) = x^n + a_1x^{n-1} + \cdots + a_{n-1}x + a_n,\ a_i \in \mathbb{R},\ i = \overline{1,n}\}$$

and $f : [a,b] \subset \mathbb{R} \to \mathbb{R}$ be a function so that $f^{(n-1)}$ is absolutely continuous and $f^{(n)} \in L_\infty[a,b]$. Assume that the following inequality holds:

$$L\left(f, f^{(1)}, \ldots, f^{(n-1)}, f^{(n)}; a, b\right) \leq R\left(\left\|f^{(n)}\right\|_{[a,b],\infty}; a, b\right), \qquad (5.3)$$

where $L(\cdot,\cdot,\ldots,\cdot;a,b) : \mathbb{R}^{(n+1)} \to \mathbb{R}$ (the left-hand side) is a general function and $R(\cdot;a,b) : [0,\infty) \to \mathbb{R}$ (the right-hand side) is monotonic nondecreasing on $[0,\infty)$.

If $g : [a,b] \to \mathbb{R}$ is such that $g^{(n-1)}$ is absolutely continuous and

$$-\infty < \gamma \leq g^{(n)} \leq \Gamma < \infty \quad on \quad [a,b], \qquad (5.4)$$

where γ, Γ are real numbers, then

$$\sup_{P_n\in\mathscr{P}_n^\circ} L\left(g - \frac{\gamma+\Gamma}{2}P_n, g^{(1)} - \frac{\gamma+\Gamma}{2}P_n^{(1)}, \ldots, g^{(n)} - \frac{\gamma+\Gamma}{2}P_n^{(n)}; a, b\right)$$
$$\leq R\left(\frac{\Gamma-\gamma}{2}; a, b\right).$$

In order to exemplify the above principle, the author has given various examples classified as "inequalities of the 0th-degree," "1st-degree" and, in general, "n^{th}-degree," where n is the maximal order of the derivative involved in the original inequality. To motivate our further exploration, we mention here only some results from the class of "0th-degree" and for the Riemann–Stieltjes integral as follows:

Theorem 5.2. *Let $f:[a,b]\to\mathbb{R}$ be a continuous function with the property that*

$$-\infty < m \le f(x) \le M < \infty \quad \text{for any } x\in[a,b] \tag{5.5}$$

and u a function of bounded variation such that $u(a)=u(b)$. Then

$$\left|\int_a^b f(t)\,du(t)\right| \le \frac{1}{2}(M-m)\bigvee_a^b(u). \tag{5.6}$$

The constant $\frac{1}{2}$ is sharp in (5.6).

Theorem 5.3. *Let $f,l:[a,b]\to\mathbb{R}$ be two continuous functions on $[a,b]$. If f satisfies (5.5) and u is of bounded variation such that*

$$\int_a^b l(t)\,du(t)=0, \tag{5.7}$$

then

$$\left|\int_a^b f(t)\,l(t)\,du(t)\right| \le \frac{1}{2}(M-m)\,\|l\|_{[a,b],\infty}\bigvee_a^b(u). \tag{5.8}$$

The constant $\frac{1}{2}$ is sharp in (5.8).

As a corollary of the above, we stated in [1] that the following Grüss type inequality obtained in [2] can be stated:

Corollary 5.4. *Let f,g be continuous on $[a,b]$, f satisfies (5.5) and u is of bounded variation and such that $u(b)\neq u(a)$. Then*

$$\begin{aligned}
&\left|\frac{1}{u(b)-u(a)}\int_a^b f(t)\,g(t)\,du(t)\right. \\
&\quad\left.-\frac{1}{u(b)-u(a)}\int_a^b f(t)\,du(t)\cdot\frac{1}{u(b)-u(a)}\int_a^b g(t)\,du(t)\right| \\
&\quad\le \frac{1}{2}(M-m)\frac{1}{|u(b)-u(a)|} \\
&\quad\left\|g-\frac{1}{u(b)-u(a)}\int_a^b g(s)\,du(s)\right\|_{[a,b],\infty}\bigvee_a^b(u).
\end{aligned} \tag{5.9}$$

The constant $\frac{1}{2}$ is best possible in (5.9).

5.2 A Perturbed Version of the Median Principle

For two real numbers $\delta, \Delta \in \mathbb{R}$ with $\Delta \neq \delta$, we consider the set of all functions $f, g : [a,b] \to \mathbb{R}$ given by:

$$\mathscr{M}_{[a,b]}(\delta,\Delta) := \left\{ (f,g) \middle| \left| f(x) - \frac{\delta+\Delta}{2} g(x) \right| \leq \frac{1}{2} |\Delta-\delta| \, |g(x)| \quad \text{for any } x \in [a,b] \right\}. \tag{5.10}$$

We observe, for instance, if $\Delta > \delta$ and $g(x) \neq 0$ for $x \in [a,b]$, then $\mathscr{M}_{[a,b]}(\delta,\Delta)$ contains the pair of functions (f,g) satisfying the condition

$$\left| \frac{f(x)}{g(x)} - \frac{\delta+\Delta}{2} \right| \leq \frac{1}{2}(\Delta-\delta), \quad \text{for all } x \in [a,b] \tag{5.11}$$

or, equivalently, the condition:

$$\delta \leq \frac{f(x)}{g(x)} \leq \Delta, \quad \text{for all } x \in [a,b]. \tag{5.12}$$

Moreover, if we assume that $g(x) > 0$ for any $x \in [a,b]$, then the condition (5.12) is equivalent with

$$\delta g(x) \leq f(x) \leq \Delta g(x), \quad \text{for all } x \in [a,b]. \tag{5.13}$$

In practical application, the assumptions (5.12) or (5.13) are natural to impose. However, for the sake of generality, we consider, in the following, the class $\mathscr{M}_{[a,b]}(\delta,\Delta)$ which is larger and can be extended to the complex case as well.

We are able now to state a perturbed version for the median principle of 0th -degree:

Lemma 5.5. *Assume that the following inequality holds:*

$$L\left(h, h^{(1)}, \ldots, h^{(n)}; a, b\right) \leq R\left(\left|h^{(n)}\right|; a, b\right), (n \geq 0) \tag{5.14}$$

where $L(\cdot, \ldots, \cdot; a, b) : \mathbb{R}^{n+1} \to \mathbb{R}$ *(the left-hand side) is a general function and* $R(\cdot; a, b) : [0, \infty) \to \mathbb{R}$ *(the right-hand side) is monotonic nondecreasing* $[0,\infty)$*. If* f, g *are such that* $f^{(n-1)}, g^{(n-1)}$ *are absolutely continuous and* $\left(f^{(n)}, g^{(n)}\right) \in \mathscr{M}_{[a,b]}(\delta_n, \Delta_n)$*, for some real numbers* $\delta_n \neq \Delta_n$*, then*

$$L\left(f - \frac{\delta_n+\Delta_n}{2} g, \ldots, f^{(n)} - \frac{\delta_n+\Delta_n}{2} g^{(n)}; a, b\right) \leq R\left(\frac{1}{2}|\Delta_n-\delta_n| \left|g^{(n)}\right|; a, b\right). \tag{5.15}$$

Proof. Since $\left(f^{(n)},g^{(n)}\right)\in\mathscr{M}_{[a,b]}(\delta_n,\Delta_n)$, then, by (5.10),

$$\left|f^{(n)}-\frac{\delta+\Delta}{2}g^{(n)}\right|\leq\frac{1}{2}|\Delta_n-\delta_n|\left|g^{(n)}\right| \tag{5.16}$$

which implies, by making use of (5.14) for $h=f-\frac{\delta_n+\Delta_n}{2}g$, that

$$L\left(f-\frac{\delta_n+\Delta_n}{2}g,...,f^{(n)}-\frac{\delta_n+\Delta_n}{2}g^{(n)};a,b\right)\leq R\left(\left|f^{(n)}-\frac{\delta_n+\Delta_n}{2}g^{(n)}\right|;a,b\right).$$

Further, by utilizing the monotonicity property of $R(\cdot;a,b)$ and (5.16) we deduce the desired result (5.15). □

5.3 Some Examples for 0th-Degree Inequalities

The main aim of the results obtained below are to provide various examples for how the above principle can be utilized in practice in order to derive inequalities when the upper bounds are expressed in terms of the difference between the supremum and the infimum of a function. These inequalites are called the "0th-degree" inequalities.

The following result concerning the Riemann–Stieltjes integral holds.

Theorem 5.6. *Let $(f,g)\in\mathscr{M}_{[a,b]}(\delta,\Delta)$ and $u:[a,b]\to\mathbb{R}$ be such that the Riemann–Stieltjes integrals $\int_a^b f(t)\,du(t)$ and $\int_a^b g(t)\,du(t)$ exist. Then*

$$\left|\int_a^b f(t)\,du(t)-\frac{\delta+\Delta}{2}\int_a^b g(t)\,du(t)\right| \tag{5.17}$$
$$\leq\frac{1}{2}|\Delta-\delta|\times\begin{cases}\|g\|_{[a,b],\infty}\bigvee_a^b(u) & \text{provided } g\in C[a,b] \text{ and } u\in BV[a,b];\\[2ex] \int_a^b|g(x)|\,du(x) & \text{provided } g\in C[a,b] \text{ and } u\in\mathscr{M}^{\nearrow}[a,b];\\[2ex] L\|g\|_{[a,b],1} & \text{provided } g\in R[a,b] \text{ and } u\in Lip_L[u,b];\end{cases}$$

where $C[a,b]$ is the class of continuous functions on $[a,b]$, $BV[a,b]$ is the class of bounded variation functions on $[a,b]$, $\mathscr{M}^{\nearrow}[a,b]$ is the class of monotonic nondecreasing functions on $[a,b]$ and $Lip_L[u,b]$ is the class of Lipschitzian functions with the constant $L>0$. All the inequalities in (5.17) are sharp.

Proof. The following result is well known for the Riemann–Stieltjes integral: If $p:[a,b]\to\mathbb{R}$ is continuous on $[a,b]$ and $v:[a,b]\to\mathbb{R}$ is of bounded variation on $[a,b]$, then the Riemann–Stieltjes integral $\int_a^b p(x)\,dv(x)$ exists and

$$\left|\int_a^b p(x)\,dv(x)\right| \le \max_{x\in[a,b]} |p(x)| \bigvee_a^b (u), \tag{5.18}$$

where $\bigvee_a^b (v)$ denotes the total variation of u on $[a,b]$.

Now, since $(f,g) \in \mathscr{M}_{[a,b]}(\delta,\Delta)$, then on applying Lemma 5.5 for the inequality (5.18) we can state

$$\begin{aligned}\left|\int_a^b f(t)\,du(t) - \frac{\delta+\Delta}{2}\int_a^b g(t)\,du(t)\right| &= \left|\int_a^b \left[f(t) - \frac{\delta+\Delta}{2} g(t)\right] du(t)\right| \\ &\le \sup_{t\in[a,b]} \left|f(t) - \frac{\delta+\Delta}{2} g(t)\right| \bigvee_a^b (u) \\ &\le \frac{1}{2}|\Delta-\delta| \,\|g\|_{[a,b],\infty} \bigvee_a^b (u)\end{aligned}$$

and the first inequality in (5.17) is proved.

If $p : [a,b] \to \mathbb{R}$ is continuous and $v : [a,b] \to \mathbb{R}$ is monotonic nondecreasing, then the Riemann–Stieltjes integral $\int_a^b p(x)\,dv(x)$ obviously exists and

$$\left|\int_a^b p(x)\,dv(x)\right| \le \int_a^b |p(x)|\,dv(x). \tag{5.19}$$

Also, if $p : [a,b] \to \mathbb{R}$ is Riemann integrable on $[a,b]$ and $v : [a,b] \to \mathbb{R}$ is l-Lipschitzian, then the Riemann–Stieltjes integral $\int_a^b p(x)\,dv(x)$ exists and

$$\left|\int_a^b p(x)\,du(x)\right| \le L\int_a^b |p(x)|\,dx. \tag{5.20}$$

Now, on applying Lemma 5.5 for these two inequalities we obtain the desired results.

To prove the sharpness of the inequalities in (5.17), we assume, for instance, that $f(x) = \Delta g(x)$, $x \in [a,b]$.

In this situation, the inequality (5.17) is obviously equivalent with

$$\left|\int_a^b g(x)\,du(x)\right| \le \begin{cases} \|g\|_{[a,b],\infty} \bigvee_a^b (u); \\ \int_a^b g(x)\,du(x); \\ L\int_a^b |g(x)|\,dx \end{cases}$$

which are obviously sharp inequalities. □

The above result has many particular instances of interest.

Corollary 5.7. *Let $f : [a,b] \to \mathbb{R}$ be such that there exist the constants $M > m$ with*

$$-\infty < m \le f(t) \le M < \infty \quad \text{for any } t \in [a,b]. \tag{5.21}$$

Then

$$\left| \int_a^b f(t)\,du(t) - \frac{m+M}{2}[u(b)-u(a)] \right| \tag{5.22}$$

$$\le \frac{1}{2}(M-m) \times \begin{cases} \bigvee_a^b(u) & \text{if } f \in C[a,b] \text{ and } u \in BV[a,b]; \\ [u(b)-u(a)] & \text{if } f \in C[a,b] \text{ and } u \in \mathcal{M}^{\nearrow}[a,b]; \\ (b-a)L & \text{if } f \in R[a,b] \text{ and } u \in Lip_L[u,b]. \end{cases}$$

Another simple result is the following one.

Corollary 5.8. *Assume that there exist the constants l, L such that:*

$$-\infty < l \le \frac{f(t)-f(x)}{t-x} \le L < \infty \tag{5.23}$$

for $t, x \in [a,b]$ with $t \ne x$. Then

$$\left| \int_a^b f(t)\,du(t) - f(x)[u(b)-u(a)] \right. \tag{5.24}$$

$$\left. - \frac{l+L}{2}\left[(x-a)u(a) + (b-x)u(b) - \int_a^b u(t)\,dt\right] \right| \tag{5.25}$$

$$\le \frac{1}{2}(L-l) \times \begin{cases} \left[\frac{1}{2}(b-a) + \left|x - \frac{a+b}{2}\right|\right] \bigvee_a^b(u) & \text{if } u \in BV[a,b]; \\ \Big[bu(b) + au(a) - x[u(a)+u(b)] & \\ \quad + \int_a^b sgn(x-t)\,u(t)\,dt\Big] & \text{if } u \in \mathcal{M}^{\nearrow}[a,b]; \\ K\left[\frac{1}{4} + \left(\frac{x - \frac{a+b}{2}}{b-a}\right)^2\right](b-a)^2 & \text{if } u \in Lip_L[u,b]. \end{cases}$$

Proof. The condition (5.23) obviously implies that

$$\left| f(t) - f(x) - \frac{l+L}{2}(t-x) \right| \le \frac{1}{2}(L-l)\,|t-x|$$

for $t, x \in [a,b]$.

If we fix $x \in [a,b]$ and apply Theorem 5.6 for $f_x(t) = f(t) - f(x)$, $g_x(t) = t - x$ and $\delta = \gamma$, $\Delta = \Gamma$, we get

$$\left|\int_a^b f(t)\,du(t) - f(x)[u(b)-u(a)] - \frac{l+L}{2}\left[(x-a)u(a)+(b-x)u(b)-\int_a^b u(t)\,dt\right]\right| \quad (5.26)$$

$$\leq \frac{1}{2}(L-l)\times\begin{cases}\sup_{t\in[a,b]}|t-x|\bigvee_a^b(u) \text{ if } u\in BV[a,b];\\[2ex] \int_a^b |t-x|\,du(t) \qquad\qquad \text{if } u\in \mathscr{M}^{\nearrow}[a,b];\\[2ex] K\int_a^b |t-x|\,dt \qquad\qquad \text{if } u\in Lip_L[u,b].\end{cases}$$

$$= \frac{1}{2}(L-l)\times\begin{cases}\left[\frac{1}{2}(b-a)+\left|x-\frac{a+b}{2}\right|\right]\bigvee_a^b(u);\\[2ex] \Big[bu(b)+au(a)-x[u(a)+u(b)] \\ \qquad +\int_a^b \operatorname{sgn}(x-t)\,u(t)\,dt\Big];\\[2ex] K\left[\frac{1}{4}+\left(\frac{x-\frac{a+b}{2}}{b-a}\right)^2\right](b-a)^2\end{cases}$$

since a simple integration by parts shows that

$$\int_a^b |t-x|\,du(t) = \int_a^x (x-t)\,du(t) + \int_x^b (t-x)\,du(t)$$
$$= bu(b)+au(a)-x[u(a)+u(b)] + \int_a^b \operatorname{sgn}(x-t)\,u(t)\,dt.$$

On the other hand,

$$\int_a^b (t-x)\,du(t) = (b-x)u(b)+(x-a)u(a) - \int_a^b u(t)\,dt,$$

which together with (5.26) produces the desired result (5.24). □

Remark 5.9. If $f:[a,b]\to\mathbb{R}$ is continuous on $[a,b]$ and differentiable on $[a,b]$ and if $\gamma=\inf_{t\in[a,b]} f'(t)$, $\Gamma=\sup_{t\in[a,b]} f'(t)$, then by Lagrange's theorem, we have

$$\gamma \leq \frac{f(t)-f(x)}{t-x} \leq \Gamma$$

for any $x,t\in[a,b]$ with $t\neq x$, and the inequality (5.24) holds true for these γ and Γ.

The following result may be stated as well:

Corollary 5.10. *Assume that there exists the constants γ,Γ so that:*

$$\left|f(t)-f(x)-\frac{\gamma+\Gamma}{2}|t-x|^{\alpha}\right| \leq \frac{1}{2}|\Gamma-\gamma|\,|t-x|^{\alpha} \quad (5.27)$$

for a given $x \in [a,b]$ and $\alpha > 0$ for each $t \in [a,b]$. Then

$$\left| \int_a^b f(t)\,du(t) - f(x)[u(b)-u(a)] \right. \tag{5.28}$$
$$\left. - \frac{\gamma+\Gamma}{2}\left[(x-a)^{\alpha}u(a)+(b-x)^{\alpha}u(b)-\alpha\int_a^b |t-x|^{\alpha-1}u(t)\,dt\right]\right|$$
$$\leq \frac{1}{2}|\Gamma-\gamma| \times \begin{cases} \left[\frac{1}{2}(b-a)+\left|x-\frac{a+b}{2}\right|\right]^{\alpha} \bigvee_a^b(u) & \text{if } u \in BV[a,b]; \\ \Big[(x-a)^{\alpha}u(a)+(b-x)^{\alpha}u(b) \\ \quad -\alpha\int_a^b (t-x)^{\alpha-1}u(t)\,dt\Big] & \text{if } u \in \mathcal{M}^{\nearrow}[a,b]; \\ \frac{L}{\alpha+1}\left[(b-x)^{\alpha+1}+(x-a)^{\alpha+1}\right] & \text{if } u \in Lip_L[u,b]. \end{cases}$$

Proof. Follows by applying Theorem 5.6 for $\delta=\gamma$, $\Delta=\Gamma$ and $g_x(t)=|t-x|^{\alpha}$ with x fixed in $[a,b]$. The details are omitted. □

Remark 5.11. The above result contains some cases of interest, for instance when $x=a$, $x=b$ or $x=\frac{a+b}{2}$. To be more specific, we choose only one case to exemplify. Let us assume that there exist $\phi,\Phi \in \mathbb{R}$ such that

$$\left| f(t) - f\left(\frac{a+b}{2}\right) - \frac{\phi+\Phi}{2}\left|t-\frac{a+b}{2}\right|^{\alpha}\right| \leq \frac{1}{2}|\Phi-\phi|\left|t-\frac{a+b}{2}\right|^{\alpha}$$

for any $t\in[a,b]$, where $\alpha>0$ is given. Then

$$\left| \int_a^b f(t)\,du(t) - f\left(\frac{a+b}{2}\right)[u(b)-u(a)] \right. \tag{5.29}$$
$$\left. - \frac{\phi+\Phi}{2}\left[\frac{(b-a)^{\alpha}}{2^{\alpha}}[u(b)+u(a)]-\alpha\int_a^b \left|t-\frac{a+b}{2}\right|^{\alpha-1}u(t)\,dt\right]\right|$$
$$\leq \frac{1}{2}|\Phi-\phi| \times \begin{cases} \frac{(b-a)^{\alpha}}{2^{\alpha}} \bigvee_a^b(u) & \text{if } u \in BV[a,b]; \\ \Big[\frac{(b-a)^{\alpha}}{2^{\alpha}}[u(b)+u(a)] \\ \quad -\alpha\cdot\int_a^b \left|t-\frac{a+b}{2}\right|^{\alpha-1}u(t)\,dt\Big] & \text{if } u \in \mathcal{M}^{\nearrow}[a,b]; \\ \frac{L(b-a)^{\alpha+1}}{2^{\alpha}(\alpha+1)} & \text{if } u \in Lip_L[u,b]. \end{cases}$$

The following result can be stated as well.

Corollary 5.12. *Assume that there exist the constants $n,N\in\mathbb{R}$ such that:*

$$\left| f(x) - \frac{1}{b-a}\int_a^b f(s)\,ds - \frac{n+N}{2}\cdot g(x)\right| \leq \frac{1}{2}|N-n|\,|g(x)|$$

for any $x \in [a,b]$. *Then*

$$\left| \int_a^b f(t)\,du(t) - \frac{u(b)-u(a)}{b-a}\int_a^b f(s)\,ds - \frac{n+N}{2}\int_a^b g(t)\,du(t) \right| \tag{5.30}$$

$$\leq \frac{1}{2}|N-n| \times \begin{cases} \|g\|_{[a,b],\infty} \bigvee_a^b(u) & \textit{provided } g \in C[a,b] \textit{ and } u \in BV[a,b]; \\ \int_a^b |g(x)|\,du(x) & \textit{provided } g \in C[a,b] \textit{ and } u \in \mathcal{M}^{\nearrow}[a,b]; \\ L\|g\|_{[a,b],1} & \textit{provided } g \in R[a,b] \textit{ and } u \in Lip_L[u,b]. \end{cases}$$

5.4 Inequalities of the 1st-Degree

An inequality that contains at most the first derivative of the involved functions will be called an inequality of the 1st-degree.

If one would like examples of such inequalities for two functions, the following Ostrowski's inequality obtained in [3] is the most suitable

$$|C(h,l;a,b)| \leq \frac{1}{8}(b-a)(M-m)\left\|h'\right\|_{[a,b],\infty}, \tag{5.31}$$

where $C(h,l;a,b)$ is the *Čebyšev functional* given by

$$C(h,l;a,b) := \frac{1}{b-a}\int_a^b h(x)\,l(x)\,dx - \frac{1}{b-a}\int_a^b h(x)\,dx \cdot \frac{1}{b-a}\int_a^b l(x)\,dx$$

and $-\infty < m \leq h(x) \leq M < \infty$ for a.e. $x \in [a,b]$ while l is absolutely continuous and such that $l' \in L_\infty[a,b]$. The constant $\frac{1}{8}$ is sharp.

Another example of such an inequality is the Čebyšev one

$$|C(h,l;a,b)| \leq \frac{1}{12}(b-a)^2\left\|h'\right\|_{[a,b],\infty}\left\|l'\right\|_{[a,b],\infty}, \tag{5.32}$$

provided h,l are absolutely continuous and $h',l' \in L_\infty[a,b]$. The constant $\frac{1}{12}$ here is sharp.

The following result holds:

Theorem 5.13. *Let* $f:[a,b] \to \mathbb{R}$ *be a Lebesgue integrable function such that* $-\infty < m \leq f(x) \leq M < \infty$ *for a.e.* $x \in [a,b]$. *If* g *and* h *are absolutely continous and such that there exists the constants* δ *and* Δ *such that* $(g',h') \in \mathcal{M}_{[a,b]}(\delta,\Delta)$, *then we have the following perturbed version of the Ostrowski's inequality:*

$$\left| C(f,g;a,b) - \frac{\Delta+\delta}{2}C(f,h;a,b) \right| \tag{5.33}$$

$$\leq \frac{1}{16}(b-a)(M-m)(\Delta-\delta)\left\|h'\right\|_{[a,b],\infty}.$$

Proof. By applying Lemma 5.5 to the Ostrowski's inequality (5.31) we have

$$\left|C\left(f,g-\frac{\Delta+\delta}{2}h;a,b\right)\right| \le \frac{1}{8}(b-a)(M-m)\left\|\frac{1}{2}(\Delta-\delta)h'\right\|_{[a,b],\infty}$$

which is clearly equivalent with the desired result (5.33). □

Finally, we have the following perturbed version of the Čebyšev inequality

Theorem 5.14. *Let f,g,h,l be absolutely continuous on $[a,b]$. If there exists the constants φ, Φ, δ and Δ such that $(f',l') \in \mathscr{M}_{[a,b]}(\varphi,\Phi)$ and $(g',h') \in \mathscr{M}_{[a,b]}(\delta,\Delta)$, then*

$$\begin{aligned}\Bigg|C(f,g;a,b) &- \frac{\Delta+\delta}{2}C(f,h;a,b) \\ &- \frac{\Phi+\varphi}{2}C(l,g;a,b) + \frac{\Delta+\delta}{2}\cdot\frac{\Phi+\varphi}{2}C(h,l;a,b)\Bigg| \\ &\le \frac{1}{48}(b-a)^2(\Phi-\varphi)(\Delta-\delta)\left\|h'\right\|_{[a,b],\infty}\left\|l'\right\|_{[a,b],\infty}.\end{aligned} \tag{5.34}$$

Similar results can be obtained for other inequalities involving the derivatives up to an order $n \ge 1$. However, the details are left to the interested reader.

References

1. S.S. Dragomir, The median principle for inequalities and applications, in *Functional Equations, Inequalities and Applications*, Ed. by Th.M. Rassias, Kluwer Acad. Publ., 2003. Preprint, *RGMIA Res. Rep. Coll.*, **5**(2002), Supplement, Article 17. [http://www.staff.vu.edu.au/RGMIA/v5(E).asp].
2. S.S. Dragomir, Improvements of Ostrowski and generalised trapezoid inequality in terms of the upper and lower bounds of the first derivative, *Tamkang J. Math.*, **34**(3) (2003), 213-222. Preprint, *RGMIA Res. Rep. Coll.*, **5**(2002), Supplement, Article 10. [http://www.staff.vu.edu.au/RGMIA/v5(E).asp].
3. A. Ostrowski, On an integral inequality, *Aequat. Math.*, **4**(1970), 358-373.

Chapter 6
Stability of a Mixed Type Additive, Quadratic, Cubic and Quartic Functional Equation

M. Eshaghi-Gordji, S. Kaboli-Gharetapeh, M.S. Moslehian, and S. Zolfaghari

Dedicated to the memory of Professor George Isac

Abstract We find the general solution of the functional equation

$$\begin{aligned} D_f(x,y) := f(x+2y)+f(x-2y)-4[f(x+y)-f(x-y)]-f(4y)+4f(3y) \\ -6f(2y)+4f(y)+6f(x)=0. \end{aligned}$$

in the context of linear spaces. We prove that if a mapping f from a linear space X into a Banach space Y satisfies $f(0)=0$ and

$$\|D_f(x,y)\| \leq \varepsilon \quad (x,y \in X),$$

where $\varepsilon > 0$, then there exist a unique additive mapping $A : X \to Y$, a unique quadratic mapping $Q_1 : X \to Y$, a unique cubic mapping $C : X \to Y$ and a unique quartic mapping $Q_2 : X \to Y$ such that

$$\|f(x)-A(x)-Q_1(x)-C(x)-Q_2(x)\| \leq \frac{1087\varepsilon}{140} \quad \forall x \in X.$$

M. Eshaghi-Gordji
Department of Mathematics, Semnan University, P.O. Box 35195-363, Semnan, Iran, e-mail: madjid.eshaghi@gmail.com

S. Kaboli-Gharetapeh
Department of Mathematics, Payame Noor University of Mashhad, Mashhad, Iran, e-mail: simin.kaboli@gmail.com

M.S. Moslehian
Department of Pure Mathematics and Center of Excellence in Analysis on Algebraic Structures (CEAAS), Ferdowsi University of Mashhad, P.O. Box 1159, Mashhad 91775, Iran, e-mail: moslehian@ferdowsi.um.ac.ir and moslehian@ams.org

S. Zolfaghari
Department of Mathematics, Semnan University, P.O. Box 35195-363, Semnan, Iran, e-mail: zolfaghgrys@yahoo.com

P.M. Pardalos et al. (eds.), *Nonlinear Analysis and Variational Problems*,
Springer Optimization and Its Applications 35, DOI 10.1007/978-1-4419-0158-3_6,

6.1 Introduction

The first stability problem concerning group homomorphisms was raised in a famous talk given by Ulam [32] in 1940.
We are given a group G and a metric group G' with metric $\rho(\cdot,\cdot)$. Given $\varepsilon > 0$, does there exist a number $\delta > 0$ such that if $f: G \to G'$ satisfies $\rho(f(xy), f(x)f(y)) < \delta$ for all $x, y \in G$, then a homomorphism $h: G \to G'$ exists with $\rho(f(x), h(x)) < \varepsilon$ for all $x \in G$?

In the next year, 1941, Ulam's problem was affirmatively solved by Hyers [10] for Banach spaces:
Suppose that E_1 is a normed space, E_2 is a Banach space and a mapping $f: E_1 \to E_2$ satisfies the inequality

$$\|f(x+y) - f(x) - f(y)\| \leq \varepsilon \qquad (x, y \in E_1),$$

where $\varepsilon > 0$ is a constant. Then the limit $T(x) := \lim 2^{-n} f(2^n x)$ exists for each $x \in E_1$ and T is the unique additive mapping satisfying

$$\|f(x) - T(x)\| \leq \varepsilon \quad (x \in E_1).$$

In 1950, Aoki [1] generalized Hyers' theorem for approximately additive mappings. In 1978, Th.M. Rassias [25] extended Hyers' theorem by obtaining a unique linear mapping under certain continuity assumption when the Cauchy difference is allowed to be unbounded (see [20]):
Suppose that E is a real normed space, F is a real Banach space and $f: E \to F$ is a mapping such that for each fixed $x \in E$, the mapping $t \mapsto f(tx)$ is continuous on the real line. Let there exist $\varepsilon \geq 0$ and $p \in [0,1)$ such that

$$\|f(x+y) - f(x) - f(y)\| \leq \varepsilon(\|x\|^p + \|y\|^p) \quad (x, y \in E).$$

Then there exists a unique linear mapping $T: E \to F$ such that

$$\|f(x) - T(x)\| \leq \varepsilon \|x\|^p / (1 - 2^{p-1}) \quad (x \in E).$$

In 1990, Th.M. Rassias [26] during the 27th International Symposium on Functional Equations asked the question whether such a theorem can also be proved for $p \geq 1$. In 1991, Gajda [8] gave an affirmative solution to this question for $p > 1$ by following the same approach as in Rassias' paper [25]. It was proved by Gajda [8], as well as by Th.M. Rassias and Šemrl [29] that one cannot prove a Rassias type theorem when $p = 1$. In 1994, P. Găvruta [9] provided a generalization of Rassias' theorem in which he replaced the bound $\varepsilon(\|x\|^p + \|y\|^p)$ by a general control function $\varphi(x, y)$. The paper of Th.M. Rassias [25] has provided a lot of influence in the development of what we now call Hyers–Ulam–Rassias stability of functional equations. During the past decades, several stability problems for various functional equations have been investigated by many mathematicians; we refer the reader to the monographs [4, 11, 12, 13, 16, 28].

The functional equation

$$f(x+y)+f(x-y)=2f(x)+2f(y), \tag{6.1}$$

is called the quadratic functional equation and every solution of the quadratic equation (6.1) is said to be a quadratic function. It is well known that a mapping f between two real vector spaces is quadratic if and only if there exists a unique symmetric bi-additive mapping B such that $f(x)=B(x,x)$ for all x, where $B(x,y)=\frac{1}{4}(f(x+y)-f(x-y))$ (see [17]). The Hyers–Ulam stability problem for the quadratic functional equation was solved by Skof [31] and, independently, by Cholewa [5]. An analogous result for quadratic stochastic processes was obtained by Nikodem [22]. In [3], Czerwik proved the generalized Hyers–Ulam stability of the quadratic functional equation. Jung [15] dealt with stability problems for the quadratic functional equation of Pexider type. The orthogonal stability of the quadratic equation was studied by Moslehian [19] and Mirzavaziri [18].

Jun and Kim [14] introduced the following functional equation

$$f(2x+y)+f(2x-y)=2f(x+y)+2f(x-y)+12f(x) \tag{6.2}$$

and established the general solution and the generalized Hyers–Ulam–Rassias stability for functional equation (6.2). Obviously, the $f(x)=x^3$ satisfies functional equation (6.2), so it is natural to call (6.2) the cubic functional equation. Every solution of the cubic functional equation is said to be a cubic mapping. Jun and Kim proved also that a mapping f between two real vector spaces X and Y is a solution of (6.2) if and only if there exists a unique mapping $C: X\times X\times X\to Y$ such that $f(x)=C(x,x,x)$ for all $x\in X$, moreover, C is symmetric for each fixed one variable and is additive for fixed two variables. Later a number of mathematicians worked on the stability of some types of the cubic equation [2, 30].

In [24], Park and Bae considered the following functional equation:

$$f(2x+y)+f(2x-y)=4(f(x+y)+f(x-y))+24f(x)-6f(y). \tag{6.3}$$

In fact, they proved that a mapping f between two real vector spaces X and Y is a solution of (6.3) if and only if there exists a unique symmetric multi-additive mapping $B: X\times X\times X\times X\to Y$ such that $f(x)=B(x,x,x,x)$ for all x. It is easy to show that the function $f(x)=x^4$ satisfies the functional equation (6.3), which is called the quartic functional equation (see also [21, 23]). In this paper, we deal with the following functional equation derived from additive, quadratic, cubic and quartic mappings

$$\begin{aligned} f(x+2y)+f(x-2y) &= 4[f(x+y)+f(x-y)]+f(4y) \\ &\quad -4f(3y)+6f(2y)-4f(y)-6f(x), \end{aligned} \tag{6.4}$$

which possess evidently the function $f(x)=ax+bx^2+cx^3+dx^4$ as a solution (see also [6, 7]).

We also investigate the general solution and the Hyers–Ulam stability for the mixed functional equation (6.4). In 1992, Th.M. Rassias and Tabor [27] first introduced the concept and coined the term mixed stability for functional equations.

6.2 General Solution

Throughout this section, X and Y will be real vector spaces. Before proceeding to the proof of Theorem 6.3, which is the main result in this section, we shall need the following two lemmas.

Lemma 6.1. *If an even mapping $f : X \to Y$ satisfies (6.4), then f is quartic-quadratic function.*

Proof. We show that the maps $g_1 : X \to Y$ defined by $g_1(x) := f(2x) - 16f(x)$ and $h_1 : X \to Y$ defined by $h_1(x) := f(2x) - 4f(x)$ are quadratic and quartic, respectively. Regarding (6.4), by putting $x = y = 0$, we get $f(0) = 0$. In addition, as f is an even mapping, so by setting $x = 0$ in (6.4), we obtain

$$f(4y) = 4f(3y) - 4f(2y) - 4f(y). \tag{6.5}$$

Replacing x by y in (6.4), we obtain

$$f(4y) = 5f(3y) - 10f(2y) + 11f(y). \tag{6.6}$$

By comparing (6.5) with (6.6), we arrive at

$$f(3y) = 6f(2y) - 15f(y). \tag{6.7}$$

According to (6.7), (6.6) can be rewritten as

$$f(4y) = 20f(2y) - 64f(y). \tag{6.8}$$

By utilizing (6.7) and (6.8), (6.4) can be written as

$$f(x+2y) + f(x-2y) = 4f(x+y) + 4f(x-y) + 2f(2y) - 8f(y) - 6f(x). \tag{6.9}$$

Interchanging x with y in (6.9) gives the equation

$$f(2x+y) + f(2x-y) = 4f(x+y) + 4f(x-y) + 2f(2x) - 8f(x) - 6f(y). \tag{6.10}$$

By substituting $x := x + y$ in (6.9),

$$f(x+3y) + f(x-y) = 4f(x+2y) - 6f(x+y) + 2f(2y) - 8f(y) + 4f(x). \tag{6.11}$$

By alternating x and y in (6.11), we lead to

$$f(3x+y) + f(x-y) = 4f(2x+y) - 6f(x+y) + 2f(2x) - 8f(x) + 4f(y). \tag{6.12}$$

And replace $-y$ by y in (6.12) to obtain

$$f(3x-y)+f(x+y)=4f(2x-y)-6f(x-y)+4f(y)+2f(2x)-8f(x). \quad (6.13)$$

If (6.12) and (6.13) are added, we have

$$\begin{aligned} f(3x+y)+f(3x-y) &= 4f(2x+y)+4f(2x-y)-7f(x+y)-7f(x-y) \\ &\quad +8f(y)+4f(2x)-16f(x). \end{aligned} \quad (6.14)$$

Now set $x+y$ in place of y in (6.9), so

$$f(3x+2y)+f(x+2y)=4f(2x+y)-8f(x+y)+2f(2(x+y))-6f(x)+4f(y). \quad (6.15)$$

Interchanging x and y in (6.15), we get

$$f(2x+3y)+f(2x+y)=4f(x+2y)-8f(x+y)+2f(2(x+y))-6f(y)+4f(x), \quad (6.16)$$

which on substitution of $-y$ for y in (6.16) gives

$$f(2x-3y)+f(2x-y)=4f(x-2y)-8f(x-y)+2f(2(x-y))-6f(y)+4f(x). \quad (6.17)$$

By adding (6.16) and (6.17), we lead to

$$\begin{aligned} f(2x+3y)+f(2x-3y) &= 4f(x+2y)+4f(x-2y)-f(2x+y)-f(2x-y) \\ &\quad +2f(2(x+y))+2f(2(x-y))-8f(x+y) \\ &\quad -8f(x-y)-12f(y)+8f(x). \end{aligned} \quad (6.18)$$

By substituting $y:=2y$ in (6.14), we get

$$\begin{aligned} f(3x+2y)+f(3x-2y) &= 4f(2(x+y))+4f(2(x-y))-7f(x+2y) \\ &\quad -7f(x-2y)+8f(2y)+4f(2x)-16f(x). \end{aligned} \quad (6.19)$$

Now if x and y are interchanged in (6.18), we can get

$$\begin{aligned} f(3x+2y)+f(3x-2y) &= 4f(2x+y)+4f(2x-y)-f(x+2y)-f(x-2y) \\ &\quad +2f(2(x+y))+2f(2(x-y))-8f(x+y) \\ &\quad -8f(x-y)-12f(x)+8f(y). \end{aligned} \quad (6.20)$$

So if (6.19) and (6.20) are compared, utilizing (6.9) and (6.10), we conclude that,

$$\begin{aligned} &[f(2(x+y))-16f(x+y)] + [f(2(x-y))-16f(x-y)] \\ &\qquad = 2[f(2x)-16f(x)]+2[f(2y)-16f(y)], \end{aligned}$$

for all $x,y \in X$. The last equality means that

$$g_1(x+y)+g_1(x-y)=2g_1(x)+2g_1(y),$$

for all $x,y \in X$. Thus $g_1 : X \to Y$ is quadratic map. With the substitutions $x := 2x$ and $y := 2y$ in (6.10), we have

$$\begin{aligned} f(2(2x+y)) + f(2(2x-y)) &= 4f(2(x+y)) + 4f(2(x-y)) \\ &\quad -6f(2y) + 2f(4x) - 8f(2x). \end{aligned} \tag{6.21}$$

Since $g_1(2x) = 4g_1(x)$ for all $x \in X$, where $g_1 : X \to Y$ is a quadratic mapping defined above, we conclude that

$$f(4x) = 20f(2x) - 64f(x), \tag{6.22}$$

for all $x,y \in X$.

Hence, according to (6.22), equality (6.21) can be written as

$$\begin{aligned} f(2(2x+y)) + f(2(2x-y)) &= 4f(2(x+y)) + 4f(2(x-y)) \\ &\quad -6f(2y) + 32f(2x) - 128f(x). \end{aligned} \tag{6.23}$$

Now if x and y are interchanged in (6.23), we can get

$$\begin{aligned} f(2(x+2y)) + f(2(x-2y)) &= 4f(2(x+y)) + 4f(2(x-y)) \\ &\quad -6f(2x) + 32f(2y) - 128f(y). \end{aligned} \tag{6.24}$$

By multiplying by 4 in (6.9) and subtracting, the last equation from (6.24), we arrive at

$$\begin{aligned} h_1(x+2y) + h_1(x-2y) &= [f(2(x+2y)) - 4f(x+2y)] + [f(2(x-2y)) - 4f(x-2y)] \\ &= 4[f(2(x+y)) - 4f(x+y)] + 4[f(2(x-y)) - 4f(x-y)] \\ &\quad +24[f(2y) - 4f(y)] - 6[f(2x) - 4f(x)] \\ &= 4h_1(x+y) + 4h_1(x-y) + 24h_1(y) - 6h_1(x), \end{aligned}$$

for all $x,y \in X$. Therefore, $h_1 : X \to Y$ is quartic map. Besides $f(x) = \frac{1}{12}h_1(x) - \frac{1}{12}g_1(x)$ for all $x \in X$, which means that f is quartic-quadratic mapping. □

Lemma 6.2. *If an odd mapping $f : X \to Y$ satisfies (6.4), then f is a cubic-additive.*

Proof. We show that the maps $g_2 : X \to Y$ defined by $g_2(x) := f(2x) - 8f(x)$ and $h_2 : X \to Y$ defined by $h_2(x) := f(2x) - 2f(x)$ are additive and cubic, respectively. Regarding (6.4), by putting $x = y = 0$, we get $f(0) = 0$. In addition, as f is an odd mapping, so by setting $x = 0$ in (6.4), we obtain

$$f(4y) = 4f(3y) - 6f(2y) + 4f(y). \tag{6.25}$$

So by replacing x by y in (6.4), we have

$$f(4y) = 5f(3y) - 10f(2y) + 9f(y). \tag{6.26}$$

By comparing (6.25) with (6.26), we arrive at

$$f(3y) = 4f(2y) - 5f(y). \tag{6.27}$$

According to (6.27), (6.26) can be written as

$$f(4y) = 10f(2y) - 16f(y). \tag{6.28}$$

By utilizing (6.27) and (6.28), (6.4) can be written as

$$f(x+2y) + f(x-2y) = 4f(x+y) + 4f(x-y) - 6f(x). \tag{6.29}$$

By substituting $x := x+y$, $y := x-y$ in (6.29), we have

$$f(3x-y) - f(x-3y) = -6f(x+y) + 4f(2x) + 4f(2y). \tag{6.30}$$

By alternating x and y in (6.29), we lead to

$$f(2x+y) - f(2x-y) = 4f(x+y) - 4f(x-y) - 6f(y). \tag{6.31}$$

By substituting $y := x+y$ in (6.31), we obtain

$$f(3x+y) - f(x-y) = 4f(2x+y) - 6f(x+y) + 4f(y). \tag{6.32}$$

Interchanging x with y in (6.32) gives the equation

$$f(x+3y) + f(x-y) = 4f(x+2y) - 6f(x+y) + 4f(x). \tag{6.33}$$

Replacing y by $-y$ in (6.33), we get

$$f(x-3y) + f(x+y) = 4f(x-2y) - 6f(x-y) + 4f(x), \tag{6.34}$$

which on substitution of $-y$ for y in (6.32) gives

$$f(3x-y) - f(x+y) = 4f(2x-y) - 6f(x-y) - 4f(y). \tag{6.35}$$

If we subtract (6.35) from (6.34), we obtain

$$f(3x-y) - f(x-3y) = 4f(2x-y) - 4f(x-2y) + 2f(x+y) - 4f(x) - 4f(y), \tag{6.36}$$

and by comparing (6.36) with (6.30), we arrive at

$$f(x-2y) - f(2x-y) = 2f(x+y) - f(2x) - f(2y) - f(x) - f(y), \tag{6.37}$$

which, by putting $y := -y$ in (6.37), leads to

$$f(x+2y) - f(2x+y) = 2f(x-y) - f(2x) + f(2y) - f(x) + f(y). \tag{6.38}$$

If we add (6.37) to (6.38), we obtain

$$\begin{aligned} f(x+2y)+f(x-2y) &= f(2x+y)+f(2x-y)+2f(x+y) \\ &\quad +2f(x-y)-2f(2x)-2f(x). \end{aligned} \tag{6.39}$$

By comparing (6.39) with (6.29), we have

$$f(2x+y)+f(2x-y)=2f(x+y)+2f(x-y)+2f(2x)-4f(x). \tag{6.40}$$

Replacing x by y and y by x in (6.40), respectively, we get

$$f(x+2y)-f(x-2y)=2f(x+y)-2f(x-y)+2f(2y)-4f(y). \tag{6.41}$$

Replacing x by $x+y$ in (6.41), we get

$$f(x+3y)-f(x-y)=2f(x+2y)-2f(x)+2f(2y)-4f(y). \tag{6.42}$$

Interchanging x with y in (6.42) gives the equation

$$f(3x+y)+f(x-y)=2f(2x+y)-2f(y)+2f(2x)-4f(x). \tag{6.43}$$

Adding (6.42) to (6.43), we arrive at

$$f(x+3y)+f(3x+y)=2f(x+2y)+2f(2x+y)+2f(2x)+2f(2y)-6f(x)-6f(y). \tag{6.44}$$

Replacing x by $x-y$ and y by $x+y$ in (6.40), respectively, we get

$$f(3x-y)+f(x-3y)=2f(2x-2y)-4f(x-y)+2f(2x)-2f(2y). \tag{6.45}$$

Interchanging x with y in (6.45) gives the equation

$$f(x-3y)+f(3x-y)=2f(2x-2y)-4f(x-y)+2f(2x)-2f(2y), \tag{6.46}$$

and we replace $-y$ and y in (6.46) to obtain

$$f(x+3y)+f(3x+y)=2f(2x+2y)-4f(x+y)+2f(2x)+2f(2y). \tag{6.47}$$

Therefore it follows from (6.44) and (6.47) that

$$f(2x+y)+f(x+2y)=f(2x+2y)-2f(x+y)+3f(x)+3f(y). \tag{6.48}$$

Replacing y by $-x-y$ in (6.48), we get

$$f(x-y)-f(x+2y)=-3f(x+y)+3f(x)+2f(y)-f(2y). \tag{6.49}$$

Interchanging x with y in (6.49) gives the equation

$$f(x-y)+f(2x+y)=3f(x+y)-2f(x)-3f(y)+f(2x), \tag{6.50}$$

which on substitution of $-y$ for y in (6.50) gives

$$f(x+y)+f(2x-y)=3f(x-y)-2f(x)+3f(y)+f(2x). \tag{6.51}$$

By putting $y:=-y$ in (6.49), we obtain

$$f(x+y)-f(x-2y)=-3f(x-y)+3f(x)-2f(y)+f(2y). \tag{6.52}$$

Adding (6.51) to (6.52), we arrive at

$$f(2x-y)-f(x-2y)=-2f(x+y)+f(x)+f(y)+f(2x)+f(2y). \tag{6.53}$$

By adding (6.48) and (6.53) and then using (6.40) and (6.41), we lead to

$$f(2x+2y)-8f(x+y)=[f(2x)-8f(x)]+[f(2y)-8f(y)].$$

The last equality means that

$$g_2(x+y)=g_2(x)+g_2(y),$$

for all $x,y\in X$. Therefore the mapping $g_2:X\to Y$ is additive. With the substitutions $x:=2x$ and $y:=2y$ in (6.41), we get

$$f(2x+4y)-f(2x-4y)=2f(2x+2y)-2f(2x-2y)+2f(4y)-4f(2y). \tag{6.54}$$

Interchanging x with y in (6.54) gives the equation

$$f(4x+2y)+f(4x-2y)=2f(2x+2y)+2f(2x-2y)+2f(4x)-4f(2x). \tag{6.55}$$

Since $g_2(2x)=2g_2(x)$ for all $x\in X$, where $g_2:X\to Y$ is additive mapping defined above, we see

$$f(4x)=10f(2x)-16f(x), \tag{6.56}$$

for all $x,y\in X$. Hence it follows from (6.40), (6.55) and (6.56) that

$$\begin{aligned}h_2(2x+y)+h_2(2x-y) &= [f(2(2x+y))-2f(2x+y)]+[f(2(2x-y))-2f(2x-y)]\\ &= 2[f(2(x+y))-2f(x+y)]+2[f(2(x-y))-2f(x-y)]\\ &\quad +12[f(2x)-2f(x)]\\ &= 2h_2(x+y)+2h_2(x-y)+12h_2(x),\end{aligned}$$

for all $x,y\in X$. Therefore, $h_2:X\to Y$ is cubic map. Besides $f(x)=\frac{1}{6}h_2(x)-\frac{1}{6}g_2(x)$ for all $x\in X$, which means that f is cubic-additive mapping. □

Theorem 6.3. *A mapping $f:X\to Y$ satisfies (6.4) for each $x,y\in X$ if and only if there exist a unique additive mapping $A:X\to Y$, a unique symmetric mapping $Q_1:X\times X\to Y$, a unique mapping $C:X\times X\times X\to Y$ and a unique symmetric multi-additive mapping $Q_2:X\times X\times X\times X\to Y$ such that $f(x)=A(x)+Q_1(x,x)+C(x,x,x)+Q_2(x,x,x,x)$ for all $x\in X$, and that Q_1 is additive for each fixed one variable, C is symmetric for each fixed one variable and is additive for fixed two variables.*

Proof. Let f satisfy (6.4). We decompose f into the even part and odd part by setting

$$f_e(x) = \frac{1}{2}(f(x)+f(-x)), f_o(x) = \frac{1}{2}(f(x)-f(-x)),$$

for all $x \in X$. By (6.4), we have

$$\begin{aligned}
&f_e(x+2y)+f_e(x-2y) = \frac{1}{2}[f(x+2y)+f(-x-2y)+f(x-2y)+f(-x+2y)]\\
&= \frac{1}{2}[f(x+2y)+f(x-2y)]+\frac{1}{2}[f(-x+(-2y))+f(-x-(-2y))]\\
&= \frac{1}{2}[4(f(x+y)+f(x-y))+f(4y)-4f(3y)+6f(2y)-4f(y)-6f(x)]\\
&\quad +\frac{1}{2}[4(f(-x-y)+f(-x-(-y)))+f(-4y)-4f(-3y)+6f(-2y)\\
&\quad -4f(-y)-6f(-x)]\\
&= 4[\frac{1}{2}(f(x+y)+f(-x-y))+\frac{1}{2}(f(-x+y)+f(x-y))]\\
&\quad +[\frac{1}{2}(f(4y)+f(-4y))]-4[\frac{1}{2}(f(3y)+f(-3y))]\\
&\quad +6[\frac{1}{2}(f(2y)-f(-2y))]-4[\frac{1}{2}(f(y)+f(-y))]-6[\frac{1}{2}(f(x)+f(-x))]\\
&= 4(f_e(x+y)+f_e(x-y))+f_e(4y)-4f_e(3y)\\
&\quad +6f_e(2y))-4f_e(y)-6f_e(x),
\end{aligned}$$

for all $x,y \in X$. This means that f_e satisfies (6.4). Similarly we can show that f_o satisfies (6.4). By Lemmas 6.1 and 6.2, f_e and f_o are quartic-quadratic and cubic-additive, respectively. Thus there exist a unique additive mapping $A : X \to Y$, a unique symmetric mapping $Q_1 : X \times X \to Y$, a unique mapping $C : X \times X \times X \to Y$ and a unique symmetric multi-additive mapping $Q_2 : X \times X \times X \times X \to Y$ such that $f(x) = A(x)+Q_1(x,x)+C(x,x,x)+Q_2(x,x,x,x)$ for all $x \in X$, and that Q_1 is additive for each fixed one variable, C is symmetric for each fixed one variable and is additive for fixed two variables. The proof of the converse is trivial. □

6.3 Stability

From this point on, let X be a real vector space and let Y be a Banach space. Before taking up the main subject, we define the difference operator $D_f : X \times X \to Y$ by

$$\begin{aligned}
D_f(x,y) := &f(x+2y)+f(x-2y)-4[f(x+y)-f(x-y)]-f(4y)+4f(3y)\\
&-6f(2y)+4f(y)+6f(x)
\end{aligned}$$

for all $x, y \in X$, where $f : X \to Y$ is a mapping. We investigate the Hyers–Ulam stability problem for functional equation (6.4) as follows.

Theorem 6.4. *Let ε be a positive real number. Suppose that an even mapping $f : X \to Y$ satisfies $f(0) = 0$ and*

$$\|D_f(x,y)\| \leq \varepsilon, \tag{6.57}$$

for all $x, y \in X$. Then the limit

$$Q_1(x) := \lim_n 4^{-n}[f(2^{n-1}x) - 16f(2^n x)],$$

exists for each $x \in X$, and $Q_1 : X \to Y$ is a unique quadratic mapping satisfying (6.4), and

$$\|f(2x) - 16f(x) - Q_1(x)\| \leq 3\varepsilon \tag{6.58}$$

for all $x \in X$.

Proof. Putting $x = 0$ in (6.57) we get

$$\|f(4y) - 4f(3y) + 4f(2y) + 4f(y)\| \leq \varepsilon. \tag{6.59}$$

Replacing x by y in (6.57), we obtain

$$\| - f(4y) + 5f(3y) - 10f(2y) + 11f(y)\| \leq \varepsilon. \tag{6.60}$$

By combining (6.59) and (6.60), we lead to

$$\|f(4x) - 20f(2x) + 64f(x)\| \leq 9\varepsilon, \tag{6.61}$$

for all $x \in X$. Put $g(x) = f(2x) - 16f(x)$, for all $x \in X$. Then by (6.61), we have

$$\|\frac{1}{4}g(2x) - g(x)\| \leq \frac{9}{4}\varepsilon. \tag{6.62}$$

By (6.62), we use iterative methods and induction on n to prove our next relation

$$\|g(x) - 4^{-n}g(2^n x)\| \leq \frac{9}{4}\varepsilon \sum_{i=0}^{n-1} \frac{1}{4^i}$$

for all $x \in X$. Now we multiply both sides of the inequality above by 4^{-m} and replace x by $2^m x$, and we get

$$\|4^{-m}g(2^m x) - 4^{-m-n}g(2^{m+n}x)\| \leq \frac{9}{4}\varepsilon \sum_{i=0}^{n-1} \frac{1}{4^{i+m}}.$$

Since the right-hand side of the last inequality tends to 0 as $m \to \infty$, the sequence $\{4^{-n}g(2^n x)\}$ is Cauchy. Then we can define

$$Q_1(x) := \lim_n 4^{-n} g(2^n x) = \lim_n 4^{-n}[f(2^{n-1}x) - 16f(2^n x)]$$

for each $x \in X$. Additionally, we have

$$\begin{aligned}\|Q_1(2x) - 4Q_1(x)\| &= \lim_n [4^{-n} g(2^{n+1}x) - 4^{-n+1} g(2^n x)] \\ &= 4\lim_n [4^{-n-1} g(2^{n+1}x) - 4^{-n} g(2^n x)] = 0, \qquad (6.63)\end{aligned}$$

for all $x \in X$. Let $D_g(x,y) := D_f(2x,2y) - 16D_f(x,y)$ for all $x \in X$. Then we have

$$\begin{aligned}D_{Q_1}(x,y) &= \lim_n \|4^{-n} D_g(2^n x, 2^n y)\| \\ &= \lim_n 4^{-n} \|D_f(2^{n-1}x, 2^{n-1}y) - 16D_f(2^n x, 2^n y)\| \\ &\leq \lim_n 4^{-n} \|D_f(2^{n-1}x, 2^{n-1}y)\| \\ &\quad + 16 \times 4^{-n} \|D_f(2^n x, 2^n y)\| \\ &= 0,\end{aligned}$$

for all $x, y \in X$. This means that Q_1 satisfies (6.4). Since Q_1 is even, then by Lemma 6.1, Q_1 satisfies (6.7) and (6.8). Thus by (6.63), we have $Q_1(3x) = 9Q_1(x)$, $Q_1(4x) = 16Q_1(x)$, for all $x \in X$. Hence by (6.4), we conclude that Q_1 is quadratic. It remains to show that Q_1 is a unique quadratic mapping, which satisfies (6.58). Suppose that there is a quadratic mapping $Q_1' : X \to Y$ satisfying (6.58). Because of $Q_1(2^n x) = 4^n Q_1(x)$, and $Q_1'(2^n x) = 4^n Q_1'(x)$ for all $x \in X$, it follows that

$$\begin{aligned}\| Q_1(x) - Q_1'(x) \| &= 4^{-n} \| Q_1(2^n x) - Q_1'(2^n x) \| \\ &\leq 4^{-n}[\| Q_1(2^n x) - f(2^n(2x)) - 16f(2^n x) \| \\ &\quad + \| Q_1'(2^n x) - f(2^n(2x)) - 16f(2^n x) \|] \\ &\leq 4^{-n}(6\varepsilon)\end{aligned}$$

for all $x \in X$. Clearly, the right-hand side of inequality above tends to 0 as $n \to \infty$. Hence, we have $Q_1(x) = Q_1'(x)$ for all $x \in X$. □

Theorem 6.5. *Let ε be a positive real number. Suppose that an even mapping $f : X \to Y$ satisfies (6.57), and $f(0) = 0$. Then the limit*

$$Q_2(x) := \lim_n 16^{-n}[f(2^{n-1}x) - 4f(2^n x)]$$

exists for each $x \in X$, and $Q_2 : X \to Y$ is a unique quartic mapping that satisfies (6.4), and

$$\|f(2x) - 4f(x) - Q_2(x)\| \leq \frac{12}{5}\varepsilon$$

for all $x \in X$.

Proof. Similar to the proof of Theorem 6.4, we can show that f satisfies (6.61). Put $h(x) = f(2x) - 4f(x)$ for all $x \in X$. Then by (6.61) we have

$$\| h(2x) - 16h(x) \| \leq 9\varepsilon. \tag{6.64}$$

By (6.64), we apply iterative methods and induction on n again to prove our next relation

$$\| h(x) - 16^{-n}h(2^n x) \| \leq \frac{9}{16}\varepsilon \sum_{i=0}^{n-1} \frac{1}{16^i}$$

for all $x \in X$. By using the Cauchy convergence criterion, we can show that the limit

$$Q_2(x) := \lim_n 16^{-n}[f(2^{n-1}x) - 4f(2^n x)]$$

exists for each $x \in X$,. The rest of the proof is similar to the proof of Theorem 6.4.

□

Theorem 6.6. *Let ε be a positive real number. Suppose that an even mapping $f : X \to Y$ satisfies (6.57) and $f(0) = 0$. Then there exist a unique quadratic mapping $Q_1 : X \to Y$ and a unique quartic mapping $Q_2 : X \to Y$ such that*

$$\| f(x) - Q_1(x) - Q_2(x) \| \leq \frac{\varepsilon}{20}, \tag{6.65}$$

for all $x \in X$.

Proof. By Theorems 6.4 and 6.5, there exist a quadratic mapping $Q_1^o : X \to Y$ and a quartic mapping $Q_2^o : X \to Y$ such that

$$\|f(2x) - 16f(x) - Q_1^o(x)\| \leq 3\varepsilon \tag{6.66}$$

and

$$\|f(2x) - 4f(x) - Q_2^o(x)\| \leq \frac{12}{5}\varepsilon. \tag{6.67}$$

Combining (6.66) and (6.67), we obtain

$$\|12f(x) + Q_1^o(x) - Q_2^o(x)\| \leq \frac{3}{5}\varepsilon.$$

By putting $Q_1(x) := -\frac{1}{12}Q_1^o(x)$, and $Q_2(x) := \frac{1}{12}Q_2^o(x)$ for all $x \in X$, we have (6.65). To prove the uniqueness property of Q_1 and Q_2, let $Q_1', Q_2' : X \to Y$ be another quadratic and quartic mappings satisfying (6.65). Put $Q_1''(x) = Q_1(x) - Q_1'(x)$, $Q_2''(x) = Q_2(x) - Q_2'(x)$ for all $x \in X$. Then by (6.65) we have

$$\begin{aligned}\lim_n 16^{-n}\|Q_1''(2^n x) - Q_2''(2^n x)\| &\leq \lim_n 16^{-n}\|f(2^n x) - Q_1(2^n x) - Q_2(2^n x)\| \\ &\quad + \lim_n 16^{-n}\|f(2^n x) - Q_1'(2^n x) - Q_2'(2^n x)\| = 0.\end{aligned}$$

Hence, we can see that $Q_1''(x) = Q_2''(x) = 0$ for all $x \in X$. □

Theorem 6.7. *Let ε be a positive real number. Suppose that an odd mapping $f : X \to Y$ satisfies (6.57). Then the limits*

$$A(x) := \lim_n 2^{-n}[f(2^{n-1}x) - 8f(2^n x)]$$

and

$$C(x) := \lim_n 8^{-n}[f(2^{n-1}x) - 2f(2^n x)],$$

exist for all $x \in X$, and $A : X \to Y$ and $C : X \to Y$ are unique additive mapping and unique cubic mapping, respectively, satisfying (6.4), and

$$\|f(2x) - 8f(x) - A(x)\| \le 9\varepsilon$$

and

$$\|f(2x) - 2f(x) - C(x)\| \le \frac{9}{7}\varepsilon$$

for all $x \in X$.

Proof. The proof is similar to the proof of Theorems 6.4 and 6.5. □

Theorem 6.8. *Let ε be a positive real number. Suppose that an odd mapping $f : X \to Y$ satisfies (6.57). Then there exist a unique additive mapping $A : X \to Y$ and a unique cubic mapping $C : X \to Y$ such that*

$$\| f(x) - A(x) - C(x) \| \le \frac{54\varepsilon}{7}, \tag{6.68}$$

for all $x \in X$.

Proof. By Theorem 6.7, there exist an additive mapping $A_0 : X \to Y$ and a cubic mapping $C_0 : X \to Y$ such that

$$\|f(2x) - 8f(x) - A_0(x)\| \le 9\varepsilon, \quad \|f(2x) - 2f(x) - C_0(x)\| \le \frac{9}{7}\varepsilon$$

for all $x \in X$. Hence, we have

$$\|6f(x) + A_0(x) - C_0(x)\| \le \frac{54}{7}\varepsilon$$

for all $x \in X$. Set $A := \frac{-A_0}{6}$, $C := \frac{C_0}{6}$. The rest of the proof is similar to the proof of Theorem 6.6. □

Now we establish the Hyers–Ulam stability of functional equation (6.4) as follows:

Theorem 6.9. *Let ε be a positive real number. Suppose that a mapping $f : X \to Y$ satisfies $f(0) = 0$ and*

$$\|D_f(x,y)\| \le \varepsilon,$$

for all $x, y \in X$. Then there exist a unique additive mapping $A : X \to Y$, a unique quadratic mapping $Q_1 : X \to Y$, a unique cubic mapping $C : X \to Y$ and a unique quartic mapping $Q_2 : X \to Y$ such that

$$\|f(x)-A(x)-Q_1(x)-C(x)-Q_2(x)\| \leq \frac{1087\varepsilon}{140}, \tag{6.69}$$

for all $x \in X$.

Proof. Let $f_e(x) = \frac{1}{2}(f(x)+f(-x))$ for all $x \in X$. Then $f_e(0) = 0$, $f_e(-x) = f_e(x)$, and

$$\|D_{f_e}(x,y)\| \leq \varepsilon,$$

for all $x, y \in X$. Hence in view of Theorem 6.6, there exist a unique quadratic mapping $Q_1 : X \to Y$ and a unique quartic mapping $Q_2 : X \to Y$ satisfying (6.65). Let $f_o(x) = \frac{1}{2}(f(x)-f(-x))$ for all $x \in X$. Then f_o is an odd mapping satisfying $\|D_{f_o}(x,y)\| \leq \varepsilon$, for each $x, y \in X$. From Theorem 6.8, there exist a unique additive mapping $A : X \to Y$ and a unique cubic mapping $C : X \to Y$ satisfying (6.68). Now it is easy to see that (6.69) holds true for all $x \in X$. □

References

1. T. Aoki, *On the stability of the linear transformationin Banach spaces*, J. Math. Soc. Japan **2** (1950) 64–66.
2. C. Baak and M.S. Moslehian, *On the stability of orthogonally cubic functional equations*, Kyungpook Math. J. **47** (2007), no. 1, 69–76.
3. S. Czerwik, *On the stability of the quadratic mapping in normed spaces*, Abh. Math. Sem. Univ. Hamburg **62** (1992), 59–64.
4. S. Czerwik, *Functional Equations and Inequalities in Several Variables*, World Scientific, Singapore, 2002.
5. P.W. Cholewa, *Remarks on the stability offunctional equations*, Aequationes Math. **27** (1984), 76–86.
6. M. Eshaghi Gordji and H. Khodaei, *Solution and stability of generalized mixed type cubic, quadratic and additive functional equation in quasi–Banach spaces*, Nonlinear Anal. **71** (2009), 5629–5643.
7. M. Eshaghi, S. Kaboli and S. Zolfaghari, *Stability of a mixed type quadratic, cubic and quartic functional equation*, preprint.
8. Z. Gajda, *On stability of additive mappings*, Int. J. Math. Math. Sci. **14** (1991), 431–434.
9. P. Găvruta, *A generalization of the Hyers–Ulam–Rassias stability of approximately additive mappings*, J. Math. Anal. Appl. **184** (1994), 431–436.
10. D.H. Hyers, *On the stability of the linear functional equation*, Proc. Nat. Acad. Sci. U.S.A. **27** (1941), 222–224.
11. D.H. Hyers, G. Isac and Th.M. Rassias, *Stability of Functional Equations in Several Variables*, Birkhäuser, Basel, 1998.
12. G. Isac and Th. M. Rassias, *On the Hyers–Ulam stability of ψ-additive mappings*, J. Approx. Theory **72** (1993), 131–137.
13. G. Isac and Th.M. Rassias, *Functional inequalities for approximately additive mappings*, Stability of mappings of Hyers–Ulam type, 117–125, Hadronic Press Collect. Orig. Artic., Hadronic Press, Palm Harbor, FL, 1994.
14. K.W. Jun and H.M. Kim, *The generalized Hyers–Ulam–Rassias stability of a cubic functional equation*, J. Math. Anal. Appl. **274** (2002), 267–278.
15. S.-M. Jung, *Stability of the quadratic equation of Pexider type*, Abh. Math. Sem. Univ. Hamburg **70** (2000), 175–190.
16. S.-M. Jung, *Hyers–Ulam–Rassias Stability of Functional Equations in Mathematical Analysis*, Hadronic Press, Palm Harbor, FL, 2001.

17. Pl. Kannappan, *Quadratic functional equation and inner product spaces*, Results Math. **27** (1995), 368–372.
18. M. Mirzavaziri and M.S. Moslehian, *A fixed point approach to stability of a quadratic equation,* Bull. Braz. Math. Soc. **37** (2006), no. 3, 361–376.
19. M.S. Moslehian, *On the orthogonal stability of the Pexiderized quadratic equation*, J. Differ. Equations Appl. **11** (2005), 999–1004.
20. M.S. Moslehian and Th.M. Rassias, *Stability of functional equations in non-Archimedian spaces*, Appl. Anal. Disc. Math., **1** (2007), no. 2, 325–334.
21. A. Najati, *On the stability of a quartic functional equation*, J. Math. Anal. Appl. **340** (2008), no. 1, 569–574.
22. K. Nikodem, *On some properties of quadratic stochastic processes*, Annales Math. Silesianae **3** (15) (1990), 59–69.
23. C. Park, *On the stability of the orthogonally quartic functional equation* Bull. Iranian Math. Soc. **31** (2005), no. 1, 63–70.
24. W.-G. Park and J.-H. Bae, *On a bi-quadratic functional equation and its stability*, Nonlinear Anal. **62** (2005), no. 4, 643–654.
25. Th.M. Rassias, *On the stability of the linear mapping in Banach spaces*, Proc. Amer. Math. Soc. **72** (1978), 297–300.
26. Th.M. Rassias, *Problem 16 ; 2, Report of the 27th International Symp. on Functional Equations*, Aequationes Math. **39** (1990), 292–293; 309.
27. Th.M. Rassias, and J. Tabor, *What is left of Hyers-Ulam stability?*, J. Natur. Geom. **1** (1992), no. 2, 65–69.
28. Th.M. Rassias (ed.), *Functional equations, inequalities and applications*, Kluwer Academic Publishers, Dordrecht, 2003.
29. Th.M. Rassias and P. Šemrl, *On the behaviour of mappings which do not satisfy Hyers–Ulam stability*, Proc. Amer. Math. Soc. **114** (1992), 989–993.
30. P.K. Sahoo, *A generalized cubic functional equation* Acta Math. Sin. (Engl. Ser.) **21** (2005), no. 5, 1159–1166.
31. F. Skof, *Proprietà locali e approssimazione di operatori*, Rend. Sem. Mat. Fis. Milano **53** (1983), 113–129.
32. S.M. Ulam, *Problems in Modern Mathematics*, Chapter VI, Science Editions, Wiley, New York, 1964.

Chapter 7
Ψ-Additive Mappings and Hyers–Ulam Stability

P. Găvruţa and L. Găvruţa

Dedicated to the memory of Professor George Isac

Abstract The notion of ψ-additive mappings was first introduced by George Isac related to the asymptotic derivative of mappings. Hyers–Ulam stability of those mappings was studied by G. Isac, Th.M. Rassias and many other authors: L. Cădariu, H.G. Dales, V. Faiziev, P. Găvruţa, R. Ger, J. Matkowski, M.S. Moslehian, S.Z. Nemeth, and V. Radu. In this paper we give a short survey about the Hyers–Ulam stability of ψ-additive mappings.

7.1 Introduction

The study of stability problems for functional equations originated from a talk of S. Ulam in 1940, when he proposed the following problem:
Let G be a group endowed with a metric d. Given $\varepsilon > 0$, does there exist a $k > 0$ such that for every function $f : G \to G$ satisfying the inequality

$$d(f(xy), f(x)f(y)) < \varepsilon, \qquad \forall x, y \in G,$$

there exists an automorphism a of G with

$$d(f(x), a(x)) < k\varepsilon, \qquad \forall x \in G \ ?$$

P. Găvruţa
Universitatea Politehnica din Timişoara, Departamentul de Matematică, Piaţa Victoriei no. 2, 300006 Timişoara, Romania, e-mail: pgavruta@yahoo.com

L. Găvruţa
Universitatea Politehnica din Timişoara, Departamentul de Matematică, Piaţa Victoriei no. 2, 300006 Timişoara, Romania, e-mail: gavruta_laura@yahoo.com

P.M. Pardalos et al. (eds.), *Nonlinear Analysis and Variational Problems*, Springer Optimization and Its Applications 35, DOI 10.1007/978-1-4419-0158-3_7,

In 1941, Hyers gave an affirmative answer to the question of Ulam for additive Cauchy equation in Banach spaces. The concept of the Hyers–Ulam stability of mappings is currently used in the spirit of Ulam's problem for approximate homomorphisms. The Hyers–Ulam stability has been mainly used to study problems concerning approximate isometries or quasi-isometries, the stability of Lorentz and conformal mappings, the stability of stationary points, the stability of convex mappings, the stability of minimum points, etc. (cf. [26],[27],[28]).

Now, let E_1 be a real normed vector space and E_2 a real Banach space and $f : E_1 \to E_2$ is an approximately additive mapping. In 1941, Hyers considered approximately additive mappings $f : E_1 \to E_2$ satisfying

$$\|f(x+y) - f(x) - f(y)\| \le \varepsilon, \quad \text{for all } x, y \in E_1.$$

Hyers proved that the limit

$$T(x) = \lim_{n\to\infty} 2^{-n} f(2^n x)$$

exists for all $x \in E_1$ and that $T : E_1 \to E_2$ is a unique additive mapping satisfying

$$\|f(x) - T(x)\| \le \varepsilon.$$

A generalized solution to Ulam's problem for additive mappings was given by T. Aoki [1] and for approximately linear mappings was given by Th.M. Rassias in 1978. Th.M. Rassias considered a mapping $f : E_1 \to E_2$ such that $t \to f(tx)$ is continuous in t for each fixed x. Assume that there exists $\theta \ge 0$ and $0 \le p < 1$ such that

$$\|f(x+y) - f(x) - f(y)\| \le \theta(\|x\|^p + \|y\|^p) \quad \text{for any } x, y \in E_1.$$

Then there exists a unique linear mapping $T : E_1 \to E_2$ such that

$$\|f(x) - T(x)\| \le \frac{2\theta}{2-2^p}\|x\|^p \quad \text{for any } x, y \in E_1.$$

The Hyers result is obtained for $p = 0$. Th.M. Rassias's proof given in [25] also applies for all real values of p that are strictly less than zero.

Rassias's stability theorem for the linear mapping (see [25]) was generalized by G. Isac and Th.M. Rassias [15] (see also [16],[17]) for ψ-additive mappings.

7.2 Results

Let E_1, E_2 be two normed vector spaces.

Definition 7.1. A mapping $f : E_1 \to E_2$ is ψ-additive if and only if there exists $\theta \ge 0$ and a function $\psi : \mathbb{R}_+ \to \mathbb{R}_+$ such that

$$\|f(x+y) - f(x) - f(y)\| \le \theta[\psi(\|x\|) + \psi(\|y\|)], \quad \text{for all } x, y \in E_1.$$

Theorem 7.2. *([15]) Consider E_1 to be a real normed vector space, E_2 a real Banach space and let $f : E_1 \to E_2$ be a mapping such that $f(tx)$ is continuous in t for each fixed x. If f is ψ-additive and ψ satisfies*

$$\begin{aligned} i) &\quad \lim_{t\to\infty}(\psi(t)/t) = 0 \\ ii) &\quad \psi(ts) \le \psi(t)\psi(s), \quad \textit{for all } t,s \in \mathbb{R}_+ \\ iii) &\quad \psi(t) < t, \quad \textit{for all } t < 1 \end{aligned}$$

then there exists a unique linear mapping $T : E_1 \to E_2$ such that $\|f(x) - T(x)\| \le (2\theta/(2-\psi(2)))\psi(\|x\|)$, for all $x \in E_1$.

A generalization of the theorem of G. Isac and Th.M. Rassias [15] was proved by P. Găvruţa [9] in the following form:

Theorem 7.3. *We denote by $(G,+)$ an Abelian group, by $(X, \|\cdot\|)$ a Banach space, and by $\varphi : G \times G \to [0,\infty)$ a mapping such that*

$$\widetilde{\varphi}(x,y) := \sum_{k=0}^{\infty} 2^{-k}\varphi(2^k x, 2^k y) < \infty \quad \textit{for all } x,y \in G.$$

Let $f : G \to X$ be such that

$$\|f(x+y) - f(x) - f(y)\| \le \varphi(x,y), \textit{for all } x,y \in G$$

Then there exists a unique mapping $T : G \to X$

$$T(x+y) = T(x) + T(y), \textit{for allx} \in G$$

and

$$\|f(x) - T(x)\| \le \frac{1}{2}\widetilde{\varphi}(x,y), \textit{for all } x \in G$$

Moreover, if G is a real normed space and X is a real Banach space and $f(tx)$ is continuous in t for each fixed x in G, then T is a linear function.

As an application of Theorem 7.3, we obtained a generalization of Theorem 7.2: Let G be a normed linear space and define $H : \mathbb{R}_+ \times \mathbb{R}_+ \to \mathbb{R}_+$ and $\varphi_0 : \mathbb{R}_+ \to \mathbb{R}_+$ such that

$$\begin{aligned} \varphi_0(\lambda) &> 0, \quad \textit{for all } \lambda > 0, \\ \varphi_0(2) &< 2, \\ \varphi_0(2\lambda) &\le \varphi_0(2)\varphi_0(\lambda), \quad \textit{for all } \lambda > 0, \\ H(\lambda t, \lambda s) &\le \varphi_0(\lambda)H(t,s), \quad \textit{for all } t,s \in \mathbb{R}_+, \lambda > 0. \end{aligned}$$

In Theorem 7.3, we take

$$\varphi(x,y) = H(\|x\|, \|y\|)$$

Then

$$\begin{aligned}\varphi(2^k x, 2^k y) &= H(2^k\|x\|, 2^k\|y\|)\\ &\leq \varphi_0(2^k)H(\|x\|,\|y\|)\\ &\leq (\varphi_0(2))^k H(\|x\|,\|y\|)\end{aligned}$$

and because $\varphi_0(2) < 2$ we have

$$\begin{aligned}\widetilde{\varphi}(x,y) &\leq \sum_{k=0}^{\infty} 2^{-k}(\varphi_0(2))^k H(\|x\|,\|y\|)\\ &= \frac{1}{1-(\varphi_0(2)/2)} H(\|x\|,\|y\|),\end{aligned}$$

and we obtain

$$\|f(x) - T(x)\| \leq \frac{1}{2}\widetilde{\varphi}(x,y) \leq \frac{1}{2-\varphi_0(2)} H(\|x\|,\|x\|)$$

or

$$\|f(x) - T(x) \leq \frac{1}{2-\varphi_0(2)} \varphi_0(\|x\|)H(1,1).$$

Also, as an application of Theorem 7.3, G. Isac and Th.M. Rassias [16] obtained a useful generalization of Theorem 7.2. Let F_ψ be the set of all functions ψ from $\mathbb{R}_+$ into $\mathbb{R}_+$ satisfying the assumptions:

$$\begin{aligned}&i)\ \lim_{t\to\infty} \frac{\psi(t)}{t} = 0,\\ &ii)\ \psi(ts) \leq \psi(t)\psi(s), \quad for\ all\ t,s \in \mathbb{R}_+,\\ &iii)\ \psi(t) < t, \quad for\ all\ t > 1.\end{aligned}$$

Let $P(\psi)$ be the convex cone generated by the set F_ψ. The function $\psi \in P(\psi)$ satisfies the assumption *i*) but generally does not satisfy the assumptions *ii*) and *iii*).

Theorem 7.4. *Let E_1 be a real normed vector space, E_2 a real Banach space and $f : E_1 \to E_2$ a continuous mapping. If f is ψ-additive with $\psi \in P(\Psi)$, then there exists a unique linear mapping $T : E_1 \to E_2$ such that*

$$\|f(x) - T(x)\| \leq \frac{2\theta}{2-\psi(2)} \psi(\|x\|), \quad for\ all\ x \in E_1.$$

Moreover, the expression of T at every point $x \in E_1$ is given by

$$T(x) = \lim_{n\to\infty} \frac{f(2^n x)}{2^n}.$$

R. Ger [11] considered stability of ψ-additive mappings and Orlicz Δ_2-conditions. He deals with a functional inequality of the form

$$\|f(x+y)-f(x)-f(y)\| \leq \varphi(\|x\|)+\psi(\|y\|) \quad (\star),$$

showing, among others, that given two self-mappings φ, ψ of the halfline $[0,\infty)$ enjoying the celebrated Orlicz Δ_2-conditions:

$$\varphi(2t) \leq k\varphi(t), \quad \psi(2t) \leq l\psi(t)$$

for all $t \in [0,\infty)$, with some constants $k,l \in [0,2)$, for every map f between a normed linear space $(X, \|\cdot\|)$ and a Banach space $(Y, \|\cdot\|)$ satisfying inequality $(\star)$, there exists exactly one additive map $a : X \to Y$ such that

$$\|f(x)-a(x)\| \leq \frac{1}{2-k}\varphi(\|x\|)+\frac{1}{2-l}\psi(\|x\|)$$

for all $x \in X$. It is clear that Ger's result is also an application of Theorem 7.3.

For several other results and a number of interesting applications in nonlinear analysis of ψ-additive mappings, one may see the papers and books cited in the References.

References

1. T. Aoki, On the stability of the linear transformation in Banach spaces, *J. Math. Soc. Japan* 2(1950) 64–66.
2. R. Badora, On approximate derivations, *Math. Ineq. Appl.* 9(1)(2006), 167–173.
3. L. Cădariu, V. Radu, On the stability of the Cauchy functional equation: a fixed point approach, ITERATION THEORY (ECIT '02), *Grazer Math. Ber.,* Berichit Nr. 346(2004), 43–52.
4. S. Czerwik, Functional Equations and Inequalities in Several Variables, *World Scientific, Singapore*, 2002.
5. H.G. Dales, M.S. Moslehian, Stability of mappings on multi-normed spaces, *Glasgow Math. J.*, 49(2007) 321–332.
6. M. Eshaghi Gordji, N. Ghobadipour, C. Park, Jordan $\star$−homomorphism on $C^\star$−algebras, arXiv:0812.2928v1, submit.
7. V.A. Faiziev, Th.M. Rassias and P.K. Sahoo, The space of (ψ,γ)–additive mappings on semigroups, *Trans. Amer. Math. Soc.* 345(11), (2002) 4455–4472.
8. V.A. Faiziev and Th.M. Rassias, The space of (ψ,γ)–pseudocharacters on semigroups, *Nonlinear Functional Analysis and Applications*, 5(1)(2000), 107–126.
9. P. Găvruţa, A generalization of the Hyers-Ulam-Rassias stability of approximately additive mappings, *J. Math. Anal. Appl.*184(1994), 431–436.
10. P. Găvruţa, On a problem of G. Isac and Th.M. Rassias concerning the stability of mappings, *J. Math. Anal. Appl.* 261(2001), 543–553.
11. R. Ger, Stability of ψ-additive mappings and Orlicz Δ_2−conditions, in: Z.Palés, Report of the General Inequalities 8 Conference, September 15-21, (2002) Noszvaj, Hungary, JIPAM 4(3)(2003), Article 49.

12. D.H. Hyers, On the stability of the linear functional equation, *Proc. Natl. Acad. Sci.* 27(1941) 222–224.
13. D.H. Hyers, G. Isac, Th.M. Rassias, Stability of Functional Equations in Several Variables, *Birkhäuser, Basel,* 1998.
14. D.H. Hyers, Th.M. Rassias, Approximate homomorphisms, *Aequationes Math.* 44(1992), 125–153.
15. G. Isac and Th.M. Rassias, On the Hyers-Ulam stability of ψ-additive mappings, *J. Approx. Theory,* 72(1993),131–137.
16. G. Isac and Th.M. Rassias, Stability of ψ-additive mappings:Applications to nonlinear analysis, *Internat. J. Math. Sci.* 19(1996), 219–228.
17. G. Isac and Th.M. Rassias, Functional inequalities for approximately additive mappings, In: Th.M. Rassias and J. Tabor (eds), *Stability of Mappings of Hyers-Ulam Type,* Hadronic Press, Florida, 1994, 117–125.
18. G. Isac and S.Z. Neméth, Scalar and Asymptotic Scalar Derivatives. Theory and Applications, Springer, 2008.
19. G. Isac, The fold complementarity problem and the order complementarity problem, *Topological Meth. Nonlin. Anal.* 8(1996),343–358.
20. Jung, S-M., Hyers-Ulam-Rassias Stability of Functional Equations in Mathematical Analysis, *Hadronic Press, Palm Harbor*, FL, 2001.
21. J. Matkowski, On subadditive and ψ-additive mappings, *Central European J. Math. 2(3)*(2004),493.
22. J. Matkowski, On subadditive functions and ψ-additive mappings, *Central European J. Math. 1(4)* (2003), 435–440.
23. V. Radu, The fixed point alternative and the stability of functional equations, *Fixed Point Theory* 4(1)(2003), 91–95.
24. J.M. Rassias, Solution of a problem of Ulam, *Journal of Approximation Theory,* vol. 57, no. 3, pp. 268—273, 1989.
25. Th.M. Rassias, On the stability of the linear mapping in Banach spaces, *Proc. Amer. Math. Soc.* 72 (1978) 297–300.
26. Th.M. Rassias and J. Tabor, What is left of Hyers-Ulam stability?, *Journal of Natural Geometry* 1(1992), 65–69.
27. Th.M. Rassias, On the stability of minimum points, *Mathematica* 45(68)(1) (2003), 93–104.
28. Th.M. Rassias, On the stability of functional equations and a problem of Ulam, *Acta Applicandae Mathematicae*, 62(1)(2000), 23–130.

Chapter 8
The Stability and Asymptotic Behavior of Quadratic Mappings on Restricted Domains

Kil-Woung Jun and Hark-Mahn Kim

Dedicated to the memory of Professor George Isac

Abstract In this paper, we investigate the generalized Hyers–Ulam stability problem for quadratic functional equations in several variables, and then obtain an asymptotic behavior of quadratic mappings on restricted domains.

8.1 Introduction

S.M. Ulam [31, 32] proposed the general Ulam stability problem: "When is it true that by slightly changing the hypotheses of a theorem one can still assert that the thesis of the theorem remains true or approximately true?" The concept of stability for a functional equation arises when we replace the functional equation by an inequality which acts as a perturbation of the equation. Thus one can ask the following question for general functional equations: If we replace a given functional equation by a functional inequality, when can one assert that the solutions of the inequality must be close to the solutions of the given equation? If the answer is affirmative, we would say that a given functional equation is stable. In 1978, P.M. Gruber [10] remarked that Ulam's problem is of particular interest in probability theory and in the case of functional equations of different types. We wish to note that stability properties of different functional equations can have applications to unrelated fields. For instance, Zhou [33] used a stability property of the functional equation

Kil-Woung Jun
Department of Mathematics, Chungnam National University, 220 Yuseong-Gu, Daejeon, 305-764, Korea, e-mail: kwjun@cnu.ac.kr

Hark-Mahn Kim
Department of Mathematics, Chungnam National University, 220 Yuseong-Gu, Daejeon, 305-764, Korea, e-mail: hmkim@cnu.ac.kr

P.M. Pardalos et al. (eds.), *Nonlinear Analysis and Variational Problems*,
Springer Optimization and Its Applications 35, DOI 10.1007/978-1-4419-0158-3_8,

$f(x-y)+f(x+y)=2f(x)$ to prove a conjecture of Z. Ditzian about the relationship between the smoothness of a mapping and the degree of its approximation by the associated Bernstein polynomials.

The case of approximately additive functions was first solved by D.H. Hyers [11], and generalized versions of the Hyers theorem for approximate additive mappings which allows the Cauchy difference to be unbounded was given by T. Aoki [3] and D.G. Bourgin [4]. In 1978, Th.M. Rassias [22] proved a theorem for the stability of the linear mapping, which allows the Cauchy difference to be controlled by a sum of powers of norms. G. Isac and Th.M. Rassias [15, 16, 17] have improved the stability results for the approximately additive functions using ψ-additive mappings and a sum of different powers of norms. During the past decades, the stability problems of several functional equations have been extensively investigated by a number of authors [13, 14, 26].

Now, a square norm on an inner product space satisfies the important parallelogram equality $\|x+y\|^2+\|x-y\|^2=2(\|x\|^2+\|y\|^2)$ for all vectors x,y in the inner product space. The following functional equation which was motivated by the identity

$$Q(x+y)+Q(x-y)=2Q(x)+2Q(y) \tag{8.1}$$

is called a quadratic functional equation, and every solution of the equation (8.1) is said to be a quadratic mapping. It is well known that a mapping Q between real vector spaces E_1,E_2 satisfies the equation (8.1) if and only if there exists a unique symmetric biadditive mapping $B: E_1\times E_1\to E_2$ such that $Q(x)=B(x,x)$ for all x [1]. The quadratic functional equation and several other functional equations are useful to characterize inner product spaces [2, 23, 29]. During the past three decades a number of papers and research monographs have been published on various generalizations and applications of the generalized Hyers–Ulam stability to a number of functional equations and mappings (see [4, 5, 8, 12, 25, 27, 28]).

In 1983, F. Skof [30] was the first author to solve the Ulam problem for additive mappings on a restricted domain. S. Jung [18] and J.M. Rassias [20] investigated the Hyers–Ulam stability for additive and quadratic mappings on restricted domains. The stability problems of several quadratic functional equations have been extensively investigated by a number of authors and there are many interesting results concerning this problem [6, 12, 19, 24].

Now we consider the following functional equations,

$$f\left(\sum_{i=1}^{d+1}x_i\right)+\sum_{1\le i<j\le d+1}f(x_i-x_j)=(d+1)\sum_{i=1}^{d+1}f(x_i), \tag{8.2}$$

$$\sum_{1\le i<j\le d+1}[f(x_i+x_j)+f(x_i-x_j)]=2d\sum_{i=1}^{d+1}f(x_i) \tag{8.3}$$

for all $(d+1)$-variables $x_1,\cdots,x_{d+1}\in E_1$, where $d\ge 1$ is a natural number. As a special case, these equations reduce to the equation (8.1) in the case $d=1$. It is easy to see that a mapping $f : E_1 \to E_2$ between vector spaces satisfies the functional equation (8.2) or (8.3) if and only if the mapping f is quadratic. In 1998, D.H. Hyers, G. Isac and Th.M. Rassias [14] introduced a new study for the asymptotic behavior of the Hyers–Ulam stability of mappings.

In this paper, we establish the new generalized Hyers–Ulam stability theorems for the general functional equations (8.2) and (8.3) with several variables and apply our results to the asymptotic behavior of quadratic mappings on restricted domains.

8.2 Approximately Quadratic Mappings

From now on, let B be a unital Banach algebra with norm $|\cdot|$, and let X and Y be left Banach B-modules with norms $\|\cdot\|$ and $||\cdot||$, respectively, unless we give any specific reference. Let $\mathbf{R}^+$ denote the set of all non-negative real numbers and d a positive integer with $d \geq 1$. A quadratic mapping $Q : X \to Y$ is called B-quadratic if

$$Q(ax) = a^2 Q(x), \quad \forall a \in B, \forall x \in X.$$

Now before taking up the main subject, given $f : X \to Y$, we define the difference operator $E_u f : X^{d+1} \to Y$ by

$$E_u f(x_1, \cdots, x_{d+1}) := f\left(\sum_{i=1}^{d+1} ux_i\right) + \sum_{1 \leq i < j \leq d+1} f(ux_i - ux_j) - (d+1) \sum_{i=1}^{d+1} u^2 f(x_i)$$

for all $u \in B(1) := \{u \in B : |u| = 1\}$, and all $(d+1)$ variables in X, which acts as a perturbation of the equation (8.2).

Theorem 8.1. *Suppose that a mapping $f : X \to Y$ satisfies*

$$\|E_u f(x_1, \cdots, x_{d+1})\| \leq \varepsilon(x_1, \cdots, x_{d+1}) \tag{8.4}$$

for all $u \in B(1)$ and all $(d+1)$-variables $x_1, \cdots, x_{d+1} \in X$, and that $\varepsilon : X^{d+1} \to \mathbf{R}^+$ is a mapping such that the series

$$\sum_{i=0}^{\infty} \frac{\varepsilon(2^i x_1, \cdots, 2^i x_{d+1})}{2^{2i}}$$

converges for all $x_1, \cdots, x_{d+1} \in X$. If f is measurable or $f(tx)$ is continuous in $t \in \mathbf{R}$ for each fixed $x \in X$, then there exists a unique B-quadratic mapping $Q : X \to Y$ which satisfies the equation (8.2) and the inequality

$$\left\| f(x) + \frac{d^2 + 3d - 6}{6} f(0) - Q(x) \right\| \leq \frac{1}{4} \sum_{i=0}^{\infty} \frac{\overline{\varepsilon}(2^i x, 2^i x)}{4^i}, \tag{8.5}$$

where $\|f(0)\| \le \frac{2\varepsilon(0,\cdots,0)}{d(d+3)}$ *and*

$$\overline{\varepsilon}(x,x) := \min\{\varepsilon(0,\cdots,\overbrace{x}^{i-th},\cdots,\overbrace{x}^{j-th},\cdots) : 1 \le i < j \le d+1\} \tag{8.6}$$

for all $x \in X$. *The mapping* Q *is defined by*

$$Q(x) = \lim_{n\to\infty} \frac{f(2^n x)}{2^{2n}} \tag{8.7}$$

for all $x \in X$.

Proof. Without loss of generality, we may assume that that $i = 1, j = 2$ in the expression (8.6). If we take $(x_1, x_2, 0, \cdots, 0)$ instead of $(x_1, \cdots, x_{d+1})$ in (8.4) with $u = 1$, we obtain

$$\Big\| f(x_1+x_2) + f(x_1-x_2) + \binom{d-1}{2} f(0) - 2f(x_1) \\ -2f(x_2) - (d+1)(d-1)f(0) \Big\| \le \varepsilon(x_1, x_2, 0, \cdots, 0),$$

which can be rewritten in the form

$$\|q(x_1+x_2) + q(x_1-x_2) - 2q(x_1) - 2q(x_2)\| \le \varepsilon(x_1, x_2, 0, \cdots, 0), \tag{8.8}$$

for all $x_1, x_2 \in X$, where $q(x) := f(x) + \frac{(d+4)(d-1)f(0)}{4}$, $x \in X$.

Now applying a standard procedure of direct method (see [5, 9]) to the last inequality (8.8), we see that there exists a unique quadratic mapping $Q : X \to Y$ which satisfies the equation (8.2) and the inequality

$$\left\| q(x) - \frac{q(0)}{3} - Q(x) \right\| \le \frac{1}{4} \sum_{i=0}^{\infty} \frac{\varepsilon(2^i x, 2^i x, \overbrace{0, \cdots, 0}^{d-1})}{2^{2i}}$$

for all $x \in X$. Under the assumption that f is measurable or $f(tx)$ is continuous in $t \in \mathbf{R}$ for each fixed $x \in X$, the quadratic function $Q : X \to Y$ satisfies

$$Q(tx) = t^2 Q(x), \quad \forall x \in X, \forall t \in \mathbf{R}.$$

That is, Q is $\mathbf{R}$-quadratic. Since Q is $\mathbf{R}$-quadratic and $Q(ux) = u^2 Q(x)$ for each element $u \in B(1)$, we see that $Q(ax) = a^2 Q(x)$ for all $a \in B$ and all $x \in X$. So $Q : X \to Y$ is also B-quadratic as desired. □

Theorem 8.2. *Suppose that a mapping* $f : X \to Y$ *satisfies*

$$\|E_u f(x_1, \cdots, x_{d+1})\| \le \varepsilon(x_1, \cdots, x_{d+1})$$

for all $u \in B(1)$ *and all* $(d+1)$*-variables* $x_1, \cdots, x_{d+1} \in X$, *and that* $\varepsilon : X^{d+1} \to \mathbf{R}^+$ *is a mapping such that the series*

$$\sum_{i=1}^{\infty} 4^i \varepsilon\left(\frac{x_1}{2^i}, \cdots, \frac{x_{d+1}}{2^i}\right)$$

converges for all $x_1, \cdots, x_{d+1} \in X$. If f is measurable or $f(tx)$ is continuous in $t \in \mathbf{R}$ for each fixed $x \in X$, then there exists a unique quadratic mapping $Q : X \to Y$ which satisfies the equation (8.2) and the inequality

$$\|f(x) - Q(x)\| \leq \frac{1}{4} \sum_{i=1}^{\infty} 4^i \overline{\varepsilon}\left(\frac{x}{2^i}, \frac{x}{2^i}\right),$$

where $\overline{\varepsilon}(x,x)$ is given as in Theorem 8.1. The mapping Q is defined by

$$Q(x) = \lim_{n\to\infty} 4^n f\left(\frac{x}{2^n}\right)$$

for all $x \in X$.

Note that one has $f(0) = 0$ in the above theorem because $\varepsilon(0, \cdots, 0) = 0$ by the convergence of the series.

Given a mapping $f : X \to Y$, we define the difference operator $D_u f : X^{d+1} \to Y$ by

$$D_u f(x_1, \cdots, x_{d+1}) := \sum_{1\leq i<j\leq d+1} [f(ux_i + ux_j) + f(ux_i - ux_j)] - 2d \sum_{i=1}^{d+1} u^2 f(x_i)$$

for all $(d+1)$-variables $x_1, \cdots, x_{d+1} \in X$, which acts as a perturbation of the equation (8.3). Furthermore, we are going to establish another theorem about the Ulam stability problem of the equation (8.3) as follows.

Theorem 8.3. *Suppose that a mapping $f : X \to Y$ satisfies*

$$\|D_u f(x_1, x_2, \cdots, x_{d+1})\| \leq \varepsilon(x_1, \cdots, x_{d+1}) \tag{8.9}$$

for all $u \in B(1)$ and all $(d+1)$-variables $x_1, \cdots, x_{d+1} \in X$, and that $\varepsilon : X^{d+1} \to \mathbf{R}^+$ is a mapping such that the series

$$\sum_{i=0}^{\infty} \frac{\varepsilon(2^i x_1, \cdots, 2^i x_{d+1})}{2^{2i}}$$

converges for all $x_1, \cdots, x_{d+1} \in X$. If f is measurable or $f(tx)$ is continuous in $t \in \mathbf{R}$ for each fixed $x \in X$, then there exists a unique quadratic mapping $Q : X \to Y$ which satisfies the equation (8.3) and the inequality

$$\left\| f(x) + \frac{(d^2+d-3)f(0)}{3} - Q(x) \right\| \leq \frac{1}{4} \sum_{i=0}^{\infty} \frac{\overline{\overline{\varepsilon}}(2^i x, 2^i x)}{2^{2i}} \tag{8.10}$$

for all $x \in X$, where a mapping $\overline{\overline{\varepsilon}} : X^2 \to Y$ is given by

$$\bar{\bar{\varepsilon}}(x,y) := min\left\{\varepsilon\left(x,0,\cdots,0,\overbrace{y}^{i},0\cdots,0\right) \;\middle|\; 2 \le i \le d+1\right\}. \tag{8.11}$$

The mapping Q is defined by

$$Q(x) = \lim_{n\to\infty} \frac{f(2^n x)}{2^{2n}}$$

for all $x \in X$.

Proof. If we take $\left(x,0,\cdots,0,\overbrace{y}^{i},0,\cdots,0\right)$ instead of $(x_1,\cdots,x_{d+1})$ in (8.9), we obtain

$$\begin{aligned}&\|f(x+y)+f(x-y)-2f(x)-2f(y)-(d^2+d-2)f(0)\|\\ &\le \varepsilon\left(x,0,\cdots,0,\overbrace{y}^{i},0,\cdots,0\right)\end{aligned}$$

for all $x,y \in X$, and all i with $2 \le i \le d+1$, which can be written in the form

$$\|q(x+y)+q(x-y)-2[q(x)+q(y)]-q(0)\| \le \bar{\bar{\varepsilon}}(x,y) \tag{8.12}$$

for all $x,y \in X$, where a mapping $q : X \to Y$ is defined by $q(x) := f(x) + \frac{(d^2+d-3)f(0)}{3}$ and a mapping $\bar{\bar{\varepsilon}} : X^2 \to Y$ is given by (8.11). Taking $y := x$ in (8.12), we get

$$\left\|\frac{q(2x)}{4} - q(x)\right\| \le \frac{1}{4}\bar{\bar{\varepsilon}}(x,x) \tag{8.13}$$

for all $x \in X$. Now applying a standard procedure of direct method [5, 9] to the last inequality (8.13), we see that there exists a unique quadratic mapping $Q : X \to Y$ which satisfies the equation (8.3) and the inequality

$$\|q(x)-Q(x)\| \le \frac{1}{4}\sum_{i=0}^{\infty} \frac{\bar{\bar{\varepsilon}}(2^i x, 2^i x)}{2^{2i}}$$

for all $x \in X$. □

Theorem 8.4. *Suppose that a mapping $f : X \to Y$ satisfies*

$$\|D_u f(x_1,x_2,\cdots,x_{d+1})\| \le \varepsilon(x_1,\cdots,x_{d+1})$$

for all $u \in B(1)$ and all $(d+1)$-variables $x_1,\cdots,x_{d+1} \in X$, and that $\varepsilon : X^{d+1} \to \mathbf{R}^+$ is a mapping such that the series

$$\sum_{i=1}^{\infty} 4^i \varepsilon\left(\frac{x_1}{2^i},\cdots,\frac{x_{d+1}}{2^i}\right)$$

converges for all $x_1, \cdots, x_{d+1} \in X$. If f is measurable or $f(tx)$ is continuous in $t \in \mathbf{R}$ for each fixed $x \in X$, then there exists a unique quadratic mapping $Q : X \to Y$ which satisfies the equation (8.3) and the inequality

$$\|f(x) - Q(x)\| \leq \frac{1}{4} \sum_{i=1}^{\infty} 4^i \bar{\bar{\varepsilon}} \left(\frac{x}{2^i}, \frac{x}{2^i} \right)$$

for all $x \in X$, where the mapping $\bar{\bar{\varepsilon}} : X^2 \to Y$ is given by (8.11). The mapping Q is defined by

$$Q(x) = \lim_{n \to \infty} 4^n f\left(\frac{x}{2^n} \right)$$

for all $x \in X$.

8.3 Quadratic Mappings on Restricted Domains

In this section, we are going to investigate the Hyers–Ulam stability problem for the equations (8.2) and (8.3) on a restricted domain. As results, we have an asymptotic property of the mappings concerning the equations (8.2) and (8.3).

Theorem 8.5. *Let $r > 0$ be fixed. Suppose that there exists a non-negative real number ε for which a mapping $f : X \to Y$ satisfies*

$$\|D_1 f(x_1, \cdots, x_{d+1})\| \leq \varepsilon \tag{8.14}$$

for all $(d+1)$-variables $x_1, \cdots, x_{d+1} \in X$ with $\sum_{i=1}^{d+1} \|x_i\| \geq r$. Then there exists a unique quadratic mapping $Q : X \to Y$ which satisfies the equation (8.3) and the inequality

$$\left\| f(x) + \frac{(d^2 + d - 3) f(0)}{2} - Q(x) \right\| \leq \frac{3\varepsilon}{2} \tag{8.15}$$

for all $x \in X$.

Proof. Taking $(x_1, \cdots, x_{d+1})$ as $(x, y, 0, \cdots, 0)$ in (8.14) with $\|x\| + \|y\| \geq r$, we obtain by the same way as (8.8)

$$\|q(x+y) + q(x-y) - 2q(x) - 2q(y) - q(0)\| \leq \varepsilon, \tag{8.16}$$

for all $x, y \in X$ with $\|x\| + \|y\| \geq r$, where $q(x) := f(x) + \frac{(d^2+d-3)f(0)}{2}$. Specially, we have $\|q(0)\| \leq \frac{\varepsilon}{3}$ by setting $y := 0$ and $x := t$ with $\|t\| \geq r$ in (8.16). Now, assume $\|x\| + \|y\| < r$. And choose a $t \in X$ with $\|t\| \geq 2r$. Then it holds clearly

$$\|x \pm t\| \geq r, \ \|y \pm t\| \geq r, \text{ and } \|2t\| + \|x + y\| \geq r.$$

Therefore from (8.16) and the following functional identity

$$\begin{aligned}
&2\big[q(x+y)+q(x-y)-2q(x)-2q(y)-q(0)\big]\\
&=\big[q(x+y+2t)+q(x-y)-2q(x+t)-2q(y+t)\big]\\
&\quad+\big[q(x+y-2t)+q(x-y)-2q(x-t)-2q(y-t)\big]\\
&\quad+\big[-q(x+y+2t)-q(x+y-2t)+2q(x+y)+2q(2t)\big]\\
&\quad+\big[2q(x+t)+2q(x-t)-4q(x)-4q(t)\big]\\
&\quad+\big[2q(y+t)+2q(y-t)-4q(y)-4q(t)\big]\\
&\quad+\big[-2q(2t)-2q(0)+4q(t)+4q(t)\big],
\end{aligned}$$

we get

$$\|q(x+y)+q(x-y)-2q(x)-2q(y)-q(0)\| \le \frac{9\varepsilon}{2} \tag{8.17}$$

for all $x,y \in X$ with $\|x\|+\|y\| < r$. Consequently, the last functional inequality holds for all $x,y \in X$ in view of (8.16) and (8.17). Now letting $y := x$ in (8.17), we obtain

$$\|q(2x)-4q(x)\| \le \frac{9\varepsilon}{2}, \ \forall x \in X.$$

Now applying a standard procedure of direct method [5, 9] to the last inequality, we see that there exists a unique quadratic mapping $Q: X \to Y$ which satisfies the equation (8.2) and the inequality

$$\|q(x)-Q(x)\| \le \frac{3\varepsilon}{2}$$

for all $x \in X$. □

In the next theorem, we have a similar stability result concerning an asymptotic property of the equation (8.3).

Theorem 8.6. *Let $r > 0$ be fixed. Suppose that there exists a non-negative real number ε for which a mapping $f: X \to Y$ satisfies*

$$\|E_1 f(x_1, x_2, \cdots, x_{d+1})\| \le \varepsilon$$

for all $(d+1)$-variables $x_1, \cdots, x_{d+1} \in X$ with $\sum_{i=1}^{d+1} \|x_i\| \ge r$. Then there exists a unique quadratic mapping $Q: X \to Y$ which satisfies the equation (8.2) and the inequality

$$\left\| f(x) + \frac{d^2+3d-6}{6} f(0) - Q(x) \right\| \le \frac{3\varepsilon}{2}$$

for all $x \in X$.

We note that if we define

$$S_{d+1} = \{(x_1, \cdots, x_{d+1}) \in X^{d+1} : \|x_i\| < r, \forall i = 1, \cdots, d+1\}$$

for some fixed $r > 0$, then we have

$$\left\{(x_1, \cdots, x_{d+1}) \in X^{d+1} : \sum_{i=1}^{d+1} \|x_i\| \geq (d+1)r\right\} \subset X^{d+1} \setminus S_{d+1}.$$

Thus the following corollary is an immediate consequence of Theorem 8.5 and Theorem 8.6.

Corollary 8.7. *If a mapping $f : X \to Y$ satisfies the functional inequality (8.14) ((8.18), respectively) for all $(x_1, \cdots, x_{d+1}) \in X^{d+1} \setminus S_{d+1}$, then there exists a unique quadratic mapping $Q : X \to Y$ which satisfies the equation (8.2) ((8.3), respectively)) and the inequality (8.15) ((8.18), respectively).*

From Theorem 8.5, we have the following corollary concerning an asymptotic property of quadratic mappings.

Corollary 8.8. *A mapping $f : X \to Y$ with $f(0) = 0$ is quadratic if and only if*

$$\begin{aligned} \textit{either} \quad & \|E_1 f(x_1, \cdots, x_{d+1})\| \to 0, \\ \textit{or} \quad & \|D_1 f(x_1, \cdots, x_{d+1})\| \to 0 \end{aligned}$$

as $\sum_{i=1}^{d+1} \|x_i\| \to \infty$.

Proof. According to our asymptotic condition, there is a sequence (ε_m) decreasing to zero such that $\|D_1 f(x_1, \cdots, x_{d+1})\| \leq \varepsilon_m$ for all $(d+1)$-variables $x_1, \cdots, x_{d+1} \in X$ with $\sum_{i=1}^{d+1} \|x_i\| \geq m$. Hence, it follows from Theorem 8.5 that there exists a unique quadratic mapping $Q_m : X \to Y$ which satisfies the equation (8.2) and the inequality

$$\|f(x) - Q_m(x)\| \leq \frac{3\varepsilon_m}{2}$$

for all $x \in X$. Let m and l be positive integers with $m > l$. Then, we obtain

$$\|f(x) - Q_m(x)\| \leq \frac{3\varepsilon_m}{2} \leq \frac{3\varepsilon_l}{2}$$

for all $x \in X$. The uniqueness of Q_l implies that $Q_m = Q_l$ for all m, l, and so

$$\|f(x) - Q_l(x)\| \leq \frac{3\varepsilon_m}{2}$$

for all $x \in X$. By letting $m \to \infty$, we conclude that f is itself quadratic.

The reverse assertion is trivial. □

References

1. J. Aczél and J. Dhombres, *Functional Equations in Several Variables*, Cambridge Univ. Press, 1989.
2. Dan Amir, *Characterizations of Inner Product Spaces*, Birkhäuser-Verlag, Basel, 1986.
3. T. Aoki, *On the stability of the linear transformation in Banach spaces,* J. Math. Soc. Japan, 2(1950), 64-66.
4. D.G. Bourgin, *Classes of transformations and bordering transformations,* Bull. Amer. Math. Soc. 57(1951), 223–237.
5. C. Borelli and G.L. Forti, *On a general Hyers–Ulam stability result,* Int. J. Math. Math. Sci. 18(1995), 229–236.
6. H.-Y. Chu, D.S. Kang and Th.M. Rassias, *On the stability of a mixed n-dimensional quadratic functional equation,* Bull. Belgian Math. Soc. Simon Stevin, 15(2008), 9–24.
7. S. Czerwik, *On the stability of the quadratic mapping in normed spaces*, Abh. Math. Sem. Univ. Hamburg, 62(1992), 59–64.
8. S. Czerwik, *Functional Equations and Inequalities in Several Variables,* World Scientific Publishing Company, Singapore, 2002.
9. G.L. Forti, *Comments on the core of the direct method for proving Hyers–Ulam stability of functional equations*, *J. Math. Anal. Appl.* 295(2004), 127–133.
10. P.M. Gruber, *Stability of isometries*, Trans. Amer. Math. Soc. 245(1978), 263–277.
11. D.H. Hyers, *On the stability of the linear functional equation*, Proc. Natl. Acad. Sci. 27(1941), 222–224.
12. D.H. Hyers and Th.M. Rassias, *Approximate homomorphisms,* Aequationes Mathematicae, 44(1992), 125–153
13. D.H. Hyers, G. Isac and Th.M. Rassias, *On the asymptoticity aspect of Hyers–Ulam stability of mappings,* Proc. Amer. Math. Soc. 126(2) (1998), 425–430.
14. D.H. Hyers, G. Isac and Th. M. Rassias *Stability of Functional Equations in Several Variables*, Birkhäuser-Verlag, Boston/Basel/Berlin, 1998.
15. G. Isac and Th. M. Rassias *On the Hyers–Ulam stability of ψ-additive mappings*, J. Approx. Theory, 72(1993), 131–137.
16. G. Isac and Th. M. Rassias *Functional inequalities for approximately additive mappings,* in: Stability of Mappings of Hyers–Ulam type, Hadronic Press, Inc. Florida, (1994), pp. 117–125.
17. G. Isac and Th. M. Rassias *Stability of ψ-additive mappings: Applications to nonlinear analysis*, Int. J. Math. Math. Sci. 19(1996), 219–228.
18. S. Jung, *On the Hyers-Ulam stability of the functional equations that have the quadratic property*, J. Math. Anal. Appl. 222(1998), 126–137.
19. C.-G. Park and Th.M. Rassias, *Hyers-Ulam stability of a generalized Apollonius type quadratic mapping,* J. Math. Anal. Appl. 322(1) (2006), 371–381.
20. J.M. Rassias, *On the Ulam stability of mixed type mappings on restricted domains,* J. Math. Anal. Appl. 276 (2002), 747–762.
21. J.M. Rassias, *Asymptotic behavior of mixed type functional equations*, Austral. J. Math. Anal. Appl. 1(1)(2004), 1–21.
22. Th.M. Rassias, On the stability of the linear mapping in Banach spaces, *Proc. Amer. Math. Soc.* 72(1978), 297–300.
23. Th.M. Rassias, *Inner product spaces and applications*, Pitman Research Notes in Mathematics Series, No. 376, Addison Wesley Longman, Harlow, 1997.
24. Th.M. Rassias, *On the stability of the quadratic functional equation and its applicattions*, Studia, Univ. Babes-Bolyai, XLIII (3) (1998), 89–124.
25. Th.M. Rassias, *On the stability of functional equations in Banach spaces*, J. Math. Anal. Appl. 251(2000), 264–284.
26. Th.M. Rassias, *On the stability of functional equations and a problem of Ulam*, Acta Appl. Math. 62(2000), 23–130.

27. Th.M. Rassias (ed.), *Functional Equations and Inequalities,* Kluwer Academic Publishers, Dordrecht, 2001.
28. Th.M.Rassias (ed.), *Functional Equations, Inequalities and Applications,* Kluwer Academic Publishers, Dordrecht, 2003.
29. D.A. Senechalle, *A characterization of inner product spaces,* Proc. Amer. Math. Soc. 19(1968), 1306–1312.
30. F. Skof, *Sull' approssimazione delle applicazioni localmente δ-additive,* Atti Accad. Sci. Torino Cl Sci. Fis. Mat. Natur. 117(1983), 377–389.
31. S.M. Ulam, *A collection of the mathematical problems*, Interscience Publ., New York, 1960.
32. S.M. Ulam, *Problems in modern mathematics*, John Wiley & Sons Inc., New York, 1964.
33. Ding-Xuan Zhou, *On a conjecture of Z. Ditzian*, J. Approx. Theory 69(1992), 167–172.

Chapter 9
A Fixed Point Approach to the Stability of a Logarithmic Functional Equation

Soon-Mo Jung and Themistocles M. Rassias

Dedicated to the memory of Professor George Isac

Abstract We will apply the fixed point method for proving the Hyers–Ulam–Rassias stability of a logarithmic functional equation of the form

$$f(\sqrt{xy}) = \frac{1}{2}f(x) + \frac{1}{2}f(y),$$

where $f : (0,\infty) \to E$ is a given function and E is a real (or complex) vector space.

9.1 Introduction

In 1940, S. M. Ulam [27] gave a wide-ranging talk before the mathematics club of the University of Wisconsin in which he discussed a number of important unsolved problems. Among those was the question concerning the stability of group homomorphisms:

> *Let G_1 be a group and let G_2 be a metric group with the metric $d(\cdot,\cdot)$. Given $\varepsilon > 0$, does there exist a $\delta > 0$ such that if a function $h : G_1 \to G_2$ satisfies the inequality $d(h(xy), h(x)h(y)) < \delta$ for all $x, y \in G_1$, then there exists a homomorphism $H : G_1 \to G_2$ with $d(h(x), H(x)) < \varepsilon$ for all $x \in G_1$?*

Soon-Mo Jung
Mathematics Section, College of Science and Technology, Hongik University, 339-701 Jochiwon, Republic of Korea, e-mail: smjung@hongik.ac.kr

Themistocles M. Rassias
Department of Mathematics, National Technical University of Athens, Zografou Campus, 15780 Athens, Greece, e-mail: trassias@math.ntua.gr

P.M. Pardalos et al. (eds.), *Nonlinear Analysis and Variational Problems*,
Springer Optimization and Its Applications 35, DOI 10.1007/978-1-4419-0158-3_9,

The Ulam problem for the case of approximately additive functions was solved by D. H. Hyers [8] under the assumption that G_1 and G_2 are Banach spaces. Indeed, Hyers proved that each solution of the inequality $\|f(x+y)-f(x)-f(y)\| \le \varepsilon$, for all x and y, can be approximated by an exact solution, say an additive function. In this case, the Cauchy additive functional equation, $f(x+y)=f(x)+f(y)$, is said to satisfy the Hyers–Ulam stability.

Th. M. Rassias [23] attempted to weaken the condition for the bound of the norm of the Cauchy difference as follows

$$\|f(x+y)-f(x)-f(y)\| \le \varepsilon(\|x\|^p+\|y\|^p)$$

and derived Hyers' theorem for the stability of the additive mapping as a special case. Thus in [23], a proof of the generalized Hyers–Ulam stability for the linear mapping between Banach spaces was obtained. A particular case of Th. M. Rassias' theorem regarding the Hyers–Ulam stability of the additive mapping was proved by T. Aoki [1]. The stability concept that was introduced by Th. M. Rassias' theorem provided some influence to a number of mathematicians to develop the notion of what is known today by the term Hyers–Ulam–Rassias stability of the linear mapping. Since then, the stability of several functional equations has been extensively investigated by several mathematicians (see, for example, [1, 4, 5, 6, 7, 10, 11, 12, 13, 16, 20, 21, 24, 25, 26] and the references therein).

The terminologies Hyers–Ulam–Rassias stability and Hyers–Ulam stability can also be applied to the case of other functional equations, differential equations, and to various integral equations.

It is not difficult to prove the Hyers–Ulam stability of the logarithmic functional equation

$$f(xy)=f(x)+f(y) \tag{9.1}$$

for the class of functions $f:(0,\infty)\to E$, where E is a real (or complex) Banach space. More precisely, if a function $f:(0,\infty)\to E$ satisfies the functional inequality

$$\|f(xy)-f(x)-f(y)\| \le \varepsilon$$

for all $x,y>0$ and for some $\varepsilon>0$, then there exists a unique solution function $F:(0,\infty)\to E$ of Eq. (9.1) such that

$$\|f(x)-F(x)\| \le \varepsilon$$

for any $x,y>0$.

In 1997, S.-M. Jung investigated the stability properties for a logarithmic functional equation of the form

$$f(x^y)=yf(x)$$

(see [15]). In this connection, we will consider the following functional equation

$$f(\sqrt{xy})=\frac{1}{2}f(x)+\frac{1}{2}f(y), \tag{9.2}$$

which can be regarded as an abstract formulation of a logarithmic identity

$$\log \sqrt{xy} = \frac{1}{2}\log x + \frac{1}{2}\log y.$$

In this paper, we will adopt the idea of L. Cădariu and V. Radu [3] and prove the Hyers–Ulam–Rassias stability and the Hyers–Ulam stability of the functional equation (9.2).

It was G. Isac and Th. M. Rassias [14] who proved for the first time in nonlinear functional analysis that new fixed point theorems can be obtained as applications of the generalized Hyers–Ulam stability approach. For an extensive treatise of various nonlinear methods and in particular fixed point theory, the reader is referred to the book [9].

Throughout this paper, a function $f : (0,\infty) \to E$ will be called a logarithmic function if f satisfies the logarithmic functional equation (9.1) for all $x, y > 0$.

9.2 Preliminaries

For a nonempty set X, we introduce the definition of the generalized metric on X. A function $d : X \times X \to [0,\infty]$ is called a generalized metric on X if and only if d satisfies

(M_1) $d(x,y) = 0$ if and only if $x = y$;
(M_2) $d(x,y) = d(y,x)$ for all $x, y \in X$;
(M_3) $d(x,z) \le d(x,y) + d(y,z)$ for all $x, y, z \in X$.

We remark that the only one difference of the generalized metric from the usual metric is that the range of the former is permitted to include the infinity.

We now introduce a fundamental result of fixed point theory. For the proof, we refer to [19]. This theorem will play an essential role in proving our main theorems.

Theorem 9.1. *Let (X,d) be a generalized complete metric space. Assume that $\Lambda : X \to X$ is a strictly contractive operator with the Lipschitz constant $L < 1$. If there exists a non-negative integer k such that $d(\Lambda^{k+1}x, \Lambda^k x) < \infty$ for some $x \in X$, then the following are true:*

(a) The sequence $\{\Lambda^n x\}$ converges to a fixed point x^* of Λ;
(b) x^* is the unique fixed point of Λ in

$$X^* = \{y \in X \,|\, d(\Lambda^k x, y) < \infty\};$$

(c) If $y \in X^*$, then

$$d(y,x^*) \le \frac{1}{1-L} d(\Lambda y, y).$$

9.3 Hyers–Ulam–Rassias Stability

In this section, we will first investigate the relationship between the logarithmic functional equations (9.1) and (9.2).

Theorem 9.2. *Let E be a real (or complex) vector space. For a given function $f : (0,\infty) \to E$, let $\hat{f} : (0,\infty) \to E$ be defined by $\hat{f}(x) = f(x) - f(1)$ for all $x > 0$. Then f is a solution function of Eq. (9.2) if and only if $\hat{f}$ is a logarithmic function.*

Proof. First, assume that f is a solution function of Eq. (9.2). It then follows from (9.2) that

$$f(\sqrt{xy}) - f(1) = \frac{1}{2}\{f(x) - f(1)\} + \frac{1}{2}\{f(y) - f(1)\}$$

for all $x, y > 0$, that is,

$$\hat{f}(\sqrt{xy}) = \frac{1}{2}\hat{f}(x) + \frac{1}{2}\hat{f}(y)$$

for any $x, y > 0$. If we set $u = \sqrt{x}$ and $v = \sqrt{y}$ in the above equation, then $\hat{f}$ satisfies

$$\hat{f}(uv) = \frac{1}{2}\hat{f}(u^2) + \frac{1}{2}\hat{f}(v^2)$$

for every $u, v > 0$. By putting $v = 1$, the last equation yields

$$\hat{f}(u^2) = 2\hat{f}(u).$$

Thus, we have

$$\hat{f}(uv) = \hat{f}(u) + \hat{f}(v)$$

for all $u, v > 0$.

Now, assume that $\hat{f}$ satisfies

$$\hat{f}(xy) = \hat{f}(x) + \hat{f}(y)$$

for all $x, y > 0$. Then, putting $y = x$, the last equation yields

$$\hat{f}(x^2) = 2\hat{f}(x).$$

Hence, we get

$$\hat{f}(x) + \hat{f}(y) = \hat{f}(xy) = \hat{f}(\sqrt{xy}^2) = 2\hat{f}(\sqrt{xy})$$

and equivalently we obtain

$$f(\sqrt{xy}) = \frac{1}{2}f(x) + \frac{1}{2}f(y)$$

for all $x, y > 0$. □

Recently, L. Cădariu and V. Radu [3] applied the fixed point method for the investigation of the Cauchy additive functional equation. Using such a clever idea, they presented a proof of the Hyers–Ulam stability of that equation ([2, 17, 22]).

By applying the idea of Cădariu and Radu, we will now prove the Hyers–Ulam–Rassias stability of the logarithmic functional equation (9.2).

Theorem 9.3. *Let* $(E, \|\cdot\|)$ *be a real (or complex) Banach space and assume that a symmetric function* $\varphi : (0,\infty)^2 \to (0,\infty)$ *is given such that there exist constants* K *and* L *with the properties:*

$$0 < K < 2, \quad L \geq 1, \tag{9.3}$$

$$\varphi(x^2, y^2) \leq K\varphi(x,y) \quad (x,y > 0), \tag{9.4}$$

and

$$\varphi(x,1) \leq L \inf_{y>0} \varphi(x,y) \quad (x > 0). \tag{9.5}$$

If a function $f : (0,\infty) \to E$ *satisfies the functional inequality*

$$\left\| f(\sqrt{xy}) - \frac{1}{2}f(x) - \frac{1}{2}f(y) \right\| \leq \varphi(x,y) \tag{9.6}$$

for any $x, y > 0$*, then there exists a unique logarithmic function* $F : (0,\infty) \to E$ *such that*

$$\|f(x) - F(x) - f(1)\| \leq \frac{2KL}{2-K}\varphi(x,x) \tag{9.7}$$

for each $x > 0$.

Proof. First, we define a set $\mathscr{X}$ by

$$\mathscr{X} = \{h : (0,\infty) \to E \mid h \text{ is a function from } (0,\infty) \text{ to } E \text{ satisfying } h(1) = 0\}$$

and introduce a generalized metric d on $\mathscr{X}$ as follows:

$$d(g,h) = \inf\{C \in [0,\infty] \mid \|g(x) - h(x)\| \leq C\varphi(x,x) \text{ for all } x > 0\}. \tag{9.8}$$

(We will here give a proof for the triangle inequality only. Assume that $d(g,h) > d(g,k) + d(k,h)$ would hold for some $g,h,k \in \mathscr{X}$. Then, due to (9.8), there should exist an $x_0 > 0$ such that

$$\begin{aligned}
\|g(x_0) - h(x_0)\| &> \{d(g,k) + d(k,h)\}\varphi(x_0,x_0) \\
&= d(g,k)\varphi(x_0,x_0) + d(k,h)\varphi(x_0,x_0) \\
&\geq \|g(x_0) - k(x_0)\| + \|k(x_0) - h(x_0)\|,
\end{aligned}$$

a contradiction.)

For the proof of the completeness of $(\mathscr{X}, d)$, we may refer the reader to [18]. Nevertheless, we will here give the proof. Let $\{h_n\}$ be a Cauchy sequence in $(\mathscr{X}, d)$.

Then, for any $\varepsilon > 0$ there exists an integer $N_\varepsilon > 0$ such that $d(h_m, h_n) \leq \varepsilon$ for all $m, n \geq N_\varepsilon$. Furthermore, it follows from (9.8) that

$$\forall\, \varepsilon > 0 \;\exists\, N_\varepsilon \in \mathbb{N} \;\forall\, m,n \geq N_\varepsilon \;\forall\, x > 0 : \; \|h_m(x) - h_n(x)\| \leq \varepsilon\varphi(x,x). \tag{9.9}$$

If x is fixed, (9.9) implies that $\{h_n(x)\}$ is a Cauchy sequence in $(E, \|\cdot\|)$. Since $(E, \|\cdot\|)$ is a complete vector space, $\{h_n(x)\}$ converges for each $x > 0$. Thus, we can define a function $h : (0, \infty) \to E$ by

$$h(x) = \lim_{n \to \infty} h_n(x),$$

and we see that $h(1) = 0$. Hence, h belongs to $\mathscr{X}$. (It has not been proved yet that $\{h_n\}$ converges to h in $(\mathscr{X}, d)$.)

If we let m increase to infinity, it then follows from (9.9) that

$$\forall\, \varepsilon > 0 \;\exists\, N_\varepsilon \in \mathbb{N} \;\forall\, n \geq N_\varepsilon \;\forall\, x > 0 : \; \|h(x) - h_n(x)\| \leq \varepsilon\varphi(x,x).$$

Moreover if we consider (9.8), then we conclude that

$$\forall\, \varepsilon > 0 \;\exists\, N_\varepsilon \in \mathbb{N} \;\forall\, n \geq N_\varepsilon : \; d(h, h_n) \leq \varepsilon,$$

that is, the Cauchy sequence $\{h_n\}$ converges to h in $(\mathscr{X}, d)$. Hence, $(\mathscr{X}, d)$ is a complete vector space.

Now, let us define an operator $\Lambda : \mathscr{X} \to \mathscr{X}$ by

$$(\Lambda h)(x) = \frac{1}{2} h\left(x^2\right) \quad (x > 0) \tag{9.10}$$

for all $h \in \mathscr{X}$. (It is obvious that $\Lambda h \in \mathscr{X}$.)

We assert that Λ is strictly contractive on $\mathscr{X}$. For any $g, h \in \mathscr{X}$, let us choose a $C_{gh} \in [0, \infty]$ satisfying $d(g,h) \leq C_{gh}$. Then, using (9.8), we have

$$\|g(x) - h(x)\| \leq C_{gh}\varphi(x,x) \quad (x > 0). \tag{9.11}$$

From (9.4), (9.10), and (9.11), we obtain

$$\begin{aligned}
\|(\Lambda g)(x) - (\Lambda h)(x)\| &= \left\| \frac{1}{2} g\left(x^2\right) - \frac{1}{2} h\left(x^2\right) \right\| \\
&\leq \frac{1}{2} C_{gh} \varphi\left(x^2, x^2\right) \\
&\leq \frac{1}{2} K C_{gh} \varphi(x,x)
\end{aligned}$$

for all $x > 0$, that is, $d(\Lambda g, \Lambda h) \leq \frac{1}{2} K C_{gh}$. Hence, we may conclude that

$$d(\Lambda g, \Lambda h) \leq \frac{1}{2} K d(g,h), \tag{9.12}$$

where by (9.3), we know that $0 < \frac{1}{2}K < 1$.

If we define $\hat{f}(x) = f(x) - f(1)$ for all $x > 0$, then $\hat{f}(1) = 0$ ($\hat{f} \in \mathscr{X}$) and from (9.6) it follows that

$$\left\| \hat{f}(\sqrt{xy}) - \frac{1}{2}\hat{f}(x) - \frac{1}{2}\hat{f}(y) \right\| \leq \varphi(x,y) \tag{9.13}$$

for any $x, y > 0$.

Moreover, it follows from (9.4), (9.5), (9.10), and (9.13) that

$$\begin{aligned} \|(\Lambda\hat{f})(x) - \hat{f}(x)\| &= \left\| \frac{1}{2}\hat{f}(x^2) - \hat{f}(x) \right\| \\ &= \left\| \frac{1}{2}\hat{f}(x^2) + \frac{1}{2}\hat{f}(1) - \hat{f}(x) \right\| \\ &\leq \varphi(x^2, 1) \\ &\leq L\varphi(x^2, x^2) \\ &\leq KL\varphi(x, x) \end{aligned}$$

for every $x > 0$. Thus, (9.8) implies that

$$d(\Lambda\hat{f}, \hat{f}) \leq KL. \tag{9.14}$$

Therefore, it follows from Theorem 9.1 (a) that there exists a function $F : (0,\infty) \to E$ such that $\Lambda^n \hat{f} \to F$ in $(\mathscr{X}, d)$ and $\Lambda F = F$. Indeed, for each $x > 0$, $(\Lambda^n \hat{f})(x) = \frac{1}{2^n}\hat{f}(x^{2^n})$ converges to $F(x)$ pointwise, that is,

$$F(x) = \lim_{n\to\infty} \frac{1}{2^n}\hat{f}\left(x^{2^n}\right)$$

for all $x > 0$. So, it follows from (9.3), (9.4), and (9.13) that

$$\begin{aligned} &\left\| F(\sqrt{xy}) - \frac{1}{2}F(x) - \frac{1}{2}F(y) \right\| \\ &= \lim_{n\to\infty} \frac{1}{2^n} \left\| \hat{f}\left(x^{2^{n-1}} y^{2^{n-1}}\right) - \frac{1}{2}\hat{f}\left(x^{2^n}\right) - \frac{1}{2}\hat{f}\left(y^{2^n}\right) \right\| \\ &\leq \lim_{n\to\infty} \frac{1}{2^n}\varphi\left(x^{2^n}, y^{2^n}\right) \\ &\leq \lim_{n\to\infty} \frac{K^n}{2^n}\varphi(x,y) \\ &= 0 \end{aligned}$$

for any $x, y > 0$. That is, F is a solution of Eq. (9.2). Hence, due to Theorem 9.2, F is a logarithmic function. Moreover, Theorem 9.1 (c), together with (9.12) and (9.14), implies that

$$d(\hat{f},F) \leq \frac{2KL}{2-K},$$

that is, in view of (9.8), the inequality (9.7) is true for each $x > 0$.

Finally, let $G : (0,\infty) \to E$ be another logarithmic function satisfying

$$\|f(x) - G(x) - f(1)\| \leq M\varphi(x,x)$$

for each $x > 0$ and for some constant $0 < M < \infty$. Then, as G is a logarithmic function, G is also a fixed point of Λ and we have $d(\hat{f},G) < \infty$. Hence

$$G \in \mathscr{X}^* = \{g \in \mathscr{X} \mid d(\hat{f},g) < \infty\}.$$

Thus, Theorem 9.1 (b) implies that $F = G$, that is, F is unique. □

9.4 Applications

By setting $\varphi(x,y) = \varepsilon$ in Theorem 9.3, we can easily deduce the Hyers–Ulam stability of Eq. (9.2).

Corollary 9.4. *Let $(E,\|\cdot\|)$ be a real (or complex) Banach space. If a function $f : (0,\infty) \to E$ satisfies the inequality*

$$\left\| f(\sqrt{xy}) - \frac{1}{2}f(x) - \frac{1}{2}f(y) \right\| \leq \varepsilon$$

for all $x,y > 0$ and for some $\varepsilon > 0$, then there exists a unique logarithmic function $F : (0,\infty) \to E$ such that

$$\|f(x) - F(x) - f(1)\| \leq 2\varepsilon$$

for any $x > 0$.

In what follows, let us define a function $\psi : (0,\infty) \to (0,\infty)$ by

$$\psi(x) = \begin{cases} \sqrt{\dfrac{\varepsilon}{2}}\,(1+x) & (0 < x < 1) \\ \sqrt{\dfrac{\varepsilon}{2}}\left(1+\dfrac{1}{x}\right) & (x \geq 1), \end{cases}$$

where ε is a positive constant. Further, we define a function $\varphi : (0,\infty)^2 \to (0,\infty)$ by $\varphi(x,y) = \psi(x)\psi(y)$ for all $x,y > 0$. Then, we have

$$\varphi(x,y) = \psi(x)\psi(y) = \begin{cases} \frac{\varepsilon}{2}(1+x)(1+y) & (0<x<1;\ 0<y<1) \\ \frac{\varepsilon}{2}(1+x)\left(1+\frac{1}{y}\right) & (0<x<1;\ y\ge 1) \\ \frac{\varepsilon}{2}\left(1+\frac{1}{x}\right)(1+y) & (x\ge 1;\ 0<y<1) \\ \frac{\varepsilon}{2}\left(1+\frac{1}{x}\right)\left(1+\frac{1}{y}\right) & (x\ge 1;\ y\ge 1) \end{cases}$$

and

$$\varphi(x^2,y^2) = \psi(x^2)\psi(y^2) = \begin{cases} \frac{\varepsilon}{2}(1+x^2)(1+y^2) & (0<x<1;\ 0<y<1) \\ \frac{\varepsilon}{2}(1+x^2)\left(1+\frac{1}{y^2}\right) & (0<x<1;\ y\ge 1) \\ \frac{\varepsilon}{2}\left(1+\frac{1}{x^2}\right)(1+y^2) & (x\ge 1;\ 0<y<1) \\ \frac{\varepsilon}{2}\left(1+\frac{1}{x^2}\right)\left(1+\frac{1}{y^2}\right) & (x\ge 1;\ y\ge 1). \end{cases}$$

Hence, we see that

$$\varphi(x^2,y^2) \le \varphi(x,y)$$

for any $x,y>0$, that is, the condition (9.4) is satisfied with $K=1$.

Moreover, we obtain

$$\varphi(x,1) = \psi(x)\psi(1) = \sqrt{2\varepsilon}\,\psi(x)$$

and

$$\begin{aligned} \inf_{y>0}\varphi(x,y) &= \inf\left\{\inf_{0<y<1}\varphi(x,y),\ \inf_{y\ge 1}\varphi(x,y)\right\} \\ &= \inf\left\{\sqrt{\frac{\varepsilon}{2}}\,\psi(x),\ \sqrt{\frac{\varepsilon}{2}}\,\psi(x)\right\} \\ &= \sqrt{\frac{\varepsilon}{2}}\,\psi(x) \end{aligned}$$

for all $x>0$. Hence, we get

$$\varphi(x,1) \le 2\inf_{y>0}\varphi(x,y)$$

for $x>0$, that is, the condition (9.5) is satisfied with $L=2$.

According to Theorem 9.3, the following corollary is true:

Corollary 9.5. *Let* $(E, \|\cdot\|)$ *be a real (or complex) Banach space. If a function* $f : (0,\infty) \to E$ *satisfies the inequality*

$$\left\| f(\sqrt{xy}) - \frac{1}{2}f(x) - \frac{1}{2}f(y) \right\| \leq \varphi(x,y)$$

for all $x, y > 0$*, then there exists a unique logarithmic function* $F : (0,\infty) \to E$ *such that*

$$\|f(x) - F(x) - f(1)\| \leq 4\varphi(x,x)$$

for any $x > 0$*.*

References

1. Aoki, T.: On the stability of the linear transformation in Banach spaces, J. Math. Soc. Japan **2**, 64–66 (1950)
2. Cădariu, L., Radu, V.: Fixed points and the stability of Jensen's functional equation, J. Inequal. Pure and Appl. Math. **4**, no. 1, Art. 4 (2003) (http://jipam.vu.edu.au)
3. Cădariu, L., Radu, V.: On the stability of the Cauchy functional equation: a fixed point approach, Grazer Math. Ber. **346**, 43–52 (2004)
4. Czerwik, S.: Functional Equations and Inequalities in Several Variables, World Scientific Publishing Co., Singapore (2002)
5. Faizev, V.A., Rassias, Th.M., Sahoo, P.K.: The space of (ϕ, α)-additive mappings on semigroups, Trans. Amer. Math. Soc. **354**, no. 11, 4455–4472 (2002)
6. Forti, G.L.: Hyers-Ulam stability of functional equations in several variables, Aequationes Math. **50**, 143–190 (1995)
7. Găvruţă, P.: A generalization of the Hyers-Ulam-Rassias stability of approximately additive mappings, J. Math. Anal. Appl. **184**, 431–436 (1994)
8. Hyers, D.H.: On the stability of the linear functional equation, Proc. Nat. Acad. Sci. USA **27**, 222–224 (1941)
9. Hyers, D.H., Isac, G., Rassias, Th.M.: Topics in Nonlinear Analysis and Applications, World Scientific Publishing Co., Singapore (1997)
10. Hyers, D.H., Isac, G., Rassias, Th.M.: Stability of Functional Equations of Several Variables, Birkhäuser, Basel, Boston (1998).
11. Hyers, D.H., Isac, G., Rassias, Th.M.: On the asymptoticity aspect of Hyers-Ulam stability of mappings, Proc. Amer. Math. Soc. **126**, no. 2, 425–430 (1998)
12. Hyers, D.H., Rassias, Th.M.: Approximate homomorphisms, Aequationes Math. **44**, 125–153 (1992)
13. Isac, G., Rassias, Th.M.: On the Hyers-Ulam stability of ϕ-additive mappings, J. Approx. Theory **72**, 131–137 (1993)
14. Isac, G., Rassias, Th.M.: Stability of additive mappings: Applications to nonlinear analysis, Internat. J. Math. Math. Sci. **19**, no. 2, 219–228 (1996)
15. Jung, S.-M.: On the superstability of the functional equation $f(x^y) = yf(x)$, Abh. Math. Sem. Univ. Hamburg **67**, 315–322 (1997)
16. Jung, S.-M.: Hyers-Ulam-Rassias Stability of Functional Equations in Mathematical Analysis, Hadronic Press, Palm Harbor, FL (2001)
17. Jung, S.-M.: A fixed point approach to the stability of isometries, J. Math. Anal. Appl. **329**, 879–890 (2007)
18. Jung, S.-M.: A fixed point approach to the stability of a Volterra integral equation, Fixed Point Theory and Applications **2007**, Article ID 57064, 9 pages, doi: 10.1155/2007/57064 (2007)

19. Margolis, B., Diaz, J.: A fixed point theorem of the alternative for contractions on a generalized complete metric space, Bull. Amer. Math. Soc. **74**, 305–309 (1968)
20. Moslehian, M.S., Rassias, Th.M.: Stability of functional equations in non-Archimedian spaces, Appl. Anal. Disc. Math. **1**, 325–334 (2007)
21. Park, C., Rassias, Th.M.: d-isometric linear mappings in linear d-normed Banach modules, J. Korean Math. Soc. **45**, no. 1, 249–271 (2008)
22. Radu, V.: The fixed point alternative and the stability of functional equations, Fixed Point Theory – An International Journal on Fixed Point Theory Computation and Applications **4**, 91–96 (2003)
23. Rassias, Th.M.: On the stability of the linear mapping in Banach spaces, Proc. Amer. Math. Soc. **72**, 297–300 (1978)
24. Rassias, Th.M.: On a modified Hyers-Ulam sequence, J. Math. Anal. Appl. **158**, 106–113 (1991)
25. Rassias, Th.M.: On the stability of functional equations and a problem of Ulam, Acta Appl. Math. **62**, 23–130 (2000)
26. Redouani, A., Elqorachi, E., Rassias, Th.M.: The superstability of d'Alembert's functional equation on step 2 nilpotent groups, Aequationes Math. **74**, 226–241 (2007)
27. Ulam, S.M.: A Collection of Mathematical Problems, Interscience Publ., New York (1960)

Chapter 10
Fixed Points and Stability of the Cauchy Functional Equation in Lie C^*-Algebras

Choonkil Park and Jianlian Cui

Dedicated to the memory of Professor George Isac

Abstract Using the fixed point method, we prove the generalized Hyers–Ulam stability of homomorphisms in C^*-algebras and Lie C^*-algebras and of derivations on C^*-algebras and Lie C^*-algebras for the 3-variable Cauchy functional equation.

10.1 Introduction and Preliminaries

The stability problem of functional equations originated from a question of Ulam [38] concerning the stability of group homomorphisms. Hyers [10] gave a first affirmative partial answer to the question of Ulam for Banach spaces. Hyers' theorem was generalized by Aoki [1] for additive mappings and by Th.M. Rassias [28] for linear mappings by considering an unbounded Cauchy difference. The paper of Th.M. Rassias [28] has provided a lot of influence in the development of what we call *generalized Hyers–Ulam stability* or as *Hyers–Ulam–Rassias stability* of functional equations. A generalization of the Th.M. Rassias theorem was obtained by Găvruta [9] by replacing the unbounded Cauchy difference by a general control function in the spirit of Th.M. Rassias' approach.

Choonkil Park
Department of Mathematics, Hanyang University, Seoul 133-791, South Korea, e-mail: baak@hanyang.ac.kr

Jianlian Cui
Department of Mathematical Sciences, Tsinghua University, Beijing 100084, P.R. China, e-mail: jcui@math.tsinghua.edu.cn

P.M. Pardalos et al. (eds.), *Nonlinear Analysis and Variational Problems*,
Springer Optimization and Its Applications 35, DOI 10.1007/978-1-4419-0158-3_10,

The functional equation

$$f(x+y)+f(x-y)=2f(x)+2f(y)$$

is called a *quadratic functional equation*. In particular, every solution of the quadratic functional equation is said to be a *quadratic function*. A generalized Hyers–Ulam stability problem for the quadratic functional equation was proved by Skof [37] for mappings $f:X\to Y$, where X is a normed space and Y is a Banach space. Cholewa [5] noticed that the theorem of Skof is still true if the relevant domain X is replaced by an Abelian group. Czerwik [6] proved the generalized Hyers–Ulam stability of the quadratic functional equation. The stability problems of several functional equations have been extensively investigated by a number of authors and there are many interesting results concerning this problem (see [7], [11]–[15], [18], [21], [24]–[26], [30]–[36]).

We recall a fundamental result in fixed point theory.

Let X be a set. A function $d:X\times X\to[0,\infty]$ is called a *generalized metric* on X if d satisfies

(1) $d(x,y)=0$ if and only if $x=y$;
(2) $d(x,y)=d(y,x)$ for all $x,y\in X$;
(3) $d(x,z)\le d(x,y)+d(y,z)$ for all $x,y,z\in X$.

Theorem 10.1. *[2, 8] Let (X,d) be a complete generalized metric space and let $J:X\to X$ be a strictly contractive mapping with Lipschitz constant $L<1$. Then for each given element $x\in X$, either*

$$d(J^nx,J^{n+1}x)=\infty$$

for all non-negative integers n or there exists a positive integer n_0 such that

(1) $d(J^nx,J^{n+1}x)<\infty, \qquad \forall n\ge n_0$;
(2) the sequence $\{J^nx\}$ converges to a fixed point y^* of J;
(3) y^* is the unique fixed point of J in the set $Y=\{y\in X\mid d(J^{n_0}x,y)<\infty\}$;
(4) $d(y,y^*)\le\frac{1}{1-L}d(y,Jy)$ for all $y\in Y$.

This paper is organized as follows: In Sections 10.2 and 10.3, using the fixed point method, we prove the generalized Hyers–Ulam stability of homomorphisms in C^*-algebras and of derivations on C^*-algebras for the 3-variable Cauchy functional equation.

In Sections 10.4 and 10.5, using the fixed point method, we prove the generalized Hyers–Ulam stability of homomorphisms in Lie C^*-algebras and of derivations on Lie C^*-algebras for the 3-variable Cauchy functional equation.

In 1996, G. Isac and Th.M. Rassias [14] were the first to provide applications of stability theory of functional equations for the proof of new fixed point theorems with applications. By using fixed point methods, the stability problems of several functional equations have been extensively investigated by a number of authors (see [4], [16], [17], [22], [23], [27]).

10.2 Stability of Homomorphisms in C^*-Algebras

Throughout this section, assume that A is a C^*-algebra with norm $\|\cdot\|_A$ and that B is a C^*-algebra with norm $\|\cdot\|_B$.

For a given mapping $f : A \to B$, we define

$$D_\mu f(x,y,z) := \mu f(x+y+z) - f(\mu x) - f(\mu y) - f(\mu z)$$

for all $\mu \in \mathbf{T}^1 := \{\nu \in \mathbf{C} : |\nu| = 1\}$ and all $x,y,z \in A$.

Note that a $\mathbf{C}$-linear mapping $H : A \to B$ is called a *homomorphism* in C^*-algebras if H satisfies $H(xy) = H(x)H(y)$ and $H(x^*) = H(x)^*$ for all $x,y \in A$.

We prove the generalized Hyers–Ulam stability of homomorphisms in C^*-algebras for the functional equation $D_\mu f(x,y,z) = 0$.

Theorem 10.2. *Let $f : A \to B$ be a mapping for which there exists a function $\varphi : A^3 \to [0,\infty)$ such that*

$$\|D_\mu f(x,y,z)\|_B \le \varphi(x,y,z), \tag{10.1}$$
$$\|f(xy) - f(x)f(y)\|_B \le \varphi(x,y,0), \tag{10.2}$$
$$\|f(x^*) - f(x)^*\|_B \le \varphi(x,x,x) \tag{10.3}$$

for all $\mu \in \mathbf{T}^1$ and all $x,y,z \in A$. If there exists an $L < 1$ such that $\varphi(x,y,z) \le 3L\varphi(\frac{x}{3}, \frac{y}{3}, \frac{z}{3})$ for all $x,y,z \in A$, then there exists a unique C^*-algebra homomorphism $H : A \to B$ such that

$$\|f(x) - H(x)\|_B \le \frac{1}{3-3L}\varphi(x,x,x) \tag{10.4}$$

for all $x \in A$.

Proof. Consider the set

$$X := \{g : A \to B\}$$

and introduce the *generalized metric* on X:

$$d(g,h) = \inf\{C \in \mathbf{R}_+ : \|g(x) - h(x)\|_B \le C\varphi(x,x,x), \quad \forall x \in A\}.$$

It is easy to show that (X,d) is complete.

Now we consider the linear mapping $J : X \to X$ such that

$$Jg(x) := \frac{1}{3}g(3x)$$

for all $x \in A$.

By Theorem 3.1 of [2],

$$d(Jg,Jh) \le Ld(g,h)$$

for all $g,h \in X$.

Letting $\mu = 1$ and $y = z = x$ in (10.1), we get

$$\|f(3x)-3f(x)\|_B \leq \varphi(x,x,x) \tag{10.5}$$

for all $x \in A$. So

$$\|f(x)-\frac{1}{3}f(3x)\|_B \leq \frac{1}{3}\varphi(x,x,x)$$

for all $x \in A$. Hence $d(f,Jf) \leq \frac{1}{3}$.

By Theorem 10.1, there exists a mapping $H : A \to B$ such that

(1) H is a fixed point of J, i.e.,

$$H(3x) = 3H(x) \tag{10.6}$$

for all $x \in A$. The mapping H is a unique fixed point of J in the set

$$Y = \{g \in X : d(f,g) < \infty\}.$$

This implies that H is a unique mapping satisfying (10.6) such that there exists $C \in (0,\infty)$ satisfying

$$\|H(x)-f(x)\|_B \leq C\varphi(x,x,x)$$

for all $x \in A$.

(2) $d(J^n f, H) \to 0$ as $n \to \infty$. This implies the equality

$$\lim_{n\to\infty} \frac{f(3^n x)}{3^n} = H(x) \tag{10.7}$$

for all $x \in A$.

(3) $d(f,H) \leq \frac{1}{1-L}d(f,Jf)$, which implies the inequality

$$d(f,H) \leq \frac{1}{3-3L}.$$

This implies that the inequality (10.4) holds.

It follows from (10.1) and (10.7) that

$$\begin{aligned}
&\|H(x+y+z)-H(x)-H(y)-H(z)\|_B \\
&= \lim_{n\to\infty} \frac{1}{3^n}\|f(3^n(x+y+z))-f(3^n x)-f(3^n y)-f(3^n z)\|_B \\
&\leq \lim_{n\to\infty} \frac{1}{3^n}\varphi(3^n x, 3^n y, 3^n z) \leq \lim_{n\to\infty} L^n \varphi(x,y,z) = 0
\end{aligned}$$

for all $x,y,z \in A$. So

$$H(x+y+z) = H(x)+H(y)+H(z) \tag{10.8}$$

for all $x,y,z \in A$. Letting $z = 0$ in (10.8), we get

$$H(x+y) = H(x)+H(y)+H(0) = H(x)+H(y)$$

for all $x, y \in A$.

Letting $y = z = x$ in (10.1), we get

$$\mu f(3x) = f(3\mu x)$$

for all $\mu \in \mathbf{T}^1$ and all $x \in A$. By a similar method to that above, we get

$$\mu H(3x) = H(3\mu x)$$

for all $\mu \in \mathbf{T}^1$ and all $x \in A$. Thus one can show that the mapping $H : A \to B$ is $\mathbf{C}$-linear.

It follows from (10.2) that

$$\begin{aligned}\|H(xy) - H(x)H(y)\|_B &= \lim_{n\to\infty} \frac{1}{9^n}\|f(9^n xy) - f(3^n x)f(3^n y)\|_B \\ &\le \lim_{n\to\infty} \frac{1}{9^n}\varphi(3^n x, 3^n y, 0) \le \lim_{n\to\infty} \frac{1}{3^n}\varphi(3^n x, 3^n y, 0) \\ &\le \lim_{n\to\infty} L^n \varphi(x, y, 0) = 0\end{aligned}$$

for all $x, y \in A$. So

$$H(xy) = H(x)H(y)$$

for all $x, y \in A$.

It follows from (10.3) that

$$\begin{aligned}\|H(x^*) - H(x)^*\|_B &= \lim_{n\to\infty} \frac{1}{3^n}\|f(3^n x^*) - f(3^n x)^*\|_B \\ &\le \lim_{n\to\infty} \frac{1}{3^n}\varphi(3^n x, 3^n x, 3^n x) \le \lim_{n\to\infty} L^n \varphi(x, x, x) = 0\end{aligned}$$

for all $x \in A$. So

$$H(x^*) = H(x)^*$$

for all $x \in A$.

Thus $H : A \to B$ is a C^*-algebra homomorphism satisfying (10.4), as desired. □

Corollary 10.3. *Let $r < 1$ and θ be non-negative real numbers, and let $f : A \to B$ be a mapping such that*

$$\|D_\mu f(x, y, z)\|_B \le \theta(\|x\|_A^r + \|y\|_A^r + \|z\|_A^r), \tag{10.9}$$

$$\|f(xy) - f(x)f(y)\|_B \le \theta(\|x\|_A^r + \|y\|_A^r), \tag{10.10}$$

$$\|f(x^*) - f(x)^*\|_B \le 3\theta\|x\|_A^r \tag{10.11}$$

for all $\mu \in \mathbf{T}^1$ and all $x, y, z \in A$. Then there exists a unique C^*-algebra homomorphism $H : A \to B$ such that

$$\|f(x) - H(x)\|_B \le \frac{3\theta}{3 - 3^r}\|x\|_A^r \tag{10.12}$$

for all $x \in A$.

Proof. The proof follows from Theorem 10.2 by taking

$$\varphi(x,y,z) := \theta(\|x\|_A^r + \|y\|_A^r + \|z\|_A^r)$$

for all $x,y,z \in A$. Then $L = 3^{r-1}$ and we get the desired result. □

Theorem 10.4. *Let $f : A \to B$ be a mapping for which there exists a function $\varphi : A^3 \to [0,\infty)$ satisfying (10.1), (10.2) and (10.3). If there exists an $L < 1$ such that $\varphi(x,y,z) \le \frac{1}{3}L\varphi(3x,3y,3z)$ for all $x,y,z \in A$, then there exists a unique C^*-algebra homomorphism $H : A \to B$ such that*

$$\|f(x) - H(x)\|_B \le \frac{L}{3-3L}\varphi(x,x,x) \tag{10.13}$$

for all $x \in A$.

Proof. We consider the linear mapping $J : X \to X$ such that

$$Jg(x) := 3g(\frac{x}{3})$$

for all $x \in A$.

It follows from (10.5) that

$$\|f(x) - 3f(\frac{x}{3})\|_B \le \varphi(\frac{x}{3},\frac{x}{3},\frac{x}{3}) \le \frac{L}{3}\varphi(x,x,x)$$

for all $x \in A$. Hence $d(f,Jf) \le \frac{L}{3}$.

By Theorem 10.1, there exists a mapping $H : A \to B$ such that

(1) H is a fixed point of J, i.e.,

$$H(3x) = 3H(x) \tag{10.14}$$

for all $x \in A$. The mapping H is a unique fixed point of J in the set

$$Y = \{g \in X : d(f,g) < \infty\}.$$

This implies that H is a unique mapping satisfying (10.14) such that there exists $C \in (0,\infty)$ satisfying

$$\|H(x) - f(x)\|_B \le C\varphi(x,x,x)$$

for all $x \in A$.

(2) $d(J^n f, H) \to 0$ as $n \to \infty$. This implies the equality

$$\lim_{n\to\infty} 3^n f(\frac{x}{3^n}) = H(x)$$

for all $x \in A$.

(3) $d(f,H) \le \frac{1}{1-L}d(f,Jf)$, which implies the inequality

$$d(f,H) \le \frac{L}{3-3L},$$

which implies that the inequality (10.13) holds.

The rest of the proof is similar to the proof of Theorem 10.2. □

Corollary 10.5. *Let $r > 2$ and θ be non-negative real numbers, and let $f : A \to B$ be a mapping satisfying (10.9), (10.10) and (10.11). Then there exists a unique C^*-algebra homomorphism $H : A \to B$ such that*

$$\|f(x) - H(x)\|_B \le \frac{3\theta}{3^r - 3}\|x\|_A^r$$

for all $x \in A$.

Proof. The proof follows from Theorem 10.4 by taking

$$\varphi(x,y,z) := \theta(\|x\|_A^r + \|y\|_A^r + \|z\|_A^r)$$

for all $x,y,z \in A$. Then $L = 3^{1-r}$ and we get the desired result. □

10.3 Stability of Derivations on C^*-Algebras

Throughout this section, assume that A is a C^*-algebra with norm $\|\cdot\|_A$.

Note that a $\mathbf{C}$-linear mapping $\delta : A \to A$ is called a *derivation* on A if δ satisfies $\delta(xy) = \delta(x)y + x\delta(y)$ for all $x,y \in A$.

We prove the generalized Hyers–Ulam stability of derivations on C^*-algebras for the functional equation $D_\mu f(x,y,z) = 0$.

Theorem 10.6. *Let $f : A \to A$ be a mapping for which there exists a function $\varphi : A^3 \to [0,\infty)$ such that*

$$\|D_\mu f(x,y,z)\|_A \le \varphi(x,y,z), \tag{10.15}$$

$$\|f(xy) - f(x)y - xf(y)\|_A \le \varphi(x,y,0) \tag{10.16}$$

for all $\mu \in \mathbf{T}^1$ and all $x,y,z \in A$. If there exists an $L < 1$ such that $\varphi(x,y,z) \le 3L\varphi(\frac{x}{3},\frac{y}{3},\frac{z}{3})$ for all $x,y,z \in A$. Then there exists a unique derivation $\delta : A \to A$ such that

$$\|f(x) - \delta(x)\|_A \le \frac{1}{3-3L}\varphi(x,x,x) \tag{10.17}$$

for all $x \in A$.

Proof. By the same reasoning as the proof of Theorem 10.2, there exists a unique $\mathbf{C}$-linear mapping $\delta : A \to A$ satisfying (10.17). The mapping $\delta : A \to A$ is given by

$$\delta(x) = \lim_{n\to\infty} \frac{f(3^n x)}{3^n}$$

for all $x \in A$.

It follows from (10.16) that

$$\begin{aligned}
&\|\delta(xy) - \delta(x)y - x\delta(y)\|_A \\
&= \lim_{n\to\infty} \frac{1}{9^n} \|f(9^n xy) - f(3^n x)\cdot 3^n y - 3^n x f(3^n y)\|_A \\
&\le \lim_{n\to\infty} \frac{1}{9^n} \varphi(3^n x, 3^n y, 0) \le \lim_{n\to\infty} \frac{1}{3^n} \varphi(3^n x, 3^n y, 0) \\
&\le \lim_{n\to\infty} L^n \varphi(x,y,0) = 0
\end{aligned}$$

for all $x, y \in A$. So

$$\delta(xy) = \delta(x)y + x\delta(y)$$

for all $x, y \in A$. Thus $\delta : A \to A$ is a derivation satisfying (10.17). □

Corollary 10.7. *Let $r < 1$ and θ be non-negative real numbers, and let $f : A \to A$ be a mapping such that*

$$\|D_\mu f(x,y,z)\|_A \le \theta(\|x\|_A^r + \|y\|_A^r + \|z\|_A^r), \tag{10.18}$$

$$\|f(xy) - f(x)y - xf(y)\|_A \le \theta(\|x\|_A^r + \|y\|_A^r) \tag{10.19}$$

for all $\mu \in \mathbf{T}^1$ and all $x, y, z \in A$. Then there exists a unique derivation $\delta : A \to A$ such that

$$\|f(x) - \delta(x)\|_A \le \frac{3\theta}{3 - 3^r} \|x\|_A^r \tag{10.20}$$

for all $x \in A$.

Proof. The proof follows from Theorem 10.6 by taking

$$\varphi(x,y,z) := \theta(\|x\|_A^r + \|y\|_A^r + \|z\|_A^r)$$

for all $x, y, z \in A$. Then $L = 3^{r-1}$ and we get the desired result. □

Theorem 10.8. *Let $f : A \to A$ be a mapping for which there exists a function $\varphi : A^3 \to [0,\infty)$ satisfying (10.15) and (10.16). If there exists an $L < 1$ such that $\varphi(x,y,z) \le \frac{1}{3} L \varphi(3x, 3y, 3z)$ for all $x, y, z \in A$, then there exists a unique derivation $\delta : A \to A$ such that*

$$\|f(x) - \delta(x)\|_A \le \frac{L}{3 - 3L} \varphi(x,x,x) \tag{10.21}$$

for all $x \in A$.

Proof. The proof is similar to the proofs of Theorems 10.4 and 10.6. □

Corollary 10.9. *Let $r > 2$ and θ be non-negative real numbers, and let $f : A \to A$ be a mapping satisfying (10.18) and (10.19). Then there exists a unique derivation $\delta : A \to A$ such that*

$$\|f(x) - \delta(x)\|_A \leq \frac{3\theta}{3^r - 3}\|x\|_A^r \qquad (10.22)$$

for all $x \in A$.

Proof. The proof follows from Theorem 10.8 by taking

$$\varphi(x,y,z) := \theta(\|x\|_A^r + \|y\|_A^r + \|z\|_A^r)$$

for all $x,y,z \in A$. Then $L = 3^{1-r}$ and we get the desired result. □

10.4 Stability of Homomorphisms in Lie C^*-Algebras

A C^*-algebra $\mathscr{C}$, endowed with the Lie product $[x,y] := \frac{xy-yx}{2}$ on $\mathscr{C}$, is called a *Lie C^*-algebra* (see [19], [20], [21]).

Definition 10.10. Let A and B be Lie C^*-algebras. A $\mathbf{C}$-linear mapping $H : A \to B$ is called a *Lie C^*-algebra homomorphism* if $H([x,y]) = [H(x), H(y)]$ for all $x, y \in A$.

Throughout this section, assume that A is a Lie C^*-algebra with norm $\|\cdot\|_A$ and that B is a Lie C^*-algebra with norm $\|\cdot\|_B$.

We prove the generalized Hyers–Ulam stability of homomorphisms in Lie C^*-algebras for the functional equation $D_\mu f(x,y,z) = 0$.

Theorem 10.11. *Let $f : A \to B$ be a mapping for which there exists a function $\varphi : A^3 \to [0,\infty)$ satisfying (10.1) such that*

$$\|f([x,y]) - [f(x), f(y)]\|_B \leq \varphi(x,y,0) \qquad (10.23)$$

for all $x,y \in A$. If there exists an $L < 1$ such that $\varphi(x,y,z) \leq 3L\varphi(\frac{x}{3}, \frac{y}{3}, \frac{z}{3})$ for all $x,y,z \in A$, then there exists a unique Lie C^*-algebra homomorphism $H : A \to B$ satisfying (10.4).

Proof. By the same reasoning as the proof of Theorem 10.2, there exists a unique $\mathbf{C}$-linear mapping $H : A \to A$ satisfying (10.4). The mapping $H : A \to B$ is given by

$$H(x) = \lim_{n\to\infty} \frac{f(3^n x)}{3^n}$$

for all $x \in A$.

It follows from (10.23) that

$$\begin{aligned}\|H([x,y]) - [H(x),H(y)]\|_B &= \lim_{n\to\infty} \frac{1}{9^n}\|f(9^n[x,y]) - [f(3^n x), f(3^n y)]\|_B \\ &\le \lim_{n\to\infty} \frac{1}{9^n}\varphi(3^n x, 3^n y, 0) \le \lim_{n\to\infty} \frac{1}{3^n}\varphi(3^n x, 3^n y, 0) \\ &\le \lim_{n\to\infty} L^n \varphi(x,y,0) = 0\end{aligned}$$

for all $x, y \in A$. So

$$H([x,y]) = [H(x), H(y)]$$

for all $x, y \in A$.

Thus $H : A \to B$ is a Lie C^*-algebra homomorphism satisfying (10.4), as desired. □

Corollary 10.12. *Let $r < 1$ and θ be non-negative real numbers, and let $f : A \to B$ be a mapping satisfying (10.9) such that*

$$\|f([x,y]) - [f(x), f(y)]\|_B \le \theta(\|x\|_A^r + \|y\|_A^r) \tag{10.24}$$

for all $x, y \in A$. Then there exists a unique Lie C^*-algebra homomorphism $H : A \to B$ satisfying (10.12).

Proof. The proof follows from Theorem 10.11 by taking

$$\varphi(x,y,z) := \theta(\|x\|_A^r + \|y\|_A^r + \|z\|_A^r)$$

for all $x, y, z \in A$. Then $L = 3^{r-1}$ and we get the desired result. □

Theorem 10.13. *Let $f : A \to B$ be a mapping for which there exists a function $\varphi : A^3 \to [0, \infty)$ satisfying (10.1) and (10.23). If there exists an $L < 1$ such that $\varphi(x,y,z) \le \frac{1}{3}L\varphi(3x, 3y, 3z)$ for all $x, y, z \in A$, then there exists a unique Lie C^*-algebra homomorphism $H : A \to B$ satisfying (10.13).*

Proof. The proof is similar to the proofs of Theorems 10.4 and 10.8. □

Corollary 10.14. *Let $r > 2$ and θ be non-negative real numbers, and let $f : A \to B$ be a mapping satisfying (10.9) and (10.24). Then there exists a unique Lie C^*-algebra homomorphism $H : A \to B$ satisfying.*

$$\frac{3\theta}{3^r - 3}\|x\|_A^r$$

for all $x \in A$.

Proof. The proof follows from Theorem 10.13 by taking

$$\varphi(x,y,z) := \theta(\|x\|_A^r + \|y\|_A^r + \|z\|_A^r)$$

for all $x, y, z \in A$. Then $L = 3^{1-r}$ and we get the desired result. □

Definition 10.15. A C^*-algebra A, endowed with the Jordan product $x \circ y := \frac{xy+yx}{2}$ for all $x, y \in A$, is called a *Jordan C^*-algebra* (see [20]).

Definition 10.16. Let A and B be Jordan C^*-algebras.

(i) A **C**-linear mapping $H : A \to B$ is called a *Jordan C^*-algebra homomorphism* if $H(x \circ y) = H(x) \circ H(y)$ for all $x, y \in A$.

(ii) A **C**-linear mapping $\delta : A \to A$ is called a *Jordan derivation* if $\delta(x \circ y) = x \circ \delta(y) + \delta(x) \circ y$ for all $x, y \in A$.

Remark 10.17. If the Lie products $[\cdot,\cdot]$ in the statements of the theorems in this section are replaced by the Jordan products $\cdot \circ \cdot$, then one obtains Jordan C^*-algebra homomorphisms instead of Lie C^*-algebra homomorphisms.

10.5 Stability of Lie Derivations on C^*-Algebras

Definition 10.18. Let A be a Lie C^*-algebra. A **C**-linear mapping $\delta : A \to A$ is called a *Lie derivation* if $\delta([x,y]) = [\delta(x), y] + [x, \delta(y)]$ for all $x, y \in A$.

Throughout this section, assume that A is a Lie C^*-algebra with norm $\|\cdot\|_A$.

We prove the generalized Hyers–Ulam stability of derivations on Lie C^*-algebras for the functional equation $D_\mu f(x,y,z) = 0$.

Theorem 10.19. *Let $f : A \to A$ be a mapping for which there exists a function $\varphi : A^3 \to [0,\infty)$ satisfying (10.15) such that*

$$\|f([x,y]) - [f(x),y] - [x, f(y)]\|_A \le \varphi(x,y,0) \tag{10.25}$$

for all $x, y \in A$. If there exists an $L < 1$ such that $\varphi(x,y,z) \le 3L\varphi(\frac{x}{3}, \frac{y}{3}, \frac{z}{3})$ for all $x, y, z \in A$, then there exists a unique Lie derivation $\delta : A \to A$ satisfying (10.17).

Proof. By the same reasoning as the proof of Theorem 10.2, there exists a unique **C**-linear mapping $\delta : A \to A$ satisfying (10.17). The mapping $\delta : A \to A$ is given by

$$\delta(x) = \lim_{n\to\infty} \frac{f(3^n x)}{3^n}$$

for all $x \in A$.

It follows from (10.25) that

$$\begin{aligned}
&\|\delta([x,y]) - [\delta(x),y] - [x,\delta(y)]\|_A \\
&= \lim_{n\to\infty} \frac{1}{9^n} \|f(9^n[x,y]) - [f(3^n x), 3^n y] - [3^n x, f(3^n y)]\|_A \\
&\le \lim_{n\to\infty} \frac{1}{9^n} \varphi(3^n x, 3^n y, 0) \le \lim_{n\to\infty} \frac{1}{3^n} \varphi(3^n x, 3^n y, 0) \\
&\le \lim_{n\to\infty} L^n \varphi(x,y,0) = 0
\end{aligned}$$

for all $x, y \in A$. So

$$\delta([x,y]) = [\delta(x), y] + [x, \delta(y)]$$

for all $x, y \in A$. Thus $\delta : A \to A$ is a derivation satisfying (10.17). □

Corollary 10.20. *Let $r < 1$ and θ be non-negative real numbers, and let $f : A \to A$ be a mapping satisfying (10.18) such that*

$$\|f([x,y]) - [f(x),y] - [x,f(y)]\|_A \leq \theta(\|x\|_A^r + \|y\|_A^r) \tag{10.26}$$

for all $x, y \in A$. Then there exists a unique Lie derivation $\delta : A \to A$ satisfying (10.20).

Proof. The proof follows from Theorem 10.19 by taking

$$\varphi(x,y,z) := \theta(\|x\|_A^r + \|y\|_A^r + \|z\|_A^r)$$

for all $x, y, z \in A$. Then $L = 3^{r-1}$ and we get the desired result. □

Theorem 10.21. *Let $f : A \to A$ be a mapping for which there exists a function $\varphi : A^3 \to [0, \infty)$ satisfying (10.15) and (10.25). If there exists an $L < 1$ such that $\varphi(x,y,z) \leq \frac{1}{3}L\varphi(3x,3y,3z)$ for all $x, y, z \in A$, then there exists a unique Lie derivation $\delta : A \to A$ satisfying (10.21).*

Proof. The proof is similar to the proofs of Theorems 10.4 and 10.8. □

Corollary 10.22. *Let $r > 2$ and θ be non-negative real numbers, and let $f : A \to A$ be a mapping satisfying (10.18) and (10.26). Then there exists a unique Lie derivation $\delta : A \to A$ satisfying (10.22).*

Proof. The proof follows from Theorem 10.21 by taking

$$\varphi(x,y,z) := \theta(\|x\|_A^r + \|y\|_A^r + \|z\|_A^r)$$

for all $x, y, z \in A$. Then $L = 3^{1-r}$ and we get the desired result. □

Remark 10.23. If the Lie products $[\cdot,\cdot]$ in the statements of the theorems in this section are replaced by the Jordan products $\cdot \circ \cdot$, then one obtains Jordan derivations instead of Lie derivations.

References

1. T. Aoki, *On the stability of the linear transformation in Banach spaces*, J. Math. Soc. Japan **2** (1950), 64–66.
2. L. Cădariu and V. Radu, *Fixed points and the stability of Jensen's functional equation*, J. Inequal. Pure Appl. Math. **4**, no. 1, Art. ID 4 (2003), 1–7.
3. L. Cădariu and V. Radu, *On the stability of the Cauchy functional equation: a fixed point approach*, Grazer Math. Ber. **346** (2004), 43–52.
4. L. Cădariu and V. Radu, *Fixed point methods for the generalized stability of functional equations in a single variable*, Fixed Point Theory and Applications **2008**, Art. ID 749392 (2008), 1–15.
5. P.W. Cholewa, *Remarks on the stability of functional equations*, Aequationes Math. **27** (1984), 76–86.
6. S. Czerwik, *On the stability of the quadratic mapping in normed spaces*, Abh. Math. Sem. Univ. Hamburg **62** (1992), 59–64.

7. S. Czerwik, *Functional Equations and Inequalities in Several Variables*, World Scientific Publishing Co., Singapore, 2002.
8. J. Diaz and B. Margolis, *A fixed point theorem of the alternative for contractions on a generalized complete metric space*, Bull. Amer. Math. Soc. **74** (1968), 305–309.
9. P. Găvruta, *A generalization of the Hyers-Ulam-Rassias stability of approximately additive mappings*, J. Math. Anal. Appl. **184** (1994), 431–436.
10. D.H. Hyers, *On the stability of the linear functional equation*, Proc. Nat. Acad. Sci. U.S.A. **27** (1941), 222–224.
11. D.H. Hyers, G. Isac and Th.M. Rassias, *Topics in Nonlinear Analysis and Applications*, World Scientific Publishing Co., Singapore, 1997.
12. D.H. Hyers, G. Isac and Th. M. Rassias, *Stability of Functional Equations in Several Variables*, Birkhäuser, Basel, 1998.
13. D.H. Hyers and Th.M. Rassias, *Approximate homomorphisms*, Aequationes Math. **44** (1992), 125–153.
14. G. Isac and Th.M. Rassias, *Stability of ψ-additive mappings: Applications to nonlinear analysis*, Int. J. Math. Math. Sci. **19** (1996), 219–228.
15. S. Jung, *Hyers-Ulam-Rassias Stability of Functional Equations in Mathematical Analysis*, Hadronic Press Inc., Palm Harbor, FL, 2001.
16. S. Jung, *A fixed point approach to the stability of isometries*, J. Math. Anal. Appl. **329** (2007), 879–890.
17. M. Mirzavaziri and M.S. Moslehian, *A fixed point approach to stability of a quadratic equation*, Bull. Braz. Math. Soc. **37** (2006), 361–376.
18. M.S. Moslehian and Th.M. Rassias, *Stability of functional equations in non-Archimedian spaces*, Appl. Anal. Disc. Math. **1** (2007), 325–334.
19. C. Park, *Lie $*$-homomorphisms between Lie C^*-algebras and Lie $*$-derivations on Lie C^*-algebras*, J. Math. Anal. Appl. **293** (2004), 419–434.
20. C. Park, *Homomorphisms between Lie JC^*-algebras and Cauchy-Rassias stability of Lie JC^*-algebra derivations*, J. Lie Theory **15** (2005), 393–414.
21. C. Park, *Homomorphisms between Poisson JC^*-algebras*, Bull. Braz. Math. Soc. **36** (2005), 79–97.
22. C. Park, *Fixed points and Hyers-Ulam-Rassias stability of Cauchy-Jensen functional equations in Banach algebras*, Fixed Point Theory and Applications **2007**, Art. ID 50175 (2007), 1–15.
23. C. Park, *Generalized Hyers-Ulam-Rassias stability of quadratic functional equations: a fixed point approach*, Fixed Point Theory and Applications **2008**, Art. ID 493751 (2008), 1–9.
24. C. Park, Y. Cho and M. Han, *Functional inequalities associated with Jordan-von Neumann type additive functional equations*, J. Inequal. Appl. **2007**, Art. ID 41820 (2007), 1–13.
25. C. Park and J. Cui, *Generalized stability of C^*-ternary quadratic mappings*, Abstract Appl. Anal. **2007**, Art. ID 23282 (2007), 1–6.
26. C. Park and A. Najati, *Homomorphisms and derivations in C^*-algebras*, Abstract Appl. Anal. **2007**, Art. ID 80630 (2007), 1–12.
27. V. Radu, *The fixed point alternative and the stability of functional equations*, Fixed Point Theory **4** (2003), 91–96.
28. Th.M. Rassias, *On the stability of the linear mapping in Banach spaces*, Proc. Amer. Math. Soc. **72** (1978), 297–300.
29. Th.M. Rassias, *New characterizations of inner product spaces*, Bull. Sci. Math. **108** (1984), 95–99.
30. Th.M. Rassias, *Problem 16; 2*, Report of the 27th International Symp. on Functional Equations, Aequationes Math. **39** (1990), 292–293; 309.
31. Th.M. Rassias, *On the stability of the quadratic functional equation and its applications*, Studia Univ. Babes-Bolyai **XLIII** (1998), 89–124.
32. Th.M. Rassias, *The problem of S.M. Ulam for approximately multiplicative mappings*, J. Math. Anal. Appl. **246** (2000), 352–378.
33. Th.M. Rassias, *On the stability of functional equations in Banach spaces*, J. Math. Anal. Appl. **251** (2000), 264–284.

34. Th.M. Rassias, *On the stability of functional equations and a problem of Ulam*, Acta Appl. Math. **62** (2000), 23–130.
35. Th.M. Rassias and P. Šemrl, *On the Hyers-Ulam stability of linear mappings*, J. Math. Anal. Appl. **173** (1993), 325–338.
36. Th.M. Rassias and K. Shibata, *Variational problem of some quadratic functionals in complex analysis*, J. Math. Anal. Appl. **228** (1998), 234–253.
37. F. Skof, *Proprietà locali e approssimazione di operatori*, Rend. Sem. Mat. Fis. Milano **53** (1983), 113–129.
38. S.M. Ulam, *Problems in Modern Mathematics*, Wiley, New York, 1960.

Chapter 11
Fixed Points and Stability of Functional Equations

Choonkil Park and Themistocles M. Rassias

Dedicated to the memory of Professor George Isac

Abstract Using the fixed point method, we prove the generalized Hyers–Ulam stability of the functional equation $f(x+y)+\frac{1}{2}f(x-y)+\frac{1}{2}f(y-x)=\frac{3}{2}f(x)+\frac{3}{2}f(y)+\frac{1}{2}f(-x)+\frac{1}{2}f(-y)$ in real Banach spaces.

11.1 Introduction and Preliminaries

The stability problem of functional equations originated from a question of Ulam [36] concerning the stability of group homomorphisms. Hyers [10] gave a first affirmative partial answer to the question of Ulam for Banach spaces. Hyers' theorem was generalized by Aoki [1] for additive mappings and by Th.M. Rassias [26] for linear mappings by considering an unbounded Cauchy difference. The paper of Th.M. Rassias [26] has provided a lot of influence in the development of what we call *generalized Hyers–Ulam stability* or as *Hyers–Ulam–Rassias stability* of functional equations. A generalization of the Th.M. Rassias theorem was obtained by Găvruta [9] by replacing the unbounded Cauchy difference by a general control function in the spirit of Th.M. Rassias' approach.

The functional equation

$$f(x+y)+f(x-y)=2f(x)+2f(y)$$

Choonkil Park
Department of Mathematics, Hanyang University, Seoul 133-791, South Korea, e-mail: baak@hanyang.ac.kr

Themistocles M. Rassias
Department of Mathematics, National Technical University of Athens, Zografou Campus, 15780 Athens, Greece, e-mail: trassias@math.ntua.gr

P.M. Pardalos et al. (eds.), *Nonlinear Analysis and Variational Problems*,
Springer Optimization and Its Applications 35, DOI 10.1007/978-1-4419-0158-3_11,

is called a *quadratic functional equation*. In particular, every solution of the quadratic functional equation is said to be a *quadratic function*. A generalized Hyers–Ulam stability problem for the quadratic functional equation was proved by Skof [35] for mappings $f : X \to Y$, where X is a normed space and Y is a Banach space. Cholewa [5] noticed that the theorem of Skof is still true if the relevant domain X is replaced by an Abelian group. Czerwik [6] proved the generalized Hyers–Ulam stability of the quadratic functional equation. The stability problems of several functional equations have been extensively investigated by a number of authors and there are many interesting results concerning this problem (see [7], [11]–[15], [18], [19], [22]–[24], [28]–[34]).

We recall a fundamental result in fixed point theory.

Let X be a set. A function $d : X \times X \to [0,\infty]$ is called a *generalized metric* on X if d satisfies

(1) $d(x,y) = 0$ if and only if $x = y$;
(2) $d(x,y) = d(y,x)$ for all $x,y \in X$;
(3) $d(x,z) \le d(x,y) + d(y,z)$ for all $x,y,z \in X$.

Theorem 11.1. *[2, 8] Let (X,d) be a complete generalized metric space and let $J : X \to X$ be a strictly contractive mapping with Lipschitz constant $L < 1$. Then for each given element $x \in X$, either*

$$d(J^n x, J^{n+1} x) = \infty$$

for all non-negative integers n or there exists a positive integer n_0 such that

(1) $d(J^n x, J^{n+1} x) < \infty, \qquad \forall n \ge n_0$;
(2) the sequence $\{J^n x\}$ converges to a fixed point y^* of J;
(3) y^* is the unique fixed point of J in the set $Y = \{y \in X \mid d(J^{n_0} x, y) < \infty\}$;
(4) $d(y,y^*) \le \frac{1}{1-L} d(y,Jy)$ for all $y \in Y$.

This paper is organized as follows: In Section 11.2, using the fixed point method, we prove the generalized Hyers-Ulam stability of the functional equation

$$f(x+y) + \frac{1}{2} f(x-y) + \frac{1}{2} f(y-x) = \frac{3}{2} f(x) + \frac{3}{2} f(y) + \frac{1}{2} f(-x) + \frac{1}{2} f(-y) \quad (11.1)$$

in real Banach spaces: an even case.

In Section 11.3, using the fixed point method, we prove the generalized Hyers–Ulam stability of the functional equation (11.1) in real Banach spaces: an odd case.

Throughout this paper, let X be a real normed vector space with norm $\|\cdot\|$, and Y a real Banach space with norm $\|\cdot\|$.

In 1996, G. Isac and Th.M. Rassias [14] were the first to provide applications of stability theory of functional equations for the proof of new fixed point theorems with applications. By using fixed point methods, the stability problems of several functional equations have been extensively investigated by a number of authors (see [4], [16], [17], [20], [21], [25]).

11.2 Fixed Points and Generalized Hyers–Ulam Stability of the Functional Equation (11.1): An Even Case

It is easily shown that an even mapping $f : X \to Y$ satisfies (11.1) if and only if the even mapping $f : X \to Y$ is a Cauchy quadratic mapping, i.e., $f(x+y)+f(x-y) = 2f(x)+2f(y)$, and that an odd mapping $f : X \to Y$ satisfies (11.1) if and only if the odd mapping mapping $f : X \to Y$ is a Cauchy additive mapping, i.e., $f(x+y) = f(x)+f(y)$.

For a given mapping $f : X \to Y$, we define

$$Cf(x,y) := f(x+y)+\frac{1}{2}f(x-y)+\frac{1}{2}f(y-x)-\frac{3}{2}f(x)-\frac{3}{2}f(y)-\frac{1}{2}f(-x) - \frac{1}{2}f(-y)$$

for all $x,y \in X$.

Using the fixed point method, we prove the generalized Hyers–Ulam stability of the functional equation $Cf(x,y) = 0$: an even case.

Theorem 11.2. *Let $f : X \to Y$ be a mapping with $f(0) = 0$ for which there exists a function $\varphi : X^2 \to [0,\infty)$ such that there exists an $L < 1$ such that $\varphi(x,y) \le \frac{1}{4}L\varphi(2x,2y)$ for all $x,y \in X$, and*

$$\|Cf(x,y)\| \le \varphi(x,y) \tag{11.2}$$

for all $x,y \in X$. Then there exists a unique Cauchy quadratic mapping $Q : X \to Y$ satisfying

$$\|f(x)+f(-x)-Q(x)\| \le \frac{L}{4-4L}(\varphi(x,x)+\varphi(-x,-x)) \tag{11.3}$$

for all $x \in X$.

Proof. Consider the set

$$S := \{g : X \to Y\}$$

and introduce the *generalized metric* on S:

$$d(g,h) = \inf\{K \in \mathbf{R}_+ : \|g(x)-h(x)\| \le K(\varphi(x,x)+\varphi(-x,-x)), \quad \forall x \in X\}.$$

It is easy to show that (S,d) is complete. (See the proof of Theorem 2.5 of [3].)

Now we consider the linear mapping $J : S \to S$ such that

$$Jg(x) := 4g\left(\frac{x}{2}\right)$$

for all $x \in X$.

It follows from the proof of Theorem 3.1 of [2] that

$$d(Jg,Jh) \le Ld(g,h)$$

for all $g,h \in S$.

Letting $y = x$ in (11.2), we get

$$\|f(2x) - 3f(x) - f(-x)\| \le \varphi(x,x) \tag{11.4}$$

for all $x \in X$. Replacing x by $-x$ in (11.4), we get

$$\|f(-2x) - 3f(-x) - f(x)\| \le \varphi(-x,-x) \tag{11.5}$$

for all $x \in X$. Let $g(x) := f(x) + f(-x)$ for all $x \in X$. Then $g : X \to Y$ is an even mapping. It follows from (11.4) and (11.5) that

$$\left\|g(x) - 4g\left(\frac{x}{2}\right)\right\| \le \frac{L}{4}(\varphi(x,x) + \varphi(-x,-x))$$

for all $x \in X$. Hence $d(g, Jg) \le \frac{L}{4}$.

By Theorem 11.1, there exists a mapping $Q : X \to Y$ satisfying the following:

(1) Q is a fixed point of J, i.e.,

$$Q\left(\frac{x}{2}\right) = \frac{1}{4}Q(x) \tag{11.6}$$

for all $x \in X$. Then $Q : X \to Y$ is an even mapping. The mapping Q is a unique fixed point of J in the set

$$M = \{g \in S : d(f,g) < \infty\}.$$

This implies that Q is a unique mapping satisfying (11.6) such that there exists a $K \in (0,\infty)$ satisfying

$$\|g(x) - Q(x)\| \le K\left(\varphi\left(\frac{x}{2},\frac{x}{2}\right) + \varphi\left(-\frac{x}{2},-\frac{x}{2}\right)\right)$$

for all $x \in X$;

(2) $d(J^n g, Q) \to 0$ as $n \to \infty$. This implies the equality

$$\lim_{n\to\infty} 4^n g\left(\frac{x}{2^n}\right) = Q(x) \tag{11.7}$$

for all $x \in X$;

(3) $d(g,Q) \le \frac{1}{1-L} d(g, Jg)$, which implies the inequality

$$d(g,Q) \le \frac{L}{4-4L}.$$

This implies that the inequality (11.3) holds.

It follows from (11.2) and (11.7) that

$$\begin{aligned}
&\|CQ(x,y)\| = \lim_{n\to\infty} 4^n \left\|Cg\left(\frac{x}{2^n},\frac{y}{2^n}\right)\right\| \\
&\le \lim_{n\to\infty} 4^n \left(\varphi\left(\frac{x}{2^n},\frac{y}{2^n}\right) + \varphi\left(-\frac{x}{2^n},-\frac{y}{2^n}\right)\right) \le \lim_{n\to\infty} L^n (\varphi(x,y) + \varphi(-x,-y)) = 0
\end{aligned}$$

for all $x, y \in X$. So $CQ(x,y) = 0$ for all $x, y \in X$. Since $Q : X \to Y$ is even, the mapping $Q : X \to Y$ is a Cauchy quadratic mapping.

Therefore, there exists a unique Cauchy quadratic mapping $Q : X \to Y$ satisfying (11.3), as desired. □

Corollary 11.3. *Let $p > 2$ and $\theta \geq 0$ be real numbers, and let $f : X \to Y$ be a mapping such that*

$$\|Cf(x,y)\| \leq \theta(||x||^p + ||y||^p) \tag{11.8}$$

for all $x, y \in X$. Then there exists a unique Cauchy quadratic mapping $Q : X \to Y$ satisfying

$$\|f(x) + f(-x) - Q(x)\| \leq \frac{4\theta}{2^p - 4} ||x||^p$$

for all $x \in X$.

Proof. The proof follows from Theorem 11.2 by taking

$$\varphi(x,y) := \theta(||x||^p + ||y||^p)$$

for all $x, y \in X$. Then we can choose $L = 2^{2-p}$ and we get the desired result. □

Remark 11.4. Let $f : X \to Y$ be a mapping for which there exists a function $\varphi : X^2 \to [0, \infty)$ satisfying (11.2) and $f(0) = 0$. By a similar method to the proof of Theorem 11.2, one can show that if there exists an $L < 1$ such that $\varphi(x,y) \leq 4L\varphi\left(\frac{x}{2}, \frac{y}{2}\right)$ for all $x, y \in X$, then there exists a unique Cauchy quadratic mapping $Q : X \to Y$ satisfying

$$\|f(x) + f(-x) - Q(x)\| \leq \frac{1}{4 - 4L}(\varphi(x,x) + \varphi(-x,-x))$$

for all $x \in X$.

For the case $0 < p < 2$, one can obtain a similar result to Corollary 11.3: Let $0 < p < 2$ and $\theta \geq 0$ be real numbers, and let $f : X \to Y$ be a mapping satisfying (11.8). Then there exists a unique Cauchy quadratic mapping $Q : X \to Y$ satisfying

$$\|f(x) + f(-x) - Q(x)\| \leq \frac{4\theta}{4 - 2^p} ||x||^p$$

for all $x \in X$.

11.3 Fixed Points and Generalized Hyers–Ulam Stability of the Functional Equation (11.1): An Odd Case

Using the fixed point method, we prove the generalized Hyers–Ulam stability of the functional equation $Cf(x,y) = 0$: an odd case.

Theorem 11.5. *Let $f : X \to Y$ be a mapping with $f(0) = 0$ for which there exists a function $\varphi : X^2 \to [0,\infty)$ such that there exists an $L < 1$ such that $\varphi(x,y) \leq \frac{1}{2}L\varphi(2x,2y)$ for all $x,y \in X$, and*

$$\|Cf(x,y)\| \leq \varphi(x,y) \tag{11.9}$$

for all $x,y \in X$. Then there exists a unique Cauchy additive mapping $A : X \to Y$ satisfying

$$\|f(x) - f(-x) - A(x)\| \leq \frac{L}{2-2L}(\varphi(x,x) + \varphi(-x,-x)) \tag{11.10}$$

for all $x \in X$.

Proof. Consider the set

$$S := \{g : X \to Y\}$$

and introduce the *generalized metric* on S:

$$d(g,h) = \inf\{K \in \mathbf{R}_+ : \|g(x) - h(x)\| \leq K(\varphi(x,x) + \varphi(-x,-x)), \quad \forall x \in X\}.$$

It is easy to show that (S,d) is complete. (See the proof of Theorem 2.5 of [3].)

Now we consider the linear mapping $J : S \to S$ such that

$$Jg(x) := 2g\left(\frac{x}{2}\right)$$

for all $x \in X$.

It follows from the proof of Theorem 3.1 of [2] that

$$d(Jg,Jh) \leq Ld(g,h)$$

for all $g,h \in S$.

Letting $y = x$ in (11.9), we get

$$\|f(2x) - 3f(x) - f(-x)\| \leq \varphi(x,x) \tag{11.11}$$

for all $x \in X$. Replacing x by $-x$ in (11.11), we get

$$\|f(-2x) - 3f(-x) - f(x)\| \leq \varphi(-x,-x) \tag{11.12}$$

for all $x \in X$. Let $g(x) := f(x) - f(-x)$ for all $x \in X$. Then $g : X \to Y$ is an odd mapping. It follows from (11.11) and (11.12) that

$$\left\| g(x) - 2g\left(\frac{x}{2}\right) \right\| \le \frac{L}{2}(\varphi(x,x) + \varphi(-x,-x))$$

for all $x \in X$. Hence $d(g, Jg) \le \frac{L}{2}$.

By Theorem 11.1, there exists a mapping $A : X \to Y$ satisfying the following:

(1) A is a fixed point of J, i.e.,

$$A\left(\frac{x}{2}\right) = \frac{1}{2}A(x) \tag{11.13}$$

for all $x \in X$. Then $A : X \to Y$ is an odd mapping. The mapping A is a unique fixed point of J in the set

$$M = \{g \in S : d(f,g) < \infty\}.$$

This implies that A is a unique mapping satisfying (11.13) such that there exists a $K \in (0, \infty)$ satisfying

$$\|g(x) - A(x)\| \le K\left(\varphi\left(\frac{x}{2}, \frac{x}{2}\right) + \varphi\left(-\frac{x}{2}, -\frac{x}{2}\right)\right)$$

for all $x \in X$;

(2) $d(J^n g, A) \to 0$ as $n \to \infty$. This implies the equality

$$\lim_{n\to\infty} 2^n g\left(\frac{x}{2^n}\right) = A(x) \tag{11.14}$$

for all $x \in X$;

(3) $d(g, A) \le \frac{1}{1-L} d(g, Jg)$, which implies the inequality

$$d(g, A) \le \frac{L}{2 - 2L}.$$

This implies that the inequality (11.10) holds. It follows from (11.9) and (11.14) that

$$\|CA(x,y)\| = \lim_{n\to\infty} 2^n \left\| Cg\left(\frac{x}{2^n}, \frac{y}{2^n}\right) \right\| \le \lim_{n\to\infty} 2^n \left(\varphi\left(\frac{x}{2^n}, \frac{y}{2^n}\right) + \varphi\left(-\frac{x}{2^n}, -\frac{y}{2^n}\right)\right)$$
$$\le \lim_{n\to\infty} L^n (\varphi(x,y) + \varphi(-x,-y)) = 0$$

for all $x, y \in X$. So $CA(x,y) = 0$ for all $x, y \in X$. Since $A : X \to Y$ is odd, the mapping $A : X \to Y$ is a Cauchy additive mapping.

Therefore, there exists a unique Cauchy additive mapping $A : X \to Y$ satisfying (11.10), as desired. □

Corollary 11.6. *Let $p > 1$ and $\theta \ge 0$ be real numbers, and let $f : X \to Y$ be a mapping satisfying (11.8). Then there exists a unique Cauchy additive mapping $A : X \to Y$ satisfying*

$$\|f(x) - f(-x) - A(x)\| \leq \frac{4\theta}{2^p - 2}||x||^p$$

for all $x \in X$.

Proof. The proof follows from Theorem 11.5 by taking

$$\varphi(x,y) := \theta(||x||^p + ||y||^p)$$

for all $x, y \in X$. Then we can choose $L = 2^{1-p}$ and we get the desired result. □

Remark 11.7. Let $f : X \to Y$ be a mapping for which there exists a function $\varphi : X^2 \to [0,\infty)$ satisfying (11.9) and $f(0) = 0$. By a similar method to the proof of Theorem 11.2, one can show that if there exists an $L < 1$ such that $\varphi(x,y) \leq 2L\varphi\left(\frac{x}{2}, \frac{y}{2}\right)$ for all $x, y \in X$, then there exists a unique Cauchy additive mapping $A : X \to Y$ satisfying

$$\|f(x) - f(-x) - A(x)\| \leq \frac{1}{2-2L}(\varphi(x,x) + \varphi(-x,-x))$$

for all $x \in X$.

For the case $0 < p < 1$, one can obtain a similar result to Corollary 11.6: Let $0 < p < 1$ and $\theta \geq 0$ be real numbers, and let $f : X \to Y$ be a mapping satisfying (11.8). Then there exists a unique Cauchy additive mapping $A : X \to Y$ satisfying

$$\|f(x) - f(-x) - A(x)\| \leq \frac{4\theta}{2 - 2^p}||x||^p$$

for all $x \in X$.

Combining Corollaries 11.3 and 11.6 yields the following.

Theorem 11.8. *Let* $p > 2$ *and* $\theta \geq 0$ *be real numbers, and let* $f : X \to Y$ *be a mapping satisfying (11.8). Then there exist a unique Cauchy quadratic mapping* $Q : X \to Y$ *and a unique Cauchy additive mapping* $A : X \to Y$ *satisfying*

$$\|2f(x) - Q(x) - A(x)\| \leq \left(\frac{4}{2^p - 4} + \frac{4}{2^p - 2}\right)\theta||x||^p$$

for all $x \in X$.

Remark 11.9. Let $f : X \to Y$ be a mapping for which there exists a function $\varphi : X^2 \to [0,\infty)$ satisfying (11.9) and $f(0) = 0$. By a similar method to the proof of Theorem 11.5, one can show that if there exists an $L < 1$ such that $\varphi(x,y) \leq 2L\varphi\left(\frac{x}{2}, \frac{y}{2}\right)$ for all $x, y \in X$, then there exists a unique Cauchy additive mapping $A : X \to Y$ satisfying

$$\|f(x) - f(-x) - A(x)\| \leq \frac{1}{2-2L}(\varphi(x,x) + \varphi(-x,-x))$$

for all $x \in X$.

Combining Remarks 11.4 and 11.7 yields the following.

Theorem 11.10. *Let $p < 1$ and $\theta \geq 0$ be real numbers, and let $f : X \to Y$ be a mapping satisfying (11.8). Then there exist a unique Cauchy quadratic mapping $Q : X \to Y$ and a unique Cauchy additive mapping $A : X \to Y$ satisfying*

$$\|2f(x) - Q(x) - A(x)\| \leq \left(\frac{4}{4-2^p} + \frac{4}{2-2^p} \right) \theta \|x\|^p$$

for all $x \in X$.

References

1. T. Aoki, *On the stability of the linear transformation in Banach spaces*, J. Math. Soc. Japan **2** (1950), 64–66.
2. L. Cădariu and V. Radu, *Fixed points and the stability of Jensen's functional equation*, J. Inequal. Pure Appl. Math. **4**, no. 1, Art. ID 4 (2003), 1–7.
3. L. Cădariu and V. Radu, *On the stability of the Cauchy functional equation: a fixed point approach*, Grazer Math. Ber. **346** (2004), 43–52.
4. L. Cădariu and V. Radu, *Fixed point methods for the generalized stability of functional equations in a single variable*, Fixed Point Theory and Applications **2008**, Art. ID 749392 (2008), 1–15.
5. P.W. Cholewa, *Remarks on the stability of functional equations*, Aequationes Math. **27** (1984), 76–86.
6. S. Czerwik, *On the stability of the quadratic mapping in normed spaces*, Abh. Math. Sem. Univ. Hamburg **62** (1992), 59–64.
7. S. Czerwik, *Functional Equations and Inequalities in Several Variables*, World Scientific Publishing Co., Singapore, 2002.
8. J. Diaz and B. Margolis, *A fixed point theorem of the alternative for contractions on a generalized complete metric space*, Bull. Amer. Math. Soc. **74** (1968), 305–309.
9. P. Găvruta, *A generalization of the Hyers-Ulam-Rassias stability of approximately additive mappings*, J. Math. Anal. Appl. **184** (1994), 431–436.
10. D.H. Hyers, *On the stability of the linear functional equation*, Proc. Nat. Acad. Sci. U.S.A. **27** (1941), 222–224.
11. D.H. Hyers, G. Isac and Th.M. Rassias, *Topics in Nonlinear Analysis and Applications*, World Scientific Publishing Co., Singapore, 1997.
12. D.H. Hyers, G. Isac and Th. M. Rassias, *Stability of Functional Equations in Several Variables*, Birkhäuser, Basel, 1998.
13. D.H. Hyers and Th.M. Rassias, *Approximate homomorphisms*, Aequationes Math. **44** (1992), 125–153.
14. G. Isac and Th.M. Rassias, *Stability of ψ-additive mappings: Applications to nonlinear analysis*, Int. J. Math. Math. Sci. **19** (1996), 219–228.
15. S. Jung, *Hyers-Ulam-Rassias Stability of Functional Equations in Mathematical Analysis*, Hadronic Press Inc., Palm Harbor, FL, 2001.
16. S. Jung, *A fixed point approach to the stability of isometries*, J. Math. Anal. Appl. **329** (2007), 879–890.
17. M. Mirzavaziri and M.S. Moslehian, *A fixed point approach to stability of a quadratic equation*, Bull. Braz. Math. Soc. **37** (2006), 361–376.
18. M.S. Moslehian and Th.M. Rassias, *Stability of functional equations in non-Archimedian spaces*, Appl. Anal. Disc. Math. **1** (2007), 325–334.

19. C. Park, *Homomorphisms between Poisson JC^*-algebras*, Bull. Braz. Math. Soc. **36** (2005), 79–97.
20. C. Park, *Fixed points and Hyers-Ulam-Rassias stability of Cauchy-Jensen functional equations in Banach algebras*, Fixed Point Theory and Applications **2007**, Art. ID 50175 (2007), 1–15.
21. C. Park, *Generalized Hyers-Ulam-Rassias stability of quadratic functional equations: a fixed point approach*, Fixed Point Theory and Applications **2008**, Art. ID 493751 (2008), 1–9.
22. C. Park, Y. Cho and M. Han, *Functional inequalities associated with Jordan-von Neumann type additive functional equations*, J. Inequal. Appl. **2007**, Art. ID 41820 (2007), 1–13.
23. C. Park and J. Cui, *Generalized stability of C^*-ternary quadratic mappings*, Abstract Appl. Anal. **2007**, Art. ID 23282 (2007), 1–6.
24. C. Park and A. Najati, *Homomorphisms and derivations in C^*-algebras*, Abstract Appl. Anal. **2007**, Art. ID 80630 (2007), 1–12.
25. V. Radu, *The fixed point alternative and the stability of functional equations*, Fixed Point Theory **4** (2003), 91–96.
26. Th.M. Rassias, *On the stability of the linear mapping in Banach spaces*, Proc. Amer. Math. Soc. **72** (1978), 297–300.
27. Th.M. Rassias, *New characterizations of inner product spaces*, Bull. Sci. Math. **108** (1984), 95–99.
28. Th.M. Rassias, *Problem 16; 2*, Report of the 27th International Symp. on Functional Equations, Aequationes Math. **39** (1990), 292–293; 309.
29. Th.M. Rassias, *On the stability of the quadratic functional equation and its applications*, Studia Univ. Babes-Bolyai **XLIII** (1998), 89–124.
30. Th.M. Rassias, *The problem of S.M. Ulam for approximately multiplicative mappings*, J. Math. Anal. Appl. **246** (2000), 352–378.
31. Th.M. Rassias, *On the stability of functional equations in Banach spaces*, J. Math. Anal. Appl. **251** (2000), 264–284.
32. Th.M. Rassias, *On the stability of functional equations and a problem of Ulam*, Acta Appl. Math. **62** (2000), 23–130.
33. Th.M. Rassias and P. Šemrl, *On the Hyers-Ulam stability of linear mappings*, J. Math. Anal. Appl. **173** (1993), 325–338.
34. Th.M. Rassias and K. Shibata, *Variational problem of some quadratic functionals in complex analysis*, J. Math. Anal. Appl. **228** (1998), 234–253.
35. F. Skof, *Proprietà locali e approssimazione di operatori*, Rend. Sem. Mat. Fis. Milano **53** (1983), 113–129.
36. S.M. Ulam, *Problems in Modern Mathematics*, Wiley, New York, 1960.

Chapter 12
Compression–Expansion Critical Point Theorems in Conical Shells

Radu Precup

Dedicated to the memory of Professor George Isac

Abstract We present compression and expansion type critical point theorems in a conical shell of a Hilbert space identified to its dual. The notion of linking is involved and the compression–expansion boundary conditions are expressed with respect to only one norm.

12.1 Introduction

One of the most common approaches to the existence, localization and multiplicity of positive solutions to nonlinear problems is based on compression and expansion conditions and Krasnoselskii type theorems.

To be more precise, let us consider a normed linear space X with norm $|.|$ and let K be a wedge of X, i.e., a closed convex subset of X, $K \neq \{0\}$, with $\lambda K \subset K$ for every $\lambda \in \mathbf{R}_+$. For two numbers $0 < R_0 < R_1$, we define the conical shell $K_{R_0 R_1}$ by

$$K_{R_0 R_1} := \{u \in K : R_0 \leq |u| \leq R_1\}.$$

Suppose that we are interested in a solution u in $K_{R_0 R_1}$ to the operator equation

$$Nu = u, \tag{12.1}$$

where N is a given nonlinear map acting in K.

Radu Precup
Department of Applied Mathematics, Babeş–Bolyai University, 400084 Cluj, Romania e-mail: r.precup@math.ubbcluj.ro

P.M. Pardalos et al. (eds.), *Nonlinear Analysis and Variational Problems*,
Springer Optimization and Its Applications 35, DOI 10.1007/978-1-4419-0158-3_12,

Here is an existence result of Krasnoselskii type which follows from index theory (see [6]), or directly, from Schauder's fixed point theorem and retract arguments (see [9]).

Theorem 12.1. *Let $\alpha, \beta > 0$, $\alpha \neq \beta$, $R_0 := \min\{\alpha, \beta\}$ and $R_1 := \max\{\alpha, \beta\}$. Assume that $N : K_{R_0R_1} \to K$ is a compact map and that the following condition is satisfied:*

$$\begin{aligned} &Nu \neq \lambda u \ \text{for} \ |u| = \alpha \ \text{and} \ \lambda > 1, \\ &Nu \neq \lambda u \ \text{for} \ |u| = \beta \ \text{and} \ 0 < \lambda < 1, \\ &\inf_{|u|=\beta} |Nu| > 0. \end{aligned} \tag{12.2}$$

Then N has a fixed point in $K_{R_0R_1}$.

Notice that condition $Nu \neq \lambda u$ for $|u| = \alpha$ and $\lambda > 1$ is known as the *Leray–Schauder boundary condition.* There is a huge literature devoted to the applications of the Leray–Schauder boundary condition to lots of classes of nonlinear problems: integral equations and boundary value problems for ordinary and partial differential equations (see [9], [10] and [12]). An interesting and atypical application to the nonlinear complementarity problem from optimization theory is due to G. Isac [7], whose notion of an *exceptional family of elements* is in connection with the Leray–Schauder boundary condition and with the so called Leray–Schauder nonlinear alternative.

If $\beta < \alpha$, then (12.2) can be seen as a *compression* property of N upon $K_{\beta\alpha}$, whereas if $\alpha < \beta$, then (12.2) represents an *expansion* property of N upon $K_{\alpha\beta}$.

A direct consequence of Theorem 12.1 is the following norm-type compression–expansion theorem.

Theorem 12.2. *Let $\alpha, \beta > 0$, $\alpha \neq \beta$, $R_0 := \min\{\alpha, \beta\}$ and $R_1 := \max\{\alpha, \beta\}$. Assume that $N : K_{R_0R_1} \to K$ is a compact map and that the following condition is satisfied:*

$$\begin{aligned} &|Nu| \leq |u| \ \text{for} \ |u| = \alpha, \\ &|Nu| \geq |u| \ \text{for} \ |u| = \beta. \end{aligned}$$

Then N has a fixed point in $K_{R_0R_1}$.

Next assume that equation (12.1) has a variational form, i.e., there exists a C^1-(energy/elevation) functional $E : X \to \mathbf{R}$ such that

$$Nu = u - E'(u).$$

Here X is a Hilbert space which is identified to its dual. Then the problem is to find solutions of equation (12.1) of a certain energy/elevation. For instance, we may think at absolute minimizers for E on $K_{R_0R_1}$, or at saddle points of E in $K_{R_0R_1}$, which in physical systems appear as stable and respectively unstable equilibria. Some ideas for an answer to this problem can be found in the recent paper [11], where only the compression case was considered together with a mountain pass argument (see [1] and [10]). Also, in [11], working in the energy space whose norm is nonmonotone, we had to define the conical shell by means of two norms (the energy norm and a monotone second one). The aim of this paper is threefold: (1) to obtain both compression and expansion type critical point theorems; (2) to use more generally the notion of linking rather than the mountain pass condition; and (3) to develop the theory in a Hilbert space endowed with only one norm.

12.2 Main Results

Throughout this section, X will be a Hilbert space, $K \subset X$ a wedge and $E \in C^1(X)$ a functional. Also we shall consider $0 < R_0 < R_1$, a closed subset $S \subset K_{R_0R_1}$ and a submanifold $Q \subset K_{R_0R_1}$ with relative boundary ∂Q, and we shall assume that S and ∂Q link (with respect to $\Gamma := \{h \in C(K_{R_0R_1}; K_{R_0R_1}) : h|_{\partial Q} = id\}$), i.e., $S \cap \partial Q = \emptyset$ and $h(Q) \cap S \neq \emptyset$ for every $h \in \Gamma$. For more information about linking, we refer the reader to [2], [5], [13] and [14].

Theorem 12.3. *Let* $\varepsilon \in \{-1; 1\}$. *Assume that*

$$u - E'(u) \in K \text{ for all } u \in K \tag{12.3}$$

and that there exists $\nu_0 > 0$ *with*

$$\varepsilon\left(E'(u), u\right) \leq \nu_0 \text{ for all } u \in K \text{ with } |u| = R_0; \tag{12.4}$$

$$\varepsilon\left(E'(u), u\right) \geq -\nu_0 \text{ for all } u \in K \text{ with } |u| = R_1. \tag{12.5}$$

In addition assume that

$$\sup_{u \in \partial Q} E(u) < \inf_{u \in S} E(u). \tag{12.6}$$

Let

$$c := \inf_{h \in \Gamma} \sup_{u \in Q} E(h(u)).$$

Then there exists a sequence (u_k) *with* $u_k \in K_{R_0R_1}$ *such that*

$$E(u_k) \to c \text{ as } k \to \infty \tag{12.7}$$

and one of the following three properties holds:

$$E'(u_k) \to 0 \text{ as } k \to \infty; \tag{12.8}$$

$$\begin{cases} |u_k| = R_0, \ \varepsilon\left(E'(u_k), u_k\right) \geq 0 \text{ and} \\ E'(u_k) - \frac{(E'(u_k), u_k)}{R_0^2} u_k \to 0 \text{ as } k \to \infty; \end{cases} \tag{12.9}$$

$$\begin{cases} |u_k| = R_1, \quad \varepsilon\left(E'\left(u_k\right), u_k\right) \leq 0 \ \ and \\ E'\left(u_k\right) - \frac{\left(E'\left(u_k\right), u_k\right)}{R_1^2} u_k \to 0 \ \ as \ \ k \to \infty. \end{cases} \tag{12.10}$$

If in addition, any sequence (u_k) *as above has a convergent subsequence (i.e., the modified Palais–Smale–Schechter (MPSS) condition holds in* $K_{R_0R_1}$*) and* E *satisfies the boundary conditions*

$$E'(u) - \varepsilon\lambda u \neq 0 \ \ for\ u \in K,\ |u| = R_0,\ \lambda > 0; \tag{12.11}$$

$$E'(u) + \varepsilon\lambda u \neq 0 \ \ for\ u \in K,\ |u| = R_1,\ \lambda > 0, \tag{12.12}$$

then there exists $u \in K_{R_0R_1}$ *with*

$$E'(u) = 0 \ \ and \ \ E(u) = c.$$

Remark 12.4. Let $N(u) := u - E'(u)$. For $\varepsilon = 1$, conditions (12.11), (12.12) can be written under the form

$$\begin{aligned} N(u) &\neq (1-\lambda)u \quad \text{for } u \in K, \ \ |u| = R_0, \ \lambda > 0; \\ N(u) &\neq (1+\lambda)u \quad \text{for } u \in K, \ \ |u| = R_1, \ \lambda > 0, \end{aligned} \tag{12.13}$$

showing a compression property of N upon $K_{R_0R_1}$. Similarly, for $\varepsilon = -1$, (12.11), (12.12) become

$$\begin{aligned} N(u) &\neq (1+\lambda)u \quad \text{for } u \in K, \ \ |u| = R_0, \ \lambda > 0; \\ N(u) &\neq (1-\lambda)u \quad \text{for } u \in K, \ \ |u| = R_1, \ \lambda > 0, \end{aligned} \tag{12.14}$$

that is, an expansion property for N.

The next critical point result can be compared to the fixed point Theorem 20.2 in [4].

Theorem 12.5. *Let* $\varepsilon \in \{-1; 1\}$. *Assume that conditions* (12.3), (12.6), (12.11) *and* (12.12) *hold. In addition assume that the restriction to* $K_{R_0R_1}$ *of the map* $N := I - E'$ *is compact and*

$$\inf\left\{ |N(u)| : u \in K,\ |u| = R_{\frac{1-\varepsilon}{2}} \right\} > 0. \tag{12.15}$$

Then there exists a point $u \in K_{R_0R_1}$ *with*

$$E'(u) = 0 \ \ and \ \ E(u) = c.$$

The following result is the compression–expansion critical point theorem accompanying the corresponding fixed point theorem of Krasnoselskii [8] (see also [6, p. 325]).

Theorem 12.6. *Assume that conditions* (12.3) *and* (12.6) *hold. In addition assume that the restriction to* $K_{R_0R_1}$ *of the map* $N := I - E'$ *is compact and one of the following conditions is satisfied:*

(a) $|N(u)| \leq |u|$ *for* $|u| = R_0$, *and* $|N(u)| \geq |u|$ *for* $|u| = R_1$,
(b) $|N(u)| \geq |u|$ *for* $|u| = R_0$, *and* $|N(u)| \leq |u|$ *for* $|u| = R_1$.

Then there exists a point $u \in K_{R_0 R_1}$ *with* $E'(u) = 0$ *and* $E(u) = c$.

Now if instead of critical points of saddle type we seek critical points of minimum type, then we obtain:

Theorem 12.7. *Let* $\varepsilon \in \{-1;1\}$. *Assume that conditions* (12.3), (12.4), (12.5) *are satisfied and that*

$$m := \inf_{K_{R_0 R_1}} E > -\infty. \tag{12.16}$$

Then there exists a sequence (u_k) *with* $u_k \in K_{R_0 R_1}$ *such that*

$$E(u_k) \to m \quad as \quad k \to \infty \tag{12.17}$$

and one of the conditions (12.8), (12.9), (12.10) *holds. If in addition, any sequence* (u_k) *as above has a convergent subsequence and* (12.11), (12.12) *are satisfied, then there exists* $u \in K_{R_0 R_1}$ *with*

$$E'(u) = 0 \quad and \quad E(u) = m.$$

Remark 12.8. If both conditions (12.6) and (12.16) are satisfied, then Theorems 12.3 and 12.7 guarantee the existence of two distinct critical points of E in $K_{R_0 R_1}$.

12.3 Proofs

The proofs need some lemmas.

Lemma 12.9 ([11]). *Let* X *be a Hilbert space,* $w, v \in X \setminus \{0\}$ *and* $\alpha, \theta \in \mathbf{R}_+$ *such that* $0 < \alpha < 1 - \theta$ *and* $(w, v) \geq -\theta |w| |v|$. *Then there exists an element* $h \in X$ *with*

$$|h| = 1, \quad (w, h) \leq -\alpha |w| \quad and \quad (v, h) < 0.$$

Assume in addition that $K \subset X$ *is a wedge. Then*

1^0 *if* $v \in K$ *and* $v - w \in K$, *then there exists a* $\lambda > 0$ *with*

$$v + \mu h \in K \quad for\ all \quad \mu \in [0, \lambda];$$

2^0 *if* $-v \in K$ *and* $-v - w \in K$, *then there exists a* $\lambda > 0$ *with*

$$-v + \mu h \in K \quad for\ all \quad \mu \in [0, \lambda].$$

In case that $1 - \theta < 2\alpha$, λ *do not depend on* v *and* w, *for* $|v| \geq R_0$ *and* $|w| \geq a > 0$.

Lemma 12.10 ([11]). *Let* X *be a Hilbert space,* $K \subset X$ *a wedge,* $D \subset K$ *a subset,* $a > 0$, *let* $G : D \to X$ *be a continuous map,* $\widehat{D} = \{u \in D : |G(u)| \geq a\}$ *and* $D_0, D_1 \subset$

$\widehat{D} \setminus \{0\}$ *be two disjoint closed sets. Assume that*

$$u - G(u) \in K \text{ for all } u \in D$$

and there exists a $\theta \in [0,1)$ *such that*

$$|(u, G(u))| \leq \theta |u| |G(u)| \text{ for all } u \in D_0 \cup D_1.$$

Then there exists $\alpha > 0$ *and a locally Lipschitz map* $H : \widehat{D} \to X$ *such that*

$$|H(u)| \leq 1, \; u + H(u) \in K, \; (G(u), H(u)) \leq -\alpha |G(u)| |H(u)| \text{ for } u \in \widehat{D}$$

and

$$\begin{aligned} (u, H(u)) &> 0 \text{ for } u \in D_0 \\ (u, H(u)) &< 0 \text{ for } u \in D_1. \end{aligned}$$

Lemma 12.11 ([3]). *Let X be a Banach space, D a closed convex set in X. Assume that* $W : D \to X$ *is a locally Lipschitz map which satisfies*

$$|W(u)| \leq C, \; \liminf_{\lambda \to 0^+} \frac{1}{\lambda} d(u + \lambda W(u), D) = 0$$

for all $u \in D$. *Then, for any* $u \in D$, *the initial value problem in Banach space*

$$\frac{d\sigma}{dt} = W(\sigma), \; \sigma(0) = u$$

has a unique solution $\sigma(u,t)$ *on* $\mathbf{R}_+$, *and* $\sigma(u,t) \in D$ *for every* $t \in \mathbf{R}_+$.

The next lemma was also given in [11]. Here we just complete its proof by the step on the locally Lipschitz extension.

Lemma 12.12. *Assume all the assumptions of Theorem* 12.3 *hold. In addition assume that there are constants* $\delta > 0$ *and* $\theta \in [0,1)$ *such that for* $u \in K_{R_0 R_1}$ *satisfying* $|E(u) - c| \leq \delta$, *one has*

$$\left|\left(E'(u), u\right)\right| \leq \theta |u| \left|E'(u)\right| \text{ if } |u| = R_0 \text{ or } |u| = R_1. \tag{12.18}$$

Then there exists a sequence of elements $u_k \in K_{R_0 R_1}$ *with*

$$E(u_k) \to c \text{ and } E'(u_k) \to 0 \text{ as } k \to \infty. \tag{12.19}$$

Proof. Assume there are no sequences satisfying (12.19). Then there would be constants $\delta, a > 0$ such that

$$\left|E'(u)\right| \geq a$$

for all u in

$$Q = \left\{u \in K_{R_0 R_1} : |E(u) - c| \leq 3\delta\right\}.$$

Clearly, we may assume $3\delta < c - \sup_{u\in\partial Q} E(u)$ and that (12.18) holds in $\widetilde{Q_0} = \{u \in Q : |u| = R_0\}$ and $\widetilde{Q_1} = \{u \in Q : |u| = R_1\}$, respectively. Denote

$$\begin{aligned} Q_0 &= \left\{u \in K_{R_0R_1} : |E(u) - c| \le 2\delta\right\} \\ Q_1 &= \left\{u \in K_{R_0R_1} : |E(u) - c| \le \delta\right\} \\ Q_2 &= K_{R_0R_1} \setminus Q_0 \\ \eta(u) &= \frac{d(u,Q_2)}{d(u,Q_1) + d(u,Q_2)}. \end{aligned}$$

We have

$$\eta(u) = 1 \text{ in } \overline{Q_1},\ \eta(u) = 0 \text{ in } \overline{Q_2},\ 0 < \eta(u) < 1 \text{ otherwise.}$$

We now apply Lemma 12.10 to $G(u) = E'(u)$, $D = K_{R_0R_1}$, $D_0 = \widetilde{Q_0}$ and $D_1 = \widetilde{Q_1}$. It follows that there exists $\alpha > 0$ and a locally Lipschitz map $H : \widehat{D} \to X$ (here $\widehat{D}$ means the set $\{u \in K_{R_0R_1} : |E'(u)| \ge a\}$) such that

$$|H(u)| \le 1,\ -\alpha\left|E'(u)\right| \ge \left(E'(u), H(u)\right) \ \text{ for } u \in \widehat{D}$$

$$\begin{aligned} (u, H(u)) &> 0 \ \text{ for } u \in \widetilde{Q_0} \\ (u, H(u)) &< 0 \ \text{ for } u \in \widetilde{Q_1} \end{aligned} \tag{12.20}$$

and

$$u + H(u) \in K \ \text{ for all } \ u \in \widehat{D}. \tag{12.21}$$

Define $W : K_{R_0R_1} \to X$ by

$$W(u) = \begin{cases} \eta(u)H(u) & \text{for } u \in \widehat{D} \\ 0 & \text{for } u \in K_{R_0R_1} \setminus \widehat{D}. \end{cases}$$

This map is locally Lipschitz and can be extended to a locally Lipschitz map on the whole of K, by setting

$$W(u) = \begin{cases} W\left(\frac{R_1}{|u|}u\right), & |u| > R_1 \\ W\left(\frac{R_0}{|u|}u\right), & \frac{R_0}{2} \le |u| < R_0 \\ \frac{2}{R_0}|u|\,W\left(\frac{R_0}{|u|}u\right), & 0 < |u| < \frac{R_0}{2} \\ 0, & u = 0. \end{cases}$$

Let σ be the semiflow generated by W as shows Lemma 12.11. Note $\sigma(u,.)$ does not exit K since for each $v \in K$, there is $\lambda > 0$ with $v + \lambda W(v) \in K$, as follows from (12.21). We claim that $\sigma(u,.)$ does not exit $K_{R_0R_1}$ for $t \in \mathbf{R}_+$ if $u \in K_{R_0R_1}$. Indeed, we have

$$\frac{d}{dt}|\sigma(u,t)|^2 = 2\left(\frac{d}{dt}\sigma(u,t), \sigma(u,t)\right)$$
$$= 2(W(\sigma(u,t)), \sigma(u,t)).$$

Assume $\sigma(u,t) \in K_{R_0R_1}$ for all $t \in [0,t_0)$ and $|\sigma(u,t_0)| = R_0$ for some $t_0 \in \mathbf{R}_+$. If $\sigma(u,t_0) \in \widetilde{Q_0}$ then (12.20) guarantees that

$$(W(\sigma(u,t)), \sigma(u,t)) > 0$$

for t in a neighborhood of t_0. If $\sigma(u,t_0) \notin \widetilde{Q_0}$, then $\eta(\sigma(u,t_0)) = 0$ in a neighborhood of t_0. Hence $d|\sigma(u,t)|^2/dt \geq 0$ in a neighborhood of t_0, which means that $|\sigma(u,t)|$ is nondecreasing on some interval $[t_0, t_0+\varepsilon)$. Similarly, if $|\sigma(u,t_0)| = R_1$, then $d|\sigma(u,t)|^2/dt \leq 0$ in a neighborhood of t_0, which means that $|\sigma(u,t)|$ is nonincreasing on some interval $[t_0, t_0+\varepsilon)$. Therefore $\sigma(u,t)$ does not exit $K_{R_0R_1}$ for $t \in \mathbf{R}_+$.

Let us denote by E_λ the level set $(E \leq \lambda)$, i.e.,

$$E_\lambda = \{u \in K_{R_0R_1} : E(u) \leq \lambda\}.$$

We have

$$\frac{dE(\sigma(u,t))}{dt} = \left(E'(\sigma(u,t)), \frac{d}{dt}\sigma(u,t)\right) \tag{12.22}$$
$$= \eta(\sigma(u,t))\left(E'(\sigma(u,t)), H(\sigma(u,t))\right)$$
$$\leq -\eta(\sigma(u,t))\,\alpha a.$$

Let $t_1 > 2\delta/(\alpha a)$ and let u be any element of $E_{c+\delta}$. If there is a $t_0 \in [0,t_1]$ with $\sigma(u,t_0) \notin Q_1$, then

$$E(\sigma(u,t_1)) \leq E(\sigma(u,t_0)) < c-\delta.$$

Hence $\sigma(u,t_1) \in E_{c-\delta}$. Otherwise, $\sigma(u,t) \in Q_1$ for all $t \in [0,t_1]$, and so $\eta(\sigma(u,t)) \equiv 1$. Then (12.22) implies

$$E(\sigma(u,t_1)) \leq E(u) - \alpha a t_1 < c+\delta-2\delta = c-\delta.$$

Thus

$$\sigma(E_{c+\delta}, t_1) \subset E_{c-\delta}. \tag{12.23}$$

Now by the definition of c, there is a $h \in \Gamma$ with

$$h(u) \in E_{c+\delta} \quad \text{for all } u \in Q. \tag{12.24}$$

We define a new map $h_1 \in \Gamma$ by

$$h_1(u) = \sigma(h(u), t_1), \; u \in K_{R_0R_1}.$$

Since η vanishes in the neighborhood of every $u \in \partial Q$, we have $\sigma(u,t) \equiv u$ for $u \in \partial Q$. Hence $h_1 \in \Gamma$. On the other hand, from (12.23) and (12.24), we have

$$E(h_1(u)) \leq c - \delta \quad \text{for all } u \in Q,$$

which contradicts the definition of c. □

Proof of Theorem 12.3. Assume that there do not exist sequences satisfying (12.7) and (12.9) or (12.10). Then there are $a, \delta > 0$ such that

$$\left| E'(u) - \frac{(E'(u), u)}{|u|^2} u \right| \geq a$$

for all $u \in K_{R_0 R_1}$ satisfying $|E(u) - c| \leq \delta$, with $|u| = R_0$ and $\varepsilon(E'(u), u) \geq 0$, or $|u| = R_1$ and $\varepsilon(E'(u), u) \leq 0$.

Let $\theta > 0$ be such that

$$0 < \theta^{-2} - 1 \leq a^2 R_0^2 \nu_0^{-2}.$$

Then, also using (12.4) and (12.5), we obtain

$$\begin{aligned}
(E'(u), u)^2 (\theta^{-2} - 1) &\leq (E'(u), u)^2 a^2 R_0^2 \nu_0^{-2} \\
&\leq (E'(u), u)^2 \left| E'(u) - \frac{(E'(u), u)}{|u|^2} u \right|^2 |u|^2 \nu_0^{-2} \\
&\leq \left| E'(u) - \frac{(E'(u), u)}{|u|^2} u \right|^2 |u|^2 .
\end{aligned}$$

It follows that

$$\begin{aligned}
(E'(u), u)^2 \theta^{-2} &\leq (E'(u), u)^2 + |E'(u)|^2 |u|^2 - (E'(u), u)^2 \\
&= |E'(u)|^2 |u|^2 .
\end{aligned}$$

Hence

$$|(E'(u), u)| \leq \theta |u| |E'(u)| .$$

Thus, for $u \in K_{R_0 R_1}$ satisfying $|E(u) - c| \leq \delta$, one has

$$|(E'(u), u)| \leq \theta |u| |E'(u)| \quad \text{if } |u| = R_0 \text{ or } |u| = R_1 .$$

Now the conclusion of the first part follows from Lemma 12.12. □

Finally we remark that (12.11), (12.12) make impossible the existence of a sequence as in (12.9) and (12.10), respectively.

Proof of Theorem 12.5. First note that (12.4) and (12.5) trivially hold since the map N is bounded on $K_{R_0 R_1}$, a consequence of its compactness. It remains to prove that the (MPSS) condition holds in $K_{R_0 R_1}$, that is, any sequence (u_k) like in Theorem

12.3 has a convergent subsequence. Let $(u_k) \subset K_{R_0 R_1}$ be such a sequence. Passing if necessary to a subsequence, we may assume that $N(u_k) \to v$ for some $v \in X$. If (12.8) is satisfied, then from $E'(u_k) = u_k - N(u_k) \to 0$ we deduce that $u_k \to v$ as we wished. Assume $\varepsilon = 1$ and (12.10). Passing to another sequence we may suppose that $\frac{(E'(u_k), u_k)}{R_1^2} \to a$ for some real number $a \leq 0$. Then (12.10) implies

$$(1-a)u_k - N(u_k) \to 0 \tag{12.25}$$

and in consequence $u_k \to \frac{1}{1-a}v$. Assume $\varepsilon = 1$ and (12.9). As above we may assume that $\frac{(E'(u_k), u_k)}{R_1^2} \to a$ this time for some real number $a \geq 0$. Notice $a \neq 1$, since otherwise (12.25) would imply that $N(u_k) \to 0$ where $|u_k| = R_0$, which contradicts (12.15). Now the conclusion follows as above. The case $\varepsilon = -1$ can be discussed similarly. □

Proof of Theorem 12.6. It is sufficient to check that (a) guarantees (12.14) and (12.15) for $\varepsilon = -1$, and (b) guarantees (12.13) and (12.15) for $\varepsilon = 1$. Thus the result follows from Theorem 12.5. □

The next lemma is used in the proof of Theorem 12.7.

Lemma 12.13. *Assume all the assumptions of Theorem* 12.7 *hold. In addition assume that there are constants* $\delta > 0$ *and* $\theta \in [0,1)$ *such that for* $u \in K_{R_0 R_1}$ *satisfying* $E(u) - m \leq \delta$, *one has*

$$\left|\left(E'(u), u\right)\right| \leq \theta\, |u|\, \left|E'(u)\right| \quad \text{if } |u| = R_0 \text{ or } |u| = R_1. \tag{12.26}$$

Then there exists a sequence of elements $u_k \in K_{R_0 R_1}$ *with*

$$E(u_k) \to m \quad \text{and} \quad E'(u_k) \to 0 \ \text{ as } k \to \infty.$$

Proof. We follow the proof of Lemma 12.12 with the only difference that one has m instead of c. Thus we obtain $\sigma(u,t)$ which does not exit $K_{R_0 R_1}$ for $t \geq 0$. We fix any $u \in Q_1 = \{v \in K_{R_0 R_1} : E(v) < m + \delta\}$ and take $t_1 > 2\delta/(\alpha a)$. Then (12.22) guarantees that $\sigma(u,t) \in Q_1$ for all $t \geq 0$. Then

$$E(\sigma(u,t_1)) \leq m + \delta - \alpha a t_1 < m - \delta,$$

contradicting the definition of m. □

Proof of Theorem 12.7. Assume there are no sequences satisfying (12.17) and (12.9) or (12.10). Then, as in the proof of Theorem 12.3, there are $\delta, \theta > 0$ such that for $u \in K_{R_0 R_1}$ satisfying $E(u) \leq m + \delta$, one has (12.26). Now the conclusion of the first part follows from Lemma 12.13.

□

References

1. A. Ambrosetti and P.H. Rabinowitz, *Dual variational methods in critical point theory and applications,* J. Funct. Anal. **14** (1973), 349-381.
2. V. Benci, *Some critical point theorems and applications*, Comm. Pure Appl. Math. **33** (1980), 147-172.
3. K. Deimling, *Ordinary Differential Equations in Banach Spaces,* Springer, Berlin, 1977.
4. K. Deimling, *Nonlinear Functional Analysis,* Springer, Berlin, 1985.
5. M. Frigon, *On a new notion of linking and application to elliptic problems at resonance,* J. Differential Equations **153** (1999), 96-120.
6. A. Granas and J. Dugundji, *Fixed Point Theory,* Springer, New York, 2003.
7. G. Isac, *Leray-Schauder Type Alternatives, Complementarity Problems and Variational Inequalities*, Kluwer, Dordrecht, 2006.
8. M.A. Krasnoselskii, *Positive Solutions of Operator Equations,* Noordhoff, Groningen, 1964.
9. D. O'Regan and R. Precup, *Theorems of Leray-Schauder Type and Applications,* Gordon and Breach, Amsterdam, 2001.
10. R. Precup, *Methods in Nonlinear Integral Equations,* Kluwer, Dordrecht, 2002.
11. R. Precup, *A compression type mountain pass theorem in conical shells*, J. Math. Anal. Appl. **338** (2008), 1116-1130.
12. R. Precup, *The Leray-Schauder boundary condition in critical point theory,* Nonlinear Anal., **71** (2009), 3218–3228.
13. M. Schechter, *Linking Methods in Critical Point Theory,* Birkhäuser, Basel, 1999.
14. M. Struwe, *Variational Methods,* Springer, Berlin, 1990.

Chapter 13
Gronwall Lemma Approach to the Hyers–Ulam–Rassias Stability of an Integral Equation

Ioan A. Rus

Dedicated to the memory of Professor George Isac

Abstract The aim of this paper is to give some Hyers–Ulam–Rassias stability results for Volterra and Fredholm integral equations. To do these, we shall use some Gronwall lemmas.

13.1 Introduction

In [10], V. Radu presents a simple and nice proof for the Hyers–Ulam stability of the Cauchy additive functional equation. S.-M. Jung, in [6], adopts the idea of V. Radu to prove the Hyers–Ulam–Rassias stability of some Volterra integral equations. The Theorem 2.1 in [6] is the following:

Theorem 13.1. *Let* $f \in C([a,b]\times\mathbb{C})$, $\varphi \in C([a,b],(0,\infty))$ *be two given functions and* $t_0 \in [a,b]$. *We suppose that:*
(i) there exists $L_f > 0$ *such that*

$$|f(t,u)-f(t,v)| \le L_f|u-v|,\ \forall\, t\in[a,b],\ \forall\, u,v\in\mathbb{C};$$

(ii) there exists $c_\varphi > 0$ *such that*

$$\left|\int_{t_0}^{t}\varphi(s)ds\right| \le c_\varphi\varphi(t),\ \forall\, t\in[a,b];$$

(iii) $0 < L_f c_\varphi < 1$.

Ioan A. Rus
Babeş-Bolyai University, Department of Applied Mathematics, Kogălniceanu Nr. 1, 400084 Cluj-Napoca, Romania, e-mail: iarus@math.ubbcluj.ro

P.M. Pardalos et al. (eds.), *Nonlinear Analysis and Variational Problems*, Springer Optimization and Its Applications 35, DOI 10.1007/978-1-4419-0158-3_13,

In the above conditions, if $x \in C([a,b],\mathbb{C})$ satisfies

$$\left| x(t) - \int_{t_0}^{t} f(s,x(s))ds \right| \leq \varphi(t),\ \forall\, t \in [a,b],$$

then there exists a unique $x^* \in C([a,b],\mathbb{C})$ such that:

(a) $x^*(t) = \int_{t_0}^{t} f(s,x^*(s))ds,\ t \in [a,b];$

and

(b) $|x(t) - x^*(t)| \leq (1 - L_f c_\varphi)^{-1}\varphi(t),\ t \in [a,b].$

Remark 13.2. In the conditions of the Theorem 13.1, the equation (a) has in $C([a,b],\mathbb{C})$ a unique solution (see H. Amann [1], C. Corduneanu [3], I.A. Rus [11], ...). So, this unique solution satisfies the inequality (b). In V. Radu's paper, the Cauchy additive functional equation has an infinity set of solutions.

The aim of this paper is to give a simple proof of the Theorem 13.1, in a more general setting, and to present a similar result for a Fredholm integral equation. To do these, we shall use some Gronwall lemmas.

13.2 Gronwall Lemmas

In this paper, we need the following results (see H. Amann [1], pp. 89–90, or C. Corduneanu [3], pp. 15–17, or P. Ver Eecke [14] pp. 224–226; see also I.A. Rus [12], [13] and [11] and the references therein).

Lemma 13.3. *Let J be an interval in $\mathbb{R}$, $t_0 \in J$ and $h,k,u \in C(J,\mathbb{R}_+)$. If*

$$u(t) \leq h(t) + \left| \int_{t_0}^{t} k(s)u(s)ds \right|,\ \forall\, t \in J,$$

then

$$u(t) \leq h(t) + \left| \int_{t_0}^{t} h(s)k(s)e^{\left|\int_s^t k(\sigma)d\sigma\right|}ds \right|,\ \forall\, t \in J.$$

Lemma 13.4. *Let $h \in C([a,b],\mathbb{R}_+)$ and $\beta > 0$ with $\beta(b-a) < 1$. If $u \in C([a,b],\mathbb{R}_+)$ satisfies*

$$u(t) \leq h(t) + \beta \int_a^b u(s)ds, \quad t \in [a,b],$$

then

$$u(t) \leq h(t) + \beta(1 - \beta(b-a))^{-1} \int_a^b h(s)ds,\ \forall\, t \in [a,b].$$

13.3 Stability of a Fixed Point Equation

Let J be an interval in $\mathbb{R}$, $(\mathbb{B}, |\cdot|)$ a (real or complex) Banach space, $A : C(J,\mathbb{B}) \to C(J,\mathbb{B})$ an operator and $\varphi \in C(J,\mathbb{R}_+)$.

Definition 13.5. The fixed point equation

$$x = A(x) \tag{13.1}$$

has the Hyers–Ulam–Rassias stability with respect to φ if there exists $c_\varphi > 0$ such that for each solution $x \in C(J,\mathbb{B})$ of

$$|x(t) - A(x)(t)| \leq \varphi(t), \ \forall\, t \in J$$

there exists a unique solution x^* of (13.1) such that

$$|x(t) - x^*(t)| \leq c_\varphi \varphi(t), \ \forall\, t \in J.$$

Remark 13.6. For the notion of Hyers–Ulam–Rassias stability for the functional equations see D.H. Hyers, G. Isac and Th.M. Rassias [5], V. Radu [10], L. Cădariu and V. Radu [2].

13.4 Stability of Volterra Integral Equations

Let $K : [a,b] \times [a,b] \times \mathbb{B} \to \mathbb{B}$, $g : [a,b] \to \mathbb{B}$ and $t_0 \in [a,b]$ be given. We consider the following integral equation

$$x(t) = g(t) + \int_{t_0}^{t} K(t,s,x(s))ds, \quad t \in [a,b]. \tag{13.2}$$

We have

Theorem 13.7. *We suppose that:*
(i) $K \in C([a,b] \times [a,b] \times \mathbb{B}, \mathbb{B})$, $g \in C([a,b],\mathbb{B})$, $\varphi \in C([a,b],\mathbb{R}_+)$ *and* $t_0 \in [a,b]$*;*
(ii) there exists $L_K > 0$ *such that*

$$|K(t,s,u) - K(t,s,v)| \leq L_K |u - v|$$

for all $t,s \in [a,b]$ *and* $u,v \in \mathbb{B}$*;*
(iii) there exists $\eta_\varphi > 0$ *such that*

$$\left| \int_{t_0}^{t} \varphi(s)ds \right| \leq \eta_\varphi \varphi(t), \ \forall\, t \in [a,b].$$

Then:
(a) The equation (13.2) has in $C([a,b],\mathbb{B})$ *a unique solution,* x^**;*

(b) If $x \in C([a,b],\mathbb{B})$ is such that

$$\left| x(t) - g(t) - \int_{t_0}^{t} K(t,s,x(s))ds \right| \leq \varphi(t)$$

for all $t \in [a,b]$, then

$$|x(t) - x^*(t)| \leq (1 + L_K \eta_\varphi e^{L_K(b-a)})\varphi(t), \ \forall\, t \in [a,b],$$

i.e., the equation (13.2) has the Hyers–Ulam–Rassias stability.

Proof. (a) It is a well-known result (see for example [3], [4], [9], [11]).

(b) We have

$$\begin{aligned} |x(t) - x^*(t)| &\leq \left| x(t) - g(t) - \int_{t_0}^{t} K(t,s,x(s))ds \right| \\ &\quad + \left| \int_{t_0}^{t} [K(t,s,x(s)) - K(t,s,x^*(s))]ds \right| \\ &\leq \varphi(t) + L_K \left| \int_{t_0}^{t} |x(s) - x^*(s)|ds \right|. \end{aligned}$$

From Lemma 13.3 it follows that

$$\begin{aligned} |x(t) - x^*(t)| &\leq \varphi(t) + L_K e^{L_K(b-a)} \left| \int_{t_0}^{t} \varphi(s)ds \right| \\ &\leq [1 + \eta_\varphi L_K e^{L_K(b-a)}]\varphi(t), \ t \in [a,b]. \end{aligned}$$

□

Remark 13.8. In the case $\mathbb{B} := c(\mathbb{R})$ and for $x \in c(\mathbb{R})$, $x = (x_0, x_1, \ldots, x_n, \ldots)$, $|x| := \sup_{k \in \mathbb{N}} |x_k|$, we have a stability result for an infinite system of Volterra integral equations.

Remark 13.9. In the case $\mathbb{B} := \mathbb{R}^m$, $|x| := \sup_{0 \leq k \leq m} |x_k|$, we have a stability result for a system of Volterra integral equations.

Remark 13.10. In the case $\mathbb{B} := \mathbb{C}$, we have a new proof and a generalization of Theorem 2.1 in [6].

13.5 Stability of Fredholm Integral Equations

For the following Fredholm integral equation

$$x(t) = g(t) + \int_{a}^{b} K(t,s,x(s))ds, \quad t \in [a,b] \tag{13.3}$$

we have

Theorem 13.11. *We suppose that:*

(i) $K \in C([a,b] \times [a,b] \times \mathbb{B}, \mathbb{B})$, $g \in C([a,b], \mathbb{B})$ *and* $\varphi \in C([a,b], \mathbb{B}_+)$;

(ii) there exists $L_K > 0$ *such that*

$$|K(t,s,u) - K(t,s,v)| \le L_K |u - v|$$

for all $t, s \in [a,b]$ *and* $u, v \in \mathbb{B}$;

(iii) $L_K(b-a) < 1$;

(iv) $\varphi(t) > 0$, $\forall\, t \in [a,b]$.

Then:

(a) The equation (13.3) has in $C([a,b], \mathbb{B})$ *a unique solution,* x^*;

(b) If $x \in C([a,b], \mathbb{B})$ *is such that*

$$\left| x(t) - g(t) - \int_a^b K(t,s,x(s))ds \right| \le \varphi(t)$$

for all $t \in [a,b]$, *then there exists* $c_\varphi > 0$ *such that*

$$|x(t) - x^*(t)| \le c_\varphi \varphi(t), \ \forall\, t \in [a,b],$$

i.e., the equation (13.3) has the Hyers–Ulam–Rassias stability.

Proof. (a) It is a well-known result (see for example [4] and [11]).

(b) We have

$$\begin{aligned} |x(t) - x^*(t)| &\le \left| x(t) - g(t) - \int_a^b K(t,s,x(s))ds \right| \\ &\quad + \left| \int_a^b [K(t,s,x(s)) - K(t,s,x^*(s))]ds \right| \\ &\le \varphi(t) + L_K \int_a^b |x(s) - x^*(s)|ds. \end{aligned}$$

From Lemma 13.4 it follows that

$$\begin{aligned} |x(t) - x^*(t)| &\le \varphi(t) + L_K(1 - L_K(b-a))^{-1} \int_a^b \varphi(s)ds \\ &\le \left[1 + L_K(1 - L_K(b-a))^{-1} \cdot \mathscr{F}(M_\varphi(b-a), m_\varphi)\right] \varphi(t), \end{aligned}$$

where m_φ and M_φ are such that

$$0 < m_\varphi \le \varphi(t) \le M_\varphi, \ \forall\, t \in [a,b].$$

□

Remark 13.12. For some particular cases, see Remark 13.8 and Remark 13.9.

References

1. H. Amann, *Ordinary Differential Equations*, Walter de Gruyter, Berlin, 1990.
2. L. Cădariu and V. Radu, *Remarks on the stability of monomial functional equations*, Fixed Point Theory, **8**(2007), No. 2, 201-218.
3. C. Corduneanu, *Principles of Differential and Integral Equations*, Chelsea Publishing Company, New York, 1971.
4. D. Guo, V. Lakshmikantham and X. Liu, *Nonlinear Integral Equations in Abstracts Spaces*, Kluwer, Dordrecht, 1996.
5. D.H. Hyers, G. Isac and Th. M. Rassias, *Stability of Functional Equations in Several Variables*, Birkhäuser, Basel, 1998.
6. S.-M. Jung, *A fixed point approach to the stability of a Volterra integral equation*, Fixed Point Theory and Applications, Vol. 2007, 9 pages.
7. V. Lakshmikantham and S. Leela, *Differential and Integral Inequalities*, Academic Press, New York, 1969.
8. D.S. Mitrinović, J.E. Pečrić and A.M. Fink, *Inequalities Involving Functions and their Integrals and Derivatives*, Kluwer, Dordrecht, 1991.
9. R. Precup, *Methods in Nonlinear Integral Equations*, Kluwer, Dordrecht, 2002.
10. V. Radu, *The fixed point alternative and the stability of functional equations*, Fixed Point Theory, **4**(2003), no. 1, 91-96.
11. I.A. Rus, *Ecuaţii diferenţiale, Ecuaţii integrale şi Sisteme dinamice*, Transilvania Press, Cluj-Napoca, 1996.
12. I.A. Rus, *Picard operators and applications*, Scientiae Mathematicae Japonicae, **58** (2003), No. 1, 191-219.
13. I.A. Rus, *Fixed points, upper and lower fixed points: abstract Gronwall lemmas*, Carpathian J. Math., **20**(2004), No. 1, 125-134.
14. P. Ver Eecke, *Applications du calcul différentiel*, Presses Universitaires de France, Paris, 1985.

Chapter 14
Brezis–Browder Principles and Applications

Mihai Turinici

Dedicated to the memory of Professor George Isac

Abstract In Part 1, we show that the version of Brezis–Browder's principle [Adv. Math., 21(1976), 355–364] for general separable sets is a logical equivalent of the Zorn–Bourbaki maximality result. Further, in Part 2, it is stressed that most (sequential) maximality principles are logical equivalents of that of Brezis–Browder; and the remaining ones, including Kang–Park's [Nonlinear Analysis, 14 (1990), 159–165] may be viewed as particular cases of an "asymptotic" type version of Brezis–Browder's principle. Finally, in Part 3, the variational statement due to Kada, Suzuki and Takahashi [Math. Japonica, 44 (1996), 381–391] is deduced from a "relative" type version of Brezis–Browder's principle. The connections with some variational statements in Suzuki [J. Math. Anal. Appl., 253 (2001), 440–458], Lin and Du [J. Math. Anal. Appl., 323 (2006), 360–370] or Al-Homidan, Ansari and Yao [Nonlin. Anal., 69 (2008), 126–139] are also investigated.

14.1 Brezis–Browder Principles in General Separable Sets

14.1.1 Introduction

Let M be some nonempty set; and $(\leq)$, some *quasi-order* (i.e., reflexive and transitive relation) over it. Further, let $x \mapsto \varphi(x)$ stand for a function from M to $R_+ := [0, \infty[$. Call the point $z \in M$, $(\leq, \varphi)$*-maximal* when

(a11) $w \in M$ and $z \leq w$ imply $\varphi(z) = \varphi(w)$.

Mihai Turinici
"A. Myller" Mathematical Seminar; "A. I. Cuza" University, 11, Copou Boulevard, 700506 Iaşi, Romania, e-mail: mturi@uaic.ro

P.M. Pardalos et al. (eds.), *Nonlinear Analysis and Variational Problems*, Springer Optimization and Its Applications 35, DOI 10.1007/978-1-4419-0158-3_14,

A basic result involving such points is the 1976 Brezis–Browder ordering principle [12] (in short: BB principle).

Theorem 14.1. *Suppose that*

(11a) $(M, \leq)$ *is sequentially inductive:*
each ascending sequence in M has an upper bound
(11b) φ *is* $(\leq)$*-decreasing* $(x \leq y \Longrightarrow \varphi(x) \geq \varphi(y))$.

Then, for each $u \in M$ *there exists a* $(\leq, \varphi)$*-maximal* $v \in M$ *with* $u \leq v$.

This principle, including the well known Ekeland's [19] [20], found some useful applications to convex and nonconvex analysis; we refer to the quoted papers for a survey of these. As a consequence, many extensions of Theorem 14.1 were proposed. Here, we shall concentrate on the *structural* way of extension. This, roughly speaking, consists of $(R_+, \geq)$ being substituted by an ordered structure $(P, \leq)$ endowed with (cardinal-) countable regularity properties for its chains. Some pioneering results in the area were obtained in the standard countable case by Gajek and Zagrodny [23]; further refinements of these may be found in the 2007 paper by Zhu and Li [85]. It is our aim in the following to show that such techniques are applicable beyond the (standard) countable framework; details will be given in Subsection 14.1.4 (the transitive case) and Subsection 14.1.5 (the amorphous case). The specific tool for deducing these is a variant of the Zorn–Bourbaki maximality principle for general separable sets, given in Subsection 14.1.3. All preliminary material involving such objects is presented in Subsection 14.1.2. Some possible applications of these facts (in the general case) are obtainable under the lines in the quoted papers; see also Zhu, Fan and Zhang [84].

14.1.2 General Separable Sets

(A) Let W stand for the class of ordinal numbers, introduced in a "factorial" way; cf. Kuratowski and Mostowski [42, Ch 7, Sect 2]. Precisely, call the partially ordered structure $(P, \leq)$, *well ordered* if each (nonempty) part of P admits a first element. Given the couple $(P, \leq)$, $(Q, \leq)$ of such objects, put

$$(P, \leq) \equiv (Q, \leq) \text{ iff there exists a strictly increasing bijection: } P \to Q.$$

This is an equivalence relation; the order type of $(P, \leq)$ (denoted $\mathrm{ord}(P, \leq)$) is just its equivalence class; also referred to as an *ordinal*.

Note that W is not a set, as results from the Burali–Forti paradox; cf. Sierpinski [55, Ch 14, Sect 2]. However, when one restricts to a *Grothendieck universe* $\mathscr{G}$ (taken as in Hasse and Michler [28, Ch 1, Sect 2]) this contradictory character is removed for the class $W(\mathscr{G})$ of all *admissible* (modulo $\mathscr{G}$) ordinals (generated by (non-contradictory) well ordered parts of $\mathscr{G}$). In the following, we drop any reference to $\mathscr{G}$, for simplicity. So, by an *ordinal* in W one actually means a $\mathscr{G}$-admissible ordinal with respect to a "sufficiently large" Grothendieck universe $\mathscr{G}$. Clearly,

$$\xi = \text{admissible ordinal and } \eta \le \xi \text{ imply } \eta = \text{admissible ordinal.}$$

Hence, in the formulae

$$W(\alpha) = \{\xi \in W; \xi < \alpha\}, \quad W[\alpha,\beta] = \{\xi \in W; \alpha \le \xi \le \beta\},$$

the symbol W in the brackets is the "absolute" class of all ordinals.

Now, an enumeration of W is realized via the immediate successor map of a subset $M \subseteq W$

$$\mathrm{suc}(M) = \min\{\xi \in W; M < \xi\} \ \text{(hence } \mathrm{suc}(\alpha) = \alpha + 1, \ \forall \alpha \in W).$$

(Here, $M < \xi$ means: $\lambda < \xi, \forall \lambda \in M$). It begins with (the set of) all natural numbers $N := \{0,1,\ldots\}$. Their immediate successor is $\omega = \mathrm{suc}(N)$ (the first transfinite ordinal); the next in this enumeration is $\omega + 1$, and so on. Put $W_0 = W \setminus \{0\} (= \{\xi \in W; \xi > 0\})$. This set is composed of two disjoint classes of ordinals. The former of these, W_0^1, collects all *first kind* ordinals $\xi > 0$ [in the sense: $W(\xi)$ admits a last element $\max[W(\xi)] = \xi - 1$]. And the latter of these, W_0^2, collects all *second kind* ordinals $\xi > 0$ [in the sense: $W(\xi)$ does not admit a last element; or, equivalently: $\lambda < \xi \Longrightarrow \lambda + 1 < \xi$]; this is also referred to as $\xi > 0$ being a *limit* ordinal.

The basic operations with ordinals may be introduced in a synthetic way as follows. Let α, β be two ordinals; and $(A, \le)$, $(B, \le)$, well-ordered structures with $\mathrm{ord}(A, \le) = \alpha, \mathrm{ord}(B, \le) = \beta$, $A \cap B = \emptyset$. Then **a)** $\alpha + \beta = \mathrm{ord}(A \cup B, \le)$, where the associated order is given by the concatenation procedure: $x < y$ iff either $[x \in A, y \in B]$ or $[(x,y \in A, x < y), (x,y \in B, x < y)]$; **b)** $\alpha \cdot \beta = \mathrm{ord}(A \times B, \le)$, where the associated order is defined by the lexicographic procedure: $(x,y) < (u,v)$ iff either $[y < v]$ or $[x < u, y = v]$. The basic properties of these may be found, e.g., in Kuratowski and Mostowski [42, Ch 7, Sect 5].

In parallel to this, we may (construct and) enumerate the class of all admissible cardinals. Let P and Q be nonempty sets; we put

$$P \preceq Q (P \sim Q) \text{ iff there exists an injection (bijection): } P \to Q.$$

The former is a quasi-order (denoted $\mathrm{card}(P) \le \mathrm{card}(Q)$); whereas the latter is an equivalence (written as $\mathrm{card}(P) = \mathrm{card}(Q)$). Denote also

$$P \prec Q \text{ if and only if } P \preceq Q \text{ and } \neg(P \sim Q).$$

This relation is *irreflexive* ($\neg(P \prec P)$, for each P) and *transitive*; hence a *strict order* (indicated as: $\mathrm{card}(P) < \mathrm{card}(Q)$). Let $\alpha > 0$ be an (admissible) ordinal; we say that it is an (admissible) *cardinal* if $W(\xi) \prec W(\alpha)$ [i.e., $\mathrm{card}(W(\xi)) < \mathrm{card}(W(\alpha))$], for each $\xi < \alpha$. The class of all these will be denoted by Z. Now, the enumeration we are looking for is realized via the immediate successor (in Z) map

$$\mathrm{SUC}(M) = \min\{\eta \in Z; M < \eta\}, \quad M \subseteq Z.$$

Precisely, this begins with the natural numbers $N := \{0,1,...\}$. The immediate successor (in Z) of all these is $\omega = \text{SUC}(N)$ (the first transfinite cardinal). To describe the remaining ones, we may introduce via transfinite recursion the function $\lambda \mapsto \aleph_\lambda$ from W to Z as

$$\begin{aligned} &\aleph_0 = \omega; \quad \text{and, for each } \lambda > 0, \\ &\aleph_\lambda = \text{SUC}(\aleph_{\lambda-1}), \quad \text{if } \lambda - 1 \text{ exists} \\ &\aleph_\lambda = \text{SUC}\{\aleph_\xi; \xi < \lambda\}, \quad \text{if } \lambda - 1 \text{ does not exist.} \end{aligned}$$

Note that, in such a case, the order structure of $Z(\omega, \leq) = \{\xi \in Z; \omega \leq \xi\}$ is completely reducible to the one of W; further details may be found in Just and Weese [35, Ch 11, Sect 11.2].

Any nonempty part P with $\text{card}(P) < \omega$ ($\text{card}(P) = \omega$) is termed *finite* (*effectively countable*); the union of these ($\text{card}(P) \leq \omega$) is referred to as P is *countable*. When $P = W(\xi)$, all such properties will be transferred to ξ. Generally, take some ordinal γ; and put $\Gamma = \aleph_\gamma$. Any nonempty part P with $\text{card}(P) < \Gamma$ ($\text{card}(P) = \Gamma$) is termed *Γ-finite* (*effectively Γ-countable*); the union of these ($\text{card}(P) \leq \Gamma$) is referred to as P is *Γ-countable*. As before, when $P = W(\xi)$, all such properties will be transferred to ξ.

Denote by Δ the immediate successor (in Z) of Γ [$\Delta = \text{SUC}(\Gamma)$; hence $\Delta = \aleph_{\gamma+1}$, if $\Gamma = \aleph_\gamma$]. By the very definition above, one has

$$\xi \text{ is } \Gamma\text{-countable, for each } \xi < \Delta; \quad \text{but } \Delta \text{ is not } \Gamma\text{-countable.} \tag{14.1}$$

A basic consequence of this is precise in the statement below (to be found, e.g., in Kuratowski and Mostowski [42, Ch 3, Sect 4]). Let M be a nonempty set; and λ, μ be ordinals with $\lambda \leq \mu$. Any map $\xi \mapsto a_\xi$ from $W(\lambda)$ to M will be referred to as a *λ-net* in M; or, simply, *net* when λ is understood. If $\lambda \leq \mu$, we say that the λ-net $(a_\xi; \xi < \lambda)$ is *μ-subordinated* (in short: *μ-sub*); and if $\lambda < \mu$, the same net will be referred to as *μ-strongly subordinated* (in short: *μ-strongsub*).

Proposition 14.2. *The following are valid:*

i) *The ordinal Δ cannot be attained via Γ-sub net limits of Γ-countable ordinals. In other words: if $(\alpha_\xi; \xi < \theta)$ is an ascending θ-net (where $\theta \leq \Gamma$) of Γ-countable ordinals then*

(a12) $\quad \alpha = \sup_\xi(\alpha_\xi)(= \lim_\xi(\alpha_\xi))$

is Γ-countable too.

ii) *Each second kind Γ-countable ordinal is attainable via such nets. In other words: if (the Γ-countable) $\alpha < \Delta$ is of second kind, there must be a strictly ascending Γ-sub net $(\alpha_\xi; \xi < \theta)$ (where $\theta \leq \Gamma$) of Γ-countable ordinals with the property (a12).*

(B) Let M be a nonempty set; and $(\leq)$, some *order* (=antisymmetric quasi-order) on it. By a *$(\leq)$-chain* of M we shall mean any (nonempty) part A of M with $(A, \leq)$ well ordered (see above). Note that any such object may be written as $A = \{a_\xi; \xi < \lambda\}$, where the net $\xi \mapsto a_\xi$ is strictly ascending ($\xi < \eta \Longrightarrow a_\xi < a_\eta$); the uniquely determined ordinal λ is just $\text{ord}(A,\leq)$. Let μ be another ordinal. If $\lambda \leq \mu$, we say that the $(\leq)$-chain $(A,\leq)$ is μ-*subordinated* (in short: μ-*sub*); and if $\lambda < \mu$, the same chain $(A,\leq)$ will be referred to as μ-*strongly-subordinated* (in short: μ-*strongsub*). A basic particular case of these conventions is the following. Let $\gamma \geq 0$ be arbitrary fixed; and put $\Gamma = \aleph_\gamma$, $\Delta = \text{SUC}(\Gamma)$ (hence $\Delta = \aleph_{\gamma+1}$). Note that, by the very definition above (and (14.1)) A is Γ-countable iff $(A,\leq)$ is Δ-strongsub. This in particular happens when $(A,\leq)$ is Γ-sub. The following characterization of the last concept is useful for us; cf. Sierpinski [55, Ch 13, Sect 5].

Proposition 14.3. *The $(\leq)$-chain A is Γ-sub if and only if*

$$\begin{aligned} &A = \{b(\xi); \xi < \theta\}, \text{ where } \theta \in W[\omega,\Gamma] \text{ and} \\ &\xi \mapsto b(\xi) \text{ is ascending } (\xi < \eta \Longrightarrow b(\xi) \leq b(\eta)). \end{aligned} \tag{14.2}$$

Let P,Q be nonempty parts with $P \supseteq Q$. We say that P is *majorized* by Q (and write $P \propto Q$) provided Q is cofinal in P ($\forall x \in P, \exists y \in Q$ with $x \leq y$). The $(\leq)$-chain $S \subseteq M$ is called *upper Γ-countable* in case:

(b12) $S \propto T$, for some Γ-sub $(\leq)$-chain $T \subseteq S$.

Clearly, this happens if S is Γ-sub. As a completion, we have

Proposition 14.4. *The generic relation holds*

$$(\forall (\leq)\text{-chain}) \ \Gamma\text{-countable} \Longrightarrow \text{upper } \Gamma\text{-countable}. \tag{14.3}$$

Hence, the $(\leq)$-chain $S \subseteq M$ is upper Γ-countable if and only if

$$S \propto T, \text{ for some } \Gamma\text{-countable } (\leq)\text{-chain } T \subseteq S. \tag{14.4}$$

The verification is immediate, via Proposition 14.2; so, we do not give details.

Remark 14.5. The reciprocal of (14.3) is not in general true; just take any $(\leq)$-chain S of M with $\Delta \leq \text{ord}(S,\leq)=$ first kind ordinal.

(C) Let us now return to our initial setting. We say that the order structure $(M,\leq)$ is *Γ-separable* if

(c12) any $(\leq)$-chain of M is upper Γ-countable.

For example, this holds (under (14.3)) whenever

(d12) $(M,\leq)$ is *strongly Γ-separable*: any $(\leq)$-chain of M is Γ-countable.

In fact, the reciprocal holds too; so that, we may formulate

Proposition 14.6. *Under these conventions,*

$$(\forall \text{ ordered structure}) \ \Gamma\text{-separable} \Longleftrightarrow \text{strongly } \Gamma\text{-separable}. \tag{14.5}$$

Proof. Assume that $(M, \leq)$ is Γ-separable; and let $S = \{s(\xi); \xi < \lambda\}$ be some $(\leq)$-chain of M; where $\lambda := \text{ord}(S, \leq)$. If, by absurd, S is not Γ-countable, we must have $\lambda \geq \Delta$. The initial segment (of S) $U = \{s(\xi); \xi < \Delta\}$ is not Γ-countable too; cf. (14.1). On the other hand, by hypothesis, U is upper Γ-countable; so, there exists a strictly ascending Γ-sub net $(\xi_\nu; \nu < \theta)$ (where $\theta \leq \Gamma$) of ranks in $W(\Delta)$ with $U \propto \{s(\xi_\nu); \nu < \theta\}$; hence $\Delta = \lim_\nu(\xi_\nu)$. This, however, cannot be accepted, in view of Proposition 14.2. Hence, S is Γ-countable; and the proof is complete. □

Remark 14.7. From this result it follows that the notion of Γ-separable structure is a transfinite extension of the concept of ω-separable structure; which has been introduced by Zhu, Fan and Zhang [84].

Now, call the cardinal Γ, *separable-admissible* (in short: sep-admissible) for $(M, \leq)$ whenever $(M, \leq)$ is Γ-separable [or, equivalently: strongly Γ-separable]; The class of all these, denoted $\text{Sep}(M, \leq)$, is nonempty; because $\text{card}(M)$ is an element of it. In addition, it is hereditary after the cardinal magnitude; i.e., $\Gamma \in \text{Sep}(M, \leq)$ and $\Gamma \leq \Delta$ imply $\Delta \in \text{Sep}(M, \leq)$. The minimal element of this set, $\text{sep}(M, \leq) = \min \text{Sep}(M, \leq)$ is therefore well defined as a cardinal number; it will be referred to as the *separability cardinal* of $(M, \leq)$. By the remark above, we have $\text{sep}(M, \leq) \leq \text{card}(M)$; but, the converse relation may be false, in general. However, in many practical situations, $\text{card}(M)$ is a good "approximation" for $\text{sep}(M, \leq)$.

(D) In the following, we shall give some useful examples of such structures.

I) Let $\mathscr{I}(M) := \{(x,x); x \in M\}$ stand for the *identical relation* over M. By an *almost uniformity* (on M) we shall mean any family $\mathscr{U}$ of parts in $M \times M$ with $\mathscr{I}(M) \subseteq \cap\mathscr{U}$. Suppose that we fixed such an object. The basic condition we need further may be written as

(12a) $\mathscr{U}$ is Γ*-pseudometrizable*: there exists a Γ-countable subfamily $\mathscr{V} \subseteq \mathscr{U}$, cofinal in $(\mathscr{U}, \supseteq)$ $[\forall U \in \mathscr{U}, \exists V \in \mathscr{V}: U \supseteq V]$

(12b) $\mathscr{U}$ is sufficient: $\cap\mathscr{U} = \mathscr{I}(M)$.

For the next one, we need some preliminaries. Call the (ascending) net $(a_\xi; \xi < \lambda)$, $\mathscr{U}$*-Cauchy*, when: $\forall U \in \mathscr{U}, \exists \mu = \mu(U)$, such that $\mu \leq \xi \leq \eta \Longrightarrow (a_\xi, a_\eta) \in U$. Likewise, call the (ascending) sequence $(b_n; n < \omega)$, $\mathscr{U}$*-asymptotic*, in case: $\forall U \in \mathscr{U}, \exists k = k(U)$, such that $n \geq k \Longrightarrow (b_n, b_{n+1}) \in U$. The following auxiliary fact is useful for us.

Lemma 14.8. *The global conditions below are equivalent to each other*

(12c) each ascending net is $\mathscr{U}$*-Cauchy*

(12d) each ascending sequence is $\mathscr{U}$*-asymptotic.*

By definition, either of the introduced properties will be referred to as: $\mathscr{U}$ is (strongly) *regular*.

Proposition 14.9. *Assume that there exists an almost uniformity* $\mathscr{U}$ *over M which is* Γ*-pseudometrizable, sufficient and (strongly) regular. Then,* $(M, \leq)$ *is (strongly)* Γ*-separable.*

Proof. Without loss of generality, one may assume that $\mathscr{U}$ itself is Γ-countable; i.e., it may be written as $\mathscr{U} = (U_\xi; \xi < \theta)$, where $\theta < \Delta$. (Otherwise, we simply replace $\mathscr{U}$ by $\mathscr{V}$). The case $\theta < \omega$ is clear; so, it remains to discuss the alternative $\theta \geq \omega$. Let S be some $(\leq)$-chain in M. If there exists a last element $s = \max(S)$, we are done; so, without restriction, one may assume that

(12e) for each $x \in S$ there exists $y \in S$ with $x < y$.

This, and the (strong) regularity of $\mathscr{U}$, yields (cf. Turinici [67])

$$\begin{array}{l}\forall x \in S, \forall U \in \mathscr{U}, \text{ there exists } y = y(x,U) \in S(x,<) \\ \text{such that: } p,q \in S, y \leq p \leq q \Longrightarrow (p,q) \in U.\end{array} \tag{14.6}$$

Fix $a \in S$. By (12e) and (14.6), there exists (in S) $a_0 > a$ with $[p,q \in S, a_0 \leq p \leq q \Longrightarrow (p,q) \in U_0]$. Further, again by this relation, there exists (in S) $a_1 > a_0$ with $[p,q \in S, a_1 \leq p \leq q \Longrightarrow (p,q) \in U_1]$; and so on. Generally, assume that, for the ordinal $\mu < \theta$ we constructed a net $(a_\xi; \xi < \mu)$ in S with the property that, for each $\lambda < \mu$,

(12f) $\xi < \lambda$ implies $a_\xi < a_\lambda$
(12g) $p,q \in S,\ a_\lambda \leq p \leq q \Longrightarrow (p,q) \in U_\lambda$.

Two possibilities may occur.

j) μ is a first kind ordinal: $\lambda = \mu - 1$ exists. Again by (12e) and (14.6), there exists (in S) $t > a_\lambda$ with $[p,q \in S, t \leq p \leq q \Longrightarrow (p,q) \in U_\mu]$. It will suffice taking $a_\mu = t$ so that (12f)+(12g) are fulfilled with (with μ in place of λ).

jj) μ is a second kind ordinal: $\mu - 1$ does not exist. By the choice of θ, the $(\leq)$-chain (in S) $T = \{a_\xi; \xi < \mu\}$ is Γ-countable. If T is cofinal in S, we are done; because (cf. Proposition 14.4) S is upper Γ-countable. Otherwise,

(12h) $a_\xi < s$, for all $\xi < \mu$ and some $s \in S$.

Again by (12e) and (14.6), there exists (in S) $t > s$ with $[p,q \in S, t \leq p \leq q \Longrightarrow (p,q) \in U_\mu]$. It will suffice then putting $a_\mu = t$ so that (12f)+(12g) are fulfilled (with μ in place of λ).

As a consequence, the process above either stops at a certain stage $\mu < \theta$ (and then, we are done); or else (in the opposite situation) it is continuable over all of $W(\theta)$; i.e., (12f)+(12g) hold for each $\lambda < \theta$. We claim that, in such a case, $T = \{a_\xi; \xi < \theta\}$ is cofinal in S; and this, combined with the Γ-countable property of the same, completes the argument. Assume not; i.e., (12h) is true with θ in place of μ. By (12e), there exists $t \in S$ with $t > s$; hence $t \neq s$. On the other hand, by the choice of $(a_\xi; \xi < \theta)$, one has $(s,t) \in U_\xi$, for all $\xi < \theta$; hence $s = t$ (by the sufficiency condition). The obtained facts involving (s,t) are contradictory. Hence, (12h) cannot hold (with θ in place of μ); and our claim follows. □

A basic particular construction of this type may be described along the following lines. By a *pseudometric* over M we mean any map $d: M \times M \to R_+$. Call this object *reflexive* provided $d(x,x) = 0$, $\forall x \in M$. Let $D = (d_\lambda; \lambda < \alpha)$ be a family of reflexive pseudometrics; where $\alpha \geq \omega$. Then $\mathscr{U}(D) = \{U(\lambda, r); \lambda < \alpha, r > 0\}$, where $U(\lambda,r) = \{(x,y) \in M \times M; d_\lambda(x,y) < r\}$, $\lambda < \alpha$, $r > 0$, is an almost uniformity over M. The sufficiency condition for this object is characterized as: D is *sufficient* $[d_\lambda(x,y) = 0, \forall\lambda < \alpha \Longrightarrow x = y]$. On the other hand, the subfamily $\mathscr{V} = \{U(\lambda, 2^{-n}); \lambda < \alpha, n < \omega\}$ is cofinal in $(\mathscr{U}, \supseteq)$; and this, in conjunction with $\mathrm{card}(W(\alpha) \times W(\omega)) = \mathrm{card}(W(\alpha))$ (cf. Alexandrov [1, Ch 3, Sect 6]) shows that $\mathscr{U}$ is Γ-pseudometrizable, where $\Gamma = \mathrm{card}(W(\alpha))$. A translation of Proposition 14.9 in terms of $D = (d_\lambda; \lambda < \alpha)$ is immediate; we do not give details. In particular, when $\Gamma = \omega$, these developments reduce to the ones in Turinici [74].

II) Let $\mathscr{T}$ be a topology on M. Any subfamily $\mathscr{B} \subseteq \mathscr{T}$ with the property that each $D \in \mathscr{T}$ is a union of members in $\mathscr{B}$, will be referred to as a *basis* for $\mathscr{T}$. If, in addition, $\mathscr{B}$ is Γ-countable, then $\mathscr{T}$ will be called *second Γ-countable*. Term the ambient order $(\leq)$, *closed from the left* provided $M(x,\geq) = \mathrm{cl}(M(x,\geq))$, for each $x \in M$; here, "cl" stand for the *closure* operator.

Proposition 14.10. *Assume that $\mathscr{T}$ is second Γ-countable and $(\leq)$ is closed from the left. Then, $(M,\leq)$ is (strongly) Γ-separable.*

Proof. Let $\mathscr{B} = \{B_\xi; \xi < \theta\}$ (where $\omega \leq \theta < \Delta$) stand for a Γ-countable basis of $\mathscr{T}$. Further, take some choice function "Ch" of the nonempty parts in M $[\mathrm{Ch}(X) \in X$, for each $X \subseteq M, X \neq \emptyset]$. Given the arbitrary fixed $(\leq)$-chain S of M, denote $T = \{\mathrm{Ch}(B \cap S); B \in \mathscr{B}\}$ (hence $T \subseteq S$). For the moment, T is Γ-countable (because $T \preceq \mathscr{B}$). In addition, we claim that $\mathrm{cl}(T) \supseteq S$ [wherefrom, T is dense in S]. In fact, let s be some point of S; and U stand for an open neighborhood of it. By the definition of $\mathscr{B}$, U may be written as a union of members in this family; so

$$U \supseteq B \ni s \text{ (hence } U \ni \mathrm{Ch}(B \cap S)), \quad \text{for some } B \in \mathscr{B};$$

and our claim follows. If T is cofinal in S, we are done (cf. Proposition 14.4). Otherwise, there must be some $s \in S$ with $T \subseteq M(s,\geq)$; wherefrom

$$S \subseteq \mathrm{cl}(T) \subseteq \mathrm{cl}(M(s,\geq)) = M(s,\geq);$$

i.e., $\{s\}$ is cofinal in S. The proof is thereby complete. □

It remains to establish under which conditions is $\mathscr{T}$, second Γ-countable. An appropriate answer may be given in a uniform context:

(12i) there exists an (up-directed) family $D = (d_\lambda; \lambda < \alpha)$ of semimetrics (over M) whose associated topology is just $\mathscr{T}$.

Then, e.g., the condition below yields the desired property for $\mathscr{T}$ [where $\Gamma = \mathrm{card}(W(\alpha))$]:

(12j) M has a Γ-countable dense subset.

The proof is very similar to the one in Bourbaki [11, Ch 9, Sect 2.8]; see also Alexandrov [1, Ch 4, Sect 4]. Finally, note that, when $\Gamma = \omega$, the developments above reduce to the ones in Zhu and Li [85]. Some related aspects may be found in Zhu, Fan and Zhang [84].

14.1.3 Zorn–Bourbaki Principles

(A) Let M be a nonempty set; and $(\leq)$, some *order* (antisymmetric quasi-order) on it. Call the point $z \in M$, $(\leq)$-*maximal* in case

(a13) $w \in M, z \leq w \Longrightarrow z = w$; i.e., $z < x$ is false, for each $x \in M$.

(Here, $(<)$ is the *strict order* attached to $(\leq)$). Sufficient conditions for the existence of such elements may be obtained as follows. Call the (nonempty) part A of M, a *linear* $(\leq)$-*chain* provided $(A, \leq)$ is linearly ordered [$\forall x, y \in A$: either $x \leq y$ or $y \leq x$]; and a *(natural)* $(\leq)$-*chain*, when $(A, \leq)$ is well ordered [cf. Subsection 14.1.2].

Theorem 14.11. *Suppose that one of the conditions below holds*

(13a) each linear $(\leq)$*-chain (of* M*) is bounded above*
(13b) each $(\leq)$*-chain (of* M*) is bounded above.*

Then, $(\leq)$ *is a normal order, in the sense: for each* $u \in M$ *there exists a* $(\leq)$*-maximal* $v \in M$ *with* $u \leq v$.

Some remarks are in order. The first explicit formulation of Theorem 14.11 in terms of (13a) was given in 1914 by Hausdorff [29, Ch 6, Sect 1]; a slight different version of it was obtained in 1922 by Kuratowski [41]. Note that the quoted authors regarded Theorem 14.11 only as a handy tool in solving various existence problems in the setting of (AC) (= the axiom of choice); cf. Taskovic [63]. Finally, again under the lines of (13a), we must mention the 1935 contribution due to Zorn [87]; who regarded Theorem 14.11 as an axiom. The version of this result involving (13b) was stated in Bourbaki [9]; who also established its equivalence with the well ordering principle in Zermelo [83] (equivalent with (AC)). For this reason, Theorem 14.11 is referred to as the Zorn–Bourbaki principle. Note that, in the context of (AC), we have: (13b) $\Longrightarrow$ (13a) (hence (13b) $\Longleftrightarrow$ (13a)); see also Felgner [22]. Further historical aspects may be found in Moore [49, Ch 4, Sect 4] and the references therein.

(B) Now, as results from the developments in Subsection 14.1.2, the verification of (13b) for (cardinal-) countable chains only will suffice (for its validity) in many concrete cases. This suggests us considering maximality principles over ordered structures with such regularity properties. So, let $(M, \leq)$ be a (partially) ordered structure. Further, take some $\gamma \geq 0$ and put $\Gamma = \aleph_\gamma, \Delta = \mathrm{SUC}(\Gamma)$ (hence $\Delta = \aleph_{\gamma+1}$). Assume firstly that

(13c) $(M, \leq)$ is sequentially Γ-inductive:
each Γ-sub $(\leq)$-chain of M is bounded from above (modulo $(\leq)$).

Note that, by Proposition 14.3, this notion is identical with the concept of sequentially inductive (ordered) structure in Turinici [67] when $\Gamma = \omega$. Moreover, by Proposition 14.4, it may be also written as

(13d) each Γ-countable $(\leq)$-chain of M is bounded above (modulo $(\leq)$).

Secondly, assume that (cf. Subsection 14.1.2)

(13e) $(M, \leq)$ is (strongly) Γ-separable: each $(\leq)$-chain $S \subseteq M$ is majorized by some Γ-countable $(\leq)$-chain $T \subseteq S$.

Remember that, by Proposition 14.6, this also reads

(13f) each $(\leq)$-chain of M is Γ-countable.

Theorem 14.12. *Assume that (13c)+(13e) hold. Then, $(\leq)$ is a normal order (in the sense above).*

Proof. By the remarks involving (13d)+(13f), it is clear that Theorem 14.11 applies to these data; and, from this, we are done. □

For the moment, Theorem 14.12 is deductible from Theorem 14.11. The reverse inclusion is also true; just take $\Gamma = \text{card}(M)$ in the above statement. Hence Theorem 14.12 and Theorem 14.11 are logically equivalent; i.e., the enlargement of Theorem 14.11 ensured by Theorem 14.12 is technical in nature.

Now, remember that the regularity conditions in Theorem 14.11 are logically minimal so that its conclusion is retainable. (See the quoted papers for details). So, it is natural to ask whether this is also true for the conditions in Theorem 14.12. Two situations may occur.

i) Assume that in Theorem 14.12 condition (13c) does not hold. By definition, there exists a Γ-sub $(\leq)$-chain K of M which is not bounded above (in M). As a consequence, $(K, \leq)$ is not sequentially Γ-inductive; but it is (strongly) Γ-separable. This, added to $(K, \leq)$ having no $(\leq)$-maximal elements, proves the logical minimality of (13c).

ii) Assume that, in Theorem 14.12, condition (13e) does not hold. By definition, there must be a $(\leq)$-chain $L \subseteq M$ with (cf. Proposition 14.4)

$$L \propto Q \text{ is false, for each } \Gamma\text{-countable } (\leq)\text{-chain } Q \subseteq L.$$

As a consequence, the structure $(L, \leq)$ is sequentially Γ-inductive; but not (strongly) Γ-separable. This, added to $(L, \leq)$ having no $(\leq)$-maximal elements, proves the logical minimality of (13e).

Summing up, we proved

Proposition 14.13. *Either of the regularity conditions (13c) and (13e) in Theorem 14.12 is logically minimal for the conclusions given there to hold.*

In particular, when $\Gamma = \omega$, Theorem 14.12 (and Proposition 14.13 as well) is just the statement in Zhu and Li [85] proved via similar techniques.

(C) An interesting completion of these facts may be given along the following lines. Let M be a nonempty set; and $(\leq)$, a quasi-order (reflexive and transitive relation) over it. The associated relation

(b13) $(x, y \in M)$: $x \preceq y$ iff $x \leq y$ and $\neg(y \leq x)$

is *irreflexive* ($\neg(x \prec x)$, $\forall x \in M$) and transitive; hence a *strict order*. As a consequence, its completion

(c13) $(x,y \in M)$: $x \preceq y$ iff either $x \prec y$ or $x = y$

is an order on M. For $z \in M$, its $(\preceq)$-maximal property is the one in (a13) (with $(\preceq)$ in place of $(\leq)$); which, in terms of our initial quasi-order means:

(d13) $z \leq w \Longrightarrow w \leq z$ (or, equivalently: $M(z,\leq) \subseteq M(z,\geq)$);

referred to as: z is $(\leq)$-*maximal*. [This must be not confused with the strong $(\leq)$-maximality of z, which may be introduced as in (a13) (the first part): $z \leq x \Longrightarrow z = x$]. Concrete circumstances for the existence of such points are obtainable from Theorem 14.12 above. For practical reasons, it would be useful having the conditions (13c)+(13e) expressed in terms of our quasi-order. However, a complete translation of (13e) is not useful from a practical perspective. So, the only acceptable way is to consider a "hybrid" result like the one below. For each θ-net $(u_\xi; \xi < \lambda)$ in M, define its ascending/boundedness (modulo $(\leq)$) properties in the usual way.

Theorem 14.14. *Let the sep-admissible cardinal Γ for $(M,\preceq)$ be such that*

(13g) $(M,\leq)$ is sequentially Γ-inductive:
each Γ-sub ascending net in M is bounded from above (modulo $(\leq)$).

Then, for each $u \in M$ there exists $v = v(u) \in M$ with **(i)** *$u \leq v$ and* **(ii)** *$M(v,\leq) \subseteq M(v,\geq)$ (or, equivalently: $v \leq x$ is false, whenever so is $x \leq v$).*

Proof. By the very choice of Γ it is clear that (13e) holds for $(M,\preceq)$; so, it will suffice showing that (13c) also holds for this structure. Let P be a Γ-sub $(\preceq)$-chain in M; hence, P may be represented as a net $(u_\xi; \xi < \mu)$ where $\mu \leq \Gamma$ and the map $\xi \mapsto u_\xi$ is strictly ascending:

$$\xi < \eta \text{ implies } u_\xi \prec u_\eta \text{ (hence } u_\xi \leq u_\eta,\ \neg(u_\eta \leq u_\xi)).$$

When μ is a first kind ordinal, we are done; because $P \preceq u_\lambda$, where $\lambda = \mu - 1$. Assume now that μ is a second kind ordinal. By the strict ascending property above, the net $(u_\xi; \xi < \lambda)$ is ascending (modulo $(\leq)$); wherefrom, by hypothesis, $u_\xi \leq v$, for all $\xi < \lambda$ and some $v \in M$. On the other hand, again by the property in question, $v \leq u_\xi$ is impossible for all $\xi < \mu$; since for each η with $\xi < \eta < \mu$ we should have $u_\eta \leq v \leq u_\xi$ (hence $u_\eta \leq u_\xi$), in contradiction with a previous relation. Hence, v is an upper bound of P (modulo $(\preceq)$; and the claim follows. But then, all is clear from Theorem 14.12, applied to the structure $(M,\preceq)$. □

14.1.4 Main Results

With this information at hand, we may now return to the questions in Subsection 14.1.1. The natural setting for discussing them is the one of *transitive* relations. This, apart from giving us new useful forms of Theorem 14.1 allows a direct approach to the quasi-order and amorphous cases.

(A) Let (M,∇) and (P,∇) be transitive structures. The relation over M

(a14) $(x,y\in M)$ $x<y$ iff $x\nabla y$ and $\neg(y\nabla x)$

is *irreflexive* $(\neg(x<x),\forall x\in M)$ and *transitive*; hence a *strict order*. As a consequence, its completion

(b14) $(x,y\in M)$ $x\overline{<}y$ iff either $x<y$ or $x=y$

is an order on M. Denote in the same way the strict/standard order on P attached to (∇). Further, let $\varphi : M\to P$ be some (∇,∇)-increasing function

(14a) $x\nabla y \Longrightarrow \varphi(x)\nabla\varphi(y)$ [equivalently: $\neg(\varphi(x)\nabla\varphi(y)) \Longrightarrow \neg(x\nabla y)$].

This allows us introducing the relation (in M)

(c14) $(x,y\in M)$ $x\sqsubset y$ iff $x\nabla y$ and $\neg(\varphi(y)\nabla\varphi(x))$.

By the remark involving (14a), one has

$$x\sqsubset y \text{ iff } x<y \text{ and } \varphi(x)<\varphi(y); \tag{14.7}$$

wherefrom, $(\sqsubset)$ is a strict order on M. Let $(\sqsubseteq)$ stand for the associated (by (b14)) order in M. Note that, by the very definitions (and remarks) above

$$[(x\sqsubset y, y\nabla z) \text{ or } (x\nabla y, y\sqsubset z)] \text{ imply } x\sqsubset z. \tag{14.8}$$

Having these precise, call the point $z\in M$, $(\nabla,\nabla;\varphi)$-*maximal*, when

(d14) for each $w\in M$: $z\nabla w$ implies $\varphi(w)\nabla\varphi(z)$.

Note that, if (M,∇) is identical with (P,∇) and (∇) is an order (on M) this concept reduces to the one in Subsection 14.1.3 (when $\varphi=$ identity). Hence, maximality results of this type are not without interest for us. The basic step in deducing these is a characterization of our concept in terms of $(\sqsubseteq)$; cf. Turinici [75]:

Lemma 14.15. *The generic relation is available*

$$(\forall z\in M):\quad (\nabla,\nabla;\varphi)\textit{-maximal} \Longleftrightarrow (\sqsubseteq)\textit{-maximal.} \tag{14.9}$$

Now, as $(\sqsubseteq)$ is an order, the developments of Subsection 14.1.3 apply to $(M,\sqsubseteq)$; and this yields the following maximality principle to be used further. Take a certain ordinal γ and put $\Gamma=\aleph_\gamma$, $\Delta=\mathrm{SUC}(\Gamma)$ (hence $\Delta=\aleph_{\gamma+1}$).

Theorem 14.16. *Assume that*

(14b) $(M,\sqsubseteq)$ *is sequentially* Γ*-inductive*
(14c) $(M,\sqsubseteq)$ *is (strongly)* Γ*-separable.*

Then, for each $u \in M$ there exists $v \in M$ with

$$v \text{ is } (\nabla, \nabla : \varphi)\text{-maximal} \tag{14.10}$$

in such a way that

$$u = v \text{ (hence } u \text{ is } (\nabla, \nabla; \varphi)\text{-maximal), whenever } M(u, \nabla) = \emptyset \tag{14.11}$$

$$u \nabla v, \quad \text{whenever } M(u, \nabla) \neq \emptyset. \tag{14.12}$$

Proof. (Sketch). By Theorem 14.12 (applicable, via (14b)+(14c)) it follows that, for each $u \in M$ there exists $v \in M$ with

$$u \sqsubseteq v \text{ (i.e.: } u \sqsubset v \text{ or } u = v); \quad \text{and } v \text{ is } (\sqsubseteq)\text{-maximal.} \tag{14.13}$$

The latter of these yields (14.10), if one takes Lemma 14.15 into account. And the former of these gives the couple of alternatives (14.11)/(14.12). □

It remains now to give sufficient conditions (involving our initial data) under which (14b)+(14c) are fulfilled. This necessitates further conventions and auxiliary facts. Let the ordinal $\lambda > 0$ be arbitrary fixed. We say that the λ-net $(b_\xi; \xi < \lambda)$ is *ascending* (modulo (∇)) when $[\xi < \eta \Longrightarrow b_\xi \nabla b_\eta]$. Given such an object, call $u \in M$ an *upper bound* (modulo (∇)) of it when

$$b_\xi \nabla u, \text{ for all } \xi < \lambda \text{ (written as: } (b_\xi; \xi < \lambda) \nabla u).$$

If u is generic in this convention, we say that $(b_\xi; \xi < \lambda)$ is *bounded from above* (modulo (∇)). Finally, we call (M, ∇), *sequentially Γ-inductive* provided

each ascending (modulo (∇)) θ-net (where $\theta \in W_0^2[\omega, \Gamma]$)
is bounded from above (modulo (∇)).

Lemma 14.17. *Under these conventions, one has*

$$(M, \nabla) \text{ is seq. } \Gamma\text{-inductive} \Longrightarrow (M, \sqsubseteq) \text{ is seq. } \Gamma\text{-inductive.} \tag{14.14}$$

We are now in position to get an appropriate answer to the posed question.

Theorem 14.18. *Suppose that*

(14d) (M, ∇) *is sequentially Γ-inductive*
(14e) $(P, \overline{<})$ *is (strongly) Γ-separable.*

Then, conclusions of Theorem 14.16 are retainable.

Proof. By Lemma 14.17, condition (14b) holds via (14d). We claim that (14c) holds too (from (14e)); and this will complete the argument. Let S be some $(\sqsubseteq)$-chain of M; and put $V = \varphi(S)$. Clearly,

$$V \text{ is a } (\overline{<})\text{-chain in } P; \quad \text{(cf. (14.7));}$$

so, in view of (14e), V is Γ-countable (in P). On the other hand, the same relation (14.7) shows that φ is an order isomorphism between $(S,\sqsubseteq)$ and $(V,\overline{<})$; wherefrom, S is countable too; and the claim follows. □

(B) In particular, assume that the (transitive) relation (∇) is a quasi-order ($\leq$) in both M and P. By Theorem 14.18 we then derive (under (14a)):

Theorem 14.19. *Assume (14e) is true, as well as*

(14f) $(M,\leq)$ *is sequentially* Γ*-inductive.*

Then, for each $u \in M$*, there exists* $v \in M$*, with*

$$[u \leq v] \text{ and } \quad [v \leq w \Longrightarrow \varphi(w) \leq \varphi(v)]. \tag{14.15}$$

In particular, when $\Gamma = \omega$, this result reduces to the one in Turinici [74]; see also Zhu and Li [85]. But, as shown in that paper, the precise statements include the Brezis–Browder ordering principle [12] (Theorem 14.1) when $(P,\leq)$ is identical with $(R_+,\geq)$. Hence, so does Theorem 14.19; and, as such, it may be also referred to in this way. On the other hand, if $(M,\leq)$ is identical with $(P,\leq)$ and $\varphi =$ the identity, Theorem 14.19 is just Theorem 14.14. Summing up,

$$\text{Th } 14.12 \Longrightarrow \text{Th } 14.18 \Longrightarrow \text{Th } 14.19 \Longrightarrow \text{Th } 14.14 \Longrightarrow \text{Th } 14.12; \tag{14.16}$$

hence, all these are mutually equivalent. In particular, this also shows that Theorem 14.19 includes the "transfinite" version of Theorem 14.1 obtained in Turinici [67]. The question of the reciprocal inclusion being also true remains open; we conjecture that the answer is positive.

(C) Let us return to our initial framework. The basic hypothesis used in all these developments is (14a). So, the question arises: what can be said about such results when (14a) is no longer available. To this end, put

(d14) $(x,y \in M)$ $x \triangle y$ iff $x \nabla y$ and $\varphi(x) \nabla \varphi(y)$.

This is a transitive relation over M; and condition (14a) holds with $(\triangle,\nabla)$ in place of (∇,∇). An application of Theorem 14.18 to these data yields an appropriate answer to the problem we deal with.

Theorem 14.20. *Assume that (14e) holds, as well as*

(14g) $(M,\triangle)$ *is sequentially inductive.*

Then, for each $u \in M$*, there exists* $v \in M$ *with*

$$v \text{ is } (\triangle,\nabla;\varphi)\text{-maximal} \tag{14.17}$$

in such a way that

$$u = v \text{ (hence } u \text{ is } (\triangle,\nabla;\varphi)\text{-maximal), whenever } M(u,\triangle) = \emptyset \tag{14.18}$$

$$u \bigtriangleup v, \quad \text{whenever } M(u, \bigtriangleup) \neq \emptyset. \tag{14.19}$$

A quasi-order version of this (under the lines of Theorem 14.19) is immediately obtainable; we do not give details. In particular, when $\Gamma = \omega$, this result is just the one in Turinici [74]; which, in turn, extends a related statement due to Kada, Suzuki and Takahashi [36]. Further structural aspects may be found in Manka [47].

14.1.5 Some Amorphous Versions

A slight extension of these facts may be reached when the relation (∇) over M is no longer transitive. Further aspects occasioned by the obtained results are then discussed.

(A) Let $(\perp)$ stand for an *amorphous* relation over M. Denote by (∇) the transitive relation (over the same) attached to $(\perp)$

(a15) $(x, y \in M)$ $x \nabla y$ iff $x = u_1 \perp ... \perp u_k = y$ (in the sense: $u_i \perp u_{i+1}, \forall i \in \{1, ..., k-1\}$), for some $k \geq 2$ and $u_1, ..., u_k \in M$.

Take a transitive relation $(\bigtriangleup)$ over P; as well as a function $\varphi : M \to P$ with

(15a) φ is $(\perp, \bigtriangleup)$-increasing: $x \perp y \Longrightarrow \varphi(x) \bigtriangleup \varphi(y)$.

Note that, under (a15) above, one gets

$$\varphi \text{ is } (\nabla, \bigtriangleup)\text{-increasing (in the sense of (14a)).}$$

Given $z \in M$, we say that it is $(\perp, \bigtriangleup; \varphi)$*-maximal*, if

(b15) (for each $w \in M$): $z \perp w \Longrightarrow \varphi(w) \bigtriangleup \varphi(z)$.

Again by (a15), one gets the generic relation

$$\text{(for each } z \in M\text{): } (\nabla, \bigtriangleup; \varphi)\text{-maximal} \Longrightarrow (\perp, \bigtriangleup; \varphi)\text{-maximal.} \tag{14.20}$$

So, existence results involving such points are deductible from Theorem 14.18 above. The only aspect to be clarified is that of expressing (14d) in terms of $(\perp)$. This will necessitate a lot of new conventions.

Let $\alpha > 0$ be an ordinal. Remember that $\omega \cdot \alpha = \text{ord}(W(\omega) \times W(\alpha), \leq)$; where $(\leq)$ stands for the lexicographic order (cf. Subsection 14.1.2). By a (ω, α)*-net*, we shall mean any map $(n, \xi) \mapsto b(n, \xi)$ from $W(\omega) \times W(\alpha)$ to M. Given such an object, call it *ascending* (modulo $(\perp)$) when

(c15) $b(n, \xi) \perp b(n+1, \xi)$, for all $n < \omega, \xi < \alpha$.

Any (ω, α)-net $(a(n, \xi); n < \omega, \xi < \alpha)$, with

(d15) $(a(n, \xi); n < \omega)$ is a subsequence of $(b(n, \xi); n < \omega)$, for each $\xi < \alpha$

will be referred to as a *strong subnet* of $(b(n,\xi);n<\omega,\xi<\alpha)$. Further, call $u\in M$, an *upper bound* (modulo $(\perp)$) of $(b(n,\xi);n<\omega,\xi<\alpha)$, when

(e15) $b(n,\xi)\perp u, \forall n<\omega, \forall\xi<\alpha$ (in short: $(b(n,\xi);n<\omega,\xi<\alpha)\perp u$).

If this holds only on a strong subnet of $(b(n,\xi);n<\omega,\xi<\alpha)$, we say that u is an *asymptotic upper bound* (modulo $(\perp)$) of this net; written as: $(b(n,\xi);n<\omega,\xi<\alpha)\perp\perp u$. When $u\in M$ is generic in these conventions, the corresponding property will be referred to as: $(b(n,\xi);n<\omega,\xi<\alpha)$ is *bounded above* (respectively, *asymptotic bounded above*) modulo $(\perp)$. Finally, call the structure $(M,\perp)$, *sequentially Γ-inductive* if

(f15) each ascending (modulo $(\perp)$) (ω,α)-net (where $\omega\cdot\alpha\le\Gamma$) is asymptotic bounded above (modulo $(\perp)$).

(Here, the couple of ordinals $[\Gamma=\aleph_\gamma,\Delta=\aleph_{\gamma+1}]$ is the one in Subsection 14.1.4).

The following auxiliary fact is useful for us.

Lemma 14.21. *Under these conventions,*

$$(M,\perp) \text{ is seq. } \Gamma\text{-inductive} \Longrightarrow (M,\nabla) \text{ is seq. } \Gamma\text{-inductive.} \tag{14.21}$$

Now, by simply adding this to Theorem 14.18, one gets

Corollary 14.22. *Assume (14e) holds, as well as*

(15b) $(M,\perp)$ *is sequentially* Γ*-inductive.*

Then, for each $u\in M$*, there exists* $v\in M$ *with*

$$v \text{ is } (\perp,\triangle;\varphi)\text{-maximal} \tag{14.22}$$

in such a way that either (14.12) (Subsection 14.1.4) is retainable or else

$$u=v \text{ (hence } u \text{ is } (\perp,\triangle;\varphi)\text{-maximal) whenever } M(u,\nabla)=\emptyset. \tag{14.23}$$

(Here, (∇) *is the transitive relation given by (a15)).*

In particular, when $(\perp)$ is a transitive relation over M, this statement reduces to Theorem 14.18. Since the opposite inclusion also holds, we get

$$\text{Corollary 14.22} \Longleftrightarrow \text{Theorem 14.18} \ (\Longleftrightarrow \text{Theorem 14.12}). \tag{14.24}$$

Hence, this extension is technical in nature.

(B) Now, the basic assumption used here is (15a). So, we may ask what happens when such a condition is no longer true. To this end, put

(g15) $(x,y\in M)$ $x\top y$ iff $x\perp y$ and $\varphi(x)\triangle\varphi(y)$.

This is an amorphous relation over M; and condition (15a) holds with $(\top,\triangle)$ in place of $(\perp,\triangle)$. An application of Corollary 14.22 to these data gives:

Corollary 14.23. *Assume (14e) holds, as well as*

(15c) $(M,\top)$ *is sequentially* Γ*-inductive.*

Then, for each $u \in M$ *there exists* $v \in M$ *with*

$$v \text{ is } (\top,\triangle;\varphi)\text{-maximal} \tag{14.25}$$

in such a way that, either (14.12) (Subsection 14.1.4) is retainable, or else

$$u = v \text{ (hence } u \text{ is } (\top,\triangle;\varphi)\text{-maximal) whenever } M(u,\nabla) = \emptyset. \tag{14.26}$$

(Here, $(\top)$ *is the amorphous relation of (g15); and* (∇)*, its associated by (a15) transitive relation).*

A basic particular case of this corresponds to the choice $\Gamma = \omega$. Then, the above results are just the ones in Turinici [74]. But, as precise there, this version of Corollary 14.23 includes a related maximality principle in Gajek and Zagrodny [23]; hence, so does our statement. Further aspects were delineated in Sonntag and Zălinescu [56].

14.2 Pseudometric Maximal Principles

14.2.1 Introduction

Let M be a nonempty set. Take a *quasi-order* $(\leq)$ (i.e., reflexive and transitive relation) over it, as well as a function $x \mapsto \psi(x)$ from M to R. Call the point $z \in M$, $(\leq,\psi)$*-maximal* when: $w \in M$ and $z \leq w$ imply $\psi(z) = \psi(w)$. A basic result about the existence of such points is the 1976 Brezis–Browder ordering principle [12]:

Theorem 14.24. *Suppose that*

(21a) $(M,\leq)$ *is sequentially inductive:*
each ascending sequence has an upper bound (modulo $(\leq)$*)*
(21b) ψ *is* $(\leq)$*-decreasing (*$x \leq y \Longrightarrow \psi(x) \geq \psi(y)$*)*
(21c) ψ *is bounded from below (*$\inf[\psi(M)] > -\infty$*).*

Then, for each $u \in M$ *there exists a* $(\leq,\psi)$*-maximal* $v \in M$ *with* $u \leq v$*.*

This statement is nothing but a particular case of the developments in Subsection 14.1.4 (when $\Gamma = \omega$); see also Zhu and Li [85]. The *structural* way of enlarging it was discussed in the precise place. In the following, the *(pseudo) metrical* extensions of Theorem 14.24 will be considered. Among these, we quote the papers by Altman [3], Turinici [66], and Anisiu [5]; see also Bae, Cho and Yeom [6]. The obtained results are interesting from a technical viewpoint. However, we must emphasize that, in all concrete situations when a maximality principle of this type is applicable, a substitution of it by the Brezis–Browder principle is always possible. This (cf. Bao and Khanh [7]) raises the question: To what extent are these enlargements of Theorem 14.24 effective? As we shall see below (in Subsection 14.2.2 and Subsection 14.2.4) the answer is negative for most of these. On the other hand, there do exist metrical maximality principles which are not comparable with Theorem 14.24; see the 1990 paper in Kang and Park [37]. It is our second aim in this exposition to show (cf. Subsection 14.2.3) that all such statements may be viewed as particular cases of an "asymptotic" type version of Theorem 14.24 (which includes it in a technical sense). Finally, in Subsection 14.2.5, an application of these facts is given to (standard) Zorn maximality principles.

14.2.2 Logical Equivalents of Brezis–Browder's Principle

Let M be some nonempty set; and $(\leq)$, some quasi-order on it. Further, let $x \mapsto \varphi(x)$ stand for a function between M and $R_+ \cup \{\infty\} = [0, \infty]$. The following maximality principle is our starting point (cf. Turinici [76]):

Proposition 14.25. *Assume (21a) and (21b) are true, as well as*

(22a) $\quad$ *($(M, \leq)$ is almost regular (modulo φ))*
$\quad \forall x \in M, \forall \varepsilon > 0, \exists y = y(x, \varepsilon) \geq x : \varphi(y) \leq \varepsilon.$

Then, for each $u \in M$ there exists $v \in M$ with $u \leq v$ and $\varphi(v) = 0$ (hence v is $(\leq, \varphi)$-maximal).

Proof. By (22a), there must be some $z \geq u$ with $\varphi(z) < \infty$. Clearly, (21a)–(21c) apply to $M(z, \leq) := \{x \in M; z \leq x\}$ and $(\leq, \varphi)$. So, for the starting point $z \in M(z, \leq)$ there exists $v \in M(z, \leq)$ with **i)** $z \leq v$ (hence $u \leq v$) and **ii)** v is $(\leq, \varphi)$-maximal in $M(z, \leq)$. Suppose by contradiction that $\gamma := \varphi(v) > 0$; and fix some β in $]0, \gamma[$. By (22a) again, there must be $y = y(v, \beta) \geq v$ (hence $y \in M(z, \leq)$) with $\varphi(y) \leq \beta < \gamma (= \varphi(v))$. This cannot be in agreement with the second conclusion above. Hence, $\varphi(v) = 0$; and we are done. □

Clearly, Proposition 14.25 is a logical consequence of Theorem 14.24. But, the converse inclusion is also true; to verify it, we need some conventions. By a (generalized) *pseudometric* over M we shall mean any map $d : M \times M \to R_+ \cup \{\infty\}$. Suppose that we introduced such an object; which is also *reflexive* $[d(x,x) = 0, \forall x \in M]$. Call the point $z \in M$, *$(\leq, d)$-maximal*, if: $u, v \in M$ and $z \leq u \leq v$ imply $d(u, v) = 0$. Note that, if d is (in addition) *sufficient* $[d(x,y) = 0 \Longrightarrow x = y]$, the $(\leq, d)$-maximal property becomes: $w \in M, z \leq w \Longrightarrow z = w$ (and reads: z is *strongly $(\leq)$-maximal*). So, existence results involving such points may be viewed as "metrical" versions of the Zorn maximality principle (cf. Moore [49, Ch 4, Sect 4]). To get sufficient conditions for these, one may proceed as below. Let (x_n) be an ascending sequence in M. The d-Cauchy property for it is introduced in the usual way $[\forall \varepsilon > 0, \exists n(\varepsilon)$ such that $n(\varepsilon) \leq p \leq q \Longrightarrow d(x_p, x_q) \leq \varepsilon]$. Also, call (x_n), *d-asymptotic* when $d(x_n, x_{n+1}) \to 0$, as $n \to \infty$. Clearly, each (ascending) d-Cauchy sequence is d-asymptotic too. The reverse implication is also true when all such sequences are involved; i.e., the global conditions below are equivalent

(22b) each ascending sequence is d-Cauchy
(22c) each ascending sequence is d-asymptotic.

By definition, either of these will be referred to as $(M,\leq)$ is *regular* (modulo d). Note that this property implies its relaxed version

(22d) ($(M,\leq)$ is weakly regular (modulo d))
$\forall x \in M, \forall \varepsilon > 0, \exists y = y(x,\varepsilon) \geq x$: $y \leq u \leq v \Longrightarrow d(u,v) \leq \varepsilon$.

The following ordering principle is then available (cf. Kang and Park [37]):

Proposition 14.26. *Assume that $(M,\leq)$ is sequentially inductive and weakly regular (modulo d). Then, for each $u \in M$ there exists a $(\leq,d)$-maximal $v \in M$ with $u \leq v$.*

Proof. Let us introduce the function (from M to $R_+ \cup \{\infty\}$)

(a22) $\varphi_d(x) = \sup\{d(u,v); x \leq u \leq v\}$, $x \in M$.

Clearly, (21b)+(21c) hold for this object, as well as (22a) (if one takes (22d) into account). Hence, Proposition 14.25 is applicable to M and $(\leq,\varphi_d)$. This, added to $[\varphi_d(z) = 0$ iff z is $(\leq,d)$-maximal$]$, gives the desired conclusion. □

As a direct consequence of this, we get the maximality principle in Turinici [68] (see also Conserva and Rizzo [16]):

Proposition 14.27. *Assume that $(M,\leq)$ is sequentially inductive and regular (modulo d). Then, the conclusion of Proposition 14.26 is holding.*

So far, Proposition 14.27 is a logical consequence of Theorem 14.24. The reciprocal of this is also true, by simply taking $d(x,y) = |\psi(x) - \psi(y)|, x,y \in M$ (where ψ is the above one). We therefore established the inclusional chain Th 14.24 $\Longrightarrow$ Prop 14.25 $\Longrightarrow$ Prop 14.26 $\Longrightarrow$ Prop 14.27 $\Longrightarrow$ Th 14.24. Hence, all these ordering principles are nothing but logical equivalents of the Brezis–Browder principle [12] (Theorem 14.24). (This also includes the related statements in Szaz [60] and Tataru [64]; which extend the one in Dancs, Hegedus and Medvegyev [17]). Further aspects may be found in Hamel [27, Ch 4]; see also Hyers, Isac and Rassias [31, Ch 5].

14.2.3 Asymptotic Extensions

(A) The developments in the preceding subsection raise the (delicate) question of whether or not extensions of Theorem 14.24 (or its variants) exist without being reducible to it. Any attempt of solving it must begin from the sequential inductivity condition (21a). Precisely, an examination of the argument in Proposition 14.26 shows that one may impose it asymptotically (i.e., to sequences (x_n) with $\varphi_d(x_n) \to$

0) for the written conclusion to be retainable. So, it is natural to ask whether this has a general character. A positive answer to this may be given under the lines below. Let again M be some nonempty set. Take a quasi-order $(\leq)$ over it, as well as a function $\varphi : M \to R_+ \cup \{\infty\}$. The following "asymptotic" counterpart of Proposition 14.25 is now available (cf. Turinici [79]):

Theorem 14.28. *Assume that (21b) and (22a) are true, as well as*

(23a) $(M, \leq)$ is sequentially inductive (modulo φ): each ascending sequence (x_n) with $\varphi(x_n) \to 0$ has an upper bound (modulo $(\leq)$).

Then, for each $u \in M$ there exists $v \in M$ with $u \leq v$ and $\varphi(v) = 0$ (hence v is $(\leq, \varphi)$-maximal).

Proof. By (22a), it is not hard to construct an ascending (modulo $(\leq)$) sequence (u_n) with ($u \leq u_0$ and) $\varphi(u_n) \leq 2^{-n}, \forall n$ (hence $\varphi(u_n) \to 0$). Let v stand for an upper bound (modulo $(\leq)$) of this sequence (assured by (23a)). This element has all properties we need. □

Now, (21a) is a particular case of (23a). This tells us that Proposition 14.25 (hence Theorem 14.24 as well) is a particular case of Theorem 14.28. The reciprocal question (Prop 14.25 $\Longrightarrow$ Th 14.28) remains open; we conjecture that the answer is negative. To explain our position, it will be useful to consider:

Example 14.29. Let $R^2 = R \times R$ stand for the Cartesian plane; and $(\leq)$ denote the partial order induced by the convex cone R^2_+. Further, put $M = A \cup B$, where $A = \{u_n := (n, 0); n \geq 0\}$, $B = \{v_n := (n, 2^{-n}); n \geq 0\}$; and take the function (from M to $R_+ \cup \{\infty\}$): $\varphi(z) = \infty$, if $z \in A$ and $\varphi(z) = 0$, if $z \in B$. For the moment, (21b) is retainable, because each point of B is (strongly) $(\leq)$-maximal (cf. Subsection 14.2.2). Moreover, (22a) is retainable too, in view of $u_n \leq v_n$, for all $n \geq 0$. Unfortunately, the structure $(M, \leq)$ cannot satisfy (21a); for, e.g., the ascending sequence (u_n) is not bounded above; so that, Proposition 14.25 is not applicable to $(M, \leq)$ and φ. Nevertheless (in compensation to this), $(M, \leq)$ fulfills (23a); wherefrom, Theorem 14.28 applies to the same data.

Summing up, Theorem 14.28 includes in a strict sense Proposition 14.25; but, this is realized at the level of *the same* structure. For a genuine answer to the posed question, a variant of Example 14.29 involving *many* sub-structures of $(M, \leq)$ is needed. Concerning this aspect, notice that roughly speaking, (21a) acts as a *global* completeness of $(M, \leq)$ and (23a) as a *local* completeness of the same, with respect to the function φ. So, if the latter property is strictly larger than the former one (modulo these sub-structures), we are done. This tells us that a promising way of constructing such examples is related to completeness type techniques, as developed in Amato [4], Liu [46], Sempi [54], and Sullivan [57]; see also Jinag and Cho [34].

(B) The following version of this result may be noted. Let M be a nonempty set; and $\mathscr{P}(M)$ stand for the class of its subsets. According to Du [18] any function $\mu : \mathscr{P}(M) \to R_+ \cup \{\infty\}$ with

(a23) $\mu(\emptyset) = 0$; $\mu(A) \leq \mu(B)$ if $A \subseteq B$

will be called *sizing-up*. Assume that we fixed such an object; and let $(\leq)$ be a quasi-order on M. Then, the function φ_μ from M to $R_+ \cup \{\infty\}$ given as

(b23) $\varphi_\mu(x) = \mu(M(x,\leq)),\ x \in M,$

fulfills (21b). An application of Theorem 14.28 yields the following practical maximality principle:

Theorem 14.30. *Suppose that*

(23b) $((M,\leq)$ *is almost regular (modulo* φ_μ*))*
$\forall x \in M, \forall \varepsilon > 0,\ \exists y = y(x,\varepsilon) \geq x$: $\mu(M(y,\leq)) \leq \varepsilon$
(23c) $(M,\leq)$ *is sequentially inductive (modulo* φ_μ*): each ascending sequence* (x_n) *with* $\mu(M(x_n,\leq)) \to 0$ *has an upper bound (modulo* $(\leq)$*).*

Then, for each $u \in M$ *there exists* $v \in M$ *with* $u \leq v$ *and* $\mu(M(v,\leq)) = 0$ *(hence* v *is* $(\leq,\varphi_\mu)$*-maximal).*

In particular, (23c) holds under

(23d) each ascending sequence (x_n) with $\mu(\{x_n, x_{n+1}, \ldots\}) \to 0$ as $n \to \infty$ has an upper bound.

Then, Theorem 14.30 is just the basic maximal principle in Du [18]; which, among others, includes Brezis–Browder's principle [12] (Theorem 14.24). An interesting aspect refers to Theorem 14.28 being deductible from Theorem 14.30; we conjecture that the answer is positive.

(C) A basic particular case of these facts corresponds to the construction in Subsection 14.2.2. Precisely, let $d : M \times M \to R_+ \cup \{\infty\}$ be a reflexive (generalized) pseudometric (over M); and $\varphi_d : M \to R_+ \cup \{\infty\}$, its associated by (a22) function. Clearly, (21b) holds in this context; and the almost regularity (modulo φ_d) condition (22a) is just the one in (22d). Putting these together, it results in the following maximality statement involving these data.

Theorem 14.31. *Assume that* $(M,\leq)$ *is sequentially inductive (modulo* φ_d*) and weakly regular (modulo* d*). Then, for each* $u \in M$ *there exists a* $(\leq,d)$*-maximal* $v \in M$ *with* $u \leq v$.

As before, the sequential inductivity (modulo φ_d) holds under (21a); wherefrom, Theorem 14.31 includes Proposition 14.26. It would be interesting to determine whether the reciprocal inclusion is also retainable; further aspects will be treated elsewhere.

14.2.4 Convergence and Uniform Versions

(A) Let us now return to the initial setting. As a rule, the ambient set is endowed with various (sequential) convergence structures; and then, it is natural to ask of which is the corresponding version of the sequential inductivity condition in such a case. For an appropriate answer, a re-formulation of Theorem 14.24 is needed. Let $(M, \leq)$ be a quasi-ordered structure; and $\varphi : M \to R \cup \{\infty\}$ be some proper bounded from below function. The following ordering principle is our starting point.

Proposition 14.32. *Assume that (in addition)*

(24a) each ascending sequence in $\mathrm{Dom}(\varphi)$ *is bounded above.*

Then, for each $u \in \mathrm{Dom}(\varphi)$ *there exists a* $(\leq, \varphi)$*-maximal* $v \in \mathrm{Dom}(\varphi)$ *with* $u \leq v$.

Proof. Denote for simplicity $M_u = M(u, \leq)$. We show that conditions of Theorem 14.24 hold for $(M_u, \leq)$ and φ. Let (x_n) be an ascending (modulo $\leq$) sequence in M_u. By (24a) (and $M_u \subseteq \mathrm{Dom}(\varphi)$) there must be some $y \in M$ with $x_n \leq y, \forall n$; wherefrom $y \in M_u$. Hence, (21a) holds for our data, as claimed. By Theorem 14.24 it then follows that, for the starting point $u \in M_u$ there exists another one $v \in M_u$ with **j)** $u \leq v$ and **jj)** v is $(\leq, \varphi)$-maximal in M_u. The former of these yields $v \in \mathrm{Dom}(\varphi)$. And, from the latter one, it results that v is $(\leq, \varphi)$-maximal in M; because $v \leq w \in M \Longrightarrow v \leq w \in M_u \Longrightarrow \varphi(v) = \varphi(w)$. □

For the moment, Proposition 14.32 is reducible to Theorem 14.24. The converse reduction also holds; because $\mathrm{Dom}(\psi) = M$ when $\psi(M) \subseteq R$. Summing up, these ordering principles are logically equivalent.

(B) Let $\mathscr{S}(M, \leq)$ stand for the class of all ascending sequences in M. By a (sequential) *convergence structure* on $(M, \leq)$ we mean, as in Kasahara [38], any part $\mathscr{C}$ of $\mathscr{S}(M, \leq) \times M$ with the properties

(a24) $x_n = x, \forall n \in N \Longrightarrow ((x_n); x) \in \mathscr{C}$

(b24) $((x_n); x) \in \mathscr{C} \Longrightarrow ((y_n); x) \in \mathscr{C}$, for each subsequence (y_n) of (x_n).

In this case, $((x_n); x) \in \mathscr{C}$ will be denoted $x_n \xrightarrow{\mathscr{C}} x$; and referred to as: x is the $\mathscr{C}$-*limit* of (x_n). When x is generic in this convention, we say that (x_n) is $\mathscr{C}$*-convergent*. Now, given such a structure, the natural way of treating (24a) is that of "splitting" it as

(c24) $(\forall (x_n) \subseteq \mathrm{Dom}(\varphi))$ ascending $\Longrightarrow$ $\mathscr{C}$-convergent $\Longrightarrow$ bounded above.

Precisely, the following convergence type version of Proposition 14.32 is available (cf. Turinici [78]):

Proposition 14.33. *Suppose that*

(24b) each ascending sequence in $\mathrm{Dom}(\varphi)$ *is* $\mathscr{C}$*-convergent*

(24c) the $\mathscr{C}$*-limit of each ascending* $\mathscr{C}$*-convergent sequence in* $\mathrm{Dom}(\varphi)$ *is an upper bound of it.*

Then, the conclusion of Proposition 14.32 is retainable.

The proof is immediate (via Proposition 14.32); just note that (24b)+(24c) $\Longrightarrow$ (24a). Hence, Proposition 14.33 is deductible from the quoted result. But, the reciprocal deduction is also possible; to verify this, it will suffice taking the (sequential) convergence structure over $(M,\leq)$ as the *bounded from above* property $\mathscr{B}$ [introduced as: $x_n \xrightarrow{\mathscr{B}} x$ iff $x_n \leq x$, for all n]. Summing up, these ordering principles are mutually equivalent; and as such, equivalent with Brezis–Browder's principle [12] (Theorem 14.24).

(C) Let us now return to the developments in Subsection 14.2.3. The pseudometric setting introduced there is an appropriate one for discussing the usefulness of the above concept in extending Theorem 14.31 above. For reasons of simplicity we shall work with its amorphous version. Precisely, denote by $\mathscr{S}(M)$, the class of all sequences (x_n) in M. By a (sequential) *convergence structure* on M we mean any part $\mathscr{C}$ of $\mathscr{S}(M)\times M$ with the properties (a24)+(b24) above. Assume that we fixed such an object and let $(\leq,d)$ be taken as before. Call the subset Z of M, $(\leq)$-*closed* (modulo $\mathscr{C}$) when the $\mathscr{C}$-limit of each ascending sequence in Z is an element of it. Further, let us say that $(\leq)$ is *self-closed* (modulo $\mathscr{C}$) when $M(x,\leq)$ is $(\leq)$-closed (modulo $\mathscr{C}$), for each $x\in M$; or, equivalently: the $\mathscr{C}$-limit of each ascending sequence is an upper bound of it. Finally, term the (reflexive) pseudometric d, $(\leq)$-*complete* (modulo $\mathscr{C}$) when each ascending d-Cauchy sequence is $\mathscr{C}$-convergent.

We may now give an appropriate answer to the posed question.

Theorem 14.34. *Suppose that $(\leq)$ is self-closed (modulo $\mathscr{C}$), d is $(\leq)$-complete (modulo $\mathscr{C}$) and $(M,\leq)$ is weakly regular (modulo d). Then, conclusions of Theorem 14.31 are retainable.*

Proof. We claim that, under the accepted conditions, Theorem 14.31 is applicable to $(M,\leq;d)$; precisely, that $(M,\leq)$ is sequentially inductive (modulo φ_d). Let (x_n) be an ascending sequence with $\varphi_d(x_n)\to 0$. In particular, it is an ascending d-Cauchy sequence; so that (by the $(\leq)$-completeness (modulo $\mathscr{C}$) of d) $x_n \xrightarrow{\mathscr{C}} y$, for some $y\in M$. Combining with the self-closeness (modulo $\mathscr{C}$) of $(\leq)$ yields $x_n\leq y$, for all n; and this proves the claim. □

Now, a good choice for our convergence structure is $\mathscr{C}=(\xrightarrow{d})$ [introduced as: $x_n \xrightarrow{d} x$ whenever $d(x_n,x)\to 0$ as $n\to\infty$; and called the *primal convergence* structure attached to d]. For, if (in addition) d is *triangular* [$d(x,z)\leq d(x,y)+d(y,z), \forall x,y,z\in M$], Theorem 14.34 includes the statement by Kang and Park [37]; which, in turn, includes the maximality principle by Granas and Horvath [25]. Further aspects of structural nature may be found in Gajek and Zagrodny [23]; see also Brunner [14] and Turinici [74].

(D) Let $(M,\leq)$ be a quasi-ordered structure. Denote $\mathscr{I}(M)=\{(x,x);x\in M\}$ (the *diagonal* of M); and let $\mathscr{V}$ be a family of parts in $M\times M$. Under a convention similar to that in Nachbin [50, Ch 2, Sect 2], we say that $\mathscr{V}$ is a *pseudo-uniformity* over it when $\cap\mathscr{V}\supseteq\mathscr{I}(M)$. Suppose that we introduced such an object. The associated (sequential) convergence structure $(\mathscr{V})$ on $(M,\leq)$ may be described as

(d24) $\quad x_n \xrightarrow{(\mathscr{V})} x$ iff $\forall V\in\mathscr{V},\exists n(V):\ n\geq n(V)\Longrightarrow (x_n,x)\in V.$

In addition to this, we may introduce the $\mathscr{V}$-*Cauchy* property for such a sequence as $\forall V\in\mathscr{V},\exists n(V): n(V)\leq p\leq q\Longrightarrow (x_p,x_q)\in V$. Now, given $\varphi: M\to R\cup\{\infty\}$ as before, the natural way of treating (24a) is that of further "splitting" (c24) as

(e24) $(\forall (x_n) \subseteq \mathrm{Dom}(\varphi))$
ascending $\Longrightarrow$ $\mathscr{V}$-Cauchy $\Longrightarrow$ $(\mathscr{V})$-convergent $\Longrightarrow$ bounded above.

To state a result of this type, we need some conventions and auxiliary facts. Call the sequence (x_n), *$\mathscr{V}$-asymptotic* when $\forall V \in \mathscr{V}, \exists n(V) : n \geq n(V) \Longrightarrow (x_n, x_{n+1}) \in V$. Clearly, each $\mathscr{V}$-Cauchy sequence is $\mathscr{V}$-asymptotic too. The converse is also true, if *all* such sequences are involved; for example, the global conditions below are equivalent to each other

(24d) each ascending sequence in $\mathrm{Dom}(\varphi)$ is $\mathscr{V}$-Cauchy
(24e) each ascending sequence in $\mathrm{Dom}(\varphi)$ is $\mathscr{V}$-asymptotic.

By definition, either of these will be referred to as $(M, \leq)$ is *regular* (modulo $(\mathscr{V}, \varphi)$). Finally, call the ambient pseudo-uniform structure $(M, \mathscr{V})$, *sequentially complete* (modulo $(\leq, \varphi)$) when each ascending $\mathscr{V}$-Cauchy sequence in $\mathrm{Dom}(\varphi)$ is $(\mathscr{V})$-convergent. The following "uniform" type version of Proposition 14.33 is then available.

Proposition 14.35. *Suppose that (24c) (related to $(\mathscr{V})$) holds, $(M, \leq)$ is regular (modulo $(\mathscr{V}, \varphi)$) and $(M, \mathscr{V})$ is sequentially complete (modulo $(\leq, \varphi)$). Then, conclusion of Proposition 14.33 is retainable.*

The proof is immediate (via Proposition 14.33); just note that the last two conditions above imply (24b) (related to $(\mathscr{V})$). As a consequence of this, Proposition 14.35 is deductible from the quoted result; hence, *a fortiori*, from Proposition 14.32. But, the reciprocal deduction is also possible. In fact, let the premises of Proposition 14.32 be in force; and put $\mathscr{V} = \{\mathrm{gr}(\leq)\}$, where $\mathrm{gr}(\leq) := \{(x,y) \in M \times M; x \leq y\}$. Clearly, $(M, \leq)$ is regular (modulo $(\mathscr{V}, \varphi)$). On the other hand, the associated (sequential) convergence structure $(\mathscr{V})$ over $(M, \leq)$ (cf. (d24)) is just the bounded from above property $\mathscr{B}$; and this tells us that (24c) is fulfilled. Finally, $(M, \mathscr{V})$ is sequentially complete (modulo $(\leq, \varphi)$), via (24a). Summing up, these ordering principles are mutually equivalent; and as such, equivalent with the Brezis–Browder principle [12] (Theorem 14.24).

The discussed particular case is an "extremal" one. To get "standard" examples in the area, we need some conventions. Let $\mathscr{V}$ be some family of parts in $M \times M$; we call it, a *fundamental system of entourages* for a uniformity over M, when (cf. Bourbaki [10, Ch 2, Sect 1])

(f24) $(\mathscr{V}, \supseteq)$ is directed and $\cap \mathscr{V} \supseteq \mathscr{I}(M)$
(g24) $\forall V \in \mathscr{V}, \exists W \in \mathscr{V} : W \subseteq V^{-1}, W \circ W \subseteq V$.

The uniformity in question is just $\mathscr{U} = \{P \subseteq M \times M;\ P \supseteq Q$, for some $Q \in \mathscr{V}\}$. As a rule, the "uniform" terminology refers to it. However (as results directly by definition), all $\mathscr{U}$-notions are in fact $\mathscr{V}$-notions. For example, the (sequential) convergence structure $(\mathscr{U})$ (introduced via (d24)) is nothing else than $(\mathscr{V})$. Likewise, the (attached to $\mathscr{U}$) Cauchy and asymptotic properties are identical with the (corresponding) ones related to $\mathscr{V}$.

The following "standard" version of Proposition 14.35 is available. (As precise before, $(M,\leq)$ is a quasi-ordered structure; and $\varphi : M \to R\cup\{\infty\}$ is some proper bounded from below function as in (21b)).

Proposition 14.36. *Assume that*

(24f) the limit of each ascending sequence in M is an upper bound of it (modulo $(\leq)$)
(24g) ($\mathscr{V}$ is $(\leq,\varphi)$-compatible) $\forall V\in\mathscr{V},\exists\varepsilon=\varepsilon(V)>0$ such that:
$x,y\in\mathrm{Dom}(\varphi), x\leq y, \varphi(x)-\varphi(y)<\varepsilon \Longrightarrow (x,y)\in V$
(24h) ($\mathscr{V}$ is sequentially $(\leq)$-complete)
each ascending $\mathscr{V}$-Cauchy sequence is $(\mathscr{V})$-convergent.

Then, for each $u\in\mathrm{Dom}(\varphi)$ there exists a $(\leq,\varphi)$-maximal $v\in\mathrm{Dom}(\varphi)$ with $u\leq v$.

Proof. Let (x_n) be an ascending (modulo $\leq$) sequence in $\mathrm{Dom}(\varphi)$. The sequence $(\varphi(x_n))$ is decreasing and bounded from below; hence a Cauchy one ($\forall\varepsilon>0,\exists n(\varepsilon)$: $n(\varepsilon)\leq p\leq q \Longrightarrow \varphi(x_p)-\varphi(x_q)<\varepsilon$). This, along with (24g), assures us that $(M,\leq)$ is regular (modulo $(\mathscr{V},\varphi)$). On the other hand, (24f) $\Longrightarrow$ (24c); and (24h) assures us that $(M,\mathscr{V})$ is sequentially complete (modulo $(\leq,\varphi)$). Hence, Proposition 14.35 is applicable to $(M,\leq)$, φ and $\mathscr{V}$; wherefrom, all is clear. □

Some remarks are in order. A direct consequence of (24g) is

$$\text{if } x,y\in\mathrm{Dom}(\varphi) \text{ are comparable (either } x\leq y \text{ or } y\leq x) \text{ and } \varphi(x)=\varphi(y) \text{ then } (x,y)\in\cap\mathscr{V}. \tag{14.27}$$

This gives the generic inclusion

$$(\forall z\in\mathrm{Dom}(\varphi))\ (\leq,\varphi)\text{-maximal} \Longrightarrow (\leq,\mathscr{V})\text{-maximal}; \tag{14.28}$$

where the last property means: $w\in M$ and $z\leq w$ imply $(z,w)\in\cap\mathscr{V}$. So, Proposition 14.36 is, at the same time, an existence principle for $(\leq,\mathscr{V})$-maximal elements; and, as such, it may be compared with a related statement in Turinici [70]. In particular, when $\mathscr{V}$ is *separated* ($\cap\mathscr{V}=\mathscr{I}(M)$), we have

$$(\forall z\in M)\ (\leq,\mathscr{V})\text{-maximal} \Longrightarrow \text{strongly } (\leq)\text{-maximal}; \tag{14.29}$$

and Proposition 14.36 yields the maximal principle in Hamel [26, Theorem 1], obtainable via different reasoning. On the other hand, (14.27) (and the separated property of $\mathscr{V}$) also gives (via (21b))

$$x,y\in\mathrm{Dom}(\varphi),\ x\leq y, y\leq x \Longrightarrow x=y. \tag{14.30}$$

The ambient quasi-order $\leq$ is therefore *antisymmetric* (hence an *order*) on $\mathrm{Dom}(\varphi)$. In fact, it may be viewed as such over all of M; for, otherwise, passing to the order (on M)

(h24) $x\preceq y$ iff either $[x,y\in M\setminus\mathrm{Dom}(\varphi), x=y]$ or $[x,y\in\mathrm{Dom}(\varphi), x\leq y]$, the regularity conditions (24f)–(24h) are retainable; and we are done. Summing up, the "separated" variant of Proposition 14.36 is identical with the main result in Brondsted [13, Theorem 1] (if we also take into account Remark 1 in that paper). In addition, the developments above tell us that the precise variant is deductible from the Brezis–Browder ordering principle [12]; see also Turinici [65]. The question of the reciprocal deduction being also possible remains open; we conjecture that the answer is negative. A partial motivation of this position comes from the fact that, in the particular case of $M = R$ (endowed with the standard order and (metrical) uniformity) any (proper bounded from below) function φ which satisfies (21b) and the compatibility condition (24g) must be injective over $\mathrm{Dom}(\varphi)$ (if we take (14.27) into account). Further aspects may be found in Isac [32] and Mizoguchi [48]; see also Szaz [61].

14.2.5 Zorn Maximality Principles

Let us now return to the setting of Subsection 14.2.2. Precisely, given the nonempty set M, take a reflexive (generalized) pseudometric $(x,y) \mapsto d(x,y)$ over it. Remember that, when d is (in addition) sufficient, the point $v \in M$ assured by Proposition 14.26 is strongly $(\leq)$-maximal. In the absence of this property, we may ask whether the weaker counterpart of this concept is holding: $w \in M, z \leq w \Longrightarrow z \geq w$ (referred to as: z is $(\leq)$-*maximal*). To establish a maximality result of this type, we need some conventions. Let $\mathrm{dist}(.,.)$ stand for the associated (to d) point to set distance function $[\mathrm{dist}(x,Z) = \inf\{d(x,z); z \in Z\}, x \in M, Z \subseteq M]$. The working hypothesis to be considered is

(25a) $((M,\leq)$ is almost weakly regular (modulo d))
$\forall x \in M,\ \forall \varepsilon > 0,\ \exists y = y(x,\varepsilon) \geq x\text{: } y \leq u \leq v \Longrightarrow \mathrm{dist}(u, M(v,\leq)) \leq \varepsilon.$

This is nothing else but the condition (22a) with respect to the function

(a25) $\psi_d(x) = \sup\{\mathrm{dist}(u, M(v,\leq)); x \leq u \leq v\},\ \ x \in M.$

Further, let $(\overset{d}{\longleftarrow})$ stand for the *dual convergence* attached to d [introduced by the convention: $x \overset{d}{\longleftarrow} x_n$ if and only if $d(x,x_n) \to 0$ as $n \to \infty$].

Theorem 14.37. *Assume that* $(M,\leq)$ *is sequentially inductive (modulo* ψ_d*) and almost weakly regular (modulo d); and* $(\leq)$ *is self-closed (modulo* $(\overset{d}{\longleftarrow})$*). Then, for each* $u \in M$ *there exists a* $(\leq)$*-maximal* $v \in M$ *with* $u \leq v$*; i.e.,* $(\leq)$ *appears as a Zorn quasi-order.*

Proof. By the admitted hypotheses (on $(M,\leq)$) it follows via Theorem 14.28 that, for the starting point $u \in M$, there exists another one $v \in M$ with $u \leq v$ and $\psi_d(v) = 0$ (hence v is $(\leq,\psi_d)$-maximal). We now claim that the generic implication is valid: $(\forall z \in M)\ \psi_d(z) = 0 \Longrightarrow z$ is $(\leq)$-maximal. (And from this, the conclusion is clear.) For, take some $w \geq z$. Since $[\mathrm{dist}(z, M(y,\leq)) = 0, \forall y \geq w]$, it is not hard to construct an ascending sequence (x_n) in $M(w, \leq)$ with $z \overset{d}{\longleftarrow} x_n$. But then, the choice of $(\leq)$ yields $x_n \leq z, \forall n$; hence $w \leq z$, as claimed. □

The following completion of this fact may be noted. Call the (ascending) sequence (x_n), *eventually d-asymptotic* when

$$\forall n, \forall \varepsilon > 0, \exists (p,q) : n \leq p < q, d(x_p, x_q) < \varepsilon.$$

This is a weaker form of the d-asymptotic property introduced in Subsection 14.2.2. Precisely, the generic implication is clear: (for each sequence) d-asymptotic $\Longrightarrow$ eventually d-asymptotic; but, the converse is not in general valid. Let us now consider the condition

(25b) (($M, \leq$) is eventually regular (modulo d))
each ascending sequence is eventually d-asymptotic.

We claim that this is a sufficient one for (25a) above. In fact, assume this were not true; then, there must be some pair $x \in M, \varepsilon > 0$ with $[\forall y \geq x, \exists (u,v) : y \leq u \leq v, \text{dist}(u, M(v, \leq)) \geq \varepsilon]$. Put $x_0 = x$; with $y = x_0$ we get a couple (x_1, x_2) with $x_0 \leq x_1 \leq x_2$, $\text{dist}(x_1, M(x_2, \leq)) \geq \varepsilon$. Further, with $y = x_2$ there exist (x_3, x_4) with $x_2 \leq x_3 \leq x_4$, $\text{dist}(x_3, M(x_4, \leq)) \geq \varepsilon$; and so on. This finally gives us an ascending sequence (x_n) with: $d(x_{2p+1}, x_k) \geq \varepsilon$, for all $k > 2p+1$ and all $p \geq 0$. So, for the ascending sequence $(y_n = x_{2n+1})$ we must have $d(y_p, y_q) \geq \varepsilon$, for all $p, q \geq 1$ with $p < q$; in contradiction with the eventual d-asymptotic property of it, ensured by (25b); hence the claim. As a direct consequence, we have (cf. Turinici [68]):

Theorem 14.38. *Assume that $(M, \leq)$ is sequentially inductive and eventually regular (modulo d); and $(\leq)$ is self-closed (modulo $(\overset{d}{\longleftarrow})$). Then, conclusions of Theorem 14.34 are retainable.*

In particular, assume that

(25c) M is $(\leq, d)$-compact (modulo $(\overset{d}{\longleftarrow})$): each ascending sequence
has a d-Cauchy convergent (modulo $(\overset{d}{\longleftarrow})$) subsequence.

The first half of this (related to the d-Cauchy property) gives at once (25b). And the second half of the same (involving the convergence (modulo $(\overset{d}{\longleftarrow})$) property) gives (21a) if one admits the self-closeness (modulo $(\overset{d}{\longleftarrow})$) of $(\leq)$. We therefore deduced:

Theorem 14.39. *Assume that M, $(\leq)$ and d are taken so as M is $(\leq, d)$-compact and $(\leq)$ is self-closed (modulo $(\overset{d}{\longleftarrow})$). Then, for each $u \in M$ there exists a $(\leq)$-maximal $v \in M$ with $u \leq v$.*

Note that, when d is (in addition) triangular and *symmetric* $[d(x,y) = d(y,x)$, for all $x, y \in M]$, the regularity condition (25c) reads in the standard way

(25d) M is $(\leq, d)$-compact:
each ascending sequence has a convergent subsequence;

and the corresponding version of Theorem 14.39 includes the metrical portion of a related statement in Ward [81]. Some applications of these to mapping theory may be found in Park and Yie [52].

14.3 Relative KST Statements

14.3.1 Introduction

Let (M,d) be a complete metric space; and $\varphi : M \to R \cup \{\infty\}$, some function with

(31a) φ is inf-proper ($\mathrm{Dom}(\varphi) \neq \emptyset$ and $\varphi_* := \inf[\varphi(M)] > -\infty$)

(31b) φ is d-lsc ($\liminf_n \varphi(x_n) \geq \varphi(x)$, whenever $x_n \xrightarrow{d} x$).

The following 1974 statement in Ekeland [19] (referred to as Ekeland's variational principle; in short: EVP) is deductible at once from the developments above (in Subsection 14.2.2):

Theorem 14.40. *Let the precise conditions hold. Then*
a) *for each $u \in \mathrm{Dom}(\varphi)$ there exists $v = v(u) \in \mathrm{Dom}(\varphi)$ with*

$$d(u,v) \leq \varphi(u) - \varphi(v) \text{ (hence } \varphi(u) \geq \varphi(v)) \tag{14.31}$$

$$\begin{aligned} &v \text{ is E-variational (modulo } (d,\varphi)\text{):} \\ &d(v,x) > \varphi(v) - \varphi(x), \quad \text{for all } x \in M \setminus \{v\} \end{aligned} \tag{14.32}$$

aa) *if $u \in \mathrm{Dom}(\varphi)$, $\rho > 0$ fulfill $\varphi(u) - \varphi_* \leq \rho$, then (14.31) gives*

$$(\varphi(u) \geq \varphi(v) \text{ and}) \quad d(u,v) \leq \rho. \tag{14.33}$$

This principle found some basic applications to control and optimization, generalized differential calculus, critical point theory and global analysis; we refer to the 1979 paper by Ekeland [20] for a survey of these. So, it cannot be surprising that, soon after its formulation, many extensions of EVP were proposed. For example, the abstract (order) one starts from the fact that, with respect to the (quasi-)order

(a31) $(x,y \in M)\ x \leq y$ iff $d(x,y) + \varphi(y) \leq \varphi(x)$

the point $v \in M$ appearing in (14.32) is *maximal*; so that, EVP is nothing but a variant of the Zorn–Bourbaki maximality principle [87], [9]. The dimensional way of extension refers to the support space (R) of $\mathrm{Codom}(\varphi)$ being substituted by a (topological or not) vector space. An account of the results in this area may be found in the 2003 monograph by Goepfert, Riahi, Tammer and Zălinescu [24, Ch 3]; see also Isac [33], Rozoveanu [53], and Turinici [72]. Finally, the metrical one consists in the conditions imposed to the ambient metric over M being relaxed. The basic 1996 result in this direction obtained by Kada, Suzuki and Takahashi [36] may be stated as follows. By a *pseudometric* over M we shall mean any map

$(x,y) \mapsto e(x,y)$ from $M \times M$ to $R_+ := [0,\infty[$. Suppose that we fixed such an object; which, in addition, is *triangular* $[e(x,z) \leq e(x,y) + e(y,z), \forall x,y,z \in M]$. We say that it is a *w-distance* (modulo d) over M provided

(b31) $y \mapsto e(x,y)$ is d-lsc on M (see above), $\forall x \in M$
(c31) e is strongly d-sufficient: for each $\varepsilon > 0$, there exists $\delta > 0$ such that: $e(z,x), e(z,y) \leq \delta \Longrightarrow d(x,y) \leq \varepsilon$.

Theorem 14.41. *Let the conditions in Theorem 14.40 be admitted; and e be some w-distance (modulo d) over M. Then*

b) *For each* $u \in \mathrm{Dom}(\varphi)$, *there exists an E-variational (modulo* (e,φ)*)* $v = v(u) \in \mathrm{Dom}(\varphi)$ *with* $\varphi(u) \geq \varphi(v)$

bb) *For each* $\rho > 0$ *and each* $u \in \mathrm{Dom}(\varphi)$ *with* $e(u,u) = 0$, $\varphi(u) \leq \varphi_* + \rho$ *there exists an E-variational (modulo* (e,φ)*)* $v = v(u,\rho) \in \mathrm{Dom}(\varphi)$ *with* $\varphi(u) \geq \varphi(v)$ *and* $e(u,v) \leq \rho$.

In particular, when $e = d$, these regularity conditions hold; and Theorem 14.41 includes the local version of EVP based upon (14.33). The relative form of the same, based upon (14.31), also holds, but indirectly; see Bao and Khanh [7] for details. Note that the (rather involved) authors' argument relies on the nonconvex minimization theorem in Takahashi [62]; see also Ume [80]. It is our aim in the following to show (cf. Subsection 14.3.4) that a simplification of this is possible. The basic tool of our investigations is (according to Subsection 14.3.3) a pseudometric variational principle (including the one in Tataru [64]) deductible from a set of maximal principles (discussed in Subsection 14.3.2) comparable with Brezis–Browder's principle [12]. As an argument for its usefulness, we show that the variational principles in Suzuki [58], Lin and Du [44], or Al-Homidan, Ansari and Yao [2] are obtainable from such an approach.

14.3.2 Maximal Principles

(A) Let M be a nonempty set; and $(\leq)$, some *quasi-order* (i.e., reflexive and transitive relation) over it. By a *pseudometric* over M we shall mean any map $d : M \times M \to R_+$. If, in addition, d is *reflexive* $[d(x,x) = 0, \forall x \in M]$, *triangular* $[d(x,z) \leq d(x,y) + d(y,z), \forall x,y,z \in M]$ and *symmetric* $[d(x,y) = d(y,x), \forall x,y \in M]$, we say that it is a *semimetric* (on M). Suppose that we fixed such an object. Call the point $z \in M$, *$(\leq,d)$-maximal*, in case: $w \in M$ and $z \leq w$ imply $d(z,w) = 0$. To get sufficient conditions for these, one may proceed as below. Call the (ascending) sequence (x_n) (in M), *d-Cauchy* when $\forall \varepsilon > 0, \exists n(\varepsilon)$ such that $n(\varepsilon) \leq p \leq q \Longrightarrow d(x_p,x_q) \leq \varepsilon$; and *d-asymptotic*, provided $d(x_n,x_{n+1}) \to 0$, as $n \to \infty$. Clearly, each (ascending) d-Cauchy sequence is d-asymptotic too. The reverse implication is also true when all such sequences are involved; i.e., the global conditions below are equivalent each other:

(32a) each ascending sequence is d-Cauchy

(32b) each ascending sequence is d-asymptotic.

By definition, either of these will be referred to as $(M, \leq)$ is *regular* (modulo d). Further, call $(M, \leq)$, *sequentially inductive* when each ascending sequence has an upper bound (modulo $(\leq)$). The following answer to the posed question obtained in Turinici [71] is available (cf. Subsection 14.2.2):

Proposition 14.42. *Assume that*

(32c) $(M, \leq)$ *is regular (modulo d) and sequentially inductive.*

Then, for each $u \in M$ *there exists an* $(\leq, d)$*-maximal* $v \in M$ *with* $u \leq v$.

(B) An interesting completion of this fact may be given under the lines below. Let the function $\varphi : M \to R \cup \{\infty\}$ be such that

(32d) φ is proper ($\mathrm{Dom}(\varphi) \neq \emptyset$), bounded from below ($\inf[\varphi(M)] > -\infty$)
(32e) φ is $(\leq)$-decreasing ($x \leq y \Longrightarrow \varphi(x) \geq \varphi(y)$).

Proposition 14.43. *Assume that (in addition) conditions of Proposition 14.42 are fulfilled over* $\mathrm{Dom}(\varphi)$*; in the sense*

(32f) $(\mathrm{Dom}(\varphi), \leq)$ *is regular (modulo d):*
each ascending sequence in $\mathrm{Dom}(\varphi)$ *is d-Cauchy*
(32g) $(\mathrm{Dom}(\varphi), \leq)$ *is relatively sequentially inductive:*
each ascending sequence in $\mathrm{Dom}(\varphi)$ *has an upper bound (in M).*

Then, for each $u \in \mathrm{Dom}(\varphi)$ *there exists* $v \in \mathrm{Dom}(\varphi)$ *with* **i)** $u \leq v$ *and* **ii)** $v \leq x \Longrightarrow d(v,x) = 0,\ \varphi(v) = \varphi(x)$.

Proof. By (32d)+(32e), φ has finite values over $M(u, \leq) := \{x \in M; u \leq x\}$; so, the convention

(a32) $e(x,y) = \max\{d(x,y), |\varphi(x) - \varphi(y)|\}, \quad x, y \in M(u, \leq)$

is meaningful and introduces a semimetric over it. We claim that Proposition 14.42 is applicable to $M(u, \leq)$ and $(\leq, e)$; wherefrom, all is clear. Let (x_n) be an ascending sequence in $M(u, \leq)$. The real sequence $(\varphi(x_n))$ is (again by (32d)+(32e)) descending and bounded; hence a Cauchy one. In addition, (x_n) is d-Cauchy by (32f); wherefrom, it is e-Cauchy; and this tells us that $(M(u, \leq), \leq)$ is regular (modulo e). On the other hand, each upper bound in M of (x_n) (existing by (32g)) belongs to $M(u, \leq)$; wherefrom, $(M(u, \leq), \leq)$ is sequentially inductive. Hence, the claim follows. □

In particular, (32f) holds under

(32h) $x, y \in \mathrm{Dom}(\varphi)$ and $x \leq y$ imply $d(x,y) \leq \varphi(x) - \varphi(y)$;

and Proposition 14.43 reduces to the Brezis–Browder ordering principle [12]. [In fact, the converse reduction is also possible; hence, these two results are logically equivalent]. Note that, by the developments in Cârjă, Necula and Vrabie [15, Ch 2,

Sect 2.1], this last statement is reducible to the *principle of dependent choices* (see, e.g., Wolk [82]); hence, so does Proposition 14.43.

(C) A useful application of these developments is to monotone variational principles. Let the triple $(M, \leq; d)$ be introduced as above; and the function $\varphi : M \to R \cup \{\infty\}$ be taken as in (32d)+(32e).

Proposition 14.44. *Assume that (in addition)*

(32i) $\quad$ (Dom(φ) *is relatively* $(\leq)$*-complete:*
each ascending d-Cauchy sequence in Dom(φ) *converges (in M)*
(32j) $\quad$ $(\leq)$ *is relatively self-closed over* Dom(φ)*:*
the limit of each ascending sequence in Dom(φ)
is an upper bound of it (in M)

Then, for each $u \in \mathrm{Dom}(\varphi)$ *there exists* $v \in \mathrm{Dom}(\varphi)$ *with* **j)** $u \leq v$, $d(u,v) \leq \varphi(u) - \varphi(v)$ *and* **jj)** $v \leq x$, $d(v,x) \leq \varphi(v) - \varphi(x)$ *imply* $d(v,x) = 0$, $\varphi(v) = \varphi(x)$.

Proof. Let $(\preceq)$ stand for the quasi-order

(b32) $\quad (x, y \in M)$: $x \preceq y$ iff $x \leq y$, $d(x,y) + \varphi(y) \leq \varphi(x)$.

It will suffice verifying that Proposition 14.43 applies to $(M, \preceq)$ and d; precisely, that (32f)+(32g) hold with $(\preceq)$ in place of $(\leq)$. Let (x_n) be an ascending (modulo $(\preceq)$) sequence in Dom(φ)

(32k) $\quad x_n \leq x_m$, $d(x_n, x_m) \leq \varphi(x_n) - \varphi(x_m)$, if $n \leq m$.

The (real) sequence $(\varphi(x_n))$ is (by (32d)+(32e)) descending and bounded; hence a Cauchy one. This, added to (32k), tells us that (x_n) is an ascending (modulo $(\leq)$) d-Cauchy sequence in Dom(φ); so, (32f) holds. Further, (32i) yields; $x_n \to x$ as $n \to \infty$, for some $x \in M$. Combining with (32j) gives (via (32e)): $x_n \leq x$ (hence $\varphi(x_n) \geq \varphi(x)$), for all n. Taking again (32k) into account gives (passing to limit as $m \to \infty$) $d(x_n, x) \leq \varphi(x_n) - \varphi(x)$, for all n. This, added to the previous relation shows that $x_n \preceq x$, for all n; and establishes (32g). The proof is thereby complete. □

In particular, a good choice for the ambient quasi-order $(\leq)$ is

(c32) $\quad x \leq y$ if and only if $d(x,y) + \varphi(y) \leq \varphi(x)$.

Note that, in such a case, the self-closeness condition (32j) holds under

(32m) $\quad \varphi$ is descending d-lsc over Dom(φ):
$\lim_n \varphi(x_n) \geq \varphi(x)$, whenever $(x_n) \subseteq \mathrm{Dom}(\varphi)$
satisfies $x_n \to x$ and $(\varphi(x_n))$ is descending.

The resulting variant of Proposition 14.44 (under these data) is a direct extension of Ekeland's variational principle [20] (Theorem 14.40). For this reason, the obtained statement will be referred to as the *monotone* EVP. An interesting question is that of Proposition 14.44 being deductible in a direct way from EVP; we conjecture that the answer is positive. Further aspects may be found in Turinici [69]; see also Borwein and Preiss [8].

14.3.3 Transitive (Pseudometric) Versions

Let M be some nonempty set and (∇) be some *transitive* (over M) relation ($x\nabla y$ and $y\nabla z$ imply $x\nabla z$). Denote

(a33) $(x,y \in M)$ $x \leq y$ iff either $x = y$ or $x\nabla y$.

This is a *quasi-order* on M; which, in addition, fulfills

$$[(x\nabla y, y \leq z) \text{ or } (x \leq y, y\nabla z)] \Longrightarrow x\nabla z. \tag{14.34}$$

Denote, for simplicity reasons

(b33) $M(x,\nabla) = \{y \in M; x\nabla y\}$ (the *x-section* of (∇)), $x \in M$.

Further, take a function $\psi : M \to R_+$. The (∇)-decreasing property for it is introduced as (cf. Subsection 14.2.1)

(c33) $x,y \in M, x\nabla y \Longrightarrow \psi(x) \geq \psi(y)$.

Note that, by (a33) above,

$$\psi \text{ is } (\nabla)\text{-decreasing} \Longleftrightarrow \psi \text{ is } (\leq)\text{-decreasing}.$$

Further, call the point $z \in M$, (∇,ψ)-*maximal*, provided

(d33) $w \in M$ and $z\nabla w$ imply $\psi(z) = \psi(w)$.

For a non-trivial concept, we must take z as (∇)-*starting* (in the sense: $M(z,\nabla) \neq \emptyset$); for, otherwise, z is vacuously (∇,ψ)-maximal. Note that such a requirement holds whenever ∇ is *z-reflexive* (i.e., $z\nabla z$). Again by (a33), the generic property holds

$$(\text{for each } z \in M)\ (\nabla,\psi)\text{-maximal} \Longleftrightarrow (\leq,\psi)\text{-maximal}.$$

As a consequence, maximality results involving our transitive relation (∇) are deductible from the Brezis–Browder principle (cf. Subsection 14.3.2) written for its associated quasi-order $(\leq)$. The key moment of this approach is that of the sequential inductivity condition about $(M,\leq)$ being assured. It would be useful to have this property expressed in terms of (∇). Call the sequence (x_n), *ascending* (modulo (∇)) when

$$x_n\nabla x_{n+1}, \forall n \text{ (or, equivalently: } x_n\nabla x_m \text{ if } n < m).$$

Note the generic (sequential) relation

$$\text{ascending (modulo } (\nabla)) \Longrightarrow \text{ascending (modulo } (\leq)).$$

The reciprocal is not in general true. For example, the constant sequence $(x_n = a; n \in N)$ is ascending (modulo $(\leq)$); but not ascending (modulo (∇)), provided that $a\nabla a$ is false. Further, given the sequence (x_n) in M, let us say that $u \in M$ is an *upper bound* (modulo (∇)) of it provided

$$x_n \nabla u, \text{ for all } n \text{ (written as: } (x_n)\nabla u).$$

If u is generic in this convention, we say that (x_n) is *bounded above* (modulo (∇)). As before, the relation below is clear

$$\text{bounded above (modulo } (\nabla)) \Longrightarrow \text{bounded above (modulo } (\leq)).$$

(The converse is not in general valid). Finally, let the concept of *sequential inductivity* for (M,∇) be that of (32a) [with (∇) in place of $(\leq)$].

Lemma 14.45. *Under the precise setting,*

$$(M,\nabla) \text{ sequentially inductive} \Longleftrightarrow (M,\leq) \text{ sequentially inductive.}$$

We are now in position to give an appropriate answer to our question; cf. Turinici [77]:

Proposition 14.46. *Assume that the structure (M,∇) is sequentially inductive and the function ψ is (∇)-decreasing. Then, for each (∇)-starting $u \in M$ there exists a (∇,ψ)-maximal $v \in M$ with $u\nabla v$.*

Proof. Let $(\leq)$ stand for the quasi-order (a33). By the remarks above (and Lemma 14.45), the Brezis–Browder principle is applicable to these data. So, for the arbitrary fixed $u_1 \in M(u,\nabla)$ there exists $v \in M$ with

$$u_1 \leq v(\text{ i.e.: either } u_1 = v \text{ or } u_1\nabla v); \text{ and } v \text{ is } (\leq,\psi)\text{-maximal.}$$

This yields $u\nabla v$; and the proof is complete. □

Clearly, the Brezis–Browder principle follows from Proposition 14.46. The reciprocal inclusion also holds, by the argument above. Hence these two statements are logically equivalent. Nevertheless, a direct use of Proposition 14.46 is more profitable in all concrete situations involving explicitly (∇); cf. Subsection 14.3.4.

An interesting completion of Proposition 14.46 may be given under the lines below. Let us introduce the concept of (∇)-*maximal* element as

(e33) $w \in M$ and $z\nabla w$ imply $z = w$;

this is a stronger version of the concept (d33). As before, it is effective only if z is (∇)-starting; for, otherwise, z is vacuously (∇)-maximal. To get a result involving such points, we need an extra condition upon our data:

(33a) (∇) is *ψ-sufficient*: $z\nabla x,\ z\nabla y,\ \psi(z) = \psi(x) = \psi(y) \Longrightarrow x = y$.

Proposition 14.47. *Suppose that conditions of Proposition 14.46 hold, as well as (33a). Then, for each (∇)-starting $u \in M$ there exists a (∇)-maximal $w \in M$ with $u\nabla w$.*

Proof. By Proposition 14.46, there must be some (∇, ψ)-maximal $v \in M$ with $u\nabla v$. If v is (∇)-maximal, we are done (with $w = v$); so, it remains the alternative of v fulfilling the opposite property:

$$v\nabla w \text{ (hence } \psi(v) = \psi(w)\text{), for some } w \in M \setminus \{v\}.$$

In this case, w is our desired element. Assume not: $w\nabla y$, for some $y \in M$, $y \neq w$. By the preceding relation we get $v\nabla y$ (hence $\psi(v) = \psi(y)$). Summing up, $v\nabla w$, $v\nabla y$ and $\psi(v) = \psi(w) = \psi(y)$; wherefrom (by (33a)) $w = y$, contradiction. This ends the argument. □

The obtained statement is nothing but a "transitive" form of the Zorn–Bourbaki maximality principle for these structures; cf. Hazen and Morin [30]. It may be also viewed as a counterpart of the "reflexive" version of Brezis–Browder principle due to Bae, Cho and Yeom [6]. Further aspects were delineated in Gajek and Zagrodny [23]; see also Hyers, Isac and Rassias [31, Ch 5].

14.3.4 Main Results

Let M be some nonempty set. Remember that, by a *pseudometric* over it we mean any map $e : M \times M \to R_+ \cup \{\infty\}$ (cf. Subsection 14.3.1). Suppose that we fixed such an object; which, in addition, is *triangular* (see above). Let also $\varphi : M \to R \cup \{\infty\}$ be some inf-proper function. For an easy reference, we shall formulate the basic regularity condition involving our data. This will necessitate some conventions and auxiliary facts. Call the sequence (x_n) in M, *strongly e-asymptotic* when

$$\text{the series } \sum_n e(x_n, x_{n+1}) \text{ converges (in } R\text{).}$$

Further, let the *e-Cauchy* property of this object be the usual one

$$\forall\delta > 0, \exists n(\delta), \text{ such that } n(\delta) \leq p < q \Longrightarrow e(x_p, x_q) \leq \delta.$$

The generic relation below is clear (by the triangular property of e)

$$\text{(for each sequence) strongly } e\text{-asymptotic} \Longrightarrow e\text{-Cauchy;} \tag{14.35}$$

but the converse is not in general true. Nevertheless, in many conditions involving *all* such objects, this is retainable. A concrete example may be constructed under the lines below. Let us introduce an *e-convergence* structure over M by

$$x_n \xrightarrow{e} x \text{ iff } \quad e(x_n, x) \to 0 \quad \text{as } n \to \infty.$$

We consider the regularity condition

(34a) (e,φ) is weakly descending complete: for each strongly e-asymptotic sequence (x_n) in $\mathrm{Dom}(\varphi)$ with $(\varphi(x_n))$ descending there exists $x \in M$ with $x_n \xrightarrow{e} x$ and $\lim_n \varphi(x_n) \geq \varphi(x)$.

By (14.35) above, it includes its (stronger) counterpart

(34b) (e,φ) is descending complete: for each e-Cauchy sequence (x_n) in $\mathrm{Dom}(\varphi)$ with $(\varphi(x_n))$ descending there exists $x \in M$ with the properties $x_n \xrightarrow{e} x$ and $\lim_n \varphi(x_n) \geq \varphi(x)$.

A remarkable fact to be added is that the reciprocal inclusion also holds:

Lemma 14.48. *Under the precise conditions,*

$$(34a) \Longrightarrow (34b); \quad \text{hence } (34a) \Longleftrightarrow (34b).$$

Now, call $v \in \mathrm{Dom}(\varphi)$, Brezis–Browder (in short: BB)-*variational* (modulo (e,φ)) provided

(a34) $x \in M, e(v,x) \leq \varphi(v) - \varphi(x) \Longrightarrow \varphi(v) = \varphi(x)$ (hence $e(v,x) = 0$).

Some basic properties of this concept are collected in

Lemma 14.49. *Suppose that* $v \in \mathrm{Dom}(\varphi)$ *is BB-variational (modulo* (e,φ)*). Then, the following are true*

$$e(v,x) \geq \varphi(v) - \varphi(x), \quad \text{for all } x \in M \tag{14.36}$$

$$e(v,x) > \varphi(v) - \varphi(x), \quad \text{for each } x \in M \text{ with } e(v,x) > 0. \tag{14.37}$$

Finally, let $(\nabla = \nabla_\varphi)$ stand for the transitive relation (over M)

(b34) $(x,y \in M)$ $x\nabla y$ iff $e(x,y) + \varphi(y) \leq \varphi(x)$.

Remember that $u \in M$ is called (∇)-starting if $M(u,\nabla) \neq \emptyset$; i.e.,

(c34) $e(u,x) + \varphi(x) \leq \varphi(u)$, for at least one $x \in M$.

This will be also referred to as u is *starting* (modulo (e,φ)); for example, (c34) holds under

(34c) e is *reflexive* at u: $e(u,u) = 0$.

We are now in position to state a pseudometric variational principle useful in the sequel.

Theorem 14.50. *Let the general conditions upon* (e,φ) *be accepted; as well as (one of the conditions) (34a)/(34b). Then, for each starting (modulo* (e,φ)*)* $u \in \mathrm{Dom}(\varphi)$ *there exists a BB-variational (modulo* (e,φ)*)* $v = v(u) \in \mathrm{Dom}(\varphi)$ *with*

$$e(u,v) \leq \varphi(u) - \varphi(v) \text{ (hence } \varphi(u) \geq \varphi(v)\text{)}. \tag{14.38}$$

Proof. Denote for simplicity

$$M_u = \{x \in M; \varphi(x) \leq \varphi(u)\} \text{ (where } u \text{ is the above one).}$$

Clearly, $\emptyset \neq M_u \subseteq \mathrm{Dom}(\varphi)$ (by the choice of u). Further, let us introduce the function (from M_u to R_+)

$$\psi(x) = \varphi(x) - \varphi_*, \quad x \in M_u \quad \text{(where } \varphi_* \text{ is that of (31a)).}$$

Note that, by (b34) (and the definitions above)

$$(x,y \in M_u) \;\; x\nabla y \text{ iff } e(x,y) \leq \psi(x) - \psi(y)(= \varphi(x) - \varphi(y)). \tag{14.39}$$

We claim that the pair (∇, ψ) fulfills conditions of Proposition 14.46 over M_u. In fact, ψ is (∇)-decreasing ; so, it remains to show that (M_u, ∇) is sequentially inductive. Let (x_n) be an ascending (modulo (∇)) sequence in M_u:

(34d) $\quad e(x_n, x_m) \leq \psi(x_n) - \psi(x_m)(= \varphi(x_n) - \varphi(x_m))$, whenever $n < m$.

The sequence $(\psi(x_n))$ is descending in R_+; hence a Cauchy one. Moreover, both these properties are transferrable to $(\varphi(x_n))$ as well; and, by (34d), (x_n) is an e-Cauchy sequence in M_u. Putting these together, it follows (via (34b)) that there must be some $y \in M$ with

$$x_n \xrightarrow{e} y \text{ and } \lim_n \varphi(x_n) \geq \varphi(y) \;\; \text{(hence } \lim_n \psi(x_n) \geq \psi(y)). \tag{14.40}$$

This gives on the one hand $\varphi(y) \leq \varphi(u)$ (or, equivalently, $\psi(y) \leq \psi(u)$); wherefrom $y \in M_u$ [because $(x_n) \subseteq M_u$]. On the other hand, fix some rank n. By (34d) and the triangular property (of $(x,y) \mapsto e(x,y)$)

$$e(x_n, y) \leq e(x_n, x_m) + e(x_m, y) \leq \psi(x_n) - \psi(x_m) + e(x_m, y), \forall m > n.$$

This, along with (14.40), yields by a limit process (relative to m)

$$e(x_n, y) \leq \psi(x_n) - \lim_m \psi(x_m) \leq \psi(x_n) - \psi(y) \;\; \text{(i.e.: } x_n \nabla y).$$

As n was arbitrarily chosen, one deduces that $(x_n)\nabla y$; and this proves our claim. By Proposition 14.46 it then follows that, for the starting (modulo (e, ψ)) $u \in M_u$ there exists a (∇, ψ)-maximal $v \in M_u$ with $u \nabla v$. It suffices now remarking that

$$v \text{ is } (\nabla, \psi)\text{-maximal (on } M_u) \Longleftrightarrow v \text{ is BB-variational (modulo } (e, \varphi))$$

to get all the conclusions above. □

Now, the regularity condition (34a) holds under

(34e) $\quad (e, \varphi)$ is weakly complete:
for each strongly e-asymptotic sequence (x_n) in $\mathrm{Dom}(\varphi)$
there exists $x \in M$ with $x_n \xrightarrow{e} x$ and $\liminf_n \varphi(x_n) \geq \varphi(x)$.

This, in the particular case when e is (in addition) *reflexive* ($e(x,x) = 0, \forall x \in M$), tells us that Theorem 14.50 includes the variational principle in Tataru [64]. The question of the converse inclusion being also true remains open; we conjecture that the answer is positive.

Let us now return to our initial setting. An interesting completion of Theorem 14.50 may be done under the lines of Subsection 14.3.2. Precisely, let the concept of E-*variational* (modulo (e,φ)) point be introduced as in (14.36) (with e in place of d). This is stronger than the concept of BB-variational (modulo (e,φ)) element introduced via (a34). To get a corresponding form of Theorem 14.50 involving such points, we have to impose (in addition to (34a)/(34b))

(34f) e is *transitively sufficient* $(e(z,x) = e(z,y) = 0 \Longrightarrow x = y)$.

Theorem 14.51. *Let the precise conditions be in force. Then, for each starting (modulo (e,φ)) $u \in \mathrm{Dom}(\varphi)$ there exists an E-variational (modulo (e,φ)) $w = w(u) \in \mathrm{Dom}(\varphi)$ with the property (14.38).*

Proof. Let M_u and ψ be introduced as in Theorem 14.50. As already shown, conditions of Proposition 14.46 hold over M_u for the couple (∇, ψ). On the other hand, the special regularity condition (33d) also holds in M_u for (∇, ψ), via (34f). Summing up, Proposition 14.47 is applicable to our data. It gives us, for the starting (modulo (e,φ)) $u \in M_u$, a (∇)-maximal $w \in M_u$ with $u \nabla w$. It suffices now remarking that

$$w \text{ is } (\nabla)\text{-maximal (in } M_u) \Longleftrightarrow w \text{ is E-variational (modulo } (e,\varphi))$$

to derive the written conclusions. □

As before, the regularity condition (34b) holds under

(34g) (e,φ) is complete:
for each e-Cauchy sequence (x_n) in $\mathrm{Dom}(\varphi)$
there exists $x \in M$ with $x_n \xrightarrow{e} x$ and $\liminf_n \varphi(x_n) \geq \varphi(x)$.

On the other hand, (34f) holds whenever e is *sufficient* $(e(x,y) = 0 \Longrightarrow x = y)$. Note that, in such a case, Theorem 14.51 becomes the variational principle in Turinici [73]. Another interesting choice is $e =$ (standard) metric over M, when Theorem 14.51 includes directly Ekeland's variational principle (Theorem 14.40). Further aspects were delineated in Dancs, Hegedus and Medvegyev [17]; see also Fang [21] and Hamel [27, Ch 4].

14.3.5 Extended KST Statements

We are now in position to get an appropriate answer to the questions in Subsection 14.3.1. Let M be a nonempty set; and $d : M \times M \to R_+$ be a pseudometric over it; supposed to be triangular (see above) and reflexive sufficient $(d(x,y) = 0 \Longleftrightarrow$

$x = y$). This map has all properties of a metric, except symmetry; we shall term it, an *almost metric* on M. Given such an object, the d-Cauchy and d-convergence properties for a sequence in M are those in Subsection 14.3.4. Let us consider the condition

(a35) d is complete: each d-Cauchy sequence is d-convergent.

Note that, by the lack of symmetry, a d-convergent sequence need not be d-Cauchy; hence, this requirement has a technical motivation only.

Let $e : M \times M \to R_+$ be a triangular pseudometric over M. We shall say that this object is a KST-*metric* (modulo d) provided

(b35) e is Cauchy d-lsc in the second variable:
(y_n) is e-Cauchy and $y_n \xrightarrow{d} y$ imply $\liminf_n e(x, y_n) \geq e(x, y), \forall x \in M$

(c35) e is Cauchy subordinated to d:
each e-Cauchy sequence is d-Cauchy (hence d-convergent).

Further, let $\varphi : M \to R \cup \{\infty\}$ be some inf-proper, d-lsc function. The following auxiliary fact will be useful for us.

Lemma 14.52. *Assume that e is some KST-metric (modulo d) over M. Then, (e, φ) is complete (in the sense of (34g)); hence (a fortiori) descending complete (in the sense of (34b)).*

Now, combining these with Theorem 14.50, we derive

Theorem 14.53. *Let e be some KST-metric (modulo d) and φ be inf-proper, d-lsc. Then, for each starting (modulo (e, φ)) $u \in \mathrm{Dom}(\varphi)$ there exists $v = v(u) \in \mathrm{Dom}(\varphi)$ with the properties (14.37) and (14.38) (Subsection 14.3.4).*

An interesting problem to be posed is that of getting a corresponding form of this result involving E-variational (modulo (e, φ)) points. The appropriate answer to this is obtainable via Theorem 14.51. Call the triangular pseudometric $e : M \times M \to R_+$, a *strong* KST-metric (modulo d) when it is a KST-metric (modulo d) and fulfills the extra condition (34f).

Theorem 14.54. *Assume that e is some strong KST-metric (modulo d) and φ is inf-proper, d-lsc. Then, for each starting (modulo (e, φ)) $u \in \mathrm{Dom}(\varphi)$ there exists $w = w(u) \in \mathrm{Dom}(\varphi)$ fulfilling*

$$e(u, w) \leq \varphi(u) - \varphi(w) \quad \textit{(hence } \varphi(u) \geq \varphi(w)) \tag{14.41}$$

$$e(w, x) > \varphi(w) - \varphi(x), \quad \textit{for all } x \in M \setminus \{w\}. \tag{14.42}$$

In the following, we shall give some particular cases of our statements.

(I) Let d be a metric (i.e., symmetric almost metric) on M; supposed to be complete (cf. (a35)). Further, let $e : M \times M \to R_+$ be some w-distance (modulo d). By (c31), one gets at once (c35) (see above). On the other hand, (b31) yields (b35); and (c31) $\Longrightarrow$ (34f); because (as d = metric)

$$e(z,x) = e(z,y) = 0 \Longrightarrow [d(x,y) \leq \varepsilon, \forall \varepsilon > 0] \Longrightarrow x = y.$$

Summing up, any w-distance is a strong KST-metric (modulo d). Hence the variational statement in Kada, Suzuki and Takahashi [36] (subsumed to Theorem 14.41) is a particular case of Theorem 14.54. But, as explicitly stated by these authors, their contribution extends the one due to T. H. Kim, E. S. Kim and S. S. Shin [40]; hence so does our statement. This also shows that any recursion to the nonconvex minimization theorem in Takahashi [62] is avoidable in such approaches. Some related facts may be found in B. M. Lee, B. S. Lee, J. S. Jung and S. S. Chang [43]; see also Zhu, Zhong and Wang [86]. For a number of structural aspects, we refer to Nemeth [51], Ume [80], and Khanh [39].

(II) Let (M,d) be a complete metric space; and $e : M \times M \to R_+$ be a triangular pseudometric over M. According to Suzuki [58], we say that this object is a *τ-distance* (modulo d) over M when there exists a function $\eta = \eta(e)$ from $M \times R_+$ to R_+ with the properties

(35a) $t \mapsto \eta(x,t)$ is increasing and $\lim_{t\to 0} \eta(x,t) = 0 = \eta(x,0), \forall x \in M$

(35b) $\lim_n [\sup\{\eta(z_n, e(z_n, y_m)); m \geq n\}] = 0$ and $y_n \xrightarrow{d} y$ imply $\liminf_n e(x,y_n) \geq e(x,y)$, for each $x \in M$

(35c) $\lim_n [\sup\{e(x_n,y_m); m \geq n\}] = 0$ and $\lim_n \eta(x_n,t_n) = 0$ imply $\lim_n \eta(y_n,t_n) = 0$

(35d) $\lim_n \eta(z_n, e(z_n,x_n)) = 0$ and $\lim_n \eta(z_n, e(z_n,y_n)) = 0$ imply $\lim_n d(x_n,y_n) = 0$.

Clearly, any w-distance is a τ-distance too; just take $\eta(x,t) = t, \forall x \in M, \forall t \in R_+$. On the other hand (as we already remarked), any w-distance is a strong KST-metric. So, it is natural asking of what can be said about the relationships between these enlargements of our initial concept. The answer to this is given in:

Proposition 14.55. *Each τ-distance (modulo d) is a strong KST-metric (modulo d); so, we have the generic inclusions (over triangular pseudometrics)*

$$w\text{-distance} \Longrightarrow \tau\text{-distance} \Longrightarrow \text{strong KST-metric (modulo } d).$$

Proof. Let $e : M \times M \to R_+$ be a τ-distance (modulo d); and $\eta = \eta(e)$ stand for some associated map fulfilling (35a)-(35d). For the moment, e is transitively sufficient. In fact, let $x,y,z \in M$ be such that $e(z,x) = e(z,y) = 0$. By (35a), we have $\eta(z,e(z,x)) = \eta(z,e(z,y)) = 0$; and this, added to (35d), gives $d(x,y) = 0$ (hence $x = y$); wherefrom (34f) holds. Further, e is Cauchy d-lsc in the second variable. To verify this, call the sequence (x_n) in M, (η,e)-*Cauchy* provided

(d35) $\lim_n [\sup\{\eta(z_n,e(z_n,x_m)); m \geq n\}] = 0,$ for some $(z_n) \subseteq M$.

By [58, Lemma 3], the generic inclusion holds

$$\text{[for each sequence] } e\text{-Cauchy} \Longrightarrow (\eta, e)\text{-Cauchy}. \tag{14.43}$$

On the other hand, note that (35b) may be written as

(35e) (y_n) is (η, e)-Cauchy and $y_n \xrightarrow{d} y$ imply
$\liminf_n e(x, y_n) \geq e(x, y)$, for each $x \in M$.

Combining these gives the conclusion (b35) we want. Finally, we show that e is Cauchy subordinated to d. Indeed, note that by [58, Lemma 1]

$$\text{[for each sequence] } (\eta, e)\text{-Cauchy} \Longrightarrow d\text{-Cauchy}; \tag{14.44}$$

and this, via (14.43), yields (c35). The proof is thereby complete. □

Now, by simply adding this to Theorem 14.54, one gets the following variational statement.

Theorem 14.56. *Assume that e is some τ-distance (modulo d), and φ is inf-proper, d-lsc. Then, for each starting (modulo (e, φ)) $u \in \mathrm{Dom}(\varphi)$ there exists an E-variational (modulo (e, φ)) $w = w(u) \in \mathrm{Dom}(\varphi)$ with the property (14.41).*

The original proof of this result was provided in Suzuki [58]; but, it is rather involved. The proposed reasoning (based on the developments in Subsection 14.3.2) may be viewed as a refinement of this one; however, it still depends on Suzuki's arguments concerning the inclusions (14.43)+(14.44). It would be interesting to have alternate proofs of these, which should avoid Proposition 14.55 above. Some applications of these techniques to variational principles may be found in Suzuki [59].

(III) Let again d be a complete metric over M; and $e : M \times M \to R_+$ be a triangular pseudometric. According to Lin and Du [44], we say that it is a *τ-function* (modulo d) provided (34f) holds and

(35f) $x \in M$, $y_n \to y$ and $e(x, y_n) \leq M$, $\forall n$ (for some $M = M(x) > 0$)
imply $e(x, y) \leq M$

(35g) $\lim_n [\sup\{e(x_n, x_m); m > n\}] = 0$ and $\lim_n e(x_n, y_n) = 0$
imply $\lim_n d(x_n, y_n) = 0$.

By [44, Remark 1] each w-distance (modulo d) is a τ-function (modulo d). We complete this remark with

Proposition 14.57. *Each τ-function (modulo d) is a strong KST-metric (modulo d). So (combining with the above) we have the generic inclusions (over triangular pseudometrics)*

$$w\text{-distance} \Longrightarrow \tau\text{-function} \Longrightarrow \text{strong KST-metric (modulo } d).$$

Proof. Let $e : M \times M \to R_+$ be some τ-function (modulo d). By definition, it fulfills (34f); so, it remains to prove that (b35)+(c35) hold. The former of these is immediate via (35f). To verify the latter, call the sequence (x_n), *almost e-Cauchy* when $\lim_n [\sup\{e(x_n, x_m); m > n\}] = 0$. By definition,

$$\text{[for each sequence]}\ \ e\text{-Cauchy} \Longrightarrow \text{almost } e\text{-Cauchy}.$$

On the other hand, [44, Lemma 2.1] tells us that

$$\text{[for each sequence]}\ \ \text{almost } e\text{-Cauchy} \Longrightarrow d\text{-Cauchy}. \tag{14.45}$$

Combining with the above gives (c35); and the claim follows. □

As a consequence of this, the variational statement (involving τ-functions (modulo d)) obtained by the quoted authors [44, Theorem 2.1] is deductible from Theorem 14.54 above. Note that (as before) the proposed proofs are still depending on Lin–Du's reasoning involving the relation (14.45); so, it would be interesting to have alternate proofs of it. Further aspects may be found in Lin and Du [45].

(IV) Let (M,d) be a complete almost metric space (see above); and $e : M \times M \to R_+$ be a triangular pseudometric over M. According to Al-Homidan Ansari and Yao [2], we say that it is a *Q-function* (modulo d) provided (c31) and (35f) hold. Clearly, the latter condition is just (b31); and from this, (b35) is fulfilled. On the other hand, (c31) yields (c35) and (34f). [The argument is similar to the "metrical" one in **(I)**]. We therefore have the inclusion (over triangular pseudometrics)

$$Q\text{-function} \Longrightarrow \text{strong KST-metric}\ \ (\text{modulo } d).$$

As a direct consequence, the variational statement (involving Q-functions (modulo d)) obtained by the quoted authors [2, Theorem 3.1] is deductible from Theorem 14.54. The converse question is still open; we conjecture that a positive answer is ultimately available.

An interesting particular case refers to d being (in addition) symmetric; hence, a metric on M. By the previous remark involving (b31), we have

j) each τ-function is a Q-function (modulo d) (cf. [2, Remark 2.1])
jj) the concepts of Q-function and w-distance are *identical*.

This, along with Proposition 14.57, gives us the inclusions

$$w\text{-distance} \Longrightarrow \tau\text{-function} \Longrightarrow w\text{-distance}\ (\text{modulo } d);$$

i.e., the concepts of w-distance and τ-function are identical (within the metric framework). This conclusion must be treated with care, in view of [2, Example 2.1]. Further aspects will be delineated elsewhere.

Acknowledgment The author is indebted to the referee for a number of useful suggestions.

This research was supported by Grant PN II PCE ID_387 from the National Authority for Scientific Research, Romania.

References

1. P. S. Alexandrov, *An Introduction to Set Theory and Topology* (Russian), Nauka, Moscow, 1977.

2. S. Al-Homidan, Q. H. Ansari and J.-C. Yao, *Some generalizations of Ekeland-type variational principle with applications to equilibrium problems and fixed point theory,* Nonlin. Anal., 69 (2008), 126–139.
3. M. Altman, *A generalization of the Brezis-Browder principle on ordered sets,* Nonlin. Anal., 6 (1981), 157–165.
4. P. Amato, *Un metodo per ridurre questioni di punto fisso a questioni di completamento e viceversa,* Boll. Un. Mat. Ital., B (6), 3 (1984), 463–476.
5. M. C. Anisiu, *On maximality principles related to Ekeland's theorem,* Seminar Funct. Analysis Numer. Meth. (Faculty of Math. Research Seminars), Preprint No. 1 (8 pp), "Babeş-Bolyai" Univ., Cluj-Napoca (România), 1987.
6. J. S. Bae, E. W. Cho and S. H. Yeom, *A generalization of the Caristi-Kirk fixed point theorem and its applications to mapping theorems,* J. Korean Math. Soc., 31 (1994), 29–48.
7. T. Q. Bao and P. Q. Khanh, *Are several recent generalizations of Ekeland's variational principle more general than the original principle?,* Acta Math. Vietnamica, 28 (2003), 345–350.
8. J. M. Borwein and D. Preiss, *A smooth variational principle with applications to subdifferentiability and to diferentiability of convex functions,* Trans. Amer. Math. Soc., 303 (1987), 517–527.
9. N. Bourbaki, *Sur le theoreme de Zorn,* Archiv Math., 2 (1949/1950), 434–437.
10. N. Bourbaki, *General Topology (Chs 1-4),* Springer, Berlin, 1989.
11. N. Bourbaki, *General Topology (Chs 5-10),* Springer, Berlin, 1989.
12. H. Brezis and F. E. Browder, *A general principle on ordered sets in nonlinear functional analysis,* Advances Math., 21 (1976), 355–364.
13. A. Brondsted, *On a lemma of Bishop and Phelps,* Pacific J. Math., 55 (1974), 335–341.
14. N. Brunner, *Topologische Maximalprinzipien,* Zeitschr. Math. Logik Grundl. Math., 33 (1987), 135–139.
15. O. Cârjă, M. Necula and I. I. Vrabie, *Viability, Invariance and Applications,* North Holland Mathematics Studies vol. 207, Elsevier B. V., Amsterdam, 2007.
16. V. Conserva and S. Rizzo, *Maximal elements in a class of order complete metric subspaces,* Math. Japonica, 37 (1992), 515–518.
17. S. Dancs, M. Hegedus and P. Medvegyev, *A general ordering and fixed-point principle in complete metric space,* Acta Sci. Math. (Szeged), 46 (1983), 381–388.
18. W. S. Du, *On some nonlinear problems induced by an abstract maximal element principle,* J. Math. Anal. Appl., 347 (2008), 391–399.
19. I. Ekeland, *On the variational principle,* J. Math. Anal. Appl., 47 (1974), 324–353.
20. I. Ekeland, *Nonconvex minimization problems,* Bull. Amer. Math. Soc. (New Series), 1 (1979), 443–474.
21. J. X. Fang, *The variational principle and fixed point theorems in certain topological spaces,* J. Math. Anal. Appl., 202 (1996), 398–412.
22. U. Felgner, *Die Existenz wohlgeordneter konfinaler Teilmengen in Ketten und das Auswahlaxiom,* Math. Zeitschrift, 111 (1969), 221–232.
23. L. Gajek and D. Zagrodny, *Countably orderable sets and their application in optimization,* Optimization, 26 (1992), 287–301.
24. A. Goepfert, H. Riahi, C. Tammer and C. Zălinescu, *Variational Methods in Partially Ordered Spaces,* Canad. Math. Soc. Books Math. vol. 17, Springer, New York, 2003.
25. A. Granas and C. D. Horvath, *On the order-theoretic Cantor theorem,* Taiwanese J. Math., 4 (2000), 203–213.
26. A. Hamel, *Equivalents to Ekeland's variational principle in uniform spaces,* Nonlin. Anal., 62 (2005), 913–924.
27. A. Hamel, *Variational Principles on Metric and Uniform Spaces,* (Habilitation Thesis), Martin-Luther University, Halle-Wittenberg (Germany), 2005.
28. M. Hasse and L. Michler, *Theorie der Kategorien,* VEB Deutscher Verlag Wissenschaften, Berlin, 1966.
29. F. Hausdorff, *Grundzuege der Mengenlehre,* von Veit & Comp., Leipzig, 1914.
30. G. B. Hazen and T. L. Morin, *Optimality conditions in nonconical multiple objective programming,* J. Optim. Th. Appl., 40 (1983), 25–60.

31. D. H. Hyers, G. Isac and T. M. Rassias, *Topics in Nonlinear Analysis and Applications,* World Sci. Publ., Singapore, 1997.
32. G. Isac, *Sur l'existence de l'optimum de Pareto,* Rivista Mat. Univ. Parma (Serie IV), 9 (1983), 303–325.
33. G. Isac, *The Ekeland's principle and Pareto ε-efficiency,* in "Multi-Objective Programming and Goal Programming" (M. Tamiz ed.), pp. 148–163, L. Notes Econ. Math. Systems vol. 432, Springer, Berlin, 1996.
34. G. J. Jinag and Y. J. Cho, *Cantor order and completeness,* Int. J. Pure Appl. Math., 2 (2002), 393–398.
35. W. Just and M. Weese, *Discovering Modern Set Theory (vol 1),* Amer. Math. Soc., Providence, RI, 1996.
36. O. Kada, T. Suzuki and W. Takahashi, *Nonconvex minimization theorems and fixed point theorems in complete metric spaces,* Math. Japonica, 44 (1996), 381–391.
37. B. G. Kang and S. Park, *On generalized ordering principles in nonlinear analysis,* Nonlin. Anal., 14 (1990), 159–165.
38. S. Kasahara, *On some generalizations of the Banach contraction theorem,* Publ. Res. Inst. Math. Sci. Kyoto Univ., 12 (1976), 427–437.
39. P. Q. Khanh, *On Caristi-Kirk's theorem and Ekeland's variational principle for Pareto extrema,* Bull. Polish Acad. Sci. (Math.), 37 (1988), 33–39.
40. T. H. Kim, E. S. Kim and S. S. Shin, *Minimization theorems relating to fixed point theorems on complete metric spaces,* Math. Japonica, 45 (1997), 97–102.
41. C. Kuratowski, *Une methode d'elimination des nombres transfinis des raisonnements mathematiques,* Fund. Math., 3 (1922), 76–108.
42. K. Kuratowski and A. Mostowski, *Set Theory,* PWN (Polish Scientific Publishers), Warsaw, 1968.
43. G. M. Lee, B. S. Lee, G. S. Jung and S. S. Chang, *Minimization theorems and fixed point theorems in generating spaces of quasi-metric family,* Fuzzy Sets Syst., 101 (1999), 143–152.
44. L. J. Lin and W. S. Du, *Ekeland's variational principle, minimax theorems and existence of nonconvex equilibria in complete metric spaces,* J. Math. Anal. Appl., 323 (2006), 360–370.
45. L. J. Lin and W. S. Du, *On maximal element theorems, variants of Ekeland's variational principle and their applications,* Nonlin. Anal., 68 (2008), 1246–1262.
46. Z. Liu, *Order completeness and stationary points,* Rostock Math. Kolloq., 50 (1997), 85–88.
47. R. Manka, *Turinici's fixed point theorem and the Axiom of Choice,* Reports Math. Logic, 22 (1988), 15–19.
48. N. Mizoguchi, *A generalization of Brondsted's result and its applications,* Proc. Amer. Math. Soc., 108 (1990), 707–714.
49. G. H. Moore, *Zermelo's Axiom of Choice: Its Origin, Development and Influence,* Springer, New York, 1982.
50. L. Nachbin, *Topology and Order,* D. van Nostrand Comp. Inc., Princeton, NJ, 1965.
51. A. B. Nemeth, *A nonconvex vector minimization problem,* Nonlin. Anal., 10 (1986), 669–678.
52. J. A. Park and S. Yie, *Surjectivity of generalized locally expansive maps,* J. Korean Math. Soc., 24 (1987), 179–185.
53. P. Rozoveanu, *Ekeland's variational principle for vector valued functions,* Math. Reports [St. Cerc. Mat.], 2(52) (2000), 351–366.
54. C. Sempi, *Hausdorff distance and the completion of probabilistic metric spaces,* Boll. Un. Mat. Ital., B (7), 6 (1992), 317–327.
55. W. Sierpinski, *Cardinal and Ordinal Numbers,* PWN (Polish Scientific Publishers), Warsaw, 1965.
56. Y. Sonntag and C. Zălinescu, *Comparison of existence results for efficient points,* J. Optim. Th. Appl., 105 (2000), 161–188.
57. F. Sullivan, *Ordering and completeness of metric spaces,* Nieuw Archief Wisk., 29 (1981), 178–193.
58. T. Suzuki, *Generalized distance and existence theorems in complete metric spaces,* J. Math. Anal. Appl., 253 (2001), 440–458.

59. T. Suzuki, *The strong Ekeland variational principle,* J. Math. Anal. Appl., 320 (2006), 787–794.
60. A. Szaz, *An Altman type generalization of the Brezis-Browder ordering principle,* Math. Moravica, 5 (2001), 1–6.
61. A. Szaz, *An improved Altman type generalization of the Brezis Browder ordering principle,* Math. Communications, 12 (2007), 155–161.
62. W. Takahashi, *Existence theorems generalizing fixed point theorems for multivalued mappings,* in "Fixed Point Theory and Applications" (J. B. Baillon, ed.), pp. 397-406, Pitman Res. Notes Math. vol. 252, Longman Sci. Tech., Harlow, 1991.
63. M. R. Taskovic, *The Axiom of Choice, fixed point theorems and inductive ordered sets,* Proc. Amer. Math. Soc., 116 (1992), 897–904.
64. D. Tataru, *Viscosity solutions of Hamilton-Jacobi equations with unbounded nonlinear terms,* J. Math. Anal. Appl., 163 (1992), 345–392.
65. M. Turinici, *A generalization of Brezis-Browder's ordering principle,* An. Şt. Univ. "A. I. Cuza" Iaşi (S I-a: Mat), 28 (1982), 11–16.
66. M. Turinici, *A generalization of Altman's ordering principle,* Proc. Amer. Math. Soc., 90 (1984), 128–132.
67. M. Turinici, *Pseudometric extensions of the Brezis-Browder ordering principle,* Math. Nachr., 130 (1987), 91–103.
68. M. Turinici, *Metric variants of the Brezis-Browder ordering principle,* Demonstr. Math., 22 (1989), 213–228.
69. M. Turinici, *A monotone version of the variational Ekeland's principle,* An. Şt. Univ. "A. I. Cuza" Iaşi (S. I-a: Mat.) 36 (1990), 329–352.
70. M. Turinici, *Vector extensions of the variational Ekeland's result,* An. Şt. Univ. "A. I. Cuza" Iaşi (S I-a: Mat), 40 (1994), 225–266.
71. M. Turinici, *Variational principles on semi-metric streuctures,* Libertas Math., 20 (2000), 161–171.
72. M. Turinici, *Minimal points in product spaces,* An. St. Univ. "Ovidius" Constanţa (Ser. Math.), 10 (2002), 109–122.
73. M. Turinici, *Pseudometric versions of the Caristi-Kirk fixed point theorem,* Fixed Point Theory (Cluj-Napoca), 5 (2004), 147–161.
74. M. Turinici, *Brezis-Browder principles in separable ordered sets,* Libertas Math., 26 (2006), 15–30.
75. M. Turinici, *Brezis-Browder principles in general separable sets,* Libertas Math., 26 (2006), 31–47.
76. M. Turinici, *Maximality principles: Theory and practice,* Sci. Papers UASVM Iaşi, 49 (2006), 323–360.
77. M. Turinici, *Remarks on some pseudometric variational statements,* An. Univ. Vest Tim. (Ser.: Mat.-Inform.), 45 (2007), 197–214.
78. M. Turinici, *Functional Variational Principles and Coercivity over Normed Spaces,* Optimization, DOI 10.1080/02331930801950993.
79. M. Turinici, *Brezis-Browder principle revisited,* Note Mat., in press.
80. J. S. Ume, *Variational principles, minimization theorems and fixed-point theorems on generalized metric spaces,* J. Optim. Th. Appl., 118 (2003), 619–633.
81. L. E. Ward Jr., *Partially ordered topological spaces,* Proc. Amer. Math. Soc., 5 (1954), 144–161.
82. E. S. Wolk, *On the principle of dependent choices and some forms of Zorn's lemma,* Canad. Math. Bull., 26 (1983), 365–367.
83. E. Zermelo, *Beweis dass jede Menge wohlgeordnet werden kann,* Math. Annalen, 50 (1904), 514–516.
84. J. Zhu, X. Fan and S. Zhang, *Fixed points of increasing operators and solutions of nonlinear impulsive integro-differential equations in Banach space,* Nonlin. Anal., 42 (2000), 599–611.
85. J. Zhu and S. J. Li, *Generalization of ordering principles and applications,* J. Optim. Th. Appl., 132 (2007), 493–507.

86. J. Zhu, C. K. Zhong and G. P. Wang, *An extension of Ekeland's variational principle in fuzzy metric space and its applications,* Fuzzy Sets Syst., 108 (1999), 353–363.
87. M. Zorn, *A remark on method in transfinite algebra,* Bull. Amer. Math. Soc., 41 (1935), 667–670.

Part II
Variational Problems

Chapter 15
A Generalized Quasi-Equilibrium Problem

Mircea Balaj and Donal O'Regan

Dedicated to the memory of Professor George Isac

Abstract In this paper, using the Kakutani–Fan–Glicksberg fixed point theorem, we obtain an existence theorem for a generalized vector quasi-equilibrium problem of the following type: for a suitable choice of the sets X, Z and V and of the mappings $T : X \multimap X$, $R : X \multimap X$, $Q : X \multimap Z$, $F : X \times X \times Z \multimap V$, $C : X \multimap V$, find $\widetilde{x} \in X$ such that $\widetilde{x} \in T(\widetilde{x})$ and $(\forall) y \in R(\widetilde{x}), (\alpha) z \in Q(\widetilde{x}), \rho\big(F((\widetilde{x}, y, z), C(\widetilde{x})\big)$, where ρ is a given binary relation on 2^V and α is any of the quantifiers $\forall$, $\exists$. Finally, several particular cases are discussed and some applications are given.

15.1 Introduction

As a direct generalization of the classical variational inequalities, the scalar equilibrium problem encompasses as special cases many important problems including optimization problems, Nash equilibrium problems, complementarity problems and fixed point problems (see [6]). Naturally, the scalar equilibrium problem has been extended and generalized for set-valued mappings in several directions by many researchers, see for example [2], [4], [8], [12], [15], [16], [26]. In the past decade, many authors studied several types of vector quasi-equilibrium problems (see for example [3], [10], [14], [19], [27]).

In this paper, we consider the following generalized quasi-equilibrium problem:

Mircea Balaj
Department of Mathematics, University of Oradea, 410087 Oradea, Romania,
email: mbalaj@uoradea.ro, e-mail: mbalaj@uoradea.ro

Donal O'Regan
Department of Mathematics, National University of Ireland, Galway, Ireland.

P.M. Pardalos et al. (eds.), *Nonlinear Analysis and Variational Problems*,
Springer Optimization and Its Applications 35, DOI 10.1007/978-1-4419-0158-3_15,

Let X be a nonempty compact convex subset of a locally convex Hausdorff topological vector space E and Z,V be two nonempty sets. Let ρ be a binary relation on 2^V, $T : X \multimap X$, $R : X \multimap X$, $Q : X \multimap Z$, $F : X \times X \times Z \multimap V$ and $C : X \multimap V$ be five mappings and α be any of the quantifiers $\forall$, $\exists$.

$(GQEP)(\alpha;\rho)$ Find $\widetilde{x} \in X$ such that $\widetilde{x} \in T(\widetilde{x})$ and

$$(\forall)y \in R(\widetilde{x}),\ (\alpha)z \in Q(\widetilde{x}),\ \rho\big(F((\widetilde{x},y,z),C(\widetilde{x})\big).$$

To motivate the problem setting, let us look at several special cases of $(GQEP)(\alpha;\rho)$. Consider the following particular relations on 2^V:

For $A,B \subseteq V$,

(i) $\rho_1\ (A,B) \Leftrightarrow A \subseteq B$,
(ii) $\rho_2\ (A,B) \Leftrightarrow A \cap B \neq \emptyset$,
(iii) $\rho_3\ (A,B) \Leftrightarrow A \nsubseteq B$,
(iv) $\rho_4(A,B)\ \Leftrightarrow A \cap B = \emptyset$,

(a) Let $T = R = 1_X$, $\phi : X \times X \times Z \to V$, $F(x,y,z) = \{\phi(x,y,z)\}$. Then, problem $(GQEP)(\forall;\rho_3)$ reduces to the following:

Find $\widetilde{x} \in X$ such that $\phi(\widetilde{x},y,z) \notin C(\widetilde{x})$, for all $y \in X$ and $z \in Q(\widetilde{x})$. This problem is considered by Ansari [1] and Lee et al. [18].

(b) In this same case, problem $(GQEP)(\exists;\rho_3)$ becomes:

Find $\widetilde{x} \in X$ such that for each $y \in X$, there exists $z \in Q(\widetilde{x})$, satisfying $\phi(\widetilde{x},y,z) \notin C(\widetilde{x})$.

The problem above has been studied by Ding and Park [7], Fang and Huang [9], and Lee and Kum [20].

(c) In the particular case $T = R = 1_X$, the problem $(GQEP)(\exists;\rho_3)$ is investigated by Fu and Wan [11].

(d) Lin [21] studied the problems $(GQEP)(\forall;\rho_1)$, $(GQEP)(\forall;\rho_4)$, $(GQEP)(\exists;\rho_2)$ and $(GQEP)(\exists;\rho_3)$ when $R = T$.

15.2 Preliminaries

Let X and Y be nonempty sets. A multivalued mapping (or simply, a mapping) $T : X \multimap Y$ is a function from X into the power set of Y. For $y \in Y$, the set $T^-(y) = \{x \in X : y \in T(x)\}$ is called the fiber of T on y. For $A \subseteq X$, $T(A) = \bigcup_{x\in A} T(x)$ is the image of A under T. If X and Y are topological spaces, a mapping $T : X \multimap Y$ is said to be: (i) *upper semicontinuous* (in short, u.s.c) (respectively, *lower semicontinuous* (in short, l.s.c.)) if for every closed subset B of Y the set $\{x \in X : T(x) \cap B \neq \emptyset\}$ (respectively, $\{x \in X : T(x) \subseteq B\}$) is closed; (ii) *closed* if its graph (that is, the set $GrT = \{(x,y) \in X \times Y :\ y \in T(x)\}$) is a closed subset of $X \times Y$.

The following lemma collects known facts about u.s.c. or l.s.c. mappings (see [17] for assertions (i), (ii), and (iii), respectively [28] for assertion (iv)).

Lemma 15.1. *Let X and Y be topological spaces and $T : X \multimap Y$ be a mapping.*

(i) If Y is compact and T is closed, then T is u.s.c..
(ii) If Y is regular and T is u.s.c. with closed values, then T is closed.
(iii) If X is a compact and T is u.s.c. with compact values, then $T(X)$ is compact.
(iv) T is l.s.c. if and only if for any $x \in X$, $y \in T(x)$ and any net $\{x_\alpha\}$ converging to x, there exists a net $\{y_\alpha\}$ converging to y, with $y_\alpha \in T(x_\alpha)$ for each α.

For a subset A of a topological vector space, the standard notations $co\,A$ and $\overline{A}$ designate the convex hull and closure of A, respectively. Instead of $co\{x_1,x_2\}$ we shall use the notation $[x_1,x_2]$.

Definition 15.2. ([21]) Let X be a convex set in a vector space, Z be a vector space and C be a convex cone in Z. A mapping $F : X \multimap Z$ is called *C-quasiconvex* (respectively, *C-quasiconvex-like*) if for all $x_1,x_2 \in X$ and $x \in [x_1,x_2]$, there exists an index $i \in \{1,2\}$ such that
$F(x_i) \subseteq F(x) + C$ (respectively, $F(x) \subseteq F(x_i) - C$).

We extend the concepts introduced above to a mapping of two variables as follows:

Definition 15.3. Let X and Y be convex sets in vector spaces, Z be a vector space and C be a convex cone in Z. A mapping $F : X \times Y \multimap Z$ is called:

(i) *(x,y)-C-quasiconvex* if for all $(x_1,y_1),(x_2,y_2) \in X \times Y$ and $x \in [x_1,x_2]$, there exist $y \in [y_1,y_2]$ and $i \in \{1,2\}$ such that $F(x_i,y_i) \subseteq F(x,y) + C$;
(ii) *(x,y)-C-quasiconvex-like* if for all $(x_1,y_1),(x_2,y_2) \in X \times Y$ and $x \in [x_1,x_2]$, there exist $y \in [y_1,y_2]$ and $i \in \{1,2\}$ such that $F(x,y) \subseteq F(x_i,y_i) - C$;
(iii) *(x,y)-quasiconcave* if for all $(x_1,y_1),(x_2,y_2) \in X \times Y$ and $x \in [x_1,x_2]$, there exist $y \in [y_1,y_2]$ and $i \in \{1,2\}$ such that $F(x,y) \subseteq F(x_i,y_i)$.

Example 15.4. Let X , Y be two real intervals and $f : Y \to X$ a function having the Darboux property. It is easy to see that the mapping $F : X \times Y \multimap \mathbb{R}$ defined by

$$F(x,y) = \begin{cases} [1,+\infty) & \text{if } (y,x) \in Grf, \\ [0,+\infty) & \text{if } (y,x) \notin Grf. \end{cases}$$

is (x,y)-quasiconcave and consequently (x,y)–C-quasiconvex-like, for any convex cone $C \subseteq \mathbb{R}$.

Definition 15.5. ([13]) For a subset K of a vector space E and $x \in E$, the *outward set* of K at x is denoted and defined as follows:
$\mathbf{O}(K;x) = \bigcup_{\lambda \geq 1}(\lambda x + (1-\lambda)K)$.

15.3 Main Result

In order to establish the main result, we need the following two lemmas:

Lemma 15.6. *([25]) Let X be a topological space, Y be a topological vector space and $S,T : X \multimap Y$ be two mappings. If S is u.s.c. with nonempty compact values and T is closed, then $S+T$ is a closed mapping.*

Lemma 15.7. *Let X be a topological space and Y be a Hausdorff topological vector space. If $f : X \to \mathbb{R}$ is a continuous function and $T : X \multimap Y$ a compact closed mapping, then the mapping $fT : X \multimap Y$ defined by $(fT)(x) = f(x)T(x)$ is closed.*

Proof. Let $(x,y) \in \overline{Gr(fT)}$. Then there exists a net $\{(x_t,y_t)\}_{t\in\Delta}$ in $Gr(fT)$ converging to (x,y). For each $t \in \Delta$ we have $y_t = f(x_t)z_t$, for some $z_t \in T(x_t)$. Since $\overline{T(X)}$ is compact there is a subnet $\{z_{t_\alpha}\}$ of $\{z_t\}$ converging to a point $z \in \overline{T(X)}$. Since the mapping T is closed, $z \in T(x)$. Hence, $y_{t_\alpha} \to f(x)z \in (fT)(x)$. The space Y is Hausdorff so $y = f(x)z$. It follows that $(x,y) \in Gr(fT)$ hence the mapping fT is closed.

□

In the next theorem, X is a nonempty compact convex subset of a locally convex Hausdorff topological vector space E, Z and V are two nonempty sets, $T : X \multimap X$, $R : X \multimap X$, $Q : X \multimap Z$, $F : X \times X \times Z \multimap V$ and $C : X \multimap V$ are five mappings, ρ is a binary relation on 2^V, and α is any of the quantifiers $\forall$, $\exists$. Denote by ρ^c the complementary relation of ρ (that is, for any $A,B \subseteq V$ exactly one of the following assertions $\rho(A,B), \rho(A,B)$ holds) and by $\overline{\alpha}$ the other of the quantifiers $\forall$, $\exists$.

Theorem 15.8. *Suppose that the following conditions are satisfied:*

(i) T is u.s.c. with nonempty compact convex values;
(ii) R is nonempty convex valued;
(iii) for each $y \in X$, the set $R^-(y) \cap \{x \in X : (\overline{\alpha})z \in Q(x), \rho^c(F(x,y,z),C(x))\}$ is open in X;
(iv) for each $x \in X$, the set $\{y \in X : (\overline{\alpha})z \in Q(x), \rho^c(F(x,y,z),C(x))\}$ is convex;
(v) for each $x \in X$, $\mathbf{O}(T(x);x) \cap R(x) \subseteq \{y \in X : (\alpha)z \in Q(x),\ \rho(F(x,y,z),C(x))\}$.

Then there exists $\widetilde{x} \in X$ such that $\widetilde{x} \in T(\widetilde{x})$ and $(\forall)y \in R(\widetilde{x})$, $(\alpha)z \in Q(\widetilde{x})$, $\rho(F(\widetilde{x},y,z),C(\widetilde{x}))$.

Proof. For $y \in X$, let

$$G_y = \{x \in X : y \in R(x) \text{ and } (\overline{\alpha})z \in Q(x), \rho^c(F(x,y,z),C(x))\} =$$

$$R^-(y) \cap \{x \in X : (\overline{\alpha})z \in Q(x), \rho^c(F(x,y,z),C(x))\}.$$

By (iii), the sets G_y are open. Let $G_0 = \{x \in X : x \notin T(x)\}$. Since the mapping T is closed, it follows readily that G_0 is open.

Suppose that the conclusion is false. Then for each $x \in X$, either $x \in G_0$ or $x \in G_y$, for some $y \in R(X)$. Thus, $X = G_0 \cup \bigcup_{y\in R(X)} G_y$. Since X is compact, there exists a finite set $\{y_1,\cdots,y_n\} \subseteq R(X)$ such that $X = G_0 \cup \bigcup_{i=1}^n G_{y_i}$. For the sake of simplicity we write G_i instead of G_{y_i}. Let $\{\beta_0,\beta_1,\ \ldots,\ \beta_n\}$ be a partition of unity on X subordinated to the open cover $\{G_0,G_1,\cdots,G_n\}$. Recall that this means that

$$\begin{cases} \beta_i : X \to [0,1] \text{ is continuous, for each } i \in \{0,1,\ldots,n\}; \\ \beta_i(x) > 0 \Rightarrow x \in G_i; \\ \Sigma_{i=0}^n \beta_i(x) = 1 \text{ for each } x \in X. \end{cases}$$

Define the mapping $S : X \multimap X$ by

$$S(x) = \beta_0(x)T(x) + \beta_1(x)y_1 + \cdot + \beta_n(x)y_n.$$

It is clear that S has nonempty compact convex values. Since the mapping $x \mapsto \beta_1(x)y_1 + \cdots + \beta_n(x)y_n$ is closed, combining Lemmas 15.6 and 15.7 we infer that S is closed, hence u.s.c. By Kakutani–Fan–Glicksberg fixed point theorem, there exists $x_0 \in X$ such that $x_0 \in S(x_0)$. We shall prove that each of the cases $\beta_0(x_0) = 0$, $\beta_0(x_0) = 1$ and $\beta_0(x_0) \in (0,1)$ leads to a contradiction. Let $I = \{i \in \{1,\cdots,n\} : \beta_i(x_0) > 0\}$. For each $i \in I$, $x_0 \in G_i$, hence $x_0 \in R^-(y_i)$ and

$$(\overline{\alpha})z \in Q(x_0), \rho^c(F(x_0,y_i,z),C(x_0)).$$

Since $\{y_i : i \in I\} \subseteq R(x_0)$ and $R(x_0)$ is convex, $co\{y_i : i \in I\} \subseteq R(x_0)$.

If $\beta_0(x_0) = 0$, then

$$x_0 = \Sigma_{i\in I}\beta_i(x_0)y_i \in co\{y_i : i \in I\}.$$

Since $x_0 \in \mathbf{O}(T(x_0);x_0)$, we have $x_0 \in \mathbf{O}(T(x_0);x_0) \cap R(x_0)$ and by (v), $(\alpha)z \in Q(x_0), \rho(F(x_0,x_0,z),C(x_0))$. On the other hand, by (iv), we have $(\overline{\alpha})z \in Q(x_0)$, $\rho^c(F(x_0,x_0,z),C(x_0))$, a contradiction.

If $\beta_0(x_0) = 1$, it follows that $x_0 \in S(x_0) = T(x_0)$. On the other hand, since $\beta_0(x_0) > 0$, $x_0 \in G_0$, that is, $x_0 \notin T(x_0)$, a contradiction again.

If $\beta_0(x_0) \in (0,1)$, then there exists $y_0 \in T(x_0)$ such that

$$x_0 = \beta_0(x_0)y_0 + \Sigma_{i\in I}\beta_i(x_0)y_i.$$

Dividing the previous relation by $1 - \beta_0(x_0)$ and denoting by $\lambda = \frac{1}{1-\beta_0(x_0)}$ we get

$$\lambda x_0 + (1-\lambda)y_0 = \Sigma_{i\in I}\frac{\beta_i(x_0)}{1-\alpha_0(x_0)}y_i \in co\{y_i : i \in I\}.$$

The previous equality implies, $\lambda x_0 + (1-\lambda)y_0 \in \mathbf{O}(T(x_0);x_0) \cap R(x_0)$. By (v), it follows that $(\alpha)z \in Q(x_0)$, $\rho(F(x_0,\lambda x_0 + (1-\lambda)y_0,z),C(x_0))$. On the other hand, since $\lambda x_0 + (1-\lambda)y_0 \in co\{y_i : i \in I\}$, we get $(\overline{\alpha})z \in Q(x_0)$, $\rho^c(F(x_0,\lambda x_0 + (1-\lambda)y_0,z),C(x_0))$. The contradiction obtained completes the proof. □

15.4 Particular Cases of Theorem 15.8

Next we are interested in replacing conditions (iii) and (iv) in Theorem 15.8 by more convenient conditions when $\alpha = \forall$, and respectively $\alpha = \exists$ and ρ is one of the relations ρ_i considered in the first section. Let us observe that each existence result concerning relation ρ_1 (respectively, ρ_2) yields an existence theorem for ρ_4 (respectively ρ_3), if we take into account the following equivalences: $F(x,y,z) \subseteq C(x) \Leftrightarrow F(x,y,z) \cap C^c(x) = \emptyset$ and $F(x,y,z) \cap C(x) \neq \emptyset \Leftrightarrow F(x,y,z) \nsubseteq C^c(x)$. For this reason we fix our attention on relations ρ_1 and ρ_2, only.

From now on we will suppose that Z is a convex subset of a topological vector space, V is a topological vector space, C is a closed mapping with nonempty convex cone values and the mapping R has nonempty convex values and open fibers. Notice that under this last assumption, condition (iii) in Theorem 15.8 is fulfilled whenever the sets $\{x \in X : (\alpha) z \in Q(x), \rho(F(x,x,z), C(x))\}$ are closed for all $y \in X$.

Theorem 15.9. *Assume that for $\alpha = \forall$ and $\rho = \rho_1$, conditions (i), (ii), and (v) are satisfied. Moreover assume that:*

(a) Q is l.s.c. with convex values;
(b) for each $y \in X$, $F(\cdot, y, \cdot)$ is l.s.c. on $X \times Z$;
(c) for each $x \in X$, the mapping $(y,z) \multimap F(x,y,z)$ is $(y,z) - C(x)$–quasiconvex;

Then there exists $\widetilde{x} \in X$ such that $\widetilde{x} \in T(\widetilde{x})$ and $F(\widetilde{x},y,z) \subseteq C(\widetilde{x})$, for all $y \in R(\widetilde{x})$, and $z \in Q(\widetilde{x})$.

Proof. It suffices to prove that for $\alpha = \forall$ and $\rho = \rho_1$, conditions (iii) and (iv) in Theorem 15.8 are fulfilled. We show first that for $y \in X$ the set $M = \{x \in X : (\forall) z \in Q(x), F(x,y,z) \subseteq C(x)\}$ is closed. Let $x \in \overline{M}$ and $\{x_t\}_{t \in \Delta}$ be a net in M converging to x. Let $z \in Q(x)$ and $v \in F(x,y,z)$ be arbitrarily fixed. Since Q is l.s.c., by Lemma 15.1 (iv), there exists a net $\{z_t\}_{t \in \Delta}$ converging to z, with $z_t \in Q(x_t)$ for all $t \in \Delta$. By (b), we infer that there exists a net $\{v_t\}_{t \in \Delta}$ converging to v, with $v_t \in F(x_t, y, z_)$, for all $t \in \Delta$. Since $x_t \in M$, $v_t \in F(x_t, y, z_t) \subseteq C(x_t)$. The mapping C is closed, and consequently $v \in C(x)$. Thus $x \in M$, hence M is closed.

It remains to show that for any $x \in X$ the set $N = \{y \in X : (\exists) z \in Q(x), F(x,y,z) \nsubseteq C(x)\}$ is convex. Let $y_1, y_2 \in N$ and $y \in [y_1, y_2]$. For each $i \in \{1,2\}$ there exists $z_i \in Q(x)$ and $i \in \{1,2\}$ such that $F(x, y_i, z_i) \nsubseteq C(x)$. By (c), there exist $z \in [z_1, z_2]$ and $i \in \{1,2\}$ such that $F(x, y_i, z_i) \subseteq F(x,y,z) + C(x)$. Moreover, since $Q(x)$ is convex, we have $z \in Q(x)$. We claim that $F(x,y,z) \nsubseteq C(x)$. If not, we have $F(x,y,z) \subseteq C(x)$, whence

$F(x, y_i, z_i) \subseteq F(x,y,z) + C(x) \subseteq C(x)$, a contradiction.

□

Theorem 15.10. *Assume that the set Z is compact, and for $\alpha = \forall$ and $\rho = \rho_2$ conditions (i), (ii), and (v) are satisfied. Moreover assume that:*

(a) Q is l.s.c. with convex values;
(b) for each $y \in X$, $F(\cdot, y, \cdot)$ is u.s.c. with compact values on $X \times Z$;
(c) for each $x \in X$, the mapping $(y,z) \multimap F(x,y,z)$ is $(y,z) - C(x)$–quasiconvex-like;

Then there exists $\widetilde{x} \in X$ such that $\widetilde{x} \in T(\widetilde{x})$ and $F(\widetilde{x},y,z) \cap C(\widetilde{x}) \neq \emptyset$, for all $y \in R(\widetilde{x})$, and $z \in Q(\widetilde{x})$.

Proof. As in the proof of Theorem 15.9, the desired conclusion follows from Theorem 15.8 as soon as we show that for $\alpha = \forall$ and $\rho = \rho_2$, conditions (iv) and (v) in Theorem 15.8 are fulfilled. For a $y \in X$ denote by $M = \{x \in X : (\forall)z \in Q(x), F(x,y,z) \cap C(x) \neq \emptyset\}$. Let $x \in \overline{M}$ and $\{x_t\}_{t\in\Delta}$ be a net in M converging to x. If $z \in Q(x)$ is arbitrarily chosen, since Q is l.s.c., there exists a net $\{z_t\}_{t\in\Delta}$ converging to z, with $z_t \in Q(x_t)$ for all $t \in \Delta$. For each $t \in \Delta$, since $x_t \in M$, there exists $v_t \in F(x_t,y,z_t) \cap C(x_t)$. By Lemma 15.1 (iii), the set $F(X,y,Z)$ is compact, hence we may assume that the net $\{v_t\}_{t\in\Delta}$ converges to a point v. The set $F(X,y,Z)$ is compact so it is regular, and by Lemma 15.1 (ii), the mapping $F(\cdot,y,\cdot)$ is closed. Hence $v \in F(x,y,z)$. Since C is closed, it follows that $v \in C(x)$. Thus $x \in M$, hence M is closed.

For $x \in X$ we show that the set $N = \{y \in X : (\exists)z \in Q(x), F(x,y,z) \cap C(x) = \emptyset\}$ is convex. Let $y_1, y_2 \in N$ and $y \in [y_1,y_2]$. For each $i \in \{1,2\}$ there exists $z_i \in Q(x)$ and $i \in \{1,2\}$ such that $F(x,y_i,z_i) \cap C(x) = \emptyset$. By (c), there exist $z \in [z_1,z_2]$ and $i \in \{1,2\}$ such that $F(x,y,z) \subseteq F(x,y_i,z_i) - C(x)$. Moreover, since $Q(x)$ is convex, we have $z \in Q(x)$. We claim that $F(x,y,z) \cap C(x) = \emptyset$. If not,

$$\emptyset \neq F(x,y,z) \cap C(x) \subseteq \big(F(x,y_i,z_i) - C(x)\big) \cap C(x),$$

hence $\emptyset \neq F(x,y_i,z_i) \cap \big(C(x) + C(x)\big) = F(x,y_i,z_i) \cap C(x)$, a contradiction. □

Theorem 15.11. *Assume that for $\alpha = \exists$ and $\rho = \rho_1$, conditions (i), (ii), and (v) are satisfied. Moreover assume that:*

(a) Q is compact and closed;
(b) for each $y \in X$, $F(\cdot,y,\cdot)$ is l.s.c. on $X \times Z$;
(c) for each fixed $x \in X$ and $z \in Q(x)$ the mapping $y \multimap F(x,y,z)$ is $C(x)$-quasiconvex;

Then there exists $\widetilde{x} \in X$ such that $\widetilde{x} \in T(\widetilde{x})$ and for each $y \in R(\widetilde{x})$ there exists $z \in Q(\widetilde{x})$ satisfying $F(\widetilde{x},y,z) \subseteq C(\widetilde{x})$.

Proof. For an arbitrary $y \in X$ let
$M = \{x \in X : (\exists)z \in Q(x), F(x,y,z) \subseteq C(x)\}$.

Let $x \in \overline{M}$ and $\{x_t\}_{t\in\Delta}$ be a net in M converging to x. For each $t \in \Delta$, there exists $z_t \in Q(x_t)$ such that $F(x_t,y,z_t) \subseteq C(x_t)$. Since Q is compact, we may assume that the net $\{z_t\}_{t\in\Delta}$ converges to a point z. The mapping Q is closed, and consequently $z \in Q(x)$. Let $v \in F(x,y,z)$ be arbitrarily chosen. By (b), there exists a net $\{v_t\}_{t\in\Delta}$ converging to v, with $v_t \in F(x_t,y,z_t)$, for all $t \in \Delta$. We have $v_t \in C(x_t)$ for all $t \in \Delta$, and since C is closed, $v \in C(x)$. Thus $x \in M$, hence M is closed.

We show next that for any $x \in X$ the set $\{y \in X : (\forall)z \in Q(x) : F(x,y,x) \nsubseteq C(x)\} = \bigcap_{z\in Q(x)} \{y \in X : F(x,y,x) \nsubseteq C(x)\}$ is convex. For this it suffices to prove that sets $N_z = \{y \in X : F(x,y,x) \nsubseteq C(x)\}$ are convex, for all $z \in Q(x)$. Let $y_1, y_2 \in N_z$ and $y \in [y_1,y_2]$. Suppose that $y \notin N_z$, that is, $F(x,y,x) \subseteq C(x)\}$. Since $F(x,\cdot,z)$ is $C(x)$-quasiconvex, there exists $i \in \{1,2\}$ such that $F(x,y_i,z) \subseteq F(x,y,z) + C(x)$. It follows that

$$F(x,y_i,z) \subseteq F(x,y,z) + C(x) \subseteq C(x) + C(x) = C(x),$$

and this contradicts $y_i \in N_z$.

The desired conclusion follows now from Theorem 15.8, in case $\alpha = \exists$ and $\rho = \rho_1$. $\square$

In a similar manner one can obtain from Theorem 15.8 the following

Theorem 15.12. *Assume that the set Z is compact, and for $\alpha = \exists$ and $\rho = \rho_2$ conditions (i), (ii), and (v) are satisfied. Moreover assume that:*

(a) Q is compact and closed;
(b) for each $y \in X$, $F(\cdot, y, \cdot)$ is u.s.c. with compact values on $X \times Z$;
(c) for each fixed $x \in X$ and $z \in Q(x)$, the mapping $y \multimap F(x,y,z)$ is $C(x)$-quasiconvex-like;

Then there exists $\widetilde{x} \in X$ such that $\widetilde{x} \in T(\widetilde{x})$ and for each $y \in R(\widetilde{x})$ there exists $z \in Q(\widetilde{x})$ satisfying $F(\widetilde{x}, y, z) \cap C(\widetilde{x}) \neq \emptyset$.

The following example shows that although the assumptions of Theorems 15.9–15.12 are not always easy to verify in practice, these results are important tools in establishing the existence of solutions for the corresponding problems.

Example 15.13. Let $X = Z = [0,3]$, $V = \mathbb{R}$ and the mappings $F : [0,3] \times [0,3] \times [0,3] \multimap \mathbb{R}$, $C : [0,3] \multimap \mathbb{R}$, $T : [0,3] \multimap [0,3]$, $R : [0,3] \multimap [0,3]$ and $Q : [0,3] \multimap [0,3]$ be defined by

$$F(x,y,z) = [x+y+z, +\infty), \;\; C(x) = [2x, +\infty),$$

$$T(x) = \begin{cases} [-2x+2, -2x+3] & \text{if } x \in [0,1], \\ [x-1, x] & \text{if } x \in (1,3]. \end{cases}$$

$$R(x) = \begin{cases} [0,1] & \text{if } x \in [0,1], \\ [0,3] & \text{if } x \in (1,2), \\ (1,3] & \text{if } x \in [2,3], \end{cases}$$

$$Qx) = \begin{cases} [x,1] & \text{if } x \in [0,1], \\ [1,x] & \text{if } x \in (1,3]. \end{cases}$$

Observe that $F(x,y,z) \subseteq C(x)$ if and only if $x \leq y+z$. One can easily check that

$$R^-(y) = \begin{cases} [0,2) & \text{if } y \in [0,1], \\ (1,3] & \text{if } y \in (1,3] \end{cases}$$

and

$$\mathbf{O}(T(x);x) \cap R(x) = \begin{cases} \emptyset & \text{if } x = 0, \\ [0,x] & \text{if } x \in \left(0, \frac{2}{3}\right], \\ [0,1] & \text{if } x \in \left(\frac{2}{3}, 1\right), \\ \{1\} & \text{if } x = 1 \\ [x,3] & \text{if } x \in (1,3]. \end{cases}$$

One can readily verify that the mappings F, C, T, R and Q satisfy all the requirements of Theorem 15.9. The set of all $\widetilde{x}$ verifying the conclusion of Theorem 15.9 is $\left[\frac{2}{3}, 1\right] \cup \{2\}$.

15.5 Applications

If $C(x) = \{0\}$, for all $x \in X$, problem $(GQEP)(\alpha;\rho_2)$ reduces to the following quasi-variational inclusion problem:

Find $\widetilde{x} \in X$ such that $\widetilde{x} \in T(\widetilde{x})$, $(\forall)y \in R(\widetilde{x})$, $(\alpha)z \in Q(\widetilde{x})$, $0 \in F(\widetilde{x},z,v)$.

This problem has been recently studied by Lin et al. in [22–25], but our results and methods of proof are different. The following two theorems can be easily obtained from Theorems 15.10 and 15.12 with the above- mentioned method.

Theorem 15.14. *Suppose that the mappings T,R and Q satisfy the following conditions:*

(i) T is u.s.c. with nonempty compact convex values, R has nonempty convex values and open fibers, Q is l.s.c with convex values;
(ii) for each $y \in X$, $F(\cdot,y,\cdot)$ is u.s.c. with compact values on $X \times Z$;
(iii) for each $x \in X$, the mapping $(y,z) \multimap F(x,y,z)$ is $(y,z)-$quasiconcave;
(iv) for each $x \in X$, $y \in \mathbf{O}(T(x);x) \cap R(x)$ and $z \in Q(x)$, $0 \in F(x,y,z)$.

Then there exists $\widetilde{x} \in X$ such that $\widetilde{x} \in T(\widetilde{x})$ and $0 \in F(\widetilde{x},y,z)$, for all $y \in R(\widetilde{x})$, and $z \in Q(\widetilde{x})$.

Theorem 15.15. *Suppose that the set Z is compact and the mappings T,R and Q satisfy the following conditions:*

(i) T is u.s.c. with nonempty compact convex values, R has nonempty convex values and open fibers, Q is compact and closed;
(ii) for each $y \in X$, $F(\cdot,y,\cdot)$ is u.s.c. with compact values on $X \times Z$;
(iii) for each $x \in X$, the mapping $(y,z) \multimap F(x,y,z)$ is $(y,z)-$quasiconcave;
(iv) for each $x \in X$, $y \in \mathbf{O}(T(x);x) \cap R(x)$ there exists $z \in Q(x)$ such that $0 \in F(x,y,z)$.

Then there exists $\widetilde{x} \in X$ such that $\widetilde{x} \in T(\widetilde{x})$ and for each $y \in R(\widetilde{x})$ there exists $z \in Q(\widetilde{x})$ satisfying $0 \in F(\widetilde{x},y,z)$.

Recall that if E is a normed vector space, the mapping $J : E \multimap E^*$ defined by

$$J(x) = \{x^* \in E^* : \|x\| = \|x^*\| = \langle x,x^*\rangle\}$$

is called the duality mapping of E. When E is a smooth normed vector space, then J is single-valued.

Theorem 15.16. *Let X be a nonempty compact convex subset of a smooth normed vector space, $T : X \multimap X$ be a u.s.c. mapping with nonempty compact convex values and $R : X \multimap X$ be a mapping with nonempty convex values and open fibers. If $\mathbf{O}(T(x);x) \cap R(x) \setminus \{x\} = \emptyset$ for all $x \in X$, then there exists $\widetilde{x} \in X$ such that $\widetilde{x} \in T(\widetilde{x}) \cap R(\widetilde{x})$.*

Proof. Since R has nonempty convex values and open fibers, by the selection theorem of Yannelis and Prabhakar [29], R has a continuous selection, that is, there exists a continuous $f: X \to X$ such that $f(x) \in R(x)$, for all $x \in X$. We apply Theorem 15.8 in the following case:

$Z = X$, $Q(x) = \{f(x)\}$, $F(x,y,z) = \{\langle x-y, J(z-x)\rangle\}$, $C(x) = [0,+\infty)$,
α any of the quantifiers $\forall$, $\exists$ and ρ any of the relations ρ_1, ρ_2 .

Notice that in this case $\rho(F(x,y,z),C(x)) \Leftrightarrow \langle x-y, J(z-x)\rangle \geq 0$. It is easy to check that all the requirements of Theorem 15.8 are satisfied. Thus, by Theorem 15.8, there exists $\widetilde{x} \in X$ such that $\widetilde{x} \in T(\widetilde{x})$ and
$\langle \widetilde{x} - y, J(f(\widetilde{x}) - x)\rangle \geq 0$ for all $y \in R(\widetilde{x})$.

Taking $y = f(\widetilde{x})$ we get

$$0 \leq \langle \widetilde{x} - f(\widetilde{x}), J(f(\widetilde{x}) - x)\rangle = -\|\widetilde{x} - f(\widetilde{x})\|,$$

hence, $\widetilde{x} = f(\widetilde{x})$. Since $\widetilde{x} \in T(\widetilde{x})$ and $f(\widetilde{x}) \in R(\widetilde{x})$, it follows $T(\widetilde{x}) \cap R(\widetilde{x}) \neq \emptyset$. □

References

1. Q. H. Ansari, *A note on generalized vector variational-like inequalities*, Optimization 41 (1997), 197-205.
2. Q. H. Ansari, W. Oettli and D. Schläger, *A generalization of vectorial equilibria*, Math. Methods Oper. Res. 47 (1997), 147-152.
3. Q. H. Ansari and F. Flores-Bazán, *Generalized vector quasi-equilibrium problems with applications*, J. Math. Anal. Appl. 277 (2003), 246-256.
4. Q. H. Ansari, S. Schaible and J. C. Yao, *The systems of vector equilibrium problems and its applications*, J. Optim. Theory Appl., 107 (2000), 547-557.
5. M. Bianchi, N. Hadjisavvas and S. Schaible, *Vector equilibrium problems with generalized monotone bifunctions* J. Optim. Theory Appl. 92 (1997), 527-542.
6. E. Blum and W. Oettli, *From optimization and variational inequalities problems to equilibrium problems* Math. Student 63 (1994), 123-146.
7. X. P. Ding and J. Y. Park, *Generalized vector equilibrium problems in generalized convex spaces*, J. Optimiz. Theory Appl. 120 (2004), 327-353.
8. M. Fakhar and J. Zafarani, *Generalized vector equilibrium problems for pseudomonotone multivalued bifunctions*, J. Optim. Theory Appl. 126 (2005), no. 1, 109–124.
9. M. Fang and N.-J. Huang, *KKM type theorems with applications to generalized vector equilibrium problems in FC-spaces*, Nonlinear Anal. 67 (2007), 809-817.
10. J. Y. Fu, *Generalized vector quasi-equilibrium problems*, Math. Methods Oper. Res. 52 (2000), 57-64.
11. J. Y. Fu and A.-H. Wan, *Generalized vector equilibrium problems with set-valued mappings*, Math. Methods Oper. Res. 56 (2002), 259-268.
12. F. Giannessi (ed.), *Vector Variational Inequalities and Vector Equilibria. Mathematical Theories*, Kluwer Academic Publishers, Dordrecht, 2000.
13. B. R. Halpern, G. M. Bergman, *A fixed-point theorem for inward and outward maps*, Trans. Amer. Math. Soc. 130 (1968), 353-358.
14. S. H. Hou, H. Yu and G. Y. Chen, *On vector quasi-equilibrium problems with set-valued maps* J. Optim. Theory Appl. 119 (2003), 485-498.
15. I. V. Konnov and J. C. Yao, *Existence of solutions for generalized vector equilibrium problems*, J. Math. Anal. Appl. 233 (1999), 328-335.
16. A. Kristály and C. Varga, *Set-valued versions of Ky Fan's inequality with application to variational inclusion theory*, J. Math. Anal. Appl. 282 (2003), 8-20.

17. M. Lassonde, *Fixed points for Kakutani factorizable multifunctions*, J. Math. Anal. Appl. 152 (1990), 46-60.
18. G. M. Lee, D. S. Kim and B. S. Lee, *Generalized vector variational inequality*, Appl. Math. Lett. 9 (1996), 39-42.
19. S. J. Li, K. L. Teo and X. Q. Yang, *Generalized vector quasi-equilibrium problems*, Math. Methods Oper. Res. 61 (2005), 385-397.
20. G. M. Lee, and S. H. Kum, *On implicit vector variational inequalities*, J. Optim. Theory Appl. 104 (2000), 409-425.
21. L. J. Lin, *System of generalized vector quasi-equilibrium problems with applications to fixed point theorems for a family of nonexpansive multivalued mappings*, J. Global Optim. 34 (2006), 5-32.
22. L. J. Lin, *Variational inclusions problems with applications to Ekeland's variational principle, fixed point and optimization problems*, J. Global Optim, 39 (2007), 509-527.
23. L. J. Lin and C. S. Chuang, *Systems of nonempty intersection theorems with applications*, Nonlinear Analysis, 69 (2008), 4063–4073, DOI. 10.1016/.2007.10.037.
24. L. J. Lin and C. I. Tu, *The study of variational inclusions problems and variational disclusions problems with applications*, Nonlinear Anal., 69 (2008), 1981-1998.
25. L. J. Lin, S. Y. Wang and C. S. Chuang, *Existence theorems of systems of variational inclusion problems with applications*, J. Global Optimization, 40 (2008), 751-764.
26. L. J. Lin, Z. T. Yu and G. Kassay, *Existence of equilibria for multivalued mappings and its application to vectorial equilibria*, J. Optim. Theory Appl. 114 (2002), 189-208.
27. S. Park, *Fixed points and quasi-equilibrium problems. Nonlinear operator theory*, Math. Comput. Modelling 32 (2000), 1297-1304.
28. N. X. Tan and P. N. Tinh, *On the existence of equilibrium points of vector functions*, Numerical Functional Analysis Optimiz., 19 (1998), 141-156.
29. N. C. Yannelis and N. D. Prabhakar, *Existence of maximal elements and equilibria in linear topological spaces*, J. Math. Econom. 12 (1983), 233-245.

Chapter 16
Double-Layer and Hybrid Dynamics of Equilibrium Problems: Applications to Markets of Environmental Products

M. Cojocaru, S. Hawkins, H. Thille, and E. Thommes

Dedicated to the memory of Professor George Isac

Abstract We present here an original method of tracking the dynamics of an equilibrium problem using an evolutionary variational inequalities and hybrid dynamical systems approach. We apply our method to describe the time evolution of a differentiated product market model under incentive policies with a finite life span. In particular, we describe trajectories of a dynamic game between two producers of a standard product and of an environmental variant of the standard product. We compute and assess the behavior of both the equilibrium (optimal) strategies, as well as the disequilibrium (no-optimal) ones of each producer involved in the oligopolistic market.

16.1 Introduction

The impact of equilibrium problems is self-evident today in many applied mathematics, systems design, management and decision making fields (see for instance

M. Cojocaru
Department of Mathematics & Statistics, University of Guelph, Guelph, ON, Canada,
e-mail: mcojocar@uoguelph.ca

S. Hawkins
Department of Mathematics & Statistics, University of Guelph, Guelph, ON, Canada,
e-mail: mcojocar@uoguelph.ca

H. Thille
Department of Economics, University of Guelph, Guelph, ON, Canada,
e-mail: hthille@uoguelph.ca

E. Thommes
Department of Physics, University of Guelph, Guelph, ON, Canada,
e-mail: ethommes@uoguelph.ca

P.M. Pardalos et al. (eds.), *Nonlinear Analysis and Variational Problems*,
Springer Optimization and Its Applications 35, DOI 10.1007/978-1-4419-0158-3_16,

telecommunications, transportation, digital and electric power networks as well as economic equilibria in the fields of market models and financial equilibria). There is an important number of volumes being dedicated to the study of equilibrium problems, both from a theoretical point of view (via optimization, game theoretic, complementarity and variational inequalities techniques), as well as from an applied point of view (via networks, agent-based modeling and social scientific methods), which in themselves constitute wide areas of research interest today (see, for instance [34],[38],[14],[31],[33],[37],[19],[15]).

In contrast, the concept of dynamics of equilibrium problems is currently of much interest, judging by the amount of applications and computational approaches to modeling network and population dynamic phenomena. This concept arises from the observation that the physical structure of, for example, a network can remain unchanged, but the phenomena which occur in these networks may vary with time. The link with previous studies of classic (static) equilibrium problems can be done in a natural way: a static configuration represents a snapshot of an evolving real phenomenon. Therefore, studying the static case can be considered only a first approach to understanding the real dynamics, useful for further developments. The results obtained in the static cases have a short range of validity and are only partially representative of the developing reality. The dynamic equilibrium setting arises when the constitutive elements of the economic or physic phenomena associated with the fixed geometry of the equilibrium problem at hand (e.g., the travel demands in the traffic models, the supplies, the demands, the supply prices, the demand prices and the shipments of commodities in the spatial price models, the change in risk perceptions and vaccine coverage in a population model) are considered time-dependent (see [7],[8],[9],[16],[11],[12],[13],[5]). To this end we would simply like to point out the impressive number of works dedicated to various formulations of equilibrium problems, time and space being too short here to outline an even brief review (we refer the reader to [35],[42],[41],[11],[12],[27],[26],[23],[32],[40],[30] among many others).

We recall that variational inequalities theory has been used to formulate, qualitatively analyze, and solve a number of equilibrium problems. In particular, variational inequalities have provided a method of solving equilibrium problems otherwise unsolvable. The complementarity conditions and the conservation laws expressing the equilibrium conditions of the various problems cannot be treated with the usual methods and, then, the result due to Smith [38] and Dafermos [14], who first proved that such problems fit very well with the variational inequality theory already introduced by Stampacchia [39], has been a breakthrough scientific result.

In the first part of this chapter (Section 16.2), we show how a first model of time dependency can be introduced in classic (static) equilibrium problem, via the so-called evolutionary variational inequalities (EVI). They were introduced in the 1960s ([6],[28],[39]), and have been used in the study of partial differential equations and boundary value problems. They are part of the general variational inequalities theory, a large area of research with important applications in control theory, optimization, operations research, economics theory and transportation science (see for example [2],[15],[16],[18],[19],[20],[21],[23],[26],[32],[34] and the references therein). The form of EVI problems we consider in the current paper represents a unified formulation coming from applied problems in traffic, spatial price and financial equilibrium problems [15],[16], and was introduced first in [7] and generalized in [12, 4]. The existence and uniqueness theory for EVI problems has been answered in many contexts; here we use the result in [15]. In [9],[4], the authors give a refinement of this existence result showing under what conditions continuous solutions exist. In [3],[8],[4], the authors present computational procedures for obtaining approximate solutions of an EVI problem of the type considered here.

In the second part of this chapter (Sections 16.3 and 16.4), we generalize the EVI model of an equilibrium problem and incorporate a rigorous mathematical framework for describing the time evolution of not only equilibrium (ideal, optimal) states of a problem with varying constraint sets, but also its time evolution away from equilibrium. We can now study the evolution, in finite-time, from disequilibrium to equilibrium, of an applied equilibrium problem whose steady states are modeled by an EVI. Such a study, started in [8] and extended in [13], is made possible by the recently introduced framework of double-layer dynamics (DLD). This framework allows the study of applications involving two types of time dependency: one represented by the time-dependent equilibria (that can be predicted for a given problem via EVI theory), and the other represented by the time-dependent behavior of the application away from the predicted equilibrium curve (studied via projected systems and flows). The interpretation of the two timescales in DLD theory was discussed in [13] and it is taken further in this paper, with the help of the theory of hybrid dynamical systems [37]. Hybrid systems theory is used to build a tracking method for the disequilibrium behavior of a problem formulated in the DLD framework.

In the last part of the chapter, we apply our theoretical results to modeling oligopolistic markets of eco-products, since environmental issues are currently at the forefront of our social lives. In general, new policy and market studies are needed in order to develop and implement long-term measures that will change the population's behavior toward environment preservation. Our application looks at a model that can be used as a testing tool for policy forecast, incorporating time evolution as the life span of a given policy. We show how a policy (such as developer subsidies for producing environmental friendly variants of an existing product) can jump start the variants' market and move consumers toward higher levels of demand for such products. We close with a few concluding remarks and acknowledgments.

16.2 Dynamic Equilibrium Problems and Variational Inequalities

16.2.1 General Formulation

We recall here a unified definition of an equilibrium problem, coming either from network or a Nash game formulation, in terms of an evolutionary variational inequality. This formulation essentially consists of noting that the constraint set of a known equilibrium problem can be thought of as time-dependent $\mathbb{K} := \mathbb{K}(t)$. The type of time-dependent constraint set below was proposed in [8],[16], for the EVI arising in time-dependent traffic network problems, spatial equilibrium problems with either quantity or price formulations, and a variety of financial equilibrium problems. Later, a similar generic formulation has been extended to migration problems and dynamic games applied to population biology and to oligopolistic market equilibrium problems [11,12,5].

We consider a nonempty, convex, closed, bounded subset of the reflexive Banach space $L^p([0,T],\mathbb{R}^q)$ given by:

$$\mathbb{K} = \Big\{ u \in L^p([0,T],\mathbb{R}^q) \,|\, \lambda(t) \le u(t) \le \mu(t) \text{ a.e. in } [0,T]; \tag{16.1}$$

$$\sum_{i=1}^{q} \xi_{ji}(t)u_i(t) = \rho_j(t) \text{ a.e. in } [0,T], i \in \{1,..,q\},\ j \in \{1,...,l\} \Big\}.$$

We let $\lambda, \mu \in L^p([0,T],\mathbb{R}^q)$ and $\rho \in L^p([0,T],\mathbb{R}^l)$ be convex functions in the above definition. For chosen values of the scalars ξ_{ji}, of the dimension q, or of the boundaries λ, μ, we obtain each of the previous above-cited model constraint set formulations as follows:

- for the traffic network problem ([16]) let $\xi_{ji}(t) = const. \in \{0,1\}$ and $\lambda(t) \ge 0$ for all $t \in [0,T]$;
- for the quantity formulation of spatial price equilibrium let $q = n+m+nm$, $\xi_{ji}(t) = const. \in \{0,1\}$, $\mu(t)$ large and $\lambda(t) = 0$, for any $t \in [0,T]$;
- for the price formulation of spatial price equilibrium [16] let $q = n+m+mn$, $\xi_{ji} = 0$ and $\lambda(t) \ge 0$ for all $t \in [0,T]$;
- for the financial equilibrium problem [16] let $q = 2n$, $\xi_{ji}(t) = -1$ for $i \in \{1,...,n\}$ and $\xi_{ji}(t) = 1$ for $i \in \{n+1,..,2n\}$; $\mu(t)$ large and $\lambda(t) = 0$, for any $t \in [0,T]$;
- for the dynamic Nash vaccination games formulation (see [12])

$$p = 2,\ \xi_{ji} : [0,T] \in (0,1),\ \sum_{i=1}^{q} \xi_{ji}(t) = 1, \lambda(t) = 0, \mu(t) = 1;$$

- for the oligopolistic market equilibrium problem (see [5]) $p = 2$, $\xi_{ji}(t) = 0, \rho_j(t) = 0$, $\lambda(t) \ge 0$.

Recall that $\ll \phi, u \gg := \int_0^T \langle \phi(u)(t), u(t) \rangle dt$ is the duality mapping on $L^p([0,T],\mathbb{R}^q)$, where
$\phi \in (L^p([0,T],\mathbb{R}2^q))^*$ and $u \in L^p([0,T],\mathbb{R}^q)$. Let $F : \mathbb{K} \to (L^p([0,T],\mathbb{R}^q))^*$; the standard form of the EVI we work with is therefore:

$$\text{find } u \in \mathbb{K} \text{ such that } \ll F(u), v-u \gg \ge 0,\ \forall v \in \mathbb{K}. \tag{16.2}$$

In order to highlight a few results concerning the existence, uniqueness and regularity of solutions to problems of type (16.2), we need to recall first the definition of pseudo-monotonicity (as generalized in [13]):

Definition 16.1. Let $K \subseteq X$ be closed, convex, where X is a reflexive Banach space space. Let $\ll \cdot,\cdot \gg$ be the pairing on X and $f : K \to X$ a mapping. Then:

1. f is called **locally** r**- strongly pseudo-monotone with degree** α **at** $x^* \in K$ if, for a given $r > 0$, there exists a neighborhood $N(x^*) \subset K$ of the point x^* with the property that for any point $x \in N(x^*)\backslash B[x^*, r]$, there exists a positive scalar $\eta > 0$ so that
$$\ll f(x^*), x - x^* \gg \geq 0 \Longrightarrow \ll f(x), x - x^* \gg \geq \eta ||x - x^*||^{\alpha}.$$
2. f is called r**- strongly pseudo-monotone with degree** α **at** $x^* \in K$ if the above holds for all $x \in K\backslash B[x^*, r]$.

Remark 16.2. 1. If $r = 0$, $\eta = 0$, then f is simply called pseudo-monotone;
2. if $r = 0$, $\eta = 0$ and $\ll f(x), x - x^* \gg > 0$ then f is called strictly pseudo-monotone.
3. Strong pseudo-monotonicity with degree α is itself a generalization of the notions of local and global strong monotonicity with degree α (see [21],[25],[32].

It was known that ([16]):

Theorem 16.3. *If F satisfies either of the following conditions:*

- *F is hemicontinuous with respect to the strong topology on $\mathbb{K}$, and there exist $A \subseteq \mathbb{K}$ nonempty, compact, and $B \subseteq \mathbb{K}$ compact such that, for every $v \in \mathbb{K} \setminus A$, there exists $u \in B$ with $\ll F(v), u - v \gg \geq 0$;*
- *F is hemicontinuous with respect to the weak topology on $\mathbb{K}$;*
- *F is pseudomonotone and hemicontinuous along line segments,*

then the EVI problem (16.2) admits a solution over the constraint set $\mathbb{K}$.

Theorem 16.4. *In [26], it is shown that if F is in addition strictly monotone, then the solution to the EVI is unique. Moreover, in [8] it is shown that if $p = 2$ and F is strictly pseudo-monotone, then the solution to (16.2) is unique.*

Regularity results for solutions to a problem of class (16.2) have been proven in the context of L^p spaces for more general constraint sets $\mathbb{K}$ (see [4]). Let $\lambda, \mu \in C([0,T], \mathbb{R}^m_+)$ such that $0 \leq \lambda < \mu$, let $A \in C([0,T], \mathbb{R}^{lm}_+)$, let $b \in C([0,T], \mathbb{R}^l_+)$ be vector-functions and consider the set
$$\mathbb{K} = \Big\{x \in L^p([0,T], \mathbb{R}^m) : \lambda(t) \leq x(t) \leq \mu(t), \, A(t)x(t) = b(t)\Big\}. \tag{16.3}$$

The following result holds (see [4] for a proof):

Theorem 16.5. *Problem (16.2) has a continuous solution on a set $\mathbb{K}$ of type (16.3) is the mapping F is strongly pseudo-monotone with degree $\alpha > 1$.*

In order to compute this solution, one can use a numerical method proposed in [7],[9],[4]. To illustrate this section, we include an example of an EVI problem, together with its numerical solution.

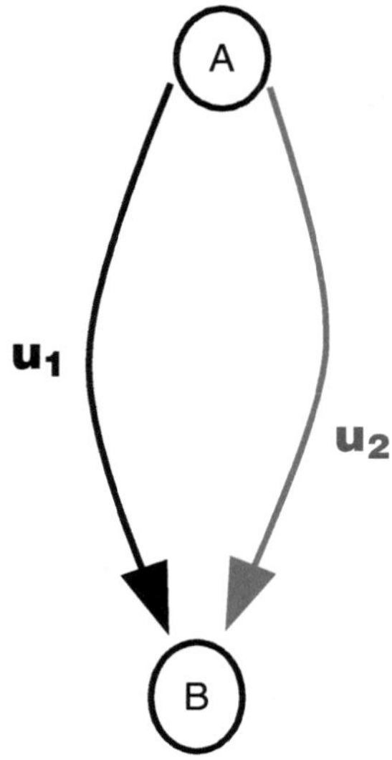

Fig. 16.1 This figure represents a simple, 1 origin-destination pair network, with two paths. The paths hold the commodity flows between the origin (1) and the destination (2).

Example 16.6. We consider a traffic network with one origin-destination pair having two links (as depicted in Figure 16.1) and the following constraint set corresponding to this network configuration

$$\mathbb{K} := \{u \in L^2([0,110],\mathbb{R}^2) \mid 0 \le u(t) \le 120,\, u_1(t)+u_2(t) = \rho_1(t) \text{ a.a. } t \in [0,110]\},$$

where

$$\rho_1(t) = \begin{cases} 4t,\ t \in [0,15] \\ 60,\ t \in (15,20] \\ 3t,\ t \in (20,40]. \\ 120,\ t \in (40,91] \\ -t+211,\ t \in (91,110] \end{cases}$$

We consider the time unit to be a minute and the time interval $[0,110]$ to correspond to 6:30 to 8:20 am during a weekday. Let the flows on each link be denoted by u_1, u_2 and the demand by ρ_1 (Figure 2 depicts the demand). We see that during the hight of rush hour, 7:10 to 8:00 am (i.e., $t \in (40,91]$) the demand is highest.
Let us also consider the cost on each link to be given by the mapping

$$F : \mathbb{K} \to L^2([0,110],\mathbb{R}^2),\ F((u_1,u_2)) = (u_1+151, u_2+60).$$

The dynamic equilibria for such a problem are given by the EVI (see also [16],[8])

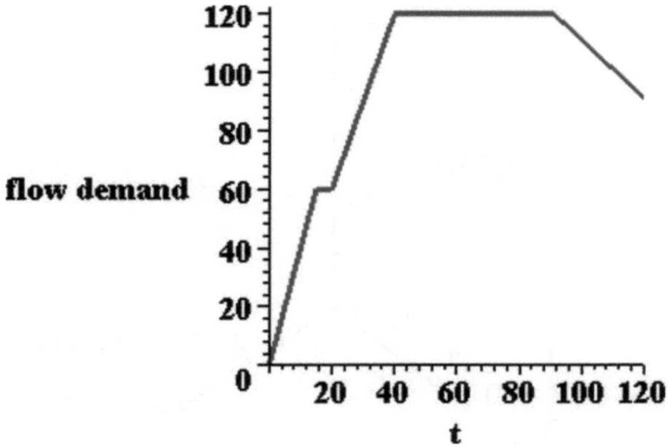

Fig. 16.2 The red curve represents the demand on the transportation network of the Example in Section 2. It is the graph of the demand function $\rho_1(t)$.

$$\int_0^{110} < F(u)(t), v(t) - u(t) > dt \geq 0, \forall v \in \mathbb{K}.$$

The mapping F is Lipschitz continuous with constant 1 and $F(u) := Au + B$ with A positive definite; the unique solution of the above EVI is piecewise continuous and has a constant value over the intervals $[15,20]$ and $[40,91]$. By the method proposed in [7] we compute an approximate solution to be

$$u^*(t) = \begin{cases} (0,4t),\ t \in [0,15] \\ (0,60),\ t \in (15,20] \\ (0,3t),\ t \in (20,\frac{91}{3}] \\ (\frac{3t-91}{2},\frac{3t+91}{2}),\ t \in (\frac{91}{3},40]. \\ (14.5,105.5),\ t \in (40,91] \\ (\frac{-t+120}{2},\frac{-t+302}{2}),\ t \in (91,110] \end{cases}$$

The graph of this solution is presented in Figure 16.3.

We note that the Wardrop equilibrium conditions are satisfied for this solution, namely all paths with positive flow in equilibrium have equal minimal costs, as can be seen below:

$$F(u^*)(t) = \begin{cases} (151,4t+60),\ t \in [0,15] \\ (151,120),\ t \in (15,20] \\ (151,3t+60),\ t \in (20,\frac{91}{3}] \\ (\frac{3t+211}{2},\frac{3t+211}{2}),\ t \in (\frac{91}{3},40] \\ (165.5,165.5),\ t \in (40,91] \\ (\frac{-t+422}{2},\frac{-t+422}{2}),\ t \in (91,110] \end{cases}.$$

We see here that users prefer the second road to the first, however, during the rush hour peak, they will use both routes, as they become equally expensive. So far, the EVI model of this problem provided the approximate equilibrium curve for

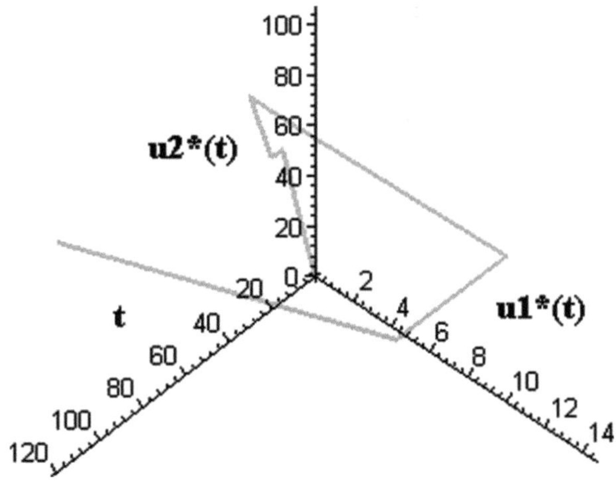

Fig. 16.3

the traffic, given a certain structure of the demand function. In general, however, the traffic may be in disequilibrium, in which case we want to know if/how will it evolve toward a steady state. This type of question is answered via the double-layer dynamics (DLD) model of this network, which we present in detail in the next section. For further discussion of this example and its disequilibrium evolution, the reader is referred to [13].

16.3 Double-Layer Dynamics and Hybrid Dynamical Systems

EVI problems of the type introduced above can be viewed as a 1-parameter family of a static variational inequality, with parameter t. From here on, we consider that our EVI (16.2) represents the model of an equilibrium problem (as for example in [16]). In this context, the parameter t will be taken to mean physical time. As t varies over $[0,T]$, the constraints of the equilibrium problem change, and so the static states describe a curve of equilibria. Double-layer dynamics (DLD) was introduced in [7],[8] as a unifying tool for deepening the study of an EVI problem with constraint sets $\mathbb{K} \subseteq L^2([0,T],\mathbb{R}^q)$, specifically for understanding the stability of the equilibrium curve arising from solving an EVI formulation of a time-dependent equilibrium problem. This has been done first in [8], where the infinite-dimensional PDS theory served for drawing conclusions about the stability of such a curve. It was then observed that the evolution of the problem, in finite time, toward a possible equilibrium state can be described with a DLD model [8, 13]. In this section we define the double-layer dynamics for equilibrium problems modeled via an EVI formulation and present for the first time a comprehensive method of tracking off-equilibrium behavior for such problems. The tracking method is based on the theory of hybrid dynamical systems.

16.3.1 DLD

Let X be a Hilbert space of arbitrary (finite or infinite) dimension and let $K \subset X$ be a non-empty, closed, convex subset. We start by introducing a nonlinear differential equation on the set K, with a discontinuities in the right-hand side. We assume the reader is familiar with the concepts of *tangent and normal cones to K at $x \in K$* ($T_K(x)$, respectively $N_K(x)$), and *the projection operator of X onto K, $P_K : X \to K$* given by $||P_K(z) - z|| = \inf_{x \in K} ||x - z||$.
The properties of the projection operator on Hilbert spaces are well-known (see for instance [43]). The directional derivative of the operator P_K is defined, for any $x \in K$ and any element $v \in X$, as the limit (for a proof see [43]):

$$\Pi_K(x,v) := \lim_{\delta \to 0^+} \frac{P_K(x+\delta v) - x}{\delta}; \text{ moreover } \Pi_K(x,v) = P_{T_K(x)}(v).$$

Let $\Pi_K : K \times X \to X$ be the operator given by $(x,v) \mapsto \Pi_K(x,v)$. Note that Π_K is discontinuous on the boundary of the set K. In [17],[22], several characterizations of Π_K are given. Assuming $F : K \to X$ to be a Lipschitz continuous vector field and $x_0 \in K$, the initial value problem

$$\frac{dx(\tau)}{d\tau} = \Pi_K(x(\tau), -F(x(\tau)),\ x(0) = x_0 \in K \tag{16.4}$$

has a unique absolutely continuous solution on the interval $[0,\infty)$, as shown in Ref. [10].

Definition 16.7. A **projected dynamical system** is given by a mapping $\phi : \mathbb{R}_+ \times K \to K$ which solves the initial value problem:

$$\dot{\phi}(\tau,x) = \Pi_K(\phi(\tau,x), -F(\phi(\tau,x))), \quad \phi(0,x) = x_0 \in K.$$

Definition 16.8. Double-layer dynamics is the projected dynamical system given by

$$\frac{du(\cdot,\tau)}{d\tau} = \Pi_{\mathbb{K}}(u(\cdot,\tau), -F(u)(\cdot,\tau)),\ u(\cdot,0) = u(\cdot) \in \mathbb{K}, \tag{16.5}$$

where $\mathbb{K} \subset X := L^2([0,T],\mathbb{R}^q)$ closed, convex, bounded subset, $F : \mathbb{K} \to L^2([0,T],\mathbb{R}^q)$ is strictly pseudo-monotone and Lipschitz continuous and there exists $u^* \in \mathbb{K}$ so that

$$\begin{gathered} u^* \text{ uniquely solves an EVI problem (16.2)} \\ \Leftrightarrow \\ u^* \text{ uniquely solves } \Pi_{\mathbb{K}}(u(\cdot,\tau), -F(u)(\cdot,\tau)) = 0. \end{gathered}$$

Note that the equivalence in Definition 16.8 above is due to the following result (see [7]):

Theorem 16.9. *Assuming F is strictly pseudo-monotone and Lipschitz continuous, the solutions of EVI problem (16.2) are the same as the critical points of PDS (16.5). The converse is also true.*

From Definition 16.8, we can see that DLD describes in fact the evolution of a curve $u \in \mathbb{K}$ with respect to time τ. Recall that we consider an EVI (16.2) as the model of an equilibrium problem. The solution of this EVI is interpreted as a curve of equilibrium states of the underlying problem over the time interval $[0,T]$. These are all the **potential** equilibrium states the problem can reach. Therefore we call $[0,T]$ **the prediction timescale**. By Definition 3.2 the equilibrium curve is stationary in the projected dynamics (16.5), hence $\tau \in [0,\infty)$ represents the evolution of the problem from disequilibrium to equilibrium. Therefore we call $[0,\infty)$ **the adjustment scale**. A DLD model includes therefore the assumptions that t and τ both represent physical time and that time flows forward, i.e., these models are, in terms of classic dynamical systems theory, not reversible in time.

The modeling questions one can answer based on DLD are of the following type: does an equilibrium problem reach one of its predicted equilibrium states **in finite time**, starting from an observed initial state $u(t_0)$, at some $t_0 \in [0,T]$?

A first answer to this question was given in [8] and then a more general one in [13]. In the next subsection, we give a new general method for estimating the adjustment behavior of an equilibrium problem over a finite time interval $[0,T]$

16.3.2 Tracking Equilibrium Dynamics: Hybrid Systems Approach

Double-layer dynamics models are in a sense generalizations of hybrid dynamical systems, given that the time evolution is continuous in both timescales, as opposed to being a combination of a discrete scale with a continuous one, as is the case for hybrid systems.

We study the question of describing the time evolution of an equilibrium problem whenever not in equilibrium. In this section, we present a method for tracking this evolution based on hybrid dynamical systems [37]. Hybrid systems are in brief dynamical systems incorporating phenomena evolving on two timescales, a discrete one and a continuous one. In fact, a hybrid system is a combinations of local flows of continuous-time dynamical systems by means of a finite number of so-called "switch and jump" mechanisms, taking place at a finite number of time instances, called "event times." Examples of such systems are numerous in control theory and complementarity systems coming from mechanics (see [37] for an introduction and examples of hybrid systems). In brief, a trajectory of a hybrid system starts with an initial given state and evolves according to a continuous dynamics generated by a (set of) constrained differential equation(s); at each discrete event time the constraints of the continuous dynamics are violated and the system has to switch to a new, different, continuous dynamics state.

Double-layer dynamics models are in a sense generalizations of hybrid dynamical systems, given that the time evolution is continuous in both timescales, as opposed to being a combination of a discrete scale with a continuous one, as is the case for hybrid systems. We proceed next to show how a hybrid dynamics approach lends itself readily as a computational approach to a double-layer dynamics model. It is known that current computational methods for deriving solutions to an EVI problem [8],[9],[3],[4] require some discretization of the time interval of interest $[0,T]$. Let us thus consider a division of the interval $[0,T]$ given by $(0 = t_0 \leq t_1 \leq \ldots \leq t_{k-1} \leq t_k = T$ so that $|t_i - t_{i+1}| = \delta > 0, \forall i \in \{1,\ldots,k\}$. In terms of DLD, whenever $[0,T]$ is discretized, we are left with solving for a solution of the EVI problem at each discretization point (which is now a finite-dimensional variational inequality for each such point), which in turn can be directly associated to a flow (16.5) on the finite-dimensional subset $\mathbb{K}(t_i)$, $\forall t_i$, $i \in \{1,\ldots,k\}$.

In a hybrid systems formalism (as developed in [37]), the division points are considered to be the discrete event times of the discrete dynamics. For each interval (t_i, t_{i+1}), the continuous dynamics will take place on the constraint set $\mathbb{K}(t_i)$ given by the projected differential equation:

$$\frac{du(t_i,\tau)}{d\tau} = P_{T_{\mathbb{K}(t_i)}(u(t_i,\tau))}(-F_i(u(t_i,\tau))), \tag{16.6}$$

$$u(t_i,t_i) \in \mathbb{K}(t_i),\; F_i : \mathbb{K}(t_i) \to \mathbb{R}^q,$$

$$\lim_{\tau \to t_{i+1}^-} u(t_i,\tau) \text{ and } \lim_{\tau \to t_i^+} u(t_i,\tau) \text{ exist}.$$

Note that at event time t_{i+1} the continuous dynamics on (t_i, t_{i+1}) switches to a new dynamics on (t_{i+1}, t_{i+2}), with a jump condition

$$u(t_{i+1},t_{i+1}) = P_{\mathbb{K}(t_{i+1})}(u(t_i, t_i + \delta)) \in \mathbb{K}(t_{i+1}).$$

Then a disequilibrium evolution of a network equilibrium problem modeled with DLD is in fact a trajectory of the hybrid system above, namely

$$u(t_0,\tau) \overset{u(t_1,t_1)=P_{\mathbb{K}(t_0)}(u(t_0,t_0+\delta))}{\longrightarrow} u(t_1,\tau) \to \ldots \overset{u(t_k,t_k)=P_{\mathbb{K}(t_k)}(u(t_{k-1},t_{k-1}+\delta))}{\longrightarrow} u(t_k,\tau).$$

Now we are ready to present our computational procedure for finding the disequilibrium (hybrid) trajectories. For each solution piece $u(t_i,\tau)$ of a continuous dynamics we use a projection like method (as in [10]) to compute $u(t_i,\tau)$, $\tau \in (t_i, t_{i+1})$ as a solution of the PrDE (16.6); it is known that such solution is in fact an absolutely continuous function on the given interval. We then project the end point of this solution piece on the next constraint set $\mathbb{K}(t_{i+1})$; the projection is then taken to be the initial point of the next solution piece of the continuous dynamics on $\mathbb{K}(t_{i+1})$. This method produces then a piecewise continuous curve which approximately tracks the disequilibrium behavior of the underlying network problem. We show in the next section how this method works when applied to a dynamic game between producers in an oligopoly environment.

16.4 Dynamics of Environmental Product Markets

In this paper, we present a first eco-product oligopolistic market model under incentive policies. The model assumes that there is a finite number of eco-product developers, denoted typically by i. Each developer may build two variants of a product: the standard or usual product (already in existence on market), and an eco-product, by which we mean a variant of the current product incorporating eco-features (for example residential developers building usual houses and eco-features houses, having perhaps one or more of: a gray water recycling system, rain water harvest system, geothermal water heating system, etc.*). Generally the prices of the eco-products are expected to be higher than the ones for the usual products. Our model is meant to be used as a tool in showing how incentive policies (to both consumers and developers) could influence (increase) the consumer demand for eco-products over a given time period.

What makes this market model novel is the incorporation of time-dependency; time is thought to influence the price, quantity and demand of an eco-product under a set policy. We develop EVI and DLD models of the competitive dynamic game at producers' level, under incentive policies. Last but not least, the reader will be able to see that there are multiple avenues to transform the model below, making it more complex, as particular product markets may require. In the next sections, we present a numerical model for markets of eco-houses; while the theoretical model can, by in large, be used for any eco-product market, some of the parameters used in the numerical illustrations refer to this particular market.

16.4.1 The Static Model

Let us consider n development sites denoted by $i \in \{1,..,n\}$; similarly, consider m to be the number of classes of consumers in an urban area, where the division can be made for instance according to either income or age, or a combination of the two. We denote by $j \in \{1,...,m\}$ a consumer class. Further, we denote by q_{ij}^k the number of k-type houses demanded at site i by group j, with $k \in \{u,e\}$, where u stands for *usual* house, and e stands for *eco*. We denote by p_i^k the price of a k-type house at site i, with $p = (..,p_i^u,..,p_i^e,..)$.

Assuming for simplicity that we have $m = 1$ (only one consumer group), we envisage a situation where consumers seeking to purchase a new home can choose from houses in two different developments, which might have different characteristics, such as distance from the city center, access to transportation networks, schools, etc. Consequently the demand for usual houses in the two developments can differ.

* The initial model was developed in conjunction with funding from the City of Guelph and Ontario Centers of Excellence, Center for Earth and Environmental Technologies on a project for emergent markets of new eco-residential developments.

We do not model these differences explicitly here, other than to allow for different "base" demands for houses in the two developments.

We therefore consider the following expressions for the demand for usual and eco-houses at site $i \in \{1,2\}$, assuming only one developer per site, as follows:

$$q_1^u = a_1^u + b_1^u p_1^u + b_1^{ue} p_1^e + (b_1^u)^{2u} p_2^u + (b_1^u)^{2e} p_2^e$$

$$q_1^e = a_1^e + b_1^{eu} p_1^u + b_1^e p_1^e + (b_1^e)^{2u} p_2^u + (b_1^e)^{2e} p_2^e$$

$$q_2^u = a_2^u + (b_2^u)^{1u} p_1^u + (b_2^u)^{1e} p_1^e + b_2^u p_2^u + b_2^{ue} p_2^e$$

$$q_2^e = a_2^e + (b_2^e)^{1u} p_1^u + (b_2^e)^{1e} p_1^e + b_2^{eu} p_2^u + b_2^e p_2^e.$$

This can be written in matrix form as

$$\underline{q} = \underline{a} + B\underline{p} \tag{16.7}$$

where

$$\underline{q} = (q_1^u, q_1^e, q_2^u, q_2^e),\ \underline{a} = (a_1^u, a_1^e, a_2^u, a_2^e),\ \underline{p} = (p_1^u, p_1^e, p_2^u, p_2^e)$$

and

$$B := \begin{pmatrix} b_1^u & b_1^{ue} & (b_1^u)^{2u} & (b_1^u)^{2e} \\ b_1^{eu} & b_1^e & (b_1^e)^{2u} & (b_1^e)^{2e} \\ (b_2^u)^{1u} & (b_2^u)^{1e} & b_2^u & b_2^{ue} \\ (b_2^e)^{1u} & (b_2^e)^{1e} & b_2^{eu} & b_2^e \end{pmatrix}. \tag{16.8}$$

The key part of consumer demand in this model is how consumers value eco-houses versus standard houses. In particular, the sensitivity of the demand for eco-houses with respect to the prices of other alternatives is central to our predictions regarding the expected market outcomes under various policy scenarios. There are a number of parameters in the system of consumer demands that govern consumer price responsiveness. First, there is the responsiveness of the demand for eco-houses at one site to the price of eco-houses at that site (b_i^e, $i \in \{1,2\}$). There are also parameters that control the responsiveness of the demand for eco-houses to the price of substitute products: usual houses in the same development (b_i^{ui}, $i \in \{1,2\}$), as well as eco- and usual houses in the other development ($(b_i^u)^{\rceil iu}, (b_i^u)^{\rceil ie}, (b_i^e)^{\rceil iu}, (b_i^e)^{\rceil ie}$, $i \in \{1,2\}$).

We consider next that the developers are in a noncooperative game toward maximizing their profits from the new developments. We define below the revenue expressions for each developer, the profit functions and the game we analyze. First, each developer's revenue can be simply estimated from the demand above to be:

$$F_i = p_i^u(\underline{q})q_i^u + p_i^e(\underline{q})q_i^e, \text{ where } i \in \{1,2\}$$

and the prices $p_i^k = p_i^k(\underline{q})$, $i \in \{1,2\}$, $k \in \{u,e\}$ are functions of the vector of quantities produced at both sites, with the expressions given by solving (16.7). Consequently, the profit functions have the expression:

$$\pi_i(\underline{q}) = F_i(\underline{q}) - c_i(\underline{q}) + \tau_i \underline{q},$$

where $c_i(\underline{q})$ is the cost function and τ_i is a policy incentive for all $i \in \{1,2\}$. There are here a few key issues to remark upon. One is that the cost structure chosen is the difference between the costs of production for usual and eco-houses. There is in principal fairly clear information about this for specific eco-features. The complication lies in defining what features are to be implemented, definition which directly impacts this cost difference. A second issue is to note that the value of τ_i is positive: the incentive here can be thought of as a tax relief to the developers for producing more environmentally advanced products. If the incentive is a subsidy, then one can consider $\tau_1 = \tau_2 > 0$.

Therefore, the developers are trying to maximize their profits as described by the following:

$$\forall i \in \{1,2\}, \quad \begin{array}{c} \max \pi_i(\underline{q}) \\ \text{s.t. } \underline{q} \geq 0 \\ q_i^u + q_i^e = cap_i \end{array}$$

where $cap_i > 0$ is given by the fact that each site has a maximum number of new residential developments it can hold, due to water, energy and natural gas capacities that can be allocated by a municipality.

Remark 16.10. Evidently the model can be generalized to $n > 2$ developers and sites, and it can be generalized to a number of $m > 1$ population groups. For the ease of the reading, we chose to present in detail a simpler case, leaving it to the reader to obtain the generalizations.

The problem of finding the optimal quantities of usual and eco-houses for all producers is then becoming a classic game, described by finding $\underline{q^*} \in K$ so that

$$\forall i \in \{1,2\}, \; \pi_i((q_i^{u*}, q_i^{e*}), \underline{\hat{q^*}_i}) \geq \pi_i((q_i^u, q_i^e), \underline{\hat{q^*}_i}), \forall (q_i^u, q_i^e) \in K_i, \tag{16.9}$$

where

$$\underline{\hat{q^*}_1} = (q_2^{u*}, q_2^{e*}), \; \underline{\hat{q^*}_2} = (q_1^{u*}, q_1^{e*}),$$

$K_i := \{(q_i^u, q_i^e) \in \mathbb{R}^2 \mid 0 \leq q_i^k \leq cap_i, k \in \{u,e\}, q_i^u + q_i^e = cap_i\}$ and $K := K_1 \times K_2$.

16.4.2 Dynamic Equilibrium Model: EVI Formulation

Following a similar method as in [5], we can think of the market model above as time dependent. Time can be thought of intervening in the size of the policy incentives $\tau_i = \tau_i(t)$ with a direct consequence upon the numbers and type of houses being developed at each site; therefore we can consider that $\underline{q} = \underline{q}(t)$. In this context, the game (16.9) above can be placed in the context of a dynamic game over a finite time interval $[0,T]$ given by finding $\underline{q^*}(t) \in K(t)$

$$\forall i \in \{1,2\}, \;\text{ for a.a. } t \in [0,T], \tag{16.10}$$

$$pi_i((q_i^{u*}(t), q_i^{e*}(t)), \hat{\underline{q}}^*_{-i}(t)) \geq \pi_i((q_i^u(t), q_i^e(t)), \hat{\underline{q}}^*_{-i}(t)), \forall (q_i^u(t), q_i^e(t)) \in K_i(t),$$

where

$$K_i(t) := \{(q_i^u(t), q_i^e(t)) \in \mathbb{R}^2 \mid 0 \leq q_i^k(t) \leq cap_i(t), k \in \{u, e\}, q_i^u(t) + q_i^e(t) = cap_i(t)\}$$

and

$$K(t) := K_1(t) \times K_2(t).$$

Under proper assumptions on the properties of the profit functions and on the marginal profit functions (see [5] for the equivalence theorem between solutions of EVI (16.2) and solutions of a dynamic game such as (16.10)), the dynamic game above can be equivalently formulated as an evolutionary variational inequality of finding $\underline{q}^*(t) \in K(t)$ for a.a. t, so that

$$\langle F(\underline{q}^*)(t), \underline{q}(t) - \underline{q}^*(t) \rangle \geq 0, \forall \underline{q}(t) \in \mathbb{K}(t) \tag{16.11}$$

where

$$F := -\frac{\partial \pi}{\partial q}, \text{ and } \frac{\partial \pi}{\partial q} = (\frac{\partial \pi_1}{\partial q_1^u}, \frac{\partial \pi_1}{\partial q_1^e}, \frac{\partial \pi_2}{\partial q_2^u}, \frac{\partial \pi_2}{\partial q_2^e}).$$

16.4.3 Example

We let the demand parameters from (16.8) be as follows:

$$B := \begin{pmatrix} -5 & -2 & -2 & 0 \\ -2 & -5 & 0 & -2 \\ 2 & 0 & -5 & -2 \\ 0 & -2 & -2 & -5 \end{pmatrix}, \underline{a} = \begin{pmatrix} 550 \\ 400 \\ 550 \\ 400 \end{pmatrix}$$

and we consider that over a period of 15 years, the developers are supported by an increase in government subsidies; the subsidies increase from an initial 500 CAD per eco-house unit to 15,000 CAD per eco-house unit, per site. We want to analyze what would be the optimal quantities and prices for usual and eco-houses at each of the two sites in this subsidy scenario. Due to the change in subsidy, it is expected that the optimal quantities as well as their prices will change over time. Hence we use an EVI model to forecast the equilibrium states of this market under this scenario.

According to our parameters above, we obtain the prices as functions of the quantities to be produced and as functions of time:

$$\underline{p}(\underline{q}(t)) = B^{-1}(\underline{q}(t) - a).$$

The profit functions for each site $i \in \{1, 2\}$ are given by

$$\pi_i(t, \underline{q}(t)) = p_i^u(\underline{q}(t)) q_i^u(t) + p_i^e(\underline{q}(t)) q_i^e(t) - c_i(\underline{q}(t)) + \tau(t) q_i^e(t),$$

where $\tau(t) = 1000t$ is the subsidy and $c_i(\underline{q}(t)) = 250q_i^u(t) + 300q_i^e(t)$ is the total cost of developing at site i.

In this case, F in (16.11) will have the form

$$F_1^u(\underline{q})(t) = -(-34q_1^u(t)/45 + 610/9 + 4q_1^e(t)/9 + 2q_2^u(t)/9 - 8q_2^e(t)/45 - 250)$$

$$F_1^e(\underline{q})(t) = -(4q_1^u(t)/9 - 34q_1^e(t)/45 + 340/9 - 8q_2^u(t)/45 + 2q_2^e(t)/9]) - 300 + \tau(t))$$

$$F_2^u(\underline{q})(t) = -(-34q_2^u(t)/45 + 610/9 + 4q_2^e(t)/9 + 2q_1^u(t)/9 - 8q_1^e(t)/45 - 250)$$

$$F_2^e(\underline{q})(t) = -(4q_2^u(t)/9 - 34q_2^e(t)/45 + 340/9 - 8q_1^u(t)/45 + 2q_1^e(t)/9]) - 300 + \tau(t))$$

and

$$\mathbb{K} = \{\underline{q} \in L^2([0,15] \to \mathbb{R}^4) \mid 0 \leq q_i^u(t), q_I^e(t) \leq 250,\ q_i^u(t) + q_i^e(t) = 250, \text{ for a.a. t}\}.$$

Given the data above, in this case we have that the optimal solution of the game is the same for both sites, therefore we only list the values for one site in Table 16.1 below.

Table 16.1 This table presents the evolution of the prices of eco-houses over time, as well the overall proportion of the eco-housing developments at both development sites.

Time	No. usual houses	No. eco-houses	Proportion of eco-houses	Price of co-house
0.5	124.29	24.9	0.16	39.42
1	124.03	25.30	0.169	39.35
2	123.50	26.02	0.174	39.21
3	147.70	51.48	0.25	36.33
4	147.13	52.15	0.26	36.20
5	146.55	52.83	0.26	36.07
6	145.98	53.50	0.26	35.94
7	145.40	54.18	0.27	35.81
8	144.83	54.85	0.27	35.68
9	144.25	55.53	0.27	35.55
10	143.68	56.20	0.28	35.42
11	143.10	56.87	0.28	35.29
12	142.53	57.55	0.28	35.15
13	141.95	58.22	0.29	35.02
14	166.24	83.75	0.33	32.13
15	165.61	84.38	0.337	32.011

These values indicate that the increase of a subsidy can decrease the price[†] of an eco-house unit, while increasing the proportion of eco- versus usual houses at each site, if the developers play their optimal strategies. For an easier comparison of the values, we represented them in the following three figures below (Figuers 16.4–16.6).

† The price unit here is 10,000 CAD.

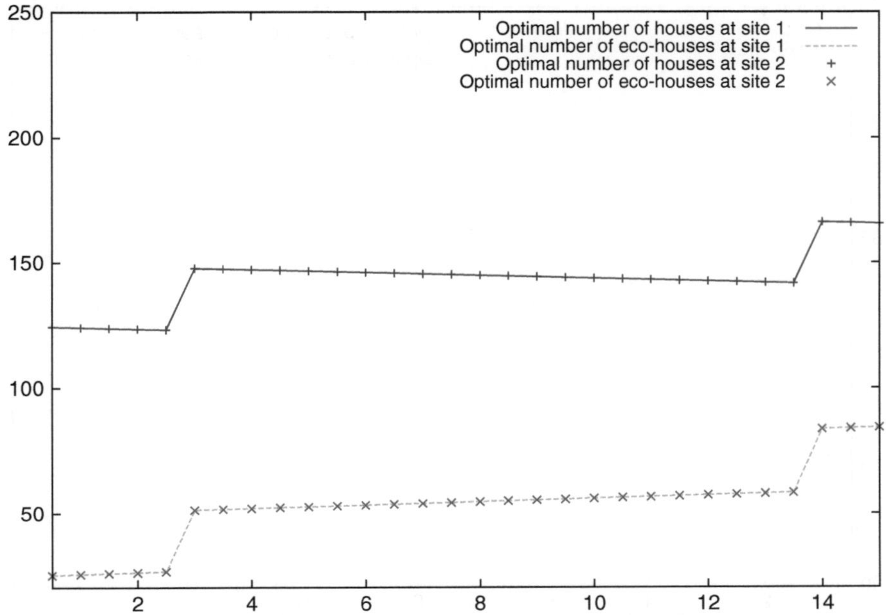

Fig. 16.4 These are the representations of the optimal usual and eco-houses over the time window, subject to the increased subsidy; note that although the subsidy increases linearly, we can identify two jumps in the optimal numbers.

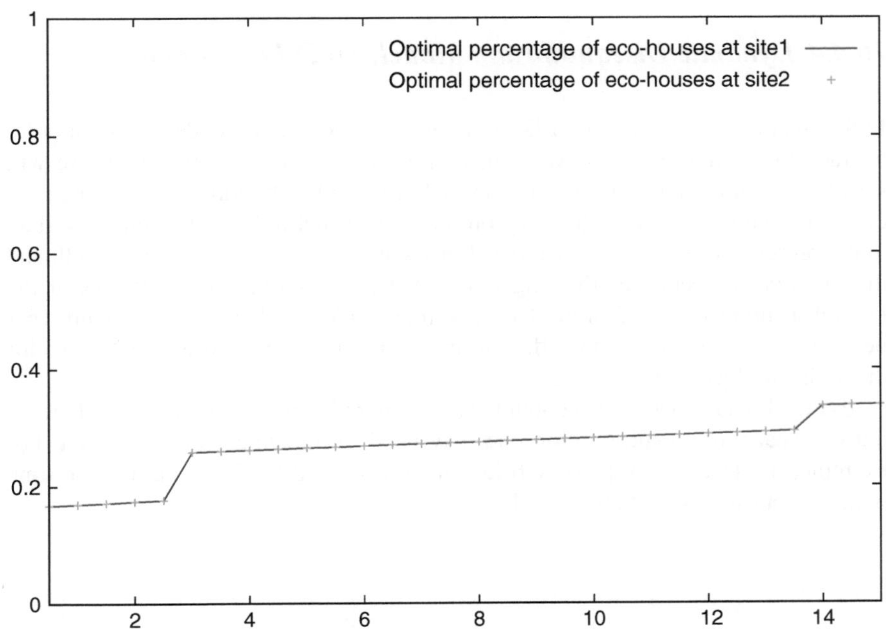

Fig. 16.5 The curves represent the proportion of eco-houses at a site as a function of time.

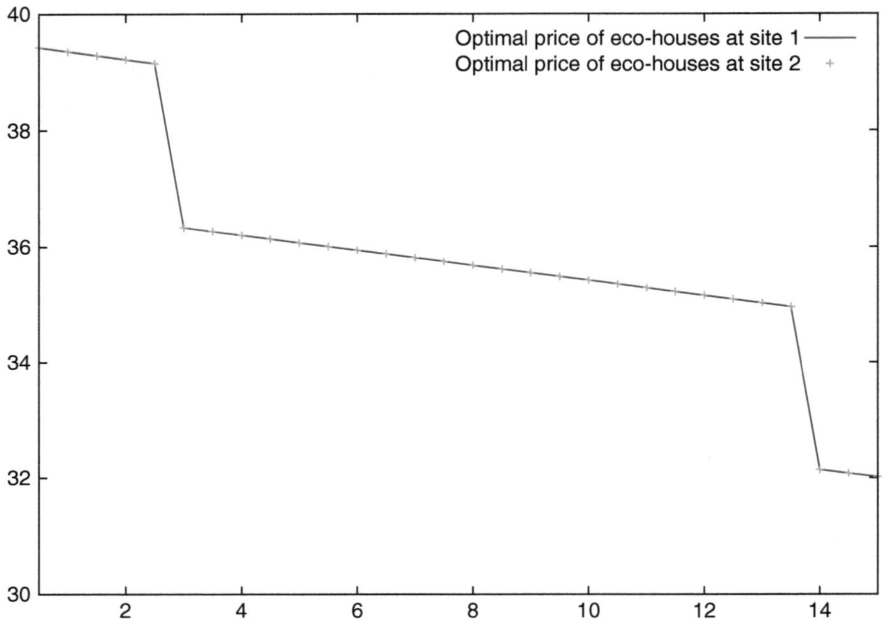

Fig. 16.6 The curves represent the price of an eco-house at a site as a function of time and subsidy; as expected, there are two bigger drops in price, consistent with the two jumps in eco-house units at a site.

16.4.4 Dynamic Disequilibrium Model: DLD Formulation

In this section, we employ a DLD formulation of the market model above in order to track the evolution of a development site away from the equilibrium. We will start by assuming that initially each developer considers building usual houses exclusively (we assume here that they produce 150 usual units). We then show that, in the presence of subsidies, each developer starts producing eco-units as well, and their output numbers are adjusting toward the predicted optimal strategies of the last subsection by the end of the 15 years time window. Table 16.2 and Figure 16.7 below show the evolution of the disequilibrium behavior (we chose here $\delta = 0.5$ for the division of $[0,15]$).

Evidently our model can account for other, more complicated scenarios, depending on what type of information one may need to implement in the market. For example, the level and type of subsidy functions could be different, and the time window considered could be varied.

Table 16.2 default

Time	Optimal no. u-houses	Optimal no. eco-houses	Adjusted no. u-houses	Adjusted no. eco-houses
1	124.03	25.30	191.51	58.48
2	123.50	26.02	182.12	67.87
3	147.70	51.48	177.97	72.02
4	147.13	52.15	176.140	73.85
5	147.13	52.15	175.32	74.67
6	145.98	53.50	174.97	75.028
7	145.40	54.18	174.812	75.18
8	144.83	54.85	174.742	75.25
9	144.25	55.53	174.712	75.28
10	143.68	56.20	174.69	75.30
11	143.10	56.87	174.69	75.30
12	142.53	57.55	174.68	75.31
13	141.95	58.22	174.68	75.31
14	166.24	83.75	174.68	75.312
15	165.61	84.38	174.68	75.312

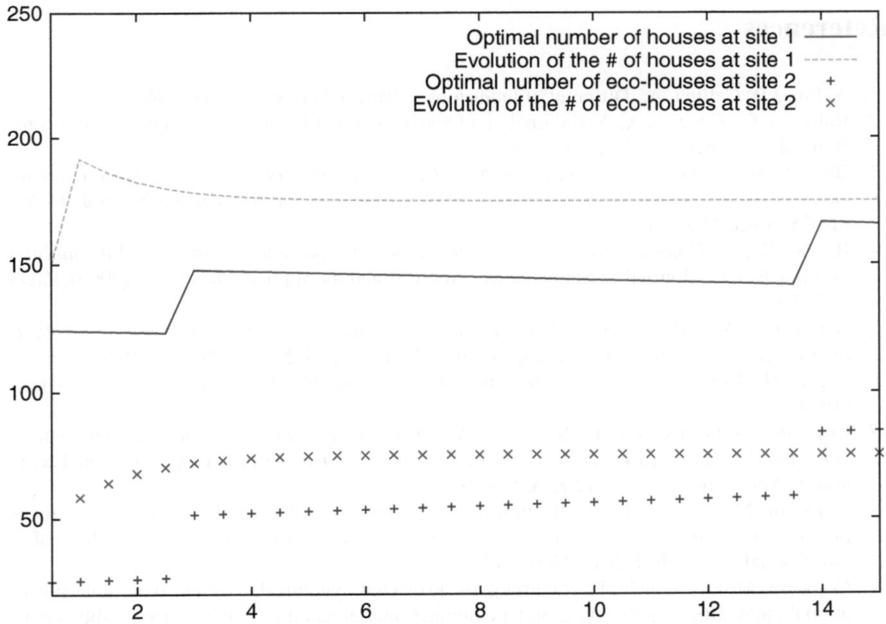

Fig. 16.7 The adjusting curves are seen to approach similar values by the end of the 15-year period.

16.5 Conclusions and Acknowledgments

We presented here a unified method of tracking the dynamics of an equilibrium problem whose constraint set varies with time. The method is using the fact that there is a natural link arising between double-layer dynamics framework of [8],[13] and hybrid dynamical systems [37]. However, we are only at the beginning of showcasing the links between DLD and hybrid systems. A systematic study of their interrelations is currently under way.

The first author would like to acknowledge the important contributions of Prof. G. Isac in the study of infinite-dimensional projected systems and their relations with variational inequalities and applications. The material in this chapter has been inspired by discussions around the relation between projected systems in Banach spaces and their relation with complementarity problems in particular. It is worth noting that these discussions have led to the current investigation into hybrid dynamical systems, among which complementarity systems are some of the most wellknown.

References

1. Aubin, J.P., Cellina, A., Differential Inclusions. Springer-Verlag, Berlin (1984).
2. Baiocchi, C., Capello, A., Variational and Quasivariational Inequalities. Applications to Free Boundary Problems, J. Wiley & Sons (1984).
3. Barbagallo, A., *Regularity results for time-dependent variational and quasivariational inequalities and computational procedures*. M^{3}AS: Mathematical Models and Methods in Applied Sciences, (2005).
4. Barbagallo, A., Cojocaru, M., *Continuity of solutions for parametric variational inequalities in Banach space*, Journal of Mathematical Analysis and its Applications, 351, 2 (2009), Pages 707–720.
5. Barbagallo, A. and Cojocaru, M.-G., *Dynamic equilibrium formulation of the oligopolistic market problem*, Mathematical and Computer Modelling, 49, 5–6 (2009), 966–976
6. Brezis, H., Inequations D'Evolution Abstraites, Comptes Rendue d'Academie des Sciences (1967).
7. Cojocaru, M.-G., Daniele, P., Nagurney, A., *Projected dynamical systems and evolutionary variational inequalities via Hilbert spaces and applications*, Journal of Optimization Theory and its Applications, 127(3) (2005), 549–563.
8. Cojocaru, M.-G., Daniele, P., Nagurney, A., *Double-layered dynamics: a unified theory of projected dynamical systems and evolutionary variational inequalities*, European Journal of Operational Research, 175, 6 (2006) 494–507.
9. Cojocaru, M.-G., Daniele, P., Nagurney, A., Projected dynamical systems, evolutionary variational inequalities, applications and a computational procedure, in Pareto Optimality, Game Theory and Equilibria. Migdalas, A., Pardalos, P. M., Pitsoulis, (Eds.), Nonconvex Optimization and its Applications Series (NOIA 2006), Kluwer Academic Publishers, Dordrecht.
10. Cojocaru, M. -G., Jonker, L. B., *Existence of solutions to projected differential equations on Hilbert spaces*, Proceedings of the American Mathematical Society 132 (2004), 183–193.
11. Cojocaru, M.-G., *Double layer dynamics theory and human migration after catastrophic events* in: Nonlinear Analysis with Applications in Economics, Energy and Transportation, Eds. E. Allevi, M. Bertocchi, A. Gnudi, I.V. Konnov, Bergamo University Press (2007), 65–86.

12. Cojocaru, M.-G., *Dynamic equilibria of group vaccination strategies in a heterogeneous population*, Journal of Global Optimization, 40 (2008), 51–63.
13. Cojocaru, M.-G., *Piecewise solutions of evolutionary variational inequalities. Consequences for the double-layer dynamics modelling of equilibrium problems*, Journal of Inequalities in Pure and Applied Mathematics, Vol. 8, Issue 2 (2007), article 63.
14. Dafermos, S., *Traffic equilibrium and variational inequalities*, Transportation Science 14 (1980), 42–54.
15. Daniele, P., Maugeri, A., Oettli, W., *Time–dependent variational inequalities*, Journal of Optimization Theory and its Applications 103 (1999), 543–555.
16. Daniele, P., Dynamic Networks and Evolutionary Variational Inequalities, Edward Elgar Publishing (2006).
17. Dupuis, P., Ishii, H. *On Lipschitz continuity of the solution mapping to the Skorokhod problem with applications*. Stochastics and Stochastics Reports 35 (1990), 31–62.
18. Dupuis, P., Nagurney, A., *Dynamical systems and variational inequalities*, Annals of Operations Research 44 (1993), 9–42.
19. J. Gwinner, J., *Time dependent variational inequalities - some recent trends*, in: Equilibrium Problems and Variational Models (Eds.: P. Daniele, F. Giannessi, and A. Maugeri), Kluwer Academic Publishers, Dordrecht (2003), 225-264.
20. Isac, G., Topological Methods in Complementarity Theory, Kluwer Academic Publishers, Dordrecht (2000).
21. Isac, G., Cojocaru, M. -G., *Variational inequalities, complementarity problems and pseudo-monotonicity. Dynamical aspects*, in: Seminar on fixed point theory (Proceedings of the International Conference on Nonlinear Operators, Differential Equations and Applications), Babes-Bolyai University of Cluj-Napoca, Vol. III (2002), 41–62.
22. Isac, G., Cojocaru, M. -G., *The projection operator in a Hilbert space and its directional derivative. Consequences for the theory of projected dynamical systems*, Journal of Function Spaces and Applications, 2(1) (2004), 71–95.
23. Isac, G., Leray-Schauder Type Alternatives, Complementarity Problems and Variational Inequalities, Springer-Verlag, Berlin (2006).
24. Karamardian, S. Schaible, S., *Seven kinds of monotone maps*, Journal of Optimization Theory and Applications, 66 (1990), 1, 37–46.
25. Krasnoselskii, M. A., Zabreiko, P. P., Geometrical Methods of Nonlinear Analysis, A Series of Comprehensive Studies in Mathematics, Vol. 263, Springer-Verlag, Berlin (1984).
26. Kinderlehrer, D., Stampacchia, D., An Introduction to Variational Inequalities and their Application, Academic Press, New York (1980).
27. Jahn, J., Introduction to the Theory of Nonlinear Optimization, Springer-Verlag, Berlin (1996).
28. Lions, J. L., Stampacchia, G., *Variational inequalities*, Communications in Pure and Applied Mathematics 22 (1967), 493–519.
29. Nagurney, A., Liu, Z., Cojocaru, M. -G., Daniele, P., *Dynamic electric power supply chains and transportation networks: an evolutionary variational inequality formulation*, Transportation Research Part E, 43 (5), (2007) 624-646.
30. Maugeri, A., Galligani E., Eds., Rendiconti del Circolo Matematico di Palermo, Serie II, Suppl. 58 (1999).
31. Nagurney A.,Network Economics: A Variational Inequality Approach, Kluwer Academic Publishers, Dordrecht (1993).
32. Nagurney, A., Zhang, D., Projected Dynamical Systems and Variational Inequalities with Applications, Kluwer Academic Publishers, Boston, MA (1996).
33. Patriksson, M., The Traffic Assignment Problem Models and Methods, VSP BV, Utrecht, The Netherlands (1994).
34. B. Ran B., Boyce, D., Modeling Dynamic Transportation Networks, Second Revised Edition, Springer, Heidelberg (1996).
35. Samuelson, P. A., *Spatial price equilibrium and linear programming*, American Economic Review, 42 (1952), 283–303.

36. Schaible, S., *Generalized monotonicity - concepts and uses*, in: Variational Inequalities and Network Equilibrium Problems, F. Gianessi, A. Maugeri (Eds.), Plenum Press, New York (1995).
37. van der Schaft, A. and H. Schumacher, An Introduction to Hybrid Dynamical Systems, LNCIS 251, Springer (2000).
38. Smith, M. J., *The existence, uniqueness and stability of traffic equilibrium*, Transportation Research 38 (1979), 295–304.
39. Stampacchia, G., *Variational inequalities, theory and applications of monotone operators*, in: Proceedings of NATO Advanced Study Institute, Oderisi, Gubbio (Eds.), Venice (1968), 101–192.
40. Steinbach, J., *On a variational inequality containing a memory term with an application in electro chemical machining*, Journal of Convex Analysis, 5, 1(1998), 6380.
41. T. Takayama, T, Judge, G. G., Spatial and Temporal Price and Allocation Models, North Holland, Amsterdam, The Netherlands (1971).
42. Wardrop, J.G., *Some theoretical aspects of road traffic research*, Proceedings of the Institute of Civil Engineers, Part II (1952), 325–378.
43. Zarantonello, E., *Projections on convex sets in Hilbert space and spectral theory*, in: Contributions to Nonlinear Functional Analysis 27, Mathematical Research Center, University of Wisconsin, Academic Press, New York (1971), 237–424.

Chapter 17
A Panoramic View on Projected Dynamical Systems

Patrizia Daniele, Sofia Giuffrè, Antonino Maugeri, and Stephane Pia

Dedicated to the memory of Professor George Isac

Abstract The theory of generalized projections both in non-pivot Hilbert spaces and strictly convex and smooth Banach spaces is developed and the related theory of projected dynamical systems is highlighted. A particular emphasis is given to the equivalence between solutions of variational inequalities and critical points of projected dynamical systems.

17.1 Introduction

The aim of this paper is to present a generalized theory of projected dynamical systems and to offer an improvement along several directions of previous results contained in the paper [15]. In [15], the authors present a detailed and self-containing outline of the projected dynamical systems and of the parallel theory of variational inequalities. Precisely, the authors note how, while variational inequalities are able

---

Patrizia Daniele
Department of Mathematics and Computer Science, University of Catania, Catania, Italy, e-mail: daniele@dmi.unict.it

Sofia Giuffrè
D.I.M.E.T., Mediterranean University, Reggio Calabria, Italy, e-mail: sofia.giuffre@unirc.it

Antonino Maugeri
Department of Mathematics and Computer Science, University of Catania, Catania, Italy, e-mail: maugeri@dmi.unict.it

Stephane Pia
Department of Mathematics and Computer Science, University of Catania, Catania, Italy, e-mail: pia@dmi.unict.it

P.M. Pardalos et al. (eds.), *Nonlinear Analysis and Variational Problems*,
Springer Optimization and Its Applications 35, DOI 10.1007/978-1-4419-0158-3_17,

to describe the equilibrium state of complex systems, projected dynamical systems allow one to study the underlying dynamics or disequilibrium behavior of such systems. The mutual dependence between variational inequalities and projected dynamical systems first was focused on by Dupuis and Nagurney (see [27]) who provided the fundamental theory for finite-dimensional projected dynamical systems and established the basic result that the set of stationary points of a projected dynamical system coincides with the set of solutions of the associate finite-dimensional variational inequality.

Isac and Cojocaru [33] initiated the systematic study of projected dynamical systems in infinite-dimensional Hilbert spaces in 2002, and [40] and [13] made explicit, for the first time, the connection between projected dynamical systems on Hilbert spaces and evolutionary variational inequalities, which Daniele, Maugeri and Oettli [23] and [24], motivated by dynamic traffic equilibrium problems, introduced in 1999.

The current paper goes beyond the framework of Hilbert spaces and expands upon the theme of the papers [8], [17], [28], [29], [30] and [31]. Precisely, following [8], [17], [30], [31], the concept of projected dynamical systems in non-pivot Hilbert spaces is introduced. The non-pivot Hilbert spaces are Hilbert spaces for which the usual identification with its topological dual does not make sense. This is the case, for example, of weighted Hilbert spaces which play a fundamental role in the study of weighted traffic equilibrium problems. The fundamental theory for such a kind of projected dynamical systems is developed and the connection between the set of critical points and the set of solutions of the weighted variational inequalities is highlighted.

Another generalization of the concept of projected dynamical systems is obtained considering the framework of strictly convex and uniformly smooth Banach spaces. In this case (see [28] and [29]), two different concepts of projected dynamical systems can be given, namely the concept of metric projected dynamical system and the one of generalized projected dynamical system. The difference arises because in the Banach space we can consider two projectors in the following way: given a closed, convex subset C of X, we can define the projectors

$$x \to P(C|x) = \left\{ y \in C : \|x-y\| = \inf_{z\in C} \|x-z\| \right\} \qquad \text{(Metric Projection)}$$

and

$$x \to \Pi_C(x) = \operatorname{argmin}_{y\in C} \left(\|x\|^2 - 2\langle J(x), y\rangle + \|y\|^2 \right) \qquad \text{(Generalized Projection)}$$

where $J : X \to X^*$ is the duality mapping between X and X^*.

Besides a study of the fundamental properties of these two projections, the basic result of the coincidence of the set of stationary points of a projected dynamical system in a non-pivot Hilbert space or in Banach spaces with the associate variational inequality is presented. In connection with these results, a focus is put on the important open problem that existence results for projected dynamical systems in strictly convex and smooth Banach spaces are not yet available, even if it is a reasonable conjecture that it is possible to prove it.

Finally, an interesting result between an equivalent formulation of projected dynamical systems in terms of unilateral differential inclusions is provided, establishing, in this way, a promising link between the theory of differential inclusions and projected dynamical systems in Banach spaces.

17.2 General Background Material

17.2.1 Spaces

17.2.1.1 Strictly Convex and Uniformly Smooth Banach Spaces

We denote by X a Banach space with dual space X^* and by $\|\cdot\|$ and $\|\cdot\|_*$ the respective norms. We denote also the duality pairing between X^* and X by $< f,x >$ for $f \in X^*$ and $x \in X$, and by $< x,f >$ the duality pairing between X and X^* for $f \in X^*$ and $x \in X$.
We define the duality mapping $J : X \to X^*$ by

$$J(x) = \{f \in X^* :< f,x >= \|f\|_*^2 = \|x\|^2\},\ \forall x \in X.$$

In the same manner we have the duality mapping $J^* : X^* \to X$ defined by:

$$J^*(f) = \{x \in X :< x,f >= \|x\|^2 = \|f\|_*^2\},\ \forall f \in X^*.$$

The existence of J and J^* is a corollary of the Hahn–Banach analytic form (see for instance [9]). We recall two definitions that we need in the sequel.

Definition 17.1 (see [25]). A space $(X,\|\cdot\|)$ is strictly convex if

$$\forall x \in X,\ \forall y \in X : \|x\| = \|y\| = 1,\ x \neq y \Rightarrow \|tx+(1-t)y\| < 1, \forall t \in]0,1[.$$

Let us denote by $S(X) = \{x \in X : \|x\| = 1\}$.

Definition 17.2 (see [25]). A Banach space X is said to be smooth at $x_0 \in S(X)$ whenever there exists a unique $f \in S(X^*)$ such that $f(x_0) = 1$. If X is smooth at each point of $S(X)$, then we say that X is smooth.

From [25], we have also the following characterization criteria: A Banach space $(X,\|\cdot\|)$ is smooth if and only if the norm $\|\cdot\|$ admits a Gâteaux derivative in each direction.

Remark 17.3. Hilbert spaces and L^p spaces $(1 < p < \infty)$ are reflexive, strictly convex and smooth.

From [10], we know that if X is reflexive, strictly convex and smooth, then J, J^* are one-to-one single-valued operators and $J^{-1} = J^*$. More precisely we have:

- X is reflexive if and only if J is surjective;
- X is smooth if and only if J is single-valued;
- X is strictly convex if and only if J is injective.

Remark 17.4. In a strictly convex and smooth Banach space, J is given by $J(x) = grad\left(\frac{\|x\|^2}{2}\right)$; for more details we refer to [3].

Example 17.5. If $X = L^p(\Omega,\mathbb{R})$ with $1 < p < \infty$ then

$$J(x) = \|x\|^{2-p}|x|^{p-1}sgn(x)$$

and

$$J^*(x) = \|x\|^{\frac{p-2}{p-1}}|x|^{\frac{1}{1-p}}sgn(x)$$

where $sgn(x) = \chi_{[x>0]} - \chi_{[x<0]}$.

17.2.1.2 The Special Case of Hilbert Spaces, the Notion of Non-pivot Hilbert Spaces

Each time we work with a Hilbert space V, it is necessary to decide whether or not we identify the topological dual space $V^* = \mathscr{L}(V,\mathbb{R})$ with V. Commonly this identification is made, one of the reasons for this being that the vectors of the polar of a set of V are in V. In some cases the identification does not make sense. For clarity of presentation, we recall below the basic results regarding the dual realization of a Hilbert space. The readers can refer to [5] for additional information.

First, consider a pre-Hilbert space V with an inner-product $((x,y))$, and its topological dual $V^* = \mathscr{L}(V,\mathbb{R})$. It is well known that V^* is a Banach space for the classical dual norm $\left(\|f\|_* = \sup\limits_{\substack{x\in V\\ x\neq\theta_V}} \frac{|f(x)|}{\|x\|}\right)$. If V is a Hilbert space, then the mapping J is linear.

Theorem 17.6 (Theorem 1 page 68, [5]). *Let V be a Hilbert space with the inner product $((x,y))$ and $J \in \mathscr{L}(V,V^*)$ the duality mapping above. Then J is a surjective isometry from V to V^*. The dual space V^* is a Hilbert space with the inner product:*

$$((f,g))_* = ((J^{-1}f, J^{-1}g)) = f(J^{-1}g).$$

Theorem 17.7 (Theorem 2 page 69, [5]). *Let V be a pre-Hilbert space. Then there exists a completion $\hat{V}$ of V, that is, an isometry j from V to the Hilbert space $\hat{V}$ such that $j(V)$ is dense in $\hat{V}$.*

Definition 17.8. Let V be a Hilbert space. We call $\{F, j\}$, where

i) F is a Hilbert space,
ii) j is an isometry from F to $\mathscr{L}(V,\mathbb{R})$

a dual realization of V. Then we set

$$\langle f,x\rangle = j\circ f(x), \forall f\in F,\ \forall x\in V,$$

where $\langle f,x\rangle$ is the duality pairing on $F\times V$.

Remark 17.9. The duality pairing is a nondegenerate bilinear form on $F\times V$ and $\|f\|_F = \sup_{\substack{x\in V\\ x\neq\theta_V}} \frac{|\langle f,x\rangle|}{\|x\|}$. These properties permit us to prove that F is isomorphic to V^*.

We deduce from Theorems 17.6 and 17.7 that $k = j^{-1}\circ J \in \mathscr{L}(V,F)$ is a surjective isometry such that

$$(x,y) = \langle k(x),y\rangle.$$

We use the following convention here: when a dual realization $\{F,j\}$ of a space has been chosen, we set $F = V^*$ and $j\circ f(x) = \langle f,x\rangle$. We say that the isometry $k : V \to V^*$ is the duality operator associated to the inner product on V and to the duality pairing on $V^*\times V$ by the relation

$$(x,y) = \langle k(x),y\rangle.$$

A special but most frequent case is to choose as a dual realization of V the couple $\{V,J\}$; in this case the Hilbert space V is called a *pivot space*. To be more precise, we introduce the following definition.

Definition 17.10. A Hilbert space H with an inner product (x,y) is called a pivot space, if we identify H^* with H. In that case

$$H^* = H,\ j = J,\ \langle x,y\rangle = (x,y).$$

Sometimes it does not make sense to identify the space itself with its topological dual, as the following example shows.
Let us consider $V = L^2(\mathbb{R},(1+|x|)) \subset L^2(\mathbb{R})$ (dense subspace of $L^2(\mathbb{R})$) endowed with the inner product:

$$(u,v)_V = \int_{\mathbb{R}} (1+|x|)u(x)v(x)dx.$$

An element $\varphi \in L^2(\mathbb{R})^*$ is also an element of V^*. If we identify φ to an element $f\in L^2(\mathbb{R})$, this function does not define a linear form on V and the expression $\varphi(v) = \langle f,v\rangle_V$ has no meaning on V. In this situation it is necessary to work in a non-pivot Hilbert space. We provide now some useful examples of non-pivot Hilbert spaces.

Let $\Omega \subset \mathbb{R}^n$ be an open subset, $a : \Omega \to R^+ \setminus \{0\}$ a continuous and strictly positive function called "weight" and $s : \Omega \to R^+ \setminus \{0\}$ a continuous and strictly positive function called "real time density." The bilinear form defined on $\mathscr{C}_0(\Omega)$ (continuous functions with compact support on Ω) by

$$\langle x, y \rangle_{a,s} = \int_\Omega x(\omega) y(\omega) a(\omega) s(\omega) d\omega$$

is an inner product. We remark here that if a is a weight, then $a^{-1} = 1/a$ is also a weight. Let us introduce the following

Definition 17.11. We call $L^2(\Omega, a, s)$ a completion of $\mathscr{C}_0(\Omega)$ for the inner product $\langle x, y \rangle_{a,s}$.

We now introduce an n-dimensional version of the previous space. If we denote by $V_i = L^2(\Omega, \mathbb{R}, a_i, s_i)$ and $V_i^* = L^2(\Omega, \mathbb{R}, a_i^{-1}, s_i)$, the space

$$V = \prod_{i=1}^m V_i \tag{17.1}$$

is a non-pivot Hilbert space with the inner product

$$(F, G)_V = (F, G)_{\mathbf{a},\mathbf{s}} = \sum_{i=1}^m \int_\Omega F_i(\omega) G_i(\omega) a_i(\omega) s_i(\omega) d\omega.$$

The space

$$V^* = \prod_{i=1}^m V_i^* \tag{17.2}$$

is clearly a non-pivot Hilbert space for the following inner product

$$(F, G)_{V^*} = (F, G)_{\mathbf{a}^{-1},\mathbf{s}} = \sum_{i=1}^m \int_\Omega \frac{F_i(\omega) G_i(\omega) s_i(\omega)}{a_i(\omega)} d\omega,$$

and the following bilinear form

$$V^* \times V \to \mathbb{R}$$

$$\langle f, x \rangle_{V^* \times V} = \langle f, x \rangle_{\mathbf{s}} = \sum_{i=1}^m \int_\Omega f_i(\omega) x_i(\omega) s_i(\omega) d\omega \tag{17.3}$$

defines a duality between V^* and V. More precisely we have (see [31] for a proof):

Proposition 17.12. *The bilinear form (17.3) defines a duality mapping between V^* and V, given by*

$$J(F) = (a_1 F_1, \ldots, a_m F_m).$$

In Section 17.2.4, we use the introduced objects to set up the weighted traffic equilibrium problem.

17.2.2 Cones and Properties

We recall in this section some definitions of cones. There are many different classes of cones, but we are interested in tangent cones which admit a mutually polar cone (a normal cone which admits as the polar cone the tangent cone). This will be used to apply decomposition theorems. In fact, we recall only the following definitions

Definition 17.13. Let $C \subset X$ be convex; we call general tangent cone to C at $\bar{x}$ the set given by:

$$T_C(\bar{x}) = \limsup_{\lambda \to 0} \frac{1}{\lambda}(C - \bar{x}).$$

Remark 17.14. Definition 17.13 is valid also if C is non-convex. If C is a convex subset of X, Definition 17.13 is equivalent to:

$$T_C(\bar{x}) = \overline{\bigcup_{\lambda > 0} \lambda(C - \bar{x})}.$$

Definition 17.15. Let $C \subset X$ be convex, we call normal cone to C at x the set given by:

$$N_C(x) = \{\xi \in X^*, <\xi, y - x> \leq 0, \forall y \in C\}.$$

Definition 17.16. Let M be a cone of X, the polar set of M, denoted by M^0, is defined by:

$$M^0 = \{\xi \in X^*, <\xi, x> \leq 0, \forall x \in M\}.$$

If X is reflexive, then the following relationships hold:

$$\begin{aligned}(T_C(x))^0 &= N_C(x), \forall x \in C \\ (N_C(x))^0 &= T_C(x), \forall x \in C.\end{aligned} \tag{17.4}$$

T_C and N_C are always closed and if C is nonempty and convex, they are nonempty and convex.
We introduce now the notion of relative interior which will be used in Section 17.5.

Definition 17.17. Let $C \subset X$ be convex. We call the relative interior of C the following set:

$$ri(C) = \{x \in C : T_C(x) = X\}.$$

Definition 17.18. Let $C \subset X$ be convex. We call the relative boundary of C the following set:

$$rb(C) = C \setminus ri(C).$$

Proposition 17.19 (Proposition 2.2 in [21]). *Let us assume that X is a reflexive Banach space and $C \subset X$ convex. If $x \in C$ we have:*

$$x \in ri(C) \Leftrightarrow N_C(x) = \{0_{X^*}\}.$$

Proof. Let it be $T_C(x) = X$ then we have:

$$N_C(x) = \{\xi \in X^* :< \xi, x > \leq 0, \forall x \in X\}$$

so we get $\forall x \in X, < \xi, x > \leq 0$ and $< \xi, x > \geq 0$ so we can deduce that $\xi = 0_{X^*}$. On the other side if $N_C(x) = \{0_{X^*}\}$ then, using the polarity, we get

$$T_C(x) = \{\xi \in X :< \xi, 0_{X^*} > \leq 0\} = X$$

and by definition $x \in ri(C)$. □

17.2.3 Projectors

17.2.3.1 Metric and Generalized Projection Operators

In Banach spaces, it is possible to define the generalized projection operator and the metric projection operator. The metric projection operator is simply the minimum norm operator and the generalized projection operator is an operator introduced by Zarantonello [48] in reflexive Banach spaces and studied deeply by Alber [2] in strictly convex and smooth Banach spaces. The purpose of this section is to recall some basic results on these operators.

Definition 17.20 (see [44]). Let X be a Banach space and C a closed convex subset of X. We call the metric projection operator from X on C the set valued mapping $P(C|\cdot) : X \to C$ defined by

$$x \to P(C|x) = \{y \in C : \|x - y\| = d_C(x)\}$$

where $d_C(x) = \inf_{z \in C} \|x - z\|$.

Note that for $x \in C$, $P(C|x)$ is the set of optimal solutions of the following minimization problem:

$$\inf_{y \in C} \|x - y\|^2. \tag{17.5}$$

From now on and unless otherwise stated, we make the following assumptions: X is a reflexive, strictly convex and smooth Banach space.
Then these additional assumptions ensure that $P(C|\cdot) = P_C(\cdot)$ is single valued and P_C is called the best approximate operator. Moreover we have the following characterization of $P_C(x)$:

$$\bar{x} = P_C(x) \Leftrightarrow < J(x - \bar{x}), y - \bar{x} > \leq 0, \ \forall y \in C. \tag{17.6}$$

As an extension of what we have on Hilbert spaces, (17.6) is called the basic variational principle for P_C in X. This characterization plays a fundamental role for our application.

Another possibility to generalize the notion of projection is to use the convex function in y $V(x,y)$ given by:

$$V(x,y) := \|x\|^2 - 2 < J(x), y > + \|y\|^2.$$

We remark that if C is a closed convex subset of X and if $x \in C$, then the problem

$$\min_{y \in C} V(x,y)$$

is uniquely solvable (apply for instance [9], Corollary III.20). Then we can give the following definition:

Definition 17.21 (see [2] or [44]). We call generalized projection of x on C the following value:

$$\Pi_C(x) := argmin_{y \in C} V(x,y).$$

Remark 17.22 (see [2]).

- The operator $\Pi_C : X \to C \subset X$ is the identity on C, i.e., for every $x \in C, \Pi_C(x) = x$.
- In a Hilbert space, $V(x,y) = \|x-y\|^2$, Π_C coincides with the projection operator P_C.

As stated in [3] we have the following characterization of $\Pi_C(x)$.

Lemma 17.23. *Assume that C is a closed convex subset of X, then:*

$$\hat{x} = \Pi_C(x) \Leftrightarrow < J(x) - J(\hat{x}), y - \hat{x} > \leq 0, \ \forall y \in C. \tag{17.7}$$

Here again the variational characterization plays a fundamental role for our application.

From Corollary 1, page 22 in [25], we know that if X is reflexive then:

$$X \text{ strictly convex} \Leftrightarrow X^* \text{ smooth},$$
$$X \text{ smooth} \Leftrightarrow X^* \text{ strictly convex}.$$

Remark 17.24. As in a Hilbert space the mapping J is linear, the two projection concepts coincide.

Remark 17.25. We can suggest a possible interpretation of the two different projection concepts. The metric projection of x to C gives the nearest point of C to x, and the generalized projection of x to C gives the point in C which makes the smallest angle with x.

17.2.3.2 A Decomposition Theorem

Theorem 17.26 ([3], Theorem 2.4). *Assume that X is a real reflexive strictly convex and smooth Banach space, and K a non-empty, closed and convex cone of X then:* $\forall x \in X$ *and* $\forall f \in X^*$ *the decompositions*

$$x = P_K(x) + J^* \Pi_{K^0} J(x) \text{ and } < \Pi_{K^0} J(x), P_K(x) >= 0$$
$$f = P_{K^0}(f) + J \Pi_K J^*(f) \text{ and } < P_{K^0}(f), \Pi_K J^*(f) >= 0 \quad (17.8)$$

hold.

Remark 17.27. If X is a non-pivot Hilbert space, the previous decomposition reduces to $x = P_K(x) + J^* P_{K^0} J(x)$ with $< P_{K^0} J(x), P_K(x) >= 0$ and $f = P_{K^0}(f) + J P_K J^*(f)$ and $< P_{K^0}(f), P_K J^*(f) >= 0$. If X is a Hilbert space and we identify X with its dual X^*, then we obtain the classical Moreau decomposition theorem $x = P_K(x) + P_{K^0}(x)$ with $< P_{K^0}(x), P_K(x) >= (P_{K^0}(x), P_K(x)) = 0$.

17.2.3.3 A Very Important Convex Set

Cojocaru, Daniele, and Nagurney in [13] showed that all the problems considered in section 17.1 can be formulated using a unified definition of a convex set as we recall below. We consider the nonempty, convex, closed, bounded subset of the strictly convex and smooth Banach space $L^p([0,T], \mathbb{R}^q)$, $1 < p < +\infty$, given by

$$\mathbb{K} = \bigcup_{t \in [0,T]} \Big\{ u \in L^p([0,T], \mathbb{R}^q) : \lambda(t) \le u(t) \le \mu(t) \text{ a.e. in } [0,T];$$
$$\sum_{i=1}^{q} \xi_{ji} u_i(t) = \rho_j(t) \text{ a.e. in } [0,T], \quad (17.9)$$
$$\xi_{ji} \in \{0,1\}, \; i \in \{1,\dots,q\} \; j \in \{1,\dots,l\} \Big\}.$$

Let $\lambda, \mu \in L^p([0,T], \mathbb{R}^q)$, $\rho \in L^p([0,T], \mathbb{R}^l)$. For chosen values of the scalars ξ_{ji}, of the dimensions q and l, and of the constraints λ, μ, we obtain each of the previous above-cited model constraint set formulations (see [13]), as follows:

- for the traffic network problem (see [24], [23]), we let $\xi_{ji} \in \{0,1\}$, $i \in \{1,\dots,q\}$, $j \in \{1,\dots,l\}$, with every column of the matrix $\Xi = [\xi_{ij}]_{1 \le i \le l,\, 1 \le j \le q}$ having only one entry different than zero and $\lambda(t) \ge 0$ for all $t \in [0,T]$;
- for the quantity formulation of spatial price equilibrium (see [18]), we let $q = n + m + nm$, $\xi_{ji} = 0$, $i \in \{1,\dots,q\}$, $j \in \{1,\dots,l\}$; $\mu(t)$ large and $\lambda(t) = 0$, for any $t \in [0,T]$;

- for the price formulation of spatial price equilibrium (see [19] and [22]), we let $q = n+m+mn$, $l = 1$, $\xi_{ji} = 0$, $i \in \{1,\dots,q\}$, $j \in \{1,\dots,l\}$, and $\lambda(t) \geq 0$ for all $t \in [0,T]$;
- for the financial equilibrium problem (see [20]), we let $q = 2mn+n$, $l = 2m$, $\xi_{ji} = \{0,1\}$ for $i \in \{1,\dots,n\}$, $j \in \{1,\dots,l\}$, with every column of the matrix Ξ having only one entry different than zero; $\mu(t)$ large and $\lambda(t) = 0$, for any $t \in [0,T]$.

17.2.4 Weighted Traffic Equilibrium Problem

Let us introduce a network $\mathcal{N}$, which means a set $\mathcal{W}$ of origin–destination pairs (origin/destination nodes) and a set $\mathcal{R}$ of routes. Each route $r \in \mathcal{R}$ links exactly one origin–destination pair $w \in \mathcal{W}$. The set of all $r \in \mathcal{R}$ which link a given $w \in \mathcal{W}$ is denoted by $\mathcal{R}(w)$. For each time $t \in (0,T)$ we consider the vector flow $F(t) \in \mathbb{R}^n$. Let us denote by Ω an open subset of $\mathbb{R}$, by $n = card(\mathcal{R})$, $\mathbf{a} = \{a_1,..,a_n\}$ and by $\mathbf{a}^{-1} = \{a_1^{-1},..,a_n^{-1}\}$ two families of weights such that for each $1 \leq i \leq n$, $a_i \in \mathcal{C}(\Omega, \mathbb{R}^+ \setminus \{0\})$. We introduce also the family of real time traffic densities $\mathbf{s} = \{s_1,..,s_n\}$ such that for each $1 \leq i \leq n$, $s_i \in \mathcal{C}(\Omega, \mathbb{R}^+ \setminus \{0\})$. To each element of $\mathbf{a}$ and $\mathbf{s}$, let us say a_i and s_i corresponds a route r, let us say r_i. Now, using the setup of section 17.2.1, we can describe the feasible flows, which have to satisfy the time-dependent capacity constraints and demand requirements, namely for all $r \in \mathcal{R}$, $w \in \mathcal{W}$ and for almost all $t \in \Omega$,

$$\lambda_r(t) \leq F_r(t) \leq \mu_r(t)$$

and

$$\sum_{r \in \mathcal{R}(w)} F_r(t) = \rho_w(t)$$

where $\lambda(t) \leq \mu(t)$ are given in $\mathbb{R}^n$, $\rho(t) \in \mathbb{R}^m$ where $m = card(\mathcal{W})$, F_r, $r \in \mathcal{R}$, denotes the flow in the route r. If $\Phi = (\Phi_{w,r})$ is the pair route incidence matrix, with $w \in \mathcal{W}$ and $r \in \mathcal{R}$, that is

$$\Phi_{w,r} := \chi_{\mathcal{R}(w)}(r),$$

the demand requirements can be written in matrix-vector notation as

$$\Phi F(t) = \rho(t).$$

The set of all feasible flows is given by

$$K := \{F \in V |\ \lambda(t) \leq F(t) \leq \mu(t), a.e.\ in\ \Omega; \\ \Phi F(t) = \rho(t),\ a.e\ in\ \Omega\}.$$

17.2.5 Time-Dependent Equilibria

We provide now the definition of equilibrium for the traffic problem. First we need to define the notion of equilibrium for a variational inequality. A variational inequality (VI) in a Hilbert space V is the problem to determine

$$x \in K: \ \langle C(x), y-x\rangle_{\mathbf{s}} \geq 0, \ \forall\, y \in K$$

where K is a closed convex subset of V and $C : K \to V^*$ is a mapping.

Definition 17.28. $H \in V$ is an equilibrium flow iff

$$H \in K, \langle C(H), F-H\rangle_{\mathbf{s}} \geq 0, \ \forall F \in K. \tag{17.10}$$

It is possible to prove the equivalence between condition (17.10) and what we will call a **weighted Wardrop condition** (17.11).

Theorem 17.29. *$H \in K$ is an equilibrium flow in the sense of (17.10) iff*

$$\forall w \in \mathcal{W}, \ \forall q, m \in \mathcal{R}(w), \ a.e. \ in \ \Omega,$$

$$\begin{aligned} & s_q(t)C_q(H(t)) < s_m(t)C_m(H(t)) \\ & \Rightarrow H_q(t) = \mu_q(t) \ or \ H_m(t) = \lambda_m(t). \end{aligned} \tag{17.11}$$

Proof. Assume that (17.11) holds. Let $w \in \mathcal{W}$ and

$$A = \{q \in \mathcal{R}(w) : \ H_q(t) < \mu_q(t)\, a.e.\, in\, \Omega\}$$

$$B = \{m \in \mathcal{R}(w) : \ H_m(t) > \lambda_m(t)\, a.e.\, in\, \Omega\}.$$

From (17.11) it follows

$$s_q(t)C_q(H(t)) \geq s_m(t)C_m(H(t))\, \forall q \in A, \forall m \in B, a.e.\, in\, \Omega.$$

Then there exists a function $\gamma_w(t) : (0,T) \to \mathbb{R}$ such that a.e. in Ω

$$\inf_{q \in A} s_q(t)C_q(H(t)) \geq \gamma_w(t) \geq \sup_{m \in B} s_m(t)C_m(H(t)).$$

Let $F \in K$ be arbitrary. For every $r \in \mathcal{R}(w)$ such that $s_r(t)C_r(H(t)) < \gamma_w(t)\, a.e.\, in\, \Omega$, it results $r \notin A$, that is $H_r(t) = \mu_r(t)\, a.e.\, in\, \Omega$. This implies $F_r(t) - H_r(t) \leq 0\ a.e.\, in\, \Omega$ and then

$$(s_r(t)C_r(H(t)) - \gamma_w(t))(F_r(t) - H_r(t)) \geq 0\, a.e.\, in\, \Omega.$$

Likewise for every $r \in \mathcal{R}(w)$ such that
$s_r(t)C_r(H(t)) > \gamma_w(t)\, a.e.\, in\, \Omega$, it results $r \notin B$ and

$$(s_r(t)C_r(H(t)) - \gamma_w(t))(F_r(t) - H_r(t)) \geq 0 a.e.\, in\, \Omega.$$

It follows

$$\sum_{r=1}^{n} s_r(t)C_r(H(t))(F_r(t)-H_r(t)) \geq$$

$$\gamma_w(t)\sum_{r=1}^{n}(F_r(t)-H_r(t)) = \gamma_w(t)(\rho_w(t)-\rho_w(t)) = 0$$

and finally we may conclude $\langle C(H), F-H\rangle_{\mathbf{s}} = \int_{\Omega}\sum_{i=1}^{n} s_i(\omega)C_i(H(\omega))(F_i(\omega) - H_i(\omega))d\omega \geq 0$, that is (17.10) holds.

Now assume that (17.11) does not hold. Then there exist $w \in \mathscr{W}$ and $q, m \in \mathscr{R}(w)$ together with a set $E \subseteq \Omega$ having positive measure such that

$$s_q(t)C_q(H(t)) < s_m(t)C_m(H(t)),$$
$$H_q(t) < \mu_q(t),\ H_m(t) > \lambda_m(t),\ a.e.\,in\,E.$$

For $t \in E$ let $\delta(t) := \min\{\mu_q(t)-H_q(t), H_m(t)-\lambda_m(t)\}$. It results $\delta(t) > 0$ a.e. on E. We construct $F : \Omega \to \mathbb{R}$ in the following way:

$$F_q(t) := H_q(t)+\delta(t),$$
$$F_m(t) := H_m(t)-\delta(t)\,a.e.\,in\,E,$$
$$F_r(t) := H_r(t)\ for\,r \neq q,m,\ a.e.\,in\,E,$$
$$F_r(t) := H_r(t)\,a.e.\,in\,\Omega\setminus E.$$

It results that $F \in K$ and

$$\langle C(H), F-H\rangle_{\mathbf{s}} = \int_{\Omega}\sum_{i=1}^{n} C_i(H(\omega))(F_i(\omega)-H_i(\omega))s_i(\omega)d\omega$$

$$= \int_{E}\delta(\omega)[s_q(\omega)C_q(H(\omega)) - s_m(\omega)C_m(H(\omega))]d\omega < 0.$$

Thus H is not an equilibrium. □

The previous equivalence shows how the optimum for the user can be conditioned by a "traffic management center" which is the organization able to establish the weights as the decision center has a global view on the overall situation.

17.3 Projected Dynamical Systems in Hilbert Spaces

17.3.1 Projected Dynamical Systems in Pivot Hilbert Spaces

Isac and Cojocaru ([32], [33]) initiated the systematic study of projected dynamical systems on infinite-dimensional Hilbert spaces in 2002 with the fundamental issue of existence of solutions to such problems answered by Cojocaru [11] in her thesis (see also Cojocaru and Jonker [12]). Let X be a pivot Hilbert space of arbitrary (finite or infinite) dimension and let $C \subset X$ be a non-empty, closed, convex subset.

Definition 17.30. We call the projected system operator in Hilbert space, the operator given by

$$\pi_C : C \times X^* \to X$$

defined by setting:

$$\pi_C(x,h) = \lim_{h \to 0} \frac{P_C(x+h) - P_C(x)}{h}.$$

The directional derivative of the operator P_C is defined, for any $x \in C$ and any element $v \in X$, as the limit (for a proof see [47]):

$$\pi_C(x,v) := \lim_{\delta \to 0^+} \frac{P_C(x+\delta v) - x}{\delta}; \text{ moreover } \pi_C(x,v) = P_{T_C(x)}(v).$$

Note that π_C is nonlinear and discontinuous on the boundary of the set C.
The following result has been shown (see [16]).

Theorem 17.31. *Let X be a Hilbert space and C be a non-empty, closed, convex subset. Let $F : C \to X$ be a Lipschitz continuous vector field and $x_0 \in C$. Then the initial value problem associated to the projected differential equation (PrDE)*

$$\frac{dx(\tau)}{d\tau} = \pi_C(x(\tau), -F(x(\tau)), \; x(0) = x_0 \in C \qquad (17.12)$$

has a unique absolutely continuous solution on the interval $[0,\infty)$.

This result is a generalization of the one in [35], where $X := \mathbb{R}^n$, C is a convex polyhedron and F has linear growth.

17.3.2 Projected Dynamical Systems in Non-pivot Hilbert Spaces

With minor modifications with respect to the pivot case, in [17] the authors extend the result obtained for PDS to the non pivot case. They first introduce non-pivot projected dynamical systems (NpPDS) and then show the existence of a solution. In analogy with [16] they introduce

Definition 17.32. A non-pivot projected differential equation (NpPrDE) is a discontinuous ODE given by:

$$\frac{dx(t)}{dt} = \pi_C(x(t), -(J^{-1} \circ F)(x(t))) = P_{T_C(x(t))}(-(J^{-1} \circ F)(x(t))). \qquad (17.13)$$

Consequently the associated Cauchy problem is given by:

$$\frac{dx(t)}{dt} = \pi_C(x(t), -(J^{-1} \circ F)(x(t))), \; x(0) = x_0 \in C. \tag{17.14}$$

Next they define a solution for a Cauchy problem of type (17.14).

Remark 17.33. In [17] a proof that

$$\pi_C(x,v) = P_{T_C(x)}(v)$$

is valid in non-pivot Hilbert spaces is given in an analogous way to the one used by Zarantonello in [47].

Definition 17.34. An absolutely continuous function $x : \mathscr{I} \subset \mathbb{R} \to X$, such that

$$\begin{cases} x(t) \in C, \; x(0) = x_0 \in C, \forall t \in \mathscr{I} \\ \dot{x}(t) = \pi_C(x(t), -(J^{-1} \circ F)(x(t))), \; \text{a.e. on } \mathscr{I} \end{cases} \tag{17.15}$$

is called a solution for the initial value problem (17.14).

Finally, assuming problem (17.14) has solutions as described above, then we are ready to introduce:

Definition 17.35. A non-pivot projected dynamical system (NpPDS) is given by a mapping $\phi : \mathbb{R}_+ \times C \to C$ which solves the initial value problem:

$$\dot{\phi}(t,x) = \pi_C(\phi(t,x), -(J^{-1} \circ F)(\phi(t,x))), \; \text{a.a. } t, \; \phi(0,x) = x_0 \in C.$$

As said at the beginning of the section, the following result has been proved in [17].

Theorem 17.36. *Let X be a non-pivot Hilbert space and C be a non-empty, closed, convex subset. Let $F : C \to X$ be a Lipschitz continuous vector field and $x_0 \in C$. Then the initial value problem associated to the projected differential equation (PrDE)*

$$\frac{dx(\tau)}{d\tau} = \pi_C(x(\tau), -F(x(\tau)), \; x(0) = x_0 \in C \tag{17.16}$$

has a unique absolutely continuous solution on the interval $[0,\infty)$.

17.4 Projected Dynamical Systems in Banach Spaces

The theory of projected dynamical systems has been introduced in finite-dimensional spaces by Dupuis and Nagurney [27] and was later extended to infinite-dimensional Hilbert spaces by Isac and Cojocaru ([32], [33]) in 2002. In this section, we put the basis for an extension of previous results to strictly convex and smooth Banach spaces. Even if actually no existence results in such spaces exist, we conjecture that it is possible to prove it. Alber and Yao in [4] already proved that an existence result for a projected dynamical system involving gradient-like functionals, but the PDS used is more focused on the relationship with variational inequalities (see Section 17.5) than on the description of the trajectory which permits to reach the equilibrium.

17.4.1 The Strictly Convex and Uniformly Smooth Case

In the paper [28], using the definitions of the metric projection operator and the generalized projection operator given in Section 17.2.3.1, the authors introduce the following concepts of projected dynamical systems in Banach spaces. We recall the following definitions:

Definition 17.37. We call the metric projected system operator, the operator given by

$$\Lambda_C^m : C \times X^* \to X$$

defined by setting:

$$\Lambda_C^m(x,h) = P_{T_C(x)}(J^*(h)).$$

So we can define, as done in [35] and in [13], the differential equation with a discontinuous right-hand side.

Definition 17.38. We call m-projected differential equation (m-PDS), the discontinuous right-hand side differential equation given by:

$$\frac{dx}{dt} = \Lambda_C^m(x, -F(x)) = P_{T_C(x)}(J^*(-F(x))). \tag{17.17}$$

Consequently, the associated Cauchy problem is given by:

$$\frac{dx}{dt} = \Lambda_C^m(x, -F(x)) = P_{T_C(x)}(J^*(-F(x))),\ x(0) = x_0 \in C. \tag{17.18}$$

Definition 17.39. A metric projected dynamical system is given by a mapping ϕ : $\mathbb{R}_+ \times K \to K$ which solves the initial value problem:

$$\dot{\phi}(t,x) = P_{T_C(\phi(t,x))}(J^*(-F(\phi(t,x)))) \text{ a.a. } t, \phi(0,x) = x_0 \in C.$$

Definition 17.40. We call the generalized projected-system operator, the operator given by

$$\Lambda_C^g : C \times X^* \to X$$

defined by setting:

$$\Lambda_C^g(x,h) = \Pi_{T_C(x)}(J^*(h)).$$

Definition 17.41. We call generalized projected differential equation (g-PDS), the discontinuous right-hand side differential equation given by:

$$\frac{dx}{dt} = \Lambda_C^g(x, -F(x)) = \Pi_{T_C(x)}(J^*(-F(x))). \tag{17.19}$$

The associated Cauchy problem is given by:

$$\frac{dx}{dt} = \Lambda_C^g(x, -F(x)) = \Pi_{T_C(x)}(J^*(-F(x))),\ x(0) = x_0 \in C. \tag{17.20}$$

Definition 17.42. A generalized projected dynamical system is given by a mapping $\phi : \mathbb{R}_+ \times C \to C$ which solves the initial value problem:

$$\dot{\phi}(t,x) = \Pi_{T_C(\phi(t,x))}(J^*(-F(\phi(t,x)))) \text{ a.a. } t, \phi(0,x) = x_0 \in C.$$

In a Hilbert space, both (17.17) and (17.19) are equal to (17.16).

17.4.2 Projected Dynamical Systems and Unilateral Differential Inclusions

To prove the existence result in Hilbert spaces, Cojocaru uses in [11] some techniques coming from differential inclusion theorems, therefore we think it is useful to include a section dedicated to equivalent formulations of a projected differential system in X, a strictly convex and smooth Banach space. We consider the two following differential inclusions:

$$-\dot{x} \in J^*(F(x) + N_{T_C(x)}(\dot{x})) \tag{17.21}$$

$$-\dot{x} \in J^*(F(x) + N_C(x)). \tag{17.22}$$

First we present the following preliminary result.

Proposition 17.43. *Let K be a non-empty closed convex cone of X. For any s and v in X, the following relations are equivalent:*

$$s = \Pi_K(v) \tag{17.23}$$

$$J(v) - J(s) \in N_K(s) \tag{17.24}$$

$$s \in K,\ J(v) - J(s) \in K^o,\ < J(v) - J(s), s >= 0 \tag{17.25}$$

$$J(v) - J(s) \in K^o,\ \textit{and}\ \forall v \in K^o,\ \|s\|^2 \leq < J(v) - v, s > . \tag{17.26}$$

Proof. Using the variational characterization of the generalized projection operator (17.7), we get that (17.23) is equivalent to:

$$s \in K,\ < J(v) - J(s), y - s > \leq 0, \forall y \in K$$

and, by definition of a normal cone, we get (17.24). Before the next step, first let us prove that $N_K(s) = K^o \cap \{s\}^\perp$.

By definition of $N_K(s)$, K^o and $\{s\}^\perp$ we get immediately that $K^o \cap \{s\}^\perp \subset N_K(s)$. Now suppose that $y \in N_K(s)$, then we have

$$< y, \eta - s > \leq 0, \ \forall \eta \in K.$$

If $< y, \eta >> 0$, as K is a cone, we get $\forall \lambda > 0$, $< y, \lambda\eta > \leq < y, s >$ which implies a contradiction. Then $< y, \eta > \leq 0$ and $y \in K^o$. As $s \in K$, we get $< y, s > \leq 0$ and as $0 \in K$, we conclude that $< y, s >= 0$ and $y \in \{s\}^\perp$. From the previous result we can conclude that

$$J(v) - J(s) \in N_K(s) \Leftrightarrow s \in K, \ J(v) - J(s) \in K^o, \ < J(v) - J(s), s >= 0.$$

Now suppose that (17.25) holds, take $v \in K^o$, as $< v, s > \leq 0 =< J(v) - J(s), s >$ we get $< v, s > \leq < J(v), s > - < J(s), s >$ and by definition of J we get:

$$\|s\|^2 \leq < J(v) - v, s >, \ \forall v \in K^o.$$

Now suppose that (17.26) holds; in particular we get

$$< v, s > \leq < J(v), s > - \|s\|^2, \forall v \in K^o.$$

If $< v, s >> 0$, we have a contradiction. In fact $< v, s >$ is bounded by $< J(v), s > - \|s\|^2$ and K^o is a cone, so we get that $< v, s > \leq 0$, $\forall v \in K^o$.
But $J(v) - J(s) \in K^o$, then $< J(v) - J(s), s > \leq 0$, if we take $v = 0$ in (17.26), we get exactly (17.25).

□

Remark 17.44. A proof of the previous result in $\mathbb{R}^n$ space can be found in [1].

Corollary 17.45. *The following statements are equivalent:*

$$\dot{x} = \Pi_{T_C(x)}(J^*(-F(x))) \tag{17.27}$$

$$-\dot{x} \in J^*(F(x) + N_{T_C(x)}(\dot{x})) \tag{17.28}$$

$$\begin{cases} -\dot{x} \in J^*(F(x) + N_C(x)) \\ -\dot{x} = J^*(F(x) + P_{N_C(x)}(-F(x)) \\ -\dot{x} = J^*(P_{N_C(x)+F(x)}(0)). \end{cases} \tag{17.29}$$

Proof. We apply Proposition 17.43 with $K = T_C(x)$, $v = J^*(-F(x))$ and $s = \dot{x}$, so we get immediately (17.27) from (17.23). From (17.24) we get

$$JJ^*(-F(x)) - J(\dot{x}) \in N_{T_C(x)}(\dot{x}).$$

As $JJ^* = Id_{X^*}$ we have the equivalence with (17.28).

From Albert's theorem we deduce that (17.27) is equivalent to

$$\dot{x} = J^*(-F(x) - P_{N_C(x)}(-F(x)));$$

so, using the variational principle for metric projections, we get:

$$< J^*(-F(x)+J(\dot{x})+F(x)), y+J(\dot{x})+F(x) > \leq 0, \ \forall y \in N_C(x)$$

and this is equivalent to

$$-\dot{x} = J^*(P_{N_C(x)+F(x)}(0)).$$

And this means that the vector $J(-\dot{x})$ is of minimum norm in $(F(x)+N_C(x))$. □

Remark 17.46. If X is a non-pivot Hilbert space, the generalized projection operator coincides with the metric projection operator and the mappings J and $J^* = J^{-1}$ are linear, therefore Corollary 17.45 can be restated in the following way:

Corollary 17.47. *The following statements are equivalent:*

$$\dot{x} = P_{T_C(x)}(J^*(-F(x))) \tag{17.30}$$

$$-\dot{x} \in J^*(F(x)) + J^*(N_{T_C(x)}(\dot{x})) \tag{17.31}$$

$$\begin{cases} -\dot{x} \in J^*(F(x)) + J^*(N_C(x)) \\ -\dot{x} = J^*(F(x)) + J^*(P_{N_C(x)}(-F(x))) \\ -\dot{x} = J^*(P_{N_C(x)+F(x)}(0)). \end{cases} \tag{17.32}$$

17.5 Bridge with Variational Inequalities

The bridge between projected dynamical systems and variational inequalities is done by a simple but very important result which states that the critical points of the PDS and the equilibrium points of VI coincide. The purpose of this section is to illustrate this point. As stated in [14], in the case in which the variational inequality is an evolutionary VI (the solution belongs to a functional space with real one-dimensional domain), we can interpret the solution of a PDS as the trajectory of a point before reaching the equilibirum and the solution of the evolutionary variational inequality as the trajectory of the equilibrium point. The purpose of this section is to clarify those points and to recall some results about variational inequalities.

We consider now the variational problem given by:

$$x \in C : < F(x), v - x > \geq 0, \ \ \forall v \in C \tag{17.33}$$

where $F : C \to X^*$.

The following existence results are known.

Definition 17.48 (see [23]). Let E be a real topological vector space, $C \subset E$ convex. Then $F : C \to E^*$ is said to be

(i) pseudomonotone iff, for all x, y $\in C$, $< F(x), y-x > \geq 0 \Rightarrow < F(y), x-y > \leq 0$;

(ii) Fan-hemicontinous iff, for all $y \in C$, the function $\xi \rightarrow < F(\xi), y-\xi >$ is upper semicontinous on C;
(iii) hemicontinous along line segments iff, for all x, $y \in C$, the function $\xi \rightarrow < F(\xi), y-x >$ is upper semicontinous on the line segment $[x,y]$.

Then we have the following result.

Theorem 17.49 (see [23]). *Let E be a real topological vector space, and let $C \subseteq E$ be convex and nonempty. Let $F : C \rightarrow E^*$ be given such that:*

(i) there exist $A \subseteq C$ compact, and $B \subseteq C$ compact, convex such that, for every $x \in C \setminus A$, there exists $y \in B$ with $< F(x), y-x >< 0$;
*either **(ii)** or **(iii)** below holds:*
(ii) F is Fan-hemicontinous;
(iii) F is pseudomonotone and hemicontinous along line segments.

Then, there exists $\bar{x} \in A$ such that $< F(\bar{x}), y-\bar{x} >\geq 0$, for all $y \in C$.

We recall the concept of a critical point for PDS:

Definition 17.50. A point $x^* \in C$ is called a critical point for equation (17.23) if

$$\pi_C(x^*, -(J^{-1} \circ F)(x^*)) = 0.$$

We have the following equivalence theorems (see [28]):

Theorem 17.51. *Assume that the hypotheses of Theorems 17.26 and 17.49 hold. Then each equilibrium point of (17.33) is a critical point of (17.17) and, if (17.17) admits critical points then they are equilibrium points of (17.33).*

Proof. Let x^* be a solution of (17.33), since J is bijective, there exists a unique $u_{x^*} \in X$ such that $-F(x^*) = J(u_{x^*})$.
So we have

$$< -J(u_{x^*}), x-x^* >\geq 0, \ \forall x \in C$$

and then

$$< -J(u_{x^*}), \lambda(x-x^*) >\geq 0, \ \forall x \in C \ \forall \lambda > 0$$

which is equivalent to writing:

$$< J(u_{x^*} - 0_X), y - 0_X >\leq 0, \ \forall y \in T_C(x^*).$$

So using the variational principle (17.6) for $P_{T_C(x^*)}$ we get

$$P_{T_C(x^*)}(u_{x^*}) = 0_X = P_{T_C(x^*)}(J^*(-F(x^*)))$$

and we deduce that x^* is a critical point of (17.17).

Now suppose that x^* is a critical point of (17.17).
We have $P_{T_C(x^*)}(J^*(-F(x^*))) = 0_X$, then we get

$$J^*(-F(x^*)) = J^* \Pi_{N_C(x^*)}(-F(x^*))$$

as $(J^*)^{-1} = J$ we get

$$-F(x^*) = \Pi_{N_C(x^*)}(-F(x^*)).$$

If $x^* \in ri(C)$: then $N_C(x^*) = 0_{X^*}$ so we get:

$$\Pi_{N_C(x^*)}(w) = \Pi_{0_{X^*}}(w) = 0_{X^*} = -F(x^*),\ \forall w \in X^*,$$

so we deduce that x^* is a solution of (17.33).

If $x^* \in rb(C)$ and $J^*(-F(x^*)) \notin T_C(x^*)$ we get $N_C(x^*) \neq 0_{X^*}$ and taking into account that $-F(x^*) = \Pi_{N_C(x^*)}(-F(x^*))$, we deduce that $-F(x^*) \in N_C(x^*)$ and so, using the definition of $N_C(x^*)$ we obtain

$$< F(x^*), x - x^* > \geq 0,\ \forall x \in C$$

which implies that x^* is a solution of (17.33).

If $x^* \in rb(C)$ and $J^*(-F(x^*)) \in T_C(x^*)$, we derive immediately

$$P_{T_C(x^*)}(J^*(-F(x^*))) = 0_X = J^*(-F(x^*))$$

but J^* is an isometry and so $-F(x^*) = 0_{X^*}$. Then again x^* is a solution of (17.33)□

Remark 17.52. In the previous proof, it is possible to avoid the use of $ri(C)$, but this notion permits one to have an easier approach to geometrical aspects of the theorem.

Theorem 17.53. *Assume that the hypotheses of Theorems (17.26) and (17.49) hold. Then each equilibrium point of (17.33) is a critical point of (17.19) and, if (17.19) admits critical points then they are equilibrium points of (17.33).*

Proof. Let x^* be a solution of (17.33), since J is bijective, there exists a unique $u_{x^*} \in X$ such that $-F(x^*) = J(u_{x^*})$.
So we have

$$< -J(u_{x^*}), x - x^* > \geq 0,\ \forall x \in C$$

and then

$$< -J(u_{x^*}), \lambda(x - x^*) > \geq 0,\ \forall x \in C\ \forall \lambda > 0,$$

which is equivalent to writing:

$$< J(u_{x^*}) - J(0_X), y - 0_X > \leq 0,\ \forall y \in T_C(x^*).$$

So, using the variational principle (17.7) for $\Pi_{T_C(x^*)}$, we get

$$\Pi_{T_C(x^*)}(u_{x^*}) = 0_X = \Pi_{T_C(x^*)}(J^*(-F(x^*)))$$

from which we deduce that x^* is a critical point of (17.19).

Now suppose that x^* is a critical point of (17.19).
$\Pi_{T_C(x^*)}(J^*(-F(x^*))) = 0_X$ therefore we get

$$J^*(-F(x^*) - P_{N_C(x^*)}(-F(x^*))) = 0_X \Leftrightarrow -F(x^*) = P_{N_C(x^*)}(-F(x^*)).$$

If $F(x^*) = 0_{X^*}$, then (17.33) is trivially verified. Now we suppose that $F(x^*) \neq 0_{X^*}$. Then as $-F(x^*) = P_{N_C(x^*)}(-F(x^*))$ we get $-F(x^*) \in N_C(x^*)$ which means

$$< -F(x^*), y - x^* > \leq 0, \ \forall y \in C$$

and this is exactly (17.33). □

Theorem 17.54. *Let X be a strictly convex and smooth Banach space and let $C \subset X$ be a non-empty, closed and convex subset. Let $F : X \to X^*$ be a vector field. Consider the variational inequality problem:*

$$x \in C : \ \langle F(x), v - x \rangle \ \geq \ 0, \ \ \forall v \in C. \tag{17.34}$$

Then if (17.23) and (17.34) admit solutions, then the critical points of (17.23) coincide with the equilibrium points of the solution set of (17.34).

Proof. It follows from the decomposition Theorem 17.26 (see also [28]). □

Remark 17.55. The previous result can be applied to non-pivot Hilbert spaces. In that case, under the conditions of Theorem 17.36 we know that the non-pivot projected dynamical system admits a solution.

17.6 Conclusion

We have shown how it is possible to extend the theory of projected dynamical systems beyond the framework of Hilbert spaces considering non-pivot Hilbert spaces and strictly convex and smooth Banach spaces. This generalization is not only motivated by love for concepts with more comprehensive domain but also by noting that the above more general spaces are really the framework in which dynamic equilibrium problems act. Almost all the results we know in the Hilbert case have been extended to the above wider framework and among the open problems we focus on the one of the existence of solutions to the Cauchy problem associated to the projected dynamical systems in the strictly convex and smooth Banach space.

References

1. V. Acary, B. Brogliato, A. Daniilidis, C. Lemaréchal, *On the Equivalence Between Complementarity Systems, Projected Systems and Unilateral Differential Inclusions*, Rapport de Recherche 5107 I.N.R.I.A, Janvier 2004.
2. Ya. I. Alber, *Metric and generalized projection operators in Banach spaces: proprieties and applications*, A. Kartsatos (Ed.), Theory and Applications of Nonlinear Operators of Monotone and Accretive Type, Marcel Dekker, New York, 1996, 15–50.
3. Ya. I. Alber, *Decomposition theorem in Banach spaces*, Field Inst. Comm. **25**, 2000, 77–99.
4. Ya. I. Alber, J-C. Yao, *On the projection dynamical systems in Banach spaces*, Taiwanese Journal of Mathematics, (to appear).

5. J-P. Aubin, *Analyse Fonctionelle appliquée*, Editions PUF, 1987.
6. A. Barbagallo, *Regularity Results For Evolutionary Variational and Quasi-Variational Inequalities and Applicatios to Dynamic Equilibrium Problems*, PhD Thesis, University of Naples, Italy, 2007.
7. A. Barbagallo, *Regularity results for time-dependent variational and quasi-variational inequalities and computational procedures*, Math. Models Methods Appl. Sci. **17**, (2007), 277–304.
8. A. Barbagallo, S. Pia, *Weighted variational inequalities in non-pivot Hilbert spaces with applications*, Computational Optimization and Applications, to appear.
9. H. Brezis, *Analyse Fonctionnelle, Théorie et Applications*, Masson, 1993.
10. V. Barbu, Th. Precapanu, *Convexity and Optimization in Banach Spaces*, Romania International Publisher, Bucarest, 1978.
11. M.G. Cojocaru, *Projected Dynamical Systems on Hilbert Spaces*, Ph. D. Thesis, Queen's University, Canada, 2002.
12. M.G. Cojocaru, L.B. Jonker, *Existence of Solutions to Projected Differential Equations in Hilbert Spaces*, Proceedings of the American Mathematical Society **132**, 2004, 183–193.
13. M.G. Cojocaru, P. Daniele, A. Nagurney, *Projected Dynamical Systems and Evolutionary Variational Inequalities Via Hilbert Spaces with Applications*, Journal of Optimization Theory and Applications **127**, no. 3, (2005), 549–563.
14. M.G. Cojocaru, P. Daniele, A. Nagurney, *Double layered Dynamics: A unified theory of Projected Dynamical Systems and Evolutionary Variational inegualities*, European Journal of Operational Research **175**, 1, (2006), 494–507.
15. M.G. Cojocaru, P. Daniele, A. Nagurney, *Projected Dynamical Systems, Evolutionary Variational Inequalities, Applications, and a Computational Procedure*, in Pareto Optimality, Game Theory and Equilibria, Springer, New York, A. Chinchuluun, P.M. Pardalos, A. Migadalas, L. Pitsoulis Eds, 2008, 392–406.
16. M.G. Cojocaru, L.B. Jonker. *Existence of solutions to Projected Differential Equations in Hilbert Spaces*, Proceeding of the American Mathematical Society **132**, 2004, 183–193.
17. M.G. Cojocaru, S. Pia. *Non-pivot and implicit projected dynamical systems on Hilbert spaces*, Preprint.
18. P. Daniele, *Time–Dependent Spatial Price Equilibrium Problem: Existence and Stability Results for the Quantity Formulation Model*, Journal of Global Optimization **28**, 2004, 283–295.
19. P. Daniele, *Evolutionary Variational Inequalities and Economic Models for Demand Supply Markets*, M3AS: Mathematical Models and Methods in Applied Sciences **4 (13)**, 2003, 471–489.
20. P. Daniele, *Variational Inequalities for Evolutionary Financial Equilibrium*, in *Innovations in Financial and Economic Networks*, Edward Elgar Publishing, Cheltenham, England, A. Nagurney, Editor, 2003, 84–108.
21. P. Daniele, S. Giuffré, G. Idone, A. Maugeri, *Infinite Dimentional Duality and Applications*, Mathematische Annalen **339** (1), 2007, 221–239.
22. P. Daniele, A. Maugeri, *On Dynamical Equilibrium Problems and Variational Inequalities*, in *Equilibrium Problems: Nonsmooth Optimization and Variational Inequality Models*, Kluwer Academic Publishers, Dordrecht, The Netherlands, F. Giannessi, A. Maugeri, and P. Pardalos, Editors, 2001, 59–69.
23. P. Daniele, A. Maugeri, W. Oettli, *Time-Dependent Traffic Equilibria*, Journal of Optimization Theory and Applications **103**, no. 3, 1999, 543–555.
24. P. Daniele, A. Maugeri, W. Oettli, *Variational Inequalities and Time–Dependent Traffic Equilibria*, C. R. Acad. Sci. Paris t. 326, serie I, 1998, 1059–1062.
25. J. Diestel, *Geometry of Banach Spaces - Selected topics*, Springer-Verlag, Berlin, 1975.
26. P. Dupuis, H. Ishii, *On Lipschitz continuity of the solution mapping to the Skorokhod problem, with applications*, Stochastics and Stochastics Reports **35**, 1990, 31–62.
27. P. Dupuis, A. Nagurney, *Dynamical Systems and Variational Inequalities*, Annals of Operations Research **44**, 1993, 9–42.
28. S. Giuffré, G. Idone, S. Pia, *Some Classes of Projected Dynamical Systems in Banach Spaces and Variational Inequalities*, Journal of Global Optimization **40**, (1–3), 2008, 119–128.

29. S. Giuffré, G. Idone, S. Pia, *Projected Dynamical Systems and Variational inequalities equivalence results*, Journal of Nonlinear and Convex Analysis **7**, no. 3, 2006, 453–463.
30. S. Giuffré, S. Pia, Wireless Communication Densities and User-Oriented Traffic Equilibrium Problem, Proceedings of IEEE 2008 Wireless Communications, Networking and Mobile Computing (WICOM 2008), Dalian, China, October 12–14, 2008.
31. S. Giuffré, S. Pia, *Weighted Traffic Equilibrium problem in non-pivot Hilbert spaces*, Nonlinear Analysis, 2009, doi: 10.1016/j.na.2009.03.044.
32. G. Isac, M.G. Cojocaru, *Variational Inequalities, Complementarity Problems and Pseudo-Monotonicity. Dynamical Aspects*, in "Seminar on Fixed Point Theory Cluj-Napoca," *Proceedings of the International Conference on Nonlinear Operators, Differential Equations and Applications*, Babes-Bolyai University of Cluj-Napoca, Vol. III, 2002, 41–62.
33. G. Isac, M.G. Cojocaru, *The Projection Operator in a Hilbert Space and its Directional Derivative. Consequences for the Theory of Projected Dynamical Systems*, Journal of Function Spaces and Applications **2**, 2004, 71–95.
34. S. Heikkila, *Monotonone Iterative Techniques for Discontinuous Nonlinear Differential Equations*, Monographs and Textbooks in Pure and Applied Mathematics **181**, Marcel Dekker, New York, 1994.
35. A. Nagurney, *Network Economics: A Variational Inequality Approach*, Second and Revised Edition, Kluwer Academic Publishers, Dordrecht, The Netherlands, 1999.
36. M.A. Noor, *Generalized multivalued quasi-variational inequalities. II*, Comput. Math. Appl. **35**, no. 5, 1998, 63–78.
37. M.A. Noor, *Generalized multivalued quasi-variational inequalities*, Comput. Math. Appl. **31**, no. 12, 1996, 1–13.
38. M.A. Noor, *Implicit dynamical systems and quasi variational inequalities*, Applied Math. and Comput. **134**, 2003, 69–81.
39. M. Patriksson, R.T. Rockafellar, *Variational Geometry and Equilibrium*, Equilibrium problems and Variational Models, Kluwer Academic Publishers, Dordrecht, 2003, 347–367.
40. F. Raciti, *Equilibria Trajectories as Stationary Solutions of Infinite Dimensional Projected Dynamical Systems*, Applied Mathematics Letters **17**, 2004, 153–158.
41. C. Ratti, R.M. Pulselli, S. Williams, D. Frenchman, *Mobile Landscapes: Using location data from cell phones for urban analysis*, Environment and Planning B. Planning and Design, 2006, 727–748.
42. C. Ratti, A. Sevtsuk, S. Huang, R. Pailer, *Mobile Landscapes: Graz in Real Time*, in Proceedings of the 3rd Symposium on LBS & TeleCartography, 28–30 November, 2007, Vienna, Austria.
43. R.T. Rockafellar, R. J-B. Wets, *Variational Analysis*, Springer-Verlag, Berlin, 1998.
44. W. Song, Z. Cao, *The generalized Decomposition Theorem in Banach spaces and Its applications*, J. Approx. Theory 129, 2004,167–181.
45. G. Tian, J. Zhou, *Quasi-variational inequality with non-compact sets*, J. Math. Anal. Appl. **160**, 1991, 583–595.
46. F. Tinti, *Numerical solution for pseudomonotone variational inequality problems by extragradient methods. Variational analysis and applications*, Nonconvex Optim. Appl. **79**, 2005, 1101–1128.
47. E.H. Zarantonello, *Projections on Convex sets in Hilbert space and spectral theory*, Contributions to Nonlinear Functional Analysis, Mathematics Research Center, Madison, April 12–14 1971 (E.H. Zarantonello, ed.) Academic Press, New York, 1971, 237–424.
48. E.H. Zarantonello, *Projectors on convex sets in reflexive Banach spaces*, Technical Summary report 1768, Winsconsin Univ. Madison Mathematics Research Center, 1977.

Chapter 18
Foundations of Set-Semidefinite Optimization

Gabriele Eichfelder and Johannes Jahn

Dedicated to the memory of Professor George Isac

Abstract In this paper, we present various foundations of a new field of research in optimization unifying semidefinite and copositive programming, which is called set-semidefinite optimization. A set-semidefinite optimization problem is a vector optimization problem with a special constraint defined by a so-called set-semidefinite ordering cone. The investigations of this chapter are based on the paper [11].

18.1 Introduction

Semidefinite programming is a rapidly growing field in optimization. There, one considers optimization problems having as a constraint that the image $G(x)$ of a matrix-valued function $G : \mathbb{R}^m \to \mathscr{S}^n$ (with $\mathscr{S}^n$ the space of the real symmetric (n,n) matrices) at some $x \in \mathbb{R}^m$ is positive semidefinite. This means that the quadratic form with the matrix $G(x)$ is non-negative on the whole space $K := \mathbb{R}^n$:

$$y^\top G(x) y \geq 0 \text{ for all } y \in K = \mathbb{R}^n.$$

Replacing the whole space by the positive orthant $K := \mathbb{R}^n_+$, we immediately arrive at the class of copositive optimization problems, where the copositivity of the matrix $G(x)$ is required, i.e.,

Gabriele Eichfelder
Department Mathematik, Universität Erlangen-Nürnberg, Martensstr. 3, D-91058 Erlangen, Germany, e-mail: Gabriele.Eichfelder@am.uni-erlangen.de

Johannes Jahn
Department Mathematik, Universität Erlangen-Nürnberg, Martensstr. 3, D-91058 Erlangen, Germany, e-mail: jahn@am.uni-erlangen.de

P.M. Pardalos et al. (eds.), *Nonlinear Analysis and Variational Problems*,
Springer Optimization and Its Applications 35, DOI 10.1007/978-1-4419-0158-3_18,

$$y^\top G(x)y \geq 0 \text{ for all } y \in K = \mathbb{R}^n_+.$$

Allowing arbitrary sets K, these two classes can be unified and extended to the class of set-semidefinite optimization problems. Instead of restricting ourselves to finite dimensions with the set of all continuous linear maps being identical with the set of matrices, we can also present the whole theory in an infinite-dimensional setting. Additionally, in opposition to the common semidefinite and copositive problems with scalar-valued objective functions, we want to allow vector-valued objective functions.

To be more specific, we have the following standard assumption.

Assumption 18.1. *Let X, Y and Z be topological linear spaces; let Z be partially ordered by a pointed convex cone C_Z; and let $f: X \to Z$ and $G: X \to L(Y,Y^*)$ (here $L(Y,Y^*)$ denotes the linear space of continuous linear maps from Y to its topological dual space Y^*) be given maps.*

In the following, we write linear forms ℓ of a topological dual space as $\langle \ell, \cdot \rangle$, and for the scalar product in $\mathbb{R}^n$ we use the notation $a^\top b$ for $a,b \in \mathbb{R}^n$.

We define a partial ordering $\preccurlyeq$ in the space $L(Y,Y^*)$ by the following ordering cone:

Definition 18.2. For an arbitrary nonempty set $K \subset Y$, the set

$$C^K_{L(Y,Y^*)} := \{A \in L(Y,Y^*) \mid \langle Ay, y \rangle \geq 0 \text{ for all } y \in K\}$$

is called *K-semidefinite cone* (for simplicity we write C^K_L). Any $A \in C^K_L$ is called *K-semidefinite*.

Thus we write equivalently $A \preccurlyeq 0_{L(Y,Y^*)}$ for $-A \in C^K_L$. For instance, for $Y = \mathbb{R}^n$ we consider the linear space $L(Y,Y^*) = \mathscr{M}^n$ of real (n,n) matrices with the scalar product $\langle A, B \rangle = \text{trace}(A \cdot B^\top)$ for all $A, B \in \mathscr{M}^n$. Then $A \in C^K_L$ is equivalent to $x^\top Ax \geq 0$ for all $x \in K$. In the following, we also consider linear subspaces of $L(Y,Y^*)$, like the linear space $\mathscr{S}^n$ of all symmetric matrices in $\mathscr{M}^n$, in the definition of C^K_L.

Using this partial ordering, we define the set-semidefinite optimization problem under Assumption 18.1

$$\min_{x \in S} f(x) \tag{SSOP}$$

with the constraint set

$$S := \{x \in X \mid G(x) \preccurlyeq 0_{L(Y,Y^*)}\}. \tag{18.1}$$

The definition of a minimal solution of the vector optimization problem (SSOP) is given in Section 18.4.

In this chapter, we proceed as follows: first we discuss in Section 18.2 applications and special cases of set-semidefinite optimization problems. Thereby we also consider the semidefinite and the copositive case, with which we started our explanations. In Section 18.3 we examine the set-semidefinite ordering cone which defines the ordering structure in (18.1). Section 18.4 includes optimality conditions and Section 18.5 presents duality results for the set-semidefinite optimization problem. In Section 18.6 we point out possible future research.

18.2 Applications of Set-Semidefinite Optimization

Before we examine theoretical properties of the problem (SSOP), we first illustrate the diversity and the importance of set-semidefinite optimization with various applications.

18.2.1 Semidefinite Optimization

The well-known semidefinite optimization problems in the real linear space $\mathscr{S}^n$ of real symmetric (n,n) matrices (for a survey see [31]) are included in the more general formulation in (SSOP). Here, $Y = \mathbb{R}^n$ and K equals the whole space $\mathbb{R}^n$, and one considers the $\mathbb{R}^n$-semidefinite cone $C^{\mathbb{R}^n}_{\mathscr{S}^n} =: \mathscr{S}^n_+$, i.e., the cone of positive semidefinite matrices, also called Löwner ordering cone ([20]). Note that in the space of symmetric matrices, the inner product $\langle \cdot,\cdot \rangle$ is defined by $\langle \cdot,\cdot \rangle := \text{trace}(A \cdot B)$ for all $A, B \in \mathscr{S}^n$. This problem class has been intensively studied in the past 10–20 years and a wide variety of applications is known. Already in 1890 Lyapunov's characterization of the stability of the solution of a linear differential equation contained a constraint using the Löwner ordering (see [27]). Quadratically constrained quadratic programs can be formulated as a linear semidefinite problem ([28, 16]). Further applications are the problem of the minimization of the maximum eigenvalue ([16, 27]), which arises for instance in stabilizing a differential equation, robust mathematical programming ([27]), control theory ([28, 27]), the consideration of NP-hard combinatorial optimization problems for obtaining lower bounds on the solutions ([13, 10]) and many more ([30]).

In the following, we discuss a problem from structural optimization ([28, 30, 16]). Here, we want to design a truss structure with k linear elastic bars, connecting a set of p nodes, such that the stiffness is maximized. We assume that the geometry of the structure, i.e., the length of the bars $l_1, \ldots, l_k > 0$, and the position of the nodes is fixed, see for instance Fig. 18.1. Assuming that a known fixed nodal load force $f_1, \ldots, f_p$ at each node is given, our aim is to determine the cross-sectional areas of the bars x_i $(i = 1, \ldots, k)$ in such a way, that the stiffest truss is found. Thereby the so-called stiffness-matrix $K(x) \in \mathscr{S}^p_+$, which depends linearly on the truss geometry given by x, is assumed to be given and to be positive definite for $x_1, \ldots, x_k > 0$. Then, maximizing the stiffness can be expressed by minimizing the elastic stored energy $E(x,f) := \frac{1}{2} f^\top K(x)^{-1} f$. Further we assume lower and upper bounds $\underline{x}_i$ (with $\underline{x}_i >$

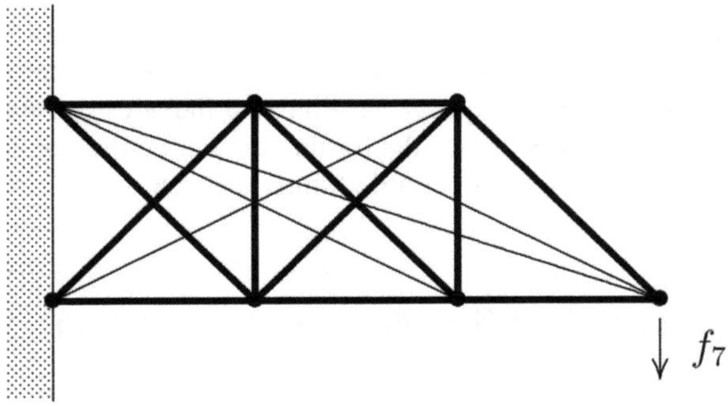

Fig. 18.1 Cantilever with 7 nodes and the load force f_7.

0), $\overline{x}_i$ on the bar cross-sectional areas x_i $(i = 1,\ldots,k)$ and an upper bound $\overline{V}$ on the total truss volume (or equivalently, weight) $\sum_{i=1}^{k} l_i x_i$. This results in the optimization problem

$$\begin{gathered}\min\ f^\top K(x)^{-1} f\\ \text{subject to the constraints}\\ \sum_{i=1}^{k} l_i x_i \le \overline{V},\\ \underline{x}_i \le x_i \le \overline{x}_i, \qquad i = 1,\ldots,k.\end{gathered}$$

By introducing an additional variable t for transforming the objective into the constraint $f^\top K(x)^{-1} f \le t$ and by using the Schur complement, we get a linear semidefinite optimization problem in x and t with some additional linear inequality constraints:

$$\begin{gathered}\min\ t\\ \text{subject to the constraints}\\ \begin{pmatrix} t & f^\top \\ f & K(x) \end{pmatrix} \in \mathscr{S}^n_+,\\ \sum_{i=1}^{k} l_i x_i \le \overline{V},\\ \underline{x}_i \le x_i \le \overline{x}_i, \qquad i = 1,\ldots,k.\end{gathered}$$

18.2.2 Copositive Optimization

A special class of set-semidefinite optimization problems is the class of copositive programs being optimization problems over the cone of copositive matrices. Thus, we consider here finite-dimensional problems in the space of symmetric matrices $\mathscr{S}^n$ with the parameter set K of the cone of K-semidefinite maps being equal to $\mathbb{R}^n_+$. These problems play an important role for instance in combinatorial optimization. Burer showed in [7] that any nonconvex quadratic program with linear and binary constraints can be modeled as a linear program over the dual of the cone of copositive matrices, i.e., over the cone of completely positive matrices. The class of copositive problems includes among others the minimum-cut graph tri-partitioning problem ([7, 24]), the quadratic assignment problem ([12, 23, 7]), the maximum stable set problem ([3, 12, 7]) or the maximum clique problem.

For instance, in the maximum clique problem one considers a simple graph $\mathscr{G}(\mathscr{V},\mathscr{E})$ with node set $\mathscr{V} = \{1,\ldots,n\}$ and edge set $\mathscr{E}$. A clique $\mathscr{C}$ is defined as a subset of the node set $\mathscr{V}$ such that every pair of nodes in $\mathscr{C}$ is connected by an edge in $\mathscr{E}$. A clique $\mathscr{C}$ is said to be maximal if there exists no larger clique which contains $\mathscr{C}$, and a (maximal) clique is said to be a maximum clique if it contains the most number of nodes among all cliques. Then, the maximum clique number $\omega(\mathscr{G})$ is given by the number of nodes in the maximum clique $\mathscr{C}$. For a survey, also over the range of applications of the maximum clique problem, see [22].

Example 18.3. In Fig. 18.2, (a) a graph with 6 nodes and the edge set $\mathscr{E} = \{\{1,2\}, \{1,3\}, \{1,5\}, \{1,6\}, \{2,3\}, \{2,4\}, \{2,6\}, \{3,4\}, \{3,5\}, \{4,5\}, \{4,6\}, \{5,6\}\}$ (an octahedron) is given. It can easily be seen that a maximum clique is for instance $\mathscr{C} = \{2,4,6\}$ (drawn in gray), i.e., the maximum clique number is $\omega(\mathscr{G}) = 3$.

This NP-hard problem can be reformulated as a copositive problem ([3, 9]):

$$\omega(\mathscr{G}) = \min\{\lambda \in \mathbb{N} \mid \lambda(E_n - A_{\mathscr{G}}) - E_n \in C^{\mathbb{R}^n_+}_{\mathscr{S}^n}\} \tag{18.2}$$

with E_n the all-ones (n,n) matrix and $A_{\mathscr{G}}$ the adjacent matrix of the graph $\mathscr{G}$, i.e., $A_{\mathscr{G}} = (a_{ij})_{n\times n}$ where $a_{ij} = 1$ if $\{i,j\} \in \mathscr{E}$, and $a_{ij} = 0$ else.

Example 18.4. We consider the problem of finding the maximum clique number of the graph of the icosahedron ([2]) given by the adjacent matrix

$$A_{\mathscr{G}} = \begin{pmatrix} 0&1&1&0&0&1&1&1&0&0&0&0\\ 1&0&1&1&1&1&0&0&0&0&0&0\\ 1&1&0&1&0&0&0&1&1&0&0&0\\ 0&1&1&0&1&0&0&0&1&1&0&0\\ 0&1&0&1&0&1&0&0&0&1&1&0\\ 1&1&0&0&1&0&1&0&0&0&1&0\\ 1&0&0&0&0&1&0&1&0&0&1&1\\ 1&0&1&0&0&0&1&0&1&0&0&1\\ 0&0&1&1&0&0&0&1&0&1&0&1\\ 0&0&0&1&1&0&0&0&1&0&1&1\\ 0&0&0&0&1&1&1&0&0&1&0&1\\ 0&0&0&0&0&0&1&1&1&1&1&0 \end{pmatrix}$$

(see Fig. 18.2, (b)). The result of the set-semidefinite optimization problem (18.2) is $\omega(\mathscr{G}) = 3$, i.e., all matrices $\lambda(E_n - A_{\mathscr{G}}) - E_n$ with $\lambda \geq 3$ are copositive, i.e., $\mathbb{R}^m_+$-semidefinite. A maximum clique is for instance $\mathscr{C} = \{3,4,9\}$.

The maximum clique problem is closely related to the maximum stable set problem. There one looks for a subset of $\mathscr{V}$, whose elements are pairwise nonadjacent, with maximum cardinality. This cardinality is then called the stability number α of the graph. For the complement graph $\bar{\mathscr{G}} := (\mathscr{V},\bar{\mathscr{E}})$, where $\bar{\mathscr{E}} := \{\{i,j\} \mid i,j \in \mathscr{V},\, i \neq j \text{ and } \{i,j\} \notin \mathscr{E}\}$, it holds $\alpha(\bar{\mathscr{G}}) = \omega(\mathscr{G})$.

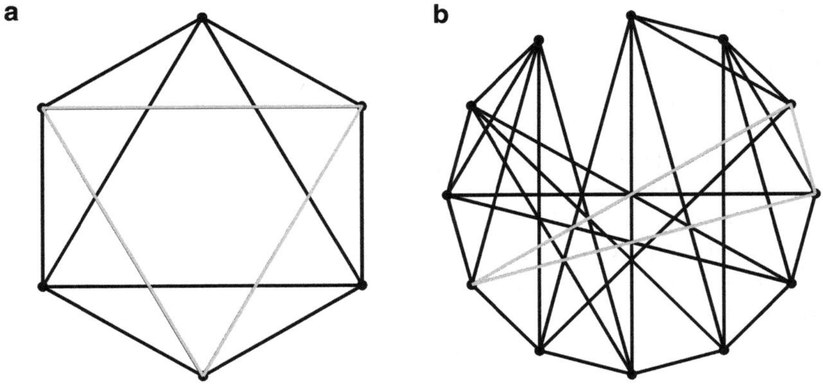

Fig. 18.2 (a) Graph of Example 18.3. (b) Graph of Example 18.4.

18.2.3 Second-Order Optimality Conditions

Now we investigate a constrained optimization problem in Banach spaces. To be more specific, let $(X, \|\cdot\|_X)$ and $(Y, \|\cdot\|_Y)$ be Banach spaces, let C be a closed convex cone in Y, and let $f : X \to \mathbb{R}$ and $g : X \to Y$ be given functions. Then we consider the optimization problem

$$\begin{aligned} &\min f(x) \\ &\text{subject to the constraint} \\ &g(x) \in -C, \\ &x \in X. \end{aligned} \tag{18.3}$$

The constraint set of this problem is denoted by $\hat{S}$. A characterization of minimal solutions of problem (18.3) can be obtained by optimality conditions like the generalized Lagrange multiplier rule. Here we discuss a necessary optimality condition of second-order (e.g., see [21, 4, 33, 8]). In the following, first-order and second-order Fréchet derivatives of f and g are denoted by f', g', f'' and g'', respectively.

Theorem 18.5 ([33, Thm. 5.5.2]). *Let $\bar{x} \in \hat{S}$ be a minimal solution of the optimization problem (18.3). Assume that the functions f and g are twice Fréchet differentiable at $\bar{x}$ and that the constraint qualification*

$$g'(\bar{x})(X) + cone(C + \{g(\bar{x})\}) = Y$$

is satisfied ($cone(\cdot)$ denotes the cone generated by a set). Let $\ell \in C^*$ (the dual cone of C) be a Lagrange multiplier for problem (18.3), i.e.

$$f'(\bar{x}) + \ell \circ g'(\bar{x}) = 0_{X^*}$$

and

$$\langle \ell, g(\bar{x}) \rangle = 0.$$

Then it follows

$$\langle (f''(\bar{x}) + \ell \circ g''(\bar{x}))(h), h \rangle \geq 0 \text{ for all } h \in T(\bar{S}, \bar{x}) \tag{18.4}$$

where

$$\bar{S} := \{x \in X \mid g(x) \in -C \text{ and } \langle \ell, g(x) \rangle = 0\} \tag{18.5}$$

and $T(\bar{S}, \bar{x})$ denotes the contingent cone (or Bouligand tangent cone) of the set $\bar{S}$ at $\bar{x}$.

A proof of Theorem 18.5 is given in [33, p. 198–199]. The necessary optimality condition (18.4) says that the second-order Fréchet derivative of the Lagrangian at the minimal solution is nonnegative on the contingent cone $T(\bar{S}, \bar{x})$, or equivalently

$$f''(\bar{x}) + \ell \circ g''(\bar{x}) \in C_L^{T(\bar{S}, \bar{x})}$$

which means that $f''(\bar{x}) + \ell \circ g''(\bar{x})$ is $T(\bar{S}, \bar{x})$-semidefinite. So, the $T(\bar{S}, \bar{x})$-semidefinite cone $C_L^{T(\bar{S}, \bar{x})}$ plays a central role in the theory of second-order optimality conditions. This shows the close connection between second-order optimality conditions and set-semidefinite cones.

Proposition 18.6. *If the map $g : X \to Y$ is $\bar{C}$-convex for*

$$\bar{C} := \{y \in C \mid \langle \ell, y \rangle = 0\} \tag{18.6}$$

with ℓ as in Thm. 18.5, i.e. for all $x_1, x_2 \in X$ and all $\lambda \in [0,1]$

$$\lambda g(x_1) + (1-\lambda) g(x_2) - g(\lambda x_1 + (1-\lambda) x_2) \in \bar{C},$$

then the set $\bar{S}$ (defined by (18.5)) is a convex set.

Proof. With the equalities (18.5) and (18.6) we obtain

$$\bar{S} = \{x \in X \mid g(x) \in -\bar{C}\}.$$

It is obvious that $\bar{C}$ is a convex cone. Then, the $\bar{C}$-convexity of the map g implies the convexity of the set $\bar{S}$. □

So, under the assumption that g is $\bar{C}$-convex the set $\bar{S}$ is convex and, therefore, the contingent cone $T(\bar{S}, \bar{x})$ is convex as well (e.g., see [16, Thm. 4.12]). In this case the set-semidefinite cone $C_L^{T(\bar{S}, \bar{x})}$ has a richer mathematical structure.

18.2.4 Semi-infinite Optimization

The classical Chebyshev approximation problem has been the starting point of semi-infinite optimization. In this class of approximation problems, a given continuous function is approximated with respect to the Chebyshev (or maximum) norm. Let M be a compact metric space and let $C(M)$ be the real linear space of continuous real-valued functions on M equipped with the Chebyshev norm $\|\cdot\|$ where

$$\|x\| = \max_{t\in M} |x(t)| \text{ for all } x \in C(M).$$

If $\hat{S}$ is a nonempty subset of $C(M)$ and $\hat{x} \in C(M)$ is a given function, then the Chebyshev approximation problem can be written as

$$\min_{x\in\hat{S}} \|x - \hat{x}\|.$$

It is obvious that this optimization problem is equivalent to the problem

$$\begin{aligned}&\min \lambda\\ &\text{subject to the constraints}\\ &|x(t) - \hat{x}(t)| \leq \lambda \text{ for all } t \in M,\\ &x \in \hat{S},\ \lambda \geq 0.\end{aligned}$$

In practice, the set $\hat{S}$ is finite-dimensional (e.g., a set of linear combinations of finitely many functions of $C(M)$) and, therefore, this optimization problem has finitely many variables and infinitely many constraints, i.e., it is a semi-infinite problem.

Later semi-infinite optimization has become an own research field in optimization. A typical semi-infinite optimization problem is of the form

$$\begin{aligned}&\min f(x)\\ &\text{subject to the constraints}\\ &g(x,t) \leq 0 \text{ for all } t \in M,\\ &x \in \hat{S}\end{aligned}$$

with a finite-dimensional set $\hat{S}$, a set M with infinitely many elements, and appropriate functions f and g. If we consider the map G in Assumption 18.1, then for an arbitrary nonempty set $K \subset Y$ the condition

$$G(x) \preccurlyeq 0_{L(Y,Y^*)} \iff G(x) \in -C_L^K \iff \langle G(x)y, y\rangle \leq 0 \text{ for all } y \in K$$

is a special semi-infinite constraint. Hence, set-semidefinite optimization problems with finitely many variables are special semi-infinite optimization problems.

It is interesting to note that the converse implication holds in special cases. If we write the system of infinitely many constraints as one abstract constraint in an appropriate normed function space with an appropriate nontrivial convex cone C, then this constraint can be written as a positive semidefinite constraint, if the ordering cone C has a closed and bounded base (see [17, Thm. 4.3]). So, under these assumptions we obtain a special constraint defined with the Löwner cone.

Connections between semi-infinite and semidefinite optimization problems are also worked out in [29].

Although we only consider quadratic forms in our inequality system, the theory of K-semidefinite optimization is quite general because K is an arbitrary set and the variable space is a topological linear space and is not restricted to a finite dimensional space.

18.3 Set-Semidefinite Cone

The key concept of the set-semidefinite optimization problem is the partial ordering defining the inequality constraint in (18.1). It is easy to see that C_L^K is in fact a convex cone for every set K and thus the related partial ordering is reflexive, transitive and compatible with the linear structure of the space. Further interesting properties will be stated in the following, including some results on the dual of this cone, which is important for achieving optimality conditions for (SSOP), and on the interior, which is for instance of interest for considering the generalized Slater condition for the problem (SSOP).

18.3.1 Properties of the Set-Semidefinite Cone

We start with some calculation rules for the set-semidefinite cone.

Lemma 18.7. *Let $K_1, K_2 \subset Y$ be given nonempty sets.*

(a) If $K_1 \subset K_2$, then

$$C_L^{K_2} \subset C_L^{K_1}.$$

(b)

$$C_L^{K_1 \cup K_2} = C_L^{K_1} \cap C_L^{K_2}.$$

(c) For $cone(K_1)$, i.e., the cone generated by K_1, we have

$$C_L^{K_1} = C_L^{cone(K_1)}.$$

Proof. (a) For $A \in C_L^{K_2}$ it is $\langle Ay, y\rangle \geq 0$ for all $y \in K_2$ and due to $K_1 \subset K_2$ it is *a fortiori* $\langle Ay, y\rangle \geq 0$ for all $y \in K_1$, i.e., $A \in C_L^{K_1}$.

(b) Using the result in (a) on the inclusions $K_1 \subset K_1 \cup K_2$ and $K_2 \subset K_1 \cup K_2$ we immediately get $C_L^{K_1 \cup K_2} \subset C_L^{K_1} \cap C_L^{K_2}$.

It remains to show $C_L^{K_1} \cap C_L^{K_2} \subset C_L^{K_1 \cup K_2}$. For $A \in C_L^{K_1} \cap C_L^{K_2}$ it is $\langle Ay, y\rangle \geq 0$ for all $y \in K_1$ and for all $y \in K_2$. Thus we have $\langle Ay, y\rangle \geq 0$ for all $y \in K_1 \cup K_2$ and therefore $A \in C_L^{K_1 \cup K_2}$.

(c) It is obvious that the condition

$$\langle Ay, y\rangle \geq 0 \text{ for all } y \in K_1$$

is equivalent to

$$\langle A(\lambda y), (\lambda y)\rangle = \lambda^2 \langle Ay, y\rangle \geq 0 \text{ for all } y \in K_1,\ \lambda > 0,$$

and thus $C_L^{K_1} = C_L^{\text{cone}(K_1)}$. □

The case that the set K is a cone is of special interest, as the copositive and the semidefinite case with $K = \mathbb{R}^n_+$ and $K = \mathbb{R}^n$ respectively demonstrates. To show that a map is K-semidefinite for a cone K, sometimes it is simpler to work with a suitable subset $B \subset K$ instead of the whole cone K. This result is already partly included in Lemma 18.7, (c), but because of its importance in solution approaches we also want to present the following more general formulation using the idea of a base of a cone.

Corollary 18.8. *Let $K \subset Y$ be a cone and let B be a subset of K such that for any $y \in K$ there exists a $\lambda \in \mathbb{R}$ and a $b \in B$ with*

$$y = \lambda b. \tag{18.7}$$

Then $A \in C_L^K$ if and only if

$$\langle Ay, y\rangle \geq 0 \textit{ for all } y \in B. \tag{18.8}$$

Proof. Because of $B \subset K$ for any $A \in C_L^K$ (18.8) follows immediately (compare Lemma 18.7, (a)). For showing the converse implication we assume that (18.8) is satisfied. Let $y \in K$ be arbitrarily chosen. Then there is a $\lambda \in \mathbb{R}$ and a $b \in B$ with (18.7) and we get

$$\langle Ay, y\rangle = \langle A(\lambda b), \lambda b\rangle = \underbrace{\lambda^2}_{\geq 0} \underbrace{\langle Ab, b\rangle}_{\geq 0} \geq 0.$$

Thus $A \in C_L^K$. □

This result can of course be generalized to arbitrary nonempty sets $K \subset Y$. To give an example for Corollary 18.8, it is demonstrated in [32] that for the case $Y = \mathbb{R}^n$, $A \in \mathscr{S}^n$ it suffices to show for a $k \in \{1, \ldots, n\}$

$$y^\top Ay \geq 0 \text{ for all } y \in \mathbb{R}^n_+ \text{ with } y_k = 1$$

to prove the copositivity, i.e. the $\mathbb{R}^n_+$-semidefiniteness, of A.

Recall that a nonempty convex subset B of a convex cone $K \neq \{0_Y\}$ is called a *base* for K, if for every $y \in K \setminus \{0_Y\}$ there exists a $\lambda > 0$ and a $b \in B$ such that the representation in (18.7) is unique. Then $\text{cone}(B) = K$. Since a base B fulfills the assumptions of Corollary 18.8, the result of this corollary also holds for a base. For example the set

$$B=\{y\in\mathbb{R}^n_+ \mid \|y\|_1=1\}$$

is a base for the cone $K=\mathbb{R}^n_+$. This set is used in [6] and already in [1] to check a symmetric matrix on $\mathbb{R}^n_+$-semidefiniteness. Each nontrivial convex cone with a base is pointed (see [15, Lemma 1.14]). Recall that a cone $K\subset Y$ is called *pointed* if $K\cap(-K)=\{0_Y\}$. If the cone K is not pointed but the linear space Y is normed, we can use in Corollary 18.8 for example the set $B=\{y\in K \mid \|y\|_Y=1\}$.

In the following lemma, it is shown that K-semidefiniteness already implies $(-K)$-semidefiniteness. We need this lemma for a result on semidefiniteness w. r. t. non-pointed cones K.

Lemma 18.9. *Let $K\subset Y$ be a given nonempty set. Then*

$$C_L^K=C_L^{-K}.$$

Proof. Let $A\in C_L^K$ be arbitrarily given. Because of the linearity of $A\in L(Y,Y^*)$ and $Ay\in Y^*$ we have

$$0\leq\langle Ay,y\rangle=\langle -Ay,-y\rangle=\langle A(-y),-y\rangle \ \text{ for all } \ y\in K.$$

This is equivalent to

$$0\leq\langle Ay,y\rangle \ \text{ for all } \ y\in -K$$

and thus to $A\in C_L^{-K}$. □

For non-pointed cones K, we can show that semidefiniteness w.r.t. a special smaller cone already implies K-semidefiniteness.

Theorem 18.10. *Let $K\subset Y$ be a cone and $b\in Y^*\setminus\{0_{Y^*}\}$ be arbitrarily given. Then*

$$K=K_1\cup K_2\cup(-K_1)$$

with the cones

$$K_1:=\{y\in K\cap(-K) \mid \langle b,y\rangle\geq 0\}$$

and

$$K_2:=\{y\in K \mid y\notin -K\}\cup\{0_Y\},$$

and we get

$$C_L^K=C_L^{K_1\cup K_2}.$$

Proof. We first show the decomposition $K=K_1\cup K_2\cup(-K_1)$ of the cone K. We have $K_1\cup(-K_1)=K\cap(-K)$ and thus

$$(K_1\cup(-K_1))\cup K_2=(K\cap(-K))\cup(K\setminus(-K))\cup\{0_Y\}=K.$$

Using this decomposition and applying Lemma 18.7, (b) and Lemma 18.9 we get

$$C_L^K=C_L^{K_1}\cap C_L^{K_2}\cap\underbrace{C_L^{-K_1}}_{=C_L^{K_1}}=C_L^{K_1}\cap C_L^{K_2}=C_L^{K_1\cup K_2}.$$ □

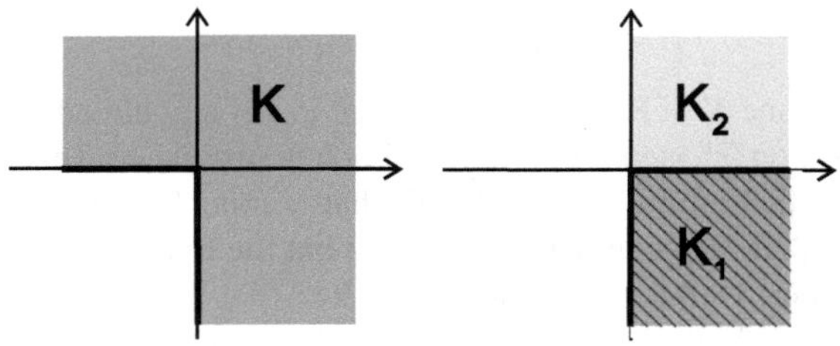

Fig. 18.3 Cones K, K_1, and K_2 of Example 18.11.

We illustrate this result with an example:

Example 18.11. Let $Y = \mathbb{R}^2$ and $K = \{y \in \mathbb{R}^2 \mid y_1 \geq 0 \;\vee\; y_2 \geq 0\}$. Then it is for $b = (1,-1)^\top$

$$K_1 = \{y \in \mathbb{R}^2 \mid y_1 \geq 0 \;\wedge\; y_2 \leq 0\}$$

and

$$K_2 = \{y \in \mathbb{R}^2 \mid y_1 > 0 \;\wedge\; y_2 > 0\} \cup \{0_2\}$$

(see Figure 18.3). Therefore, if $A \in \mathscr{M}^2$ is semidefinite w.r.t. the cone $K_1 \cup K_2$, then it is also K-semidefinite.

Next we come to some results concerning the eigenvalues and eigenvectors of a K-semidefinite linear map.

Lemma 18.12. *Let $(Y, \langle\cdot,\cdot\rangle)$ be a real Hilbert space. Let $K \subset Y$ be a given nonempty set and $A \in L(Y,Y)$ be K-semidefinite. Then for every eigenvector $y \in K$ of A the correspondent eigenvalue λ is non-negative.*

Proof. Due to $y \in K$ and $y \neq 0_Y$, we have for the associated eigenvalue λ

$$0 \leq \langle Ay, y\rangle = \langle \lambda y, y\rangle = \lambda \underbrace{\langle y, y\rangle}_{>0}$$

and thus $\lambda \geq 0$. □

This lemma generalizes the well-known result that for a semidefinite matrix (i.e., $K = \mathbb{R}^n$) all eigenvalues are non-negative. Kaplan shows in [18] (see also [19]) that if a matrix $A \in \mathscr{S}^n$ is copositive, i.e., $\mathbb{R}^n_+$-semidefinite, then A has no eigenvector $y \in \operatorname{int}(\mathbb{R}^n_+)$ with associated eigenvalue $\lambda < 0$.

Lemma 18.13. *Let $(Y, \langle\cdot,\cdot\rangle)$ be a real Hilbert space and $A \in L(Y,Y)$ arbitrarily chosen. Let $y^1, \ldots, y^k \in Y$ $(k \in \mathbb{N})$ be k eigenvectors of A with $\langle y^i, y^j\rangle \geq 0$ for all $i, j \in \{1, \ldots, k\}$, and with eigenvalues $\lambda_i \geq 0$. Let $K \subset Y$ be a nonempty set with*

$$K \subset cone(convex\ hull\{y^1,\ldots,y^k\}).$$

Then A is K-semidefinite.

Proof. Let $y \in K$ be arbitrarily chosen. Then there exist $\beta,\mu_1,\ldots,\mu_k \geq 0$ with

$$\sum_{i=1}^{k} \mu_i = 1 \text{ and } y = \beta \sum_{i=1}^{k} \mu_i y^i.$$

Because of the linearity of A, we conclude

$$\begin{aligned}
\langle Ay,y\rangle &= \left\langle A\left(\beta \sum_{i=1}^{k} \mu_i y^i\right), \beta \sum_{i=1}^{k} \mu_i y^i \right\rangle \\
&= \beta^2 \sum_{i=1}^{k} \sum_{j=1}^{k} \mu_i\mu_j \langle Ay^i, y^j\rangle \\
&= \beta^2 \sum_{i=1}^{k} \sum_{j=1}^{k} \underbrace{\mu_i\mu_j}_{\geq 0} \underbrace{\lambda_i}_{\geq 0} \underbrace{\langle y^i, y^j\rangle}_{\geq 0} \\
&\geq 0.
\end{aligned}$$

Thus A is K-semidefinite. □

18.3.2 Dual and Interior of the Set-Semidefinite Cone

For the finite-dimensional case $Y = \mathbb{R}^n$ and the matrix space $\mathcal{S}^n$ we summarize some results on dual cones. Proofs can be found in [16] for a closed convex cone K, in [26] for a given nonempty set $K \subset \mathbb{R}^n$ and in [25] for a polyhedral cone $K \subset \mathbb{R}^n$.

Lemma 18.14. *(a) Let $K \subset \mathbb{R}^n$ be a nonempty given set. Then*

$$(C^K_{\mathcal{S}^n})^* = cl\ cone(convex\ hull\{\, xx^\top \mid x \in K\}).$$

(b) Let $K \subset \mathbb{R}^n$ be a closed convex cone. Then

$$(C^K_{\mathcal{S}^n})^* = convex\ hull\{xx^\top \mid x \in K\}$$

and $(C^K_{\mathcal{S}^n})^*$ is closed.
(c) Let $K = \mathbb{R}^n$, then $C^K_{\mathcal{S}^n} = \mathcal{S}^n_+$ and $(C^K_{\mathcal{S}^n})^* = C^K_{\mathcal{S}^n}$, i.e. $\mathcal{S}^n_+$ is self-dual.

In the literature, elements of the dual cone

$$\left(C_{\mathcal{S}^n}^{\mathbb{R}^n_+}\right)^* = \text{convex hull}\{xx^\top \mid x \in \mathbb{R}^n_+\}$$

are called completely positive matrices.

For determining the interior of the cone C_L^K, let $\hat{C}_L^K$ denote the set of strict K-semidefinite maps, i.e., let

$$\hat{C}_L^K := \{A \in L(Y,Y^*) \mid \langle Ay, y\rangle > 0 \text{ for all } y \in K \setminus \{0_Y\}\},$$

for an arbitrary nonempty set $K \subset Y$. Of course it is $\hat{C}_L^K \subset C_L^K$. If the set $\hat{C}_L^K$ is nonempty for a reflexive Banach space Y, it is under some additional assumptions the interior of the cone C_L^K.

Theorem 18.15. *Let $(Y, \|\cdot\|_Y)$ be a real reflexive Banach space, let K be a closed convex cone, and let the set $\hat{C}_L^K$ be nonempty. If there are linear functionals $\ell_1, \ldots, \ell_k \in Y^*$ and real numbers $\alpha_1 \neq 0, \ldots, \alpha_k \neq 0$ for some $k \in \mathbb{N}$ so that*

$$K = \bigcup_{i=1}^{k} \mathit{cone}\,(K_i)$$

with

$$K_i := \{y \in K \cap B \mid \ell_i(y) = \alpha_i\} \textit{ for all } i \in \{1, \ldots, k\}$$

(B denotes the closed unit ball) and for every $A \in \hat{C}_L^K$ the quadratic form $\langle A\cdot, \cdot\rangle$ is weakly lower semicontinuous, then

$$\mathit{int}(C_L^K) = \hat{C}_L^K.$$

Proof. For the proof of the inclusion $\hat{C}_L^K \subset \text{int}(C_L^K)$ we fix an arbitrary map $A \in \hat{C}_L^K$. Under the assumptions of the theorem, the cone K is weakly closed and the closed unit ball B is weakly compact (see [15, Lemma 1.41]) and thus the set $K \cap B$ is also weakly compact. Consequently, for every $i \in \{1, \ldots, k\}$ the set K_i is weakly compact as well. Further it is $K_i \subset K \setminus \{0_Y\}$ for $i = 1, \ldots, k$. Since the quadratic form $\langle A\cdot, \cdot\rangle$ is weakly lower semicontinuous, by Thm. 2.3 in [16] there exists a scalar $\varepsilon > 0$ with

$$\varepsilon \leq \min_{y \in K_i} \langle Ay, y\rangle \text{ for all } i \in \{1, \ldots, k\}.$$

Let $\mathcal{N}_\varepsilon(A) := \{D \in L(Y,Y^*) \mid \|D - A\|_{L(Y,Y^*)} < \varepsilon\}$ denote the ε-neighborhood of A and let a map $D \in \mathcal{N}_\varepsilon(A)$ be arbitrarily chosen. For an arbitrary $y \in K \setminus \{0_Y\}$ we have $y \in \text{cone}(K_i) \setminus \{0_Y\}$ for some $i \in \{1, \ldots, k\}$ or

$$y = \lambda_i k_i \text{ for some } \lambda_i > 0 \text{ and some } k_i \in K_i.$$

Then we obtain because of $K_i \subset B$

$$\begin{aligned}
\frac{1}{\lambda_i^2}\langle Dy,y\rangle &= \langle Dk_i,k_i\rangle \\
&= \underbrace{\langle Ak_i,k_i\rangle}_{\geq\varepsilon} + \langle (D-A)k_i,k_i\rangle \\
&\geq \varepsilon - |\langle (D-A)k_i,k_i\rangle| \\
&\geq \varepsilon - \|(D-A)k_i\|_{Y^*}\cdot\underbrace{\|k_i\|_Y}_{\leq 1} \\
&\geq \varepsilon - \underbrace{\|D-A\|_{L(Y,Y^*)}}_{<\varepsilon}\cdot\underbrace{\|k_i\|_Y}_{\leq 1} \\
&\geq 0
\end{aligned}$$

and thus $D \in C_L^K$. Therefore we have shown $\mathscr{N}_\varepsilon(A) \subset C_L^K$ and thus $A \in \operatorname{int}(C_L^K)$. As $A \in \hat{C}_L^K$ is arbitrarily chosen we have $\hat{C}_L^K \subset \operatorname{int}(C_L^K)$.

It remains to show $\operatorname{int}(C_L^K) \subset \hat{C}_L^K$. Let $A \in \operatorname{int}(C_L^K)$ be arbitrarily chosen. Then there exists a ε-neighborhood $\mathscr{N}_\varepsilon(A)$ of A with $\mathscr{N}_\varepsilon(A) \subset C_L^K$. For arbitrarily chosen $D \in \hat{C}_L^K$ there exists a $\lambda > 0$ with

$$D^\lambda := A + \lambda\,(A-D) \in \mathscr{N}_\varepsilon(A).$$

Consequently, we have

$$A = \frac{1}{1+\lambda}D^\lambda + \frac{\lambda}{1+\lambda}D$$

and we obtain for all $y \in K\setminus\{0_Y\}$

$$\langle Ay,y\rangle = \frac{1}{1+\lambda}\underbrace{\langle D^\lambda y,y\rangle}_{\geq 0} + \frac{\lambda}{1+\lambda}\underbrace{\langle Dy,y\rangle}_{>0} > 0,$$

i.e. $A \in \hat{C}_L^K$. □

If the cone K is a subset of a finite-dimensional real Banach space, the quadratic form $\langle A\cdot,\cdot\rangle$ for $A \in \hat{C}_L^K$ is weakly lower semicontinuous. For a real Hilbert space $(Y,\langle\cdot,\cdot\rangle)$, any quadratic form $\langle A\cdot,\cdot\rangle$ for which A is Y-semidefinite is weakly lower semicontinuous (see [14, Theorem 2.1]). Besides, if $(Y,\langle\cdot,\cdot\rangle)$ is a real Hilbert space, then $\hat{C}_L^K \neq \emptyset$ as the identity map is always an element of $\hat{C}_L^K$ for any set $K \subset Y$. Notice that in the case of $Y = \mathbb{R}^n$ for instance the convex cone $K = \mathbb{R}^n_+$ fulfills the assumptions of Theorem 18.15 because of the equality

$$K = \operatorname{cone}\underbrace{\left\{y \in \mathbb{R}^n_+ \;\middle|\; \sum_{i=1}^n y_i = 1\right\}}_{=:K_1}.$$

Using various subsets $K_1,\dots,K_k$ any closed convex nontrivial cone K in $\mathbb{R}^n$ fulfills the assumptions of the previous theorem.

Corollary 18.16. *Let $Y=\mathbb{R}^n$ and $K\subset\mathbb{R}^n$ be a closed convex nontrivial cone. Then*

$$int(C_L^K)=\hat{C}_L^K.$$

Proof. As Y is finite-dimensional the set $\hat{C}_L^K$ is nonempty and the quadratic forms $\langle A\cdot,\cdot\rangle$ are weakly lower semicontinuous for all $A\in\hat{C}_L^K$. Consider the l_1 norm in $\mathbb{R}^n$, i.e., the closed unit ball B equals $\{y\in\mathbb{R}^n \mid \sum_{i=1}^n |y_i|\le 1\}$. For $l_1,\dots,l_{2^n}\in\mathbb{R}^n$ denoting all vectors of the form $(\pm 1,\dots,\pm 1)\in\mathbb{R}^n$ we define

$$K_i := \{y\in K\cap B \mid l_i(y)=1\} \text{ for all } i\in\{1,\dots,2^n\}$$

(notice that the cone generated by the set K_i equals the intersection of the i-th orthant in $\mathbb{R}^n$ with the cone K). If we set $I:=\{i\in\{1,\dots,2^n\}\mid K_i\neq\emptyset\}$, we then obtain

$$K=\bigcup_{i\in I}\text{cone}\,(K_i)$$

and the assertion follows from Theorem 18.15. □

The result of Theorem 18.15 is a generalization of a result in [5]. There it is shown that in the finite dimensional case $Y=\mathbb{R}^n$ for $K=\mathbb{R}^n_+$ it holds

$$\text{int}(C^K_{\mathcal{S}^n})=\{A\in\mathcal{S}^n \mid y^\top Ay>0 \text{ for all } y\in\mathbb{R}^n_+\setminus\{0_{\mathbb{R}^n}\}\},$$

i.e., the interior elements of the cone of copositive matrices are exactly the strict copositive matrices.

As a consequence of Corollary 18.16 we have the following results for $Y=\mathbb{R}^n$ and the matrix space $\mathcal{S}^n$, but direct proofs can be found in [16].

Corollary 18.17. *Let $K\subset\mathbb{R}^n$ be a convex cone. Then*

$$\{A\in\mathcal{S}^n \mid A \text{ is positive definite }\}\subset int(C^K_{\mathcal{S}^n}).$$

For $K=\mathbb{R}^n$ equality holds, i.e.

$$int(C^K_{\mathcal{S}^n})=\{A\in\mathcal{S}^n \mid A \text{ is positive definite }\}.$$

18.4 Optimality Conditions

In this section, we investigate the set-semidefinite optimization problem (SSOP) and formulate necessary and sufficient optimality conditions for this problem. For an arbitrary nonempty set $K\subset Y$ we consider the K-semidefinite cone C_L^K inducing the partial ordering $\preccurlyeq$ in the definition of the constraint set S in (18.1). To be more concrete, under Assumption 18.1 we examine the set-semidefinite optimization problem

$$\begin{array}{c} \min f(x) \\ \text{subject to the constraint} \\ -G(x) \in C_L^K, \\ x \in X \end{array} \tag{18.9}$$

with the constraint set $S = \{x \in X \mid G(x) \in -C_L^K\}$.

Recall that a *minimal solution* $\bar{x} \in S$ of the vector optimization problem (18.9) is defined as the preimage of a minimal element $f(\bar{x})$ of the image set $f(S)$, i.e., $\bar{x}$ is a minimal solution if

$$(\{f(\bar{x})\} - C_Z) \cap f(S) = \{f(\bar{x})\}$$

(compare [15]). Under the additional assumption that C_Z has a nonempty interior $\mathrm{int}(C_Z)$, an element $\bar{x} \in S$ is called a *weakly minimal solution* of problem (18.9), if $f(\bar{x})$ is a weakly minimal element of the image set $f(S)$, i.e.,

$$(\{f(\bar{x})\} - \mathrm{int}(C_Z)) \cap f(S) = \emptyset.$$

The following theorem gives a necessary condition for a minimal solution of problem (18.9).

Theorem 18.18 (necessary optimality condition). *Let the set-semidefinite optimization problem (18.9) be given under Assumption 18.1, and let X be a real Banach space and Y and Z real normed spaces. Let the ordering cones C_Z and C_L^K have a nonempty interior. Let $\bar{x} \in S$ be a weakly minimal solution of problem (18.9). Moreover, let f and G be Fréchet differentiable at $\bar{x}$. Then there are continuous linear functionals $t \in C_Z^*$, $U \in (C_L^K)^*$ with $(t,U) \neq 0_{Z^* \times L(Y,Y^*)^*}$ so that*

$$t \circ f'(\bar{x}) + U \circ G'(\bar{x}) = 0_{X^*} \tag{18.10}$$

and

$$\langle U, G(\bar{x}) \rangle = 0. \tag{18.11}$$

If, in addition to the given assumptions

$$G'(\bar{x})(X) + cone\left(C_L^K + \{G(\bar{x})\}\right) = L(Y,Y^*) \tag{18.12}$$

then $t \neq 0_{Z^*}$.

Proof. The first part of this theorem follows from a general Lagrange multiplier rule given in [15, Thm. 7.4], if we notice that the superset used in [15] equals the whole space X in our case. For the proof of the second part of this theorem, let the condition (18.12) be satisfied. Now assume that $t = 0_{Z^*}$. Then for an arbitrary element $A \in L(Y,Y^*)$ there is a non-negative number α, a vector $x \in X$ and a map $D \in C_L^K$ with

$$A = G'(\bar{x})(x) + \alpha(D + G(\bar{x})).$$

Hence we obtain with the equations (18.10) and (18.11) and the positivity of U

$$\langle U,A\rangle = \underbrace{\left(U\circ G'(\bar{x})\right)(x)}_{=0} + \underbrace{\alpha\langle U,D\rangle}_{\geq 0} + \underbrace{\alpha\langle U,G(\bar{x})\rangle}_{=0} \geq 0.$$

Consequently, we have $U = 0_{L(Y,Y^*)^*}$. But this contradicts the assertion that $(t,U) \neq 0_{Z^*\times L(Y,Y^*)^*}$. □

The necessary optimality condition given in Theorem 18.18 generalizes the well-known Lagrange multiplier rule. The regularity assumption (18.12) is the known Kurcyusz–Robinson–Zowe regularity condition (see [16, p.114]). It does not use the interior of the ordering cone and, therefore, it is more general than the known Slater condition.

Corollary 18.19. *Let the assumptions of Theorem 18.18 be satisfied and let $C_Z \neq Z$. If $\bar{x} \in S$ is a minimal solution of problem (18.9), then we obtain the necessary condition in Theorem 18.18.*

Proof. Under the additional assumption $C_Z \neq Z$, every minimal solution of problem (18.9) is also a weakly minimal solution of (18.9) (see [15, Lemma 4.14]). Hence, we get the same assertion as in Theorem 18.18. □

Under generalized convexity assumptions the necessary optimality condition in Theorem 18.18 is also a sufficient optimality condition. Here we recall the concept of $\tilde{C}$-quasiconvexity (see also [16, Def. 5.12]).

Definition 18.20. Let T be a nonempty subset of a real linear space X, and let $\tilde{C}$ be a nonempty subset of a real normed space V. Let $h\colon T \to V$ be a given map having a directional derivative at some $\bar{x} \in T$ in every direction $x - \bar{x}$ with arbitrary $x \in T$. The map h is called *$\tilde{C}$-quasiconvex* at $\bar{x}$, if for all $x \in T$

$$h(x) - h(\bar{x}) \in \tilde{C} \;\Rightarrow\; h'(\bar{x})(x-\bar{x}) \in \tilde{C}.$$

Now we present the sufficient optimality condition.

Theorem 18.21 (sufficient optimality condition). *Let the set-semidefinite optimization problem (18.9) be given under Assumption 18.1, let Y and Z be real normed spaces, and let the ordering cone C_Z be nontrivial. Let f and G have a directional derivative at some $\bar{x} \in S$. Moreover, let the map $(f,G)\colon X \to Z \times L(Y,Y^*)$ be $\tilde{C}$-quasiconvex at $\bar{x}$ with*

$$\tilde{C} := (-C_Z \setminus \{0_Z\}) \times \left(-C_L^K + lin(G(\bar{x}))\right).$$

If there are continuous linear functionals $t \in C_Z^{\#}$, $U \in (C_L^K)^*$ so that

$$t\circ f'(\bar{x}) + U\circ G'(\bar{x}) = 0_{X^*} \tag{18.13}$$

and

$$\langle U,G(\bar{x})\rangle = 0, \tag{18.14}$$

then $\bar{x}$ is a minimal solution of problem (18.9).

Here

$$C_Z^{\#} := \{t \in Z^* \mid \langle t,c\rangle > 0 \text{ for all } c \in C_Z \setminus \{0_Z\}\} \tag{18.15}$$

denotes the so-called quasi-interior of the dual cone C_Z^* and lin$(G(\bar{x}))$ denotes the one-dimensional linear space spanned by $G(\bar{x})$.

Proof. Assume that there is a vector $x \in X$ with

$$(f'(\bar{x})(x-\bar{x}), G'(\bar{x})(x-\bar{x})) \in \tilde{C}.$$

Then we have

$$f'(\bar{x})(x-\bar{x}) \in -C_Z \setminus \{0_Z\}$$

and

$$G'(\bar{x})(x-\bar{x}) \in -C_L^K + \mathrm{lin}(G(\bar{x})),$$

and we obtain with the definition of the quasi-interior $C_Z^{\#}$ and with the equation (18.14) for some $\alpha \in \mathbb{R}$

$$(t \circ f'(\bar{x}) + U \circ G'(\bar{x}))(x-\bar{x}) < \alpha\langle U, G(\bar{x})\rangle = 0.$$

This contradicts the equation (18.13). Hence, we have for all $x \in X$

$$(f'(\bar{x})(x-\bar{x}), G'(\bar{x})(x-\bar{x})) \notin \tilde{C}$$

and because (f,G) is $\tilde{C}$-quasiconvex this implies

$$(f(x)-f(\bar{x}), G(x)-G(\bar{x})) \notin \tilde{C} \text{ for all } x \in X.$$

This means that there is no $x \in X$ with

$$f(x)-f(\bar{x}) \in -C_Z \setminus \{0_Z\}$$

and

$$\begin{aligned} G(x)-G(\bar{x}) &\in -C_L^K + \mathrm{lin}(G(\bar{x})) \\ \Leftrightarrow \qquad G(x) &\in \{G(\bar{x})\} - C_L^K + \mathrm{lin}(G(\bar{x})) \\ &\subset -C_L^K - C_L^K + \mathrm{lin}(G(\bar{x})) \\ &= -C_L^K + \mathrm{lin}(G(\bar{x})) \end{aligned}$$

and thus in particular with $G(x) \in -C_L^K$. Consequently, there is no feasible vector $x \in S$ with

$$f(x) \in \{f(\bar{x})\} - C_Z \setminus \{0_Z\}$$

or

$$(f(\bar{x}) - C_Z) \cap f(S) = \{f(\bar{x})\}.$$

This means that $\bar{x}$ is a minimal solution of problem (18.9). □

A similar result can also be shown for the weak minimality notion (see [15, Thm. 7.20 and Cor. 7.21]). Notice that for the sufficient optimality condition we use a stronger assumption on t. Here t should be in the quasi-interior of the dual cone, whereas in Corollary 18.19 the functional t is an element of the dual cone. This theoretical gap results from the fact that there is no complete characterization of minimal solutions, if one works with linear scalarization (for instance, see Thm. 5.18 and the following discussion in [15]).

Now we shortly discuss the application of these optimality conditions to the finite-dimensional case. In Assumption 18.1, we consider the special cases $X = \mathbb{R}^m$, $Z = \mathbb{R}^k$ and $Y = \mathbb{R}^n$ and the functions $f : \mathbb{R}^m \to \mathbb{R}^k$ and $G : \mathbb{R}^m \to \mathscr{M}^n$. If f is differentiable at a weakly minimal solution $\bar{x}$ of problem (18.9) and G is elementwise differentiable at $\bar{x}$, then the equation (18.10) can be written as

$$\sum_{i=1}^{k} t_i \nabla f_i(\bar{x}) + \begin{pmatrix} \langle U \,,\, G_{x_1}(\bar{x})\rangle \\ \vdots \\ \langle U \,,\, G_{x_m}(\bar{x})\rangle \end{pmatrix} = 0_{\mathbb{R}^m}$$

for a vector $t \in C^*_{\mathbb{R}^k}$ and a matrix $U \in (C^K_{\mathscr{M}^n})^*$, and the equation (18.11) equals

$$\langle U \,,\, G(\bar{x})\rangle = 0.$$

Here we use the notation

$$G_{x_i} := \begin{pmatrix} \frac{\partial}{\partial x_i} G_{11} & \cdots & \frac{\partial}{\partial x_i} G_{1n} \\ \vdots & & \vdots \\ \frac{\partial}{\partial x_i} G_{n1} & \cdots & \frac{\partial}{\partial x_i} G_{nn} \end{pmatrix} \quad \text{for all } i \in \{1, \ldots, m\}.$$

So, the general optimality condition reduces to a KKT-type condition. For more details we refer to [11].

18.5 Nonconvex Duality

As in the previous section, we examine the vector optimization problem (SSOP) under Assumption 18.1, i.e., we consider the problem

$$\begin{array}{c} \min\ f(x) \\ \text{subject to the constraint} \\ -G(x) \in C^K_L, \\ x \in X \end{array} \tag{18.16}$$

for an arbitrary nonempty set $K \subset Y$. As in (18.1) the constraint set is denoted by S. In this section, problem (18.16) is called *primal problem*. In contrast to the presentation in [11], we do not assume that the maps f and G are convex but that a certain composite map is convex-like and, therefore, this approach also includes some nonconvex maps.

Definition 18.22. Let $\hat{S}$ be a nonempty subset of a real linear space and let E be a partially ordered real linear space with an ordering cone C. A map $h : \hat{S} \to E$ is called *convex-like*, if the set $h(\hat{S}) + C$ is convex.

It is shown in [16, Example 6.4] that every convex map is also convex-like and that there are convex-like maps which are not convex. So, this class of maps includes the class of convex maps and it slightly goes beyond this class. Therefore, we speak of nonconvex duality.

The standard assumption of this section is summarized in:

Assumption 18.23. *Let Assumption 18.1 be satisfied, let $K \subset Y$ be an arbitrary nonempty set and let the quasi-interior $C_Z^{\#}$ (see (18.15)) be nonempty.*

Under this assumption, we assign a *dual problem* to the primal problem (18.16):

$$\begin{array}{c} \max z \\ \text{subject to the constraints} \\ \langle t, z\rangle \leq \inf\limits_{x\in X} (t \circ f + U \circ G)(x), \\ z \in Z,\ t \in C_Z^{\#},\ U \in (C_L^K)^*. \end{array} \tag{18.17}$$

A triple (z,t,U) is called a *maximal solution* of this problem if it is a minimal solution w.r.t. the ordering cone $-C_Z$. A first relationship between the primal and dual problem is given in:

Theorem 18.24 (weak duality theorem). *Let Assumption 18.23 be satisfied. For every feasible triple $(\bar{z}, \bar{t}, \bar{U})$ of the dual problem (18.17)*

$$\langle \bar{t}, \bar{z}\rangle \leq (\bar{t} \circ f)(\bar{x}) \textit{ for all } \bar{x} \in S.$$

Proof. We have for an arbitrary $\bar{x} \in S$

$$\langle \bar{t}, \bar{z}\rangle \leq \inf_{x\in X} (\bar{t} \circ f + \bar{U} \circ G)(x) \leq (\bar{t} \circ f)(\bar{x}) + \underbrace{(\bar{U} \circ G)(\bar{x})}_{\leq 0} \leq (\bar{t} \circ f)(\bar{x}).$$

□

Now we present a nonconvex strong duality theorem.

Theorem 18.25 (strong duality theorem). *Let Assumption 18.23 be satisfied, let the interior int(C_L^K) be nonempty, and let $\bar{x} \in S$ be a minimal solution of the primal problem (18.16) with the additional property that for some $t \in C_Z^{\#}$*

$$(t \circ f)(\bar{x}) \leq (t \circ f)(x) \ \textit{ for all } x \in S. \tag{18.18}$$

If $(t\circ f,G)$ *is convex-like and the generalized Slater condition is satisfied, i.e., there exists a vector* $\hat{x}\in X$ *with* $-G(\hat{x})\in int(C_L^K)$*, then there is a maximal solution* $(\bar{z},t,\bar{U})$ *of the dual problem (18.17) with* $f(\bar{x})=\bar{z}$*.*

Proof. Let $\bar{x}\in S$ be a minimal solution of the primal problem (18.16) and let some $t\in C_Z^{\#}$ be given so that the inequality (18.18) is satisfied. Then we investigate the set

$$\begin{aligned} M &:= \{\, (\langle t,f(x)\rangle+\alpha,\, G(x)+y)\in\mathbb{R}\times L(Y,Y^*)\mid x\in X,\ \alpha\geq 0,\ y\in C_L^K\,\} \\ &= (t\circ f,G)(X)+(\mathbb{R}_+\times C_L^K). \end{aligned}$$

Since $(t\circ f,G)$ is convex-like, the set M is convex. Because of $\text{int}(C_L^K)\neq\emptyset$, the set M has a nonempty interior $\text{int}(M)$ as well. With the inequality (18.18) we conclude

$$(\langle t,f(\bar{x})\rangle, 0_{L(Y,Y^*)})\notin \text{int}(M)$$

or

$$\text{int}(M)\cap\{\,(\langle t,f(\bar{x})\rangle,0_{L(Y,Y^*)})\,\}=\emptyset.$$

By the Eidelheit separation theorem (e.g., see [16, Thm. C.2]), there are real numbers μ and γ and a continuous linear functional $U\in L(Y,Y^*)^*$ with $(\mu,U)\neq(0,0_{L(Y,Y^*)^*})$ and

$$\mu\beta+\langle U,z\rangle>\gamma\geq\mu\langle t,f(\bar{x})\rangle \text{ for all } (\beta,z)\in\text{int}(M). \tag{18.19}$$

Since every convex subset of a real normed space with nonempty interior is contained in the closure of the interior of this set, we conclude from the inequality (18.19)

$$\mu(\langle t,f(x)\rangle+\alpha)+\langle U,G(x)+y\rangle\geq\gamma\geq\mu\langle t,f(\bar{x})\rangle \text{ for all } x\in X,\alpha\geq 0,y\in C_L^K. \tag{18.20}$$

For $x=\bar{x}$ and $\alpha=0$, it follows from the inequality (18.20)

$$\langle U,y\rangle\geq-\langle U,G(\bar{x})\rangle \text{ for all } y\in C_L^K. \tag{18.21}$$

With standard arguments we get immediately $U\in(C_L^K)^*$. For $y=0_{L(Y,Y^*)}$, it follows from the inequality (18.21) $\langle U,G(\bar{x})\rangle\geq 0$. Because of $-G(\bar{x})\in C_L^K$ and $U\in(C_L^K)^*$ we also have $\langle U,G(\bar{x})\rangle\leq 0$, which leads to

$$\langle U,G(\bar{x})\rangle=0.$$

For $x=\bar{x}$ and $y=0_{L(Y,Y^*)}$, we get from the inequality (18.20)

$$\mu\alpha\geq 0 \text{ for all } \alpha\geq 0$$

which implies $\mu\geq 0$. For the proof of $\mu>0$, we assume that $\mu=0$. Then it follows from the inequality (18.20) with $y=0_{L(Y,Y^*)}$

$$\langle U,G(x)\rangle\geq 0 \text{ for all } x\in X.$$

Because of the generalized Slater condition, there is an $\hat{x} \in X$ with $-G(\hat{x}) \in \text{int}(C_L^K)$, and then we have

$$\langle U, G(\hat{x})\rangle = 0.$$

Now we want to show that $U = 0_{L(Y,Y^*)^*}$. For that purpose we assume that $U \neq 0_{L(Y,Y^*)^*}$, i.e., there is a $y \in L(Y,Y^*)$ with $U(y) > 0$. Then we have

$$\langle U, \lambda y + (1-\lambda)\, G(\hat{x})\rangle > 0 \text{ for all } \lambda \in (0,1], \tag{18.22}$$

and because of $-G(\hat{x}) \in \text{int}(C_L^K)$ there is a $\bar{\lambda} \in (0,1)$ with

$$\lambda y + (1-\lambda)\, G(\hat{x}) \in -C_L^K \text{ for all } \lambda \in [0,\bar{\lambda}].$$

Then we get

$$\langle U, \lambda y + (1-\lambda)\, G(\hat{x})\rangle \leq 0 \text{ for all } \lambda \in [0,\bar{\lambda}]$$

which contradicts the inequality (18.22). Hence, with the assumption $\mu = 0$ we also obtain $U = 0_{L(Y,Y^*)^*}$, a contradiction to $(\mu, U) \neq (0, 0_{L(Y,Y^*)^*})$. Consequently, we have $\mu \neq 0$ and therefore $\mu > 0$. Then we conclude from the inequality (18.20) with $\alpha = 0$ and $y = 0_{L(Y,Y^*)^*}$

$$\mu\langle t, f(x)\rangle + \langle U, G(x)\rangle \geq \mu\langle t, f(\bar{x})\rangle \text{ for all } x \in X$$

and

$$\langle t, f(x)\rangle + \frac{1}{\mu}\langle U, G(x)\rangle \geq \langle t, f(\bar{x})\rangle \text{ for all } x \in X.$$

If we define $\bar{U} := \frac{1}{\mu} U \in (C_L^K)^*$ we obtain with $\langle \bar{U}, G(\bar{x})\rangle = 0$

$$\inf_{x\in X} \langle t, f(x)\rangle + \langle \bar{U}, G(x)\rangle \geq \langle t, f(\bar{x})\rangle + \langle \bar{U}, G(\bar{x})\rangle.$$

Hence we have

$$\langle t, f(\bar{x})\rangle = \langle t, f(\bar{x})\rangle + \langle \bar{U}, G(\bar{x})\rangle = \inf_{x\in X} \langle t, f(x)\rangle + \langle \bar{U}, G(x)\rangle,$$

i.e., for $\bar{z} := f(\bar{x})$ the triple $(\bar{z}, t, \bar{U})$ satisfies the constraints of the dual problem (18.17). With the weak duality theorem (Theorem 18.24), the triple $(\bar{z}, t, \bar{U})$ is a maximal point of $\langle t, \cdot\rangle$ on the feasible set of the dual problem. Then by a known scalarization theorem (see [15, Thm. 5.18, (b)]), the triple $(\bar{z}, t, \bar{U})$ is a maximal solution of the dual problem with $\bar{z} = f(\bar{x})$. □

Notice that it is also possible to work with the weak minimality notion according to [15, p. 196]. But here we restrict ourselves to the minimality notion which is of interest in practice.

Theorem 18.26 (strong converse duality theorem). *Let Assumption 18.23 be satisfied and, in addition, let Z be locally convex, let the map f be convex-like, let the set $f(S) + C_Z$ be closed, and let for arbitrary $t \in C_Z^{\#}$*

$$\inf_{x\in S}(t\circ f)(x) = \sup_{U\in(C_L^K)^*}\inf_{x\in X}(t\circ f + U\circ G)(x).$$

Then for every maximal solution $(\bar{z},\bar{t},\bar{U})$ of the dual problem (18.17) there exists a minimal solution $\bar{x}$ of the primal problem (18.16) with $\bar{z} = f(\bar{x})$.

Proof. This theorem is an equivalent formulation of Theorem 8.9, (b) in [15] in the case of set-semidefinite optimization. □

An application of this duality approach to linear problems is presented in [11]. Here we mention only the finite-dimensional case. For arbitrary matrices $A^{(1)},\ldots,$ $A^{(m)}, B \in \mathscr{M}^n$ and $C \in \mathbb{R}^{k\times m}$ the primal problem

$$\begin{gathered}\min\ Cx \\ \text{subject to the constraint} \\ A^{(1)}x_1 + \ldots + A^{(m)}x_m - B \in C^K_{\mathscr{M}^n}, \\ x \in \mathbb{R}^m\end{gathered}$$

is associated to the dual problem

$$\begin{gathered}\max\ z \\ \text{subject to the constraints} \\ \langle U, A^{(1)}\rangle = c_1^\top t, \\ \vdots \\ \langle U, A^{(m)}\rangle = c_m^\top t, \\ t^\top z = \langle U, B\rangle, \\ z\in\mathbb{R}^k,\ t\in C^{\#}_{\mathbb{R}^k},\ U\in(C^K_{\mathscr{M}^n})^*\end{gathered}$$

where c_i $(1 \le i \le m)$ denotes the i-th column of the matrix C. For details we refer to [11].

18.6 Future Research

Set-semidefinite optimization is a unified approach for different important research fields in optimization. Future challenges are numerical methods for the solution of these optimization problems. It has been already shown in [11] that penalty approaches, can be used in special cases. But it is still open whether these approaches can be efficiently applied to concrete problems. Besides these penalty approaches, other standard numerical methods in set-semidefinite optimization are desirable. Another open question is the determination of the largest set $\tilde{K}$ on which a continuous linear map is $\tilde{K}$-semidefinite. Here, first results in the finite-dimensional case allow the development of a numerical test on K-semidefiniteness by checking whether $K \subset \tilde{K}$.

Acknowledgment The authors thank Prof. Constantin Zălinescu for valuable discussions.

References

1. L.-F. Andersson, G. Chang and T. Elfving, "Criteria for copositive matrices using simplices and barycentric coordinates", *Linear Algebra Appl.* 220 (1995) 9–30.
2. I.M. Bomze and E. de Klerk, "Solving standard quadratic optimization problems via linear, semidefinite and copositive programming", *J. Global Optim.* 24 (2002) 163–185.
3. I.M. Bomze, M. Dür, E. de Klerk, C. Roos, A.J. Quist and T. Terlaky, "On copositive programming and standard quadratic optimization problems", *J. Global Optim.* 18 (2000) 301–320.
4. J.M. Borwein, "Necessary and sufficient conditions for quadratic minimality", *Numerical Funct. Anal. Optimization* 5 (1982) 127–140.
5. S. Bundfuss and M. Dür, "Criteria for copositivity and approximations of the copositive cone" (Preprint, Department of Mathematics, Darmstadt University of Technology, Darmstadt, 2006).
6. S. Bundfuss and M. Dür, "Algorithmic Copositivity Detection by Simplicial Partition", *Linear Algebra Appl.* 428 (2008) 1511–1523.
7. S. Burer, "On the copositive representation of binary and continuous nonconvex quadratic programs", *Math. Program* 120 (2009) 479–495, DOI 10.1007/s10107-008-0223-z.
8. G. Danninger, "Role of copositivity in optimality criteria for nonconvex optimization problems", *J. Optimization Theory Appl.* 75 (1992) 535–558.
9. E. de Klerk and D.V. Pasechnik, "Approximation of the stability number of a graph via copositive programming", *SIAM J. Optim.* 12 (2002) 875–892.
10. I. Dukanovic and F. Rendl, "Semidefinite programming relaxations for graph coloring and maximal clique problems", *Math. Program.* 109(2) (2007) 345–365.
11. G. Eichfelder and J. Jahn, "Set-semidefinite optimization", *J. Convex Anal.* 15 (2008) 767–801.
12. M. Fukuda, M. Yamashita and M. Kojima, "Computational Prospects on Copositive Programming", *RIMS Kokyuroku* 1526 (2006) 207–213.
13. M.X. Goemans, "Semidefinite programming in combinatorial optimization", *Math. Program.* 79 (1997) 143–161.
14. M.S. Gowda, "A characterization of positive semidefinite operators on a Hilbert space", *J. Optimization Theory Appl.* 48 (1986) 419–425.
15. J. Jahn, *Vector Optimization - Theory, Applications, and Extensions* (Springer, Berlin, 2004).
16. J. Jahn, *Introduction to the Theory of Nonlinear Optimization* (Springer, Berlin, 2007).
17. J. Jahn, *Int. J. Optimization: Theory, Methods and Appl.* 1 (2009) 123–139.
18. W. Kaplan, "A test for copositive matrices", *Linear Algebra Appl.* 313 (2000) 203–206.
19. W. Kaplan, "A copositivity probe", *Linear Algebra Appl.* 337 (2001) 237–251.
20. K. Löwner, "Über monotone Matrixfunktionen", *Math. Z.* 38 (1934) 177–216.
21. H. Maurer and J. Zowe, "First and second-order necessary and sufficient optimality conditions for infinite-dimensional programming problems", *Math. Programming* 16 (1979) 98–110.
22. P.M. Pardalos, M. Panos and J.E. Xue, "The maximum clique problem", *J. Global Optim.* 4(3) (1994) 301-328
23. J. Povh and F. Rendl, "Copositive and semidefinite relaxations of the quadratic assignment problem" (manuscript, University of Maribor, Faculty of Logistics, Celje, Slovenia, 2006).
24. J. Povh and F. Rendl, "A copositive programming approach to graph partitioning", *SIAM J. Optimization* 18 (2007) 223–241.
25. A.J. Quist, E. de Klerk, C. Roos and T. Terlaky, "Copositive relaxation for general quadratic programming", *Optim. Methods Softw.* 9 (1998) 185–208.
26. J.F. Sturm and S. Zhang, "On cones of nonnegative quadratic functions", *Math. Oper. Res.* 28 (2003) 246–267.

27. M.J. Todd, "Semidefinite optimization", *Acta Numerica* 10 (2001) 515–560.
28. L. Vandenberghe and S. Boyd, "Semidefinite programming", *SIAM Rev.* 38 (1996) 49–95.
29. L. Vandenberghe and S. Boyd, "Connections between semi-infinite and semidefinite programming", in: R. Reemtsen and J.J. Rueckmann, *Semi-Infinite Programming* (Kluwer, Boston, 1998), p. 277–294.
30. L. Vandenberghe and S. Boyd, "Applications of semidefinite programming", *Appl. Numer. Math.* 29(3) (1999) 283–299.
31. H. Wolkowicz, R. Saigal and L. Vandenberghe (eds.), *Handbook on Semidefinite Programming* (Kluwer, 2000).
32. H. Väliaho, "Quadratic-programming criteria for copositive matrices", *Linear Algebra Appl.* 119 (1989) 163–182.
33. J. Werner, *Optimization, Theory and Applications* (Vieweg, Braunschweig, 1984).

Chapter 19
On the Envelope of a Variational Inequality

F. Giannessi and A.A. Khan

Dedicated to the blessed memory of Professor George Isac

Abstract Recently, it has been shown that analyzing variational inequalities and their generalizations by means of a separation scheme leads to connection of different topics, such as regularization, penalization, duality, and so on. This has been done by introducing the definition of image of a variational and quasi-variational inequality, and then exploiting the separation approach. Here we extend the definition of image of a variational inequality and make some comments on further investigations.

19.1 Introduction

Some recent investigations (see [1]–[13]) have shown that separation arguments can be considered as a root for developing theoretical analysis of variational and quasi-variational inequalities ([2]). In particular, it has been shown that some related theories, such as gap functions, regularization, penalization and duality, etc., can be connected to each other. It is reasonable to expect some improvements in each theory by exploiting the results obtained in the other ones. Such a development has been based on the introduction of the concept of image of a variational and quasi-variational inequality. However, such a concept of image has shown an asymmetry with respect to the analogous situation in the field of constrained extrema. Here, we

F. Giannessi
Department of Mathematics, Faculty of Natural Sciences, University of Pisa, Via F. Buonarroti, 56127, Pisa, Italy.

A.A. Khan
School of Mathematical Sciences, Rochester Institute of Technology, 85 Lomb Memorial Drive, Rochester, New York 14623, USA, e-mail: aaksma@rit.edu

P.M. Pardalos et al. (eds.), *Nonlinear Analysis and Variational Problems*, Springer Optimization and Its Applications 35, DOI 10.1007/978-1-4419-0158-3_19,

suggest a way of overcoming such an asymmetry. We hope that this will lead to as many achievements as has happened in the field of constrained extrema. Notations and definitions of Image Space Analysis will be recalled shortly; details can be found in [6].

Assume we are given the positive integers m, n, the set $X \subset \mathbb{R}^n$, and the functions $F : \mathbb{R}^n \to \mathbb{R}^n$, $g : \mathbb{R}^n \to \mathbb{R}^m$. Consider the simplest form of a variational inequality in $\mathbb{R}^n$: find $y \in K := \{x \in X : \; g(x) \geq 0\}$, such that:

$$\langle F(y), x - y \rangle \geq 0, \quad \forall x \in K, \tag{19.1}$$

where $\langle \cdot, \cdot \rangle$ denotes the usual scalar product in $\mathbb{R}^n$.

The above-mentioned separation scheme starts with the obvious remark that $y \in K$ is a solution of (19.1), if and only if the system (in the unknown x):

$$u := \langle F(y), y - x \rangle > 0, \quad v := g(x) \geq 0, \quad x \in X \tag{19.2}$$

is impossible. The space where (u, v) runs is the Image Space associated with (19.1); the set

$$\mathscr{K}(y) := \{(u,v) \in \mathbb{R} \times \mathbb{R}^m : u = \langle F(y), y - x \rangle, \; v = g(x), \; x \in X\}, \; y \in X$$

is the *image* of (19.1). System (19.2) is associated with the set:

$$\mathscr{H} := \{(u,v) \in \mathbb{R} \times \mathbb{R}^m : \quad u > 0, \quad v \geq 0\}.$$

Another obvious remark is that the impossibility of (19.2) holds, if and only if:

$$\mathscr{H} \cap \mathscr{K}(y) = \emptyset. \tag{19.3}$$

To (19.1), we associate the following image problem:

$$\max(u), \quad \text{subject to} \quad (u,v) \in \mathscr{K}(y) \cap (\mathbb{R} \times \mathbb{R}^m_+). \tag{19.4}$$

In the Image Space we define an extremum problem also when in the initial space we are not given an extremum problem; this fact is not surprising, since the starting point for introducing the Image Space is a system, namely (19.2), which can be associated or not with an extremum problem. It is easy to show that y is a maximum point of (19.4), if and only if it is a solution of (19.1). Note that the feasible region of (19.4) depends on the unknown; hence, following the variational terminology, (19.4) can be called a *quasi-maximum problem.*

To prove directly (19.3) is, in general, practically impossible. A way of overcoming this drawback consists in using separation arguments to show disjunction between $\mathscr{H}$ and $\mathscr{K}(y)$; more precisely, a family of separation functions is introduced and a member of the family is searched for which includes $\mathscr{H}$ and $\mathscr{K}$ in the positive and the nonpositive level sets, respectively, [6]; as a by-product of such a separation approach, it is possible to derive wide classes of penalization and of gap functions for (19.1).

When there exists a derivable function $f : \mathbb{R}^n \to \mathbb{R}$, such that its gradient equals F, or

$$f'(x) = F(x) \tag{19.5}$$

then (19.1) expresses a first-order necessary condition for y to be a minimum point of problems

$$\min f(x), \quad \text{subject to} \quad x \in K. \tag{19.6}$$

19.2 Auxiliary Variational Inequality

Let us set $x = (x_1, \ldots, x_n)$, $F_i : \mathbb{R}^n \to \mathbb{R}$ and $F(x) = (F_1(x), \ldots, F_n(x))$. Consider a generic, but fixed, $\tilde{x} \in \mathbb{R}^n$, and the function $\Phi : \mathbb{R}^n \to \mathbb{R}$ defined by

$$\Phi(x) = \Phi(x_1, \ldots, x_m) := \int_0^1 \langle x - \tilde{x}, F(\tilde{x} + (x - \tilde{x})t) \rangle dt, \quad x \in \mathbb{R}^n, \tag{19.7}$$

and assume that the integral exists. Indeed, we should write $\Phi(\tilde{x}, x)$ to show explicitly the dependence of Φ on $\tilde{x}$ too; for the sake of simplicity, in the sequel this will be done only when strictly necessary. It is well known that:

Proposition 19.1. *If $F \in C^1(\mathbb{R}^n)$ and is symmetric, then Φ exists, is derivable, and*

$$\Phi'(x) = F(x). \tag{19.8}$$

Of course, when (19.7) holds, then Φ depends on $\tilde{x}$ in an additive way, namely it is the sum of a function of x and one of $\tilde{x}$.

In some particular cases, Φ can have a simpler form. For instance, if

$$F_i(x) = \sum_{j=1}^{n} \hat{F}_{ij}(x_j), \quad i = 1, \ldots, n, \tag{19.9}$$

then we find:

$$\Phi(x) = \sum_{i,j=1}^{n} (x_i - \tilde{x}_i) \int_{\tilde{x}_j}^{x_j} \hat{F}_{ij}(t)\, dt. \tag{19.10}$$

An obvious consequence of Proposition 19.1 is the following:

Corollary 19.2. *Under the assumptions of Proposition 19.1, (19.1) is a first-order necessary condition for y to be a minimum point of the problem:*

$$\min \Phi(x), \quad \text{subject to } x \in K, \tag{19.11}$$

which, up to a constant of addition for Φ, is (19.6).

Now, let us introduce another variational inequality, which will be associated to (19.1) and called *auxiliary variational inequality.* It consist of finding $y, z \in K$ such that:

$$\Phi(z) - \Phi(y) + \langle F(z), x - z\rangle \geq 0, \quad \forall\, x \in K. \tag{19.12}$$

It is obvious to observe that (19.12) represents an embedding* of (19.1): at $z = y$, (19.12) coincides with (19.1).

The separation scheme outlined in Section 19.1 for (19.1) can be extended to (19.12). Elements y and z are a solution of (19.12), if and only if the system (in the unknown x):

$$u := \Phi(y) - \varphi(z) + \langle F(z), z - x\rangle > 0,\ v := g(x) \geq 0,\ x \in X \tag{19.13}$$

is impossible. The set

$$\mathcal{K}(y,z) := \{(u,v) \in \mathbb{R}^{m+1} : u = \Phi(y) - \Phi(z) + \langle F(z), z - x\rangle,\, v = g(x),\, x \in X\} \tag{19.14}$$

is the image of (19.12), and will be called the *auxiliary image* of (19.1). Of course the impossibility of (19.13) holds if and only if

$$\mathcal{H} \cap \mathcal{K}(y;z) = \emptyset. \tag{19.15}$$

Now the question is to establish relationships between (19.3) and (19.15).

Proposition 19.3. *(19.15) holds if*

(i) *either y makes (19.3) true and $z = y$;*
(ii) *or y is a minimum point of (19.11) and z (in place of y) makes (19.3) true.*

Proof. **(i)** Trivial. **(ii)** The second assumption implies:

$$\langle F(z), z - x\rangle \leq 0, \quad \forall x \in K.$$

The first assumption gives:

$$\Phi(y) - \Phi(z) \leq 0.$$

The last two inequalities imply (19.15). □

Proposition 19.4. *If (19.15) holds with*

(i) *$y = z$, then (19.3) holds;*
(ii) *z as a minimum point of (19.11), then (19.3) holds with $y = z$.*

Proof. **(i)** is trivial. **(ii)** The second part of assumption means that

$$\Phi(y) - \Phi(z) \geq 0.$$

Ab absurdo, suppose that (19.3) be false; then there exists $\hat{x} \in K$, such that:

$$\langle F(z), z - \hat{x}\rangle > 0.$$

* y in (19.12) is not necessarily the same as in (19.1); it so is unless differently specified.

The last two inequalities contradict (19.15). □

Example 19.5. Let us set $n = 2$, $X = \mathbb{R}^2$, $K = \{x \in \mathbb{R}^2 : \frac{1}{2}x_1^2 - x_2 \geq 0\}$, $F(x) = (2x_1, 2(x_2 - 1))$. We find:

$$\Phi(x) = x_1^2 + x_2^2 - 2x_2 - \tilde{x}_1^2 - \tilde{x}_2^2 + 2\tilde{x}_2.$$

We see that neither (19.1), nor (19.12) are fulfilled. Note that F is the gradient of Φ, that $y = (0,0)$ is the minimum point of (19.11) at which the first-order Kuhn–Tucker necessary condition is satisfied.

Example 19.6. Let us set $n = 1$, $X = \mathbb{R}$, $F(x) = x^2$, $K = \mathbb{R}_+$. We find:

$$\Phi(x) = \frac{1}{3}x^3 - \tilde{x}^3,$$

and

$$\mathscr{K}(y;z) = \{(u,v) \in \mathbb{R}^2 : u = \frac{1}{3}y^3 + \frac{2}{3}z^3 - z^2 v,\ v \in \mathbb{R}\}, \quad y,z \in \mathbb{R}.$$

$\mathscr{K}(y;z)$ is a family of lines which admits the envelope of equation:

$$u = \frac{1}{3}y^3 - \frac{1}{3}v^3.$$

We see that $y = 0$ is the only solution of (19.1) and $y = z = 0$ the only one of (19.12). □

Example 19.7. Let us set $n = 1$, $X = \mathbb{R}$, $K = \mathbb{R}_+$, and

$$F(x) = \begin{cases} x+1, & \text{if } x < 0, \\ 1, & \text{if } 0 \leq x < 1, \\ x, & \text{if } 1 \leq x. \end{cases}$$

We find:

$$\text{at } \tilde{x} < 0, \quad \Phi(x) = \begin{cases} \frac{1}{2}x^2 + x - \frac{1}{2}\tilde{x}^2 - \tilde{x}, & \text{if } x < 0, \\ x - \frac{1}{2}\tilde{x}^2 - \tilde{x}, & \text{if } 0 \leq x < 1, \\ \frac{1}{2}x^2 + \frac{1}{2} - \frac{1}{2}\tilde{x}^2 - \tilde{x}, & \text{if } 1 \leq x; \end{cases}$$

$$\text{at } 0 \leq \tilde{x} < 1, \quad \Phi(x) = \begin{cases} \frac{1}{2}x^2 + x - \tilde{x}, & \text{if } x < 0, \\ x - \tilde{x}, & \text{if } 0 \leq x < 1, \\ \frac{1}{2}x^2 + \frac{1}{2} - \tilde{x}, & \text{if } 1 \leq x; \end{cases}$$

$$\text{at } 1 \leq \tilde{x}, \quad \Phi(x) = \begin{cases} \frac{1}{2}x^2 + x - \frac{1}{2}\tilde{x}^2 - \frac{1}{2}, & \text{if } x < 0, \\ x - \frac{1}{2}\tilde{x}^2 - \frac{1}{2}, & \text{if } 0 \leq x < 1, \\ \frac{1}{2}x^2 + \frac{1}{2} - \frac{1}{2}\tilde{x}^2 - \frac{1}{2}, & \text{if } 1 \leq x; \end{cases}$$

If we set:

$$f(x) := \begin{cases} \frac{1}{2}x^2 + x, & \text{if } x < 0, \\ x, & \text{if } 0 \le x < 1, \\ \frac{1}{2}x^2 + \frac{1}{2}x, & \text{if } 1 \le x, \end{cases}$$

then we have:

$$\Phi(x) = f(x) - f(\tilde{x}).$$

Without any loss of generality, we set $\tilde{x} = 0$. Then we find:

$$\mathscr{K}(x;z) = \{(u,v) \in \mathbb{R}^2 : u = \Phi(y) + \begin{cases} -\frac{1}{2}z^2 - z + (z+1)(z-v), & \text{if } v < 0, \\ -z + z - v, & \text{if } 0 \le v < 1, \\ -\frac{1}{2}z^2 - \frac{1}{2} + z(z-v), & \text{if } 1 \le v. \end{cases}$$

$\mathscr{K}(y;z)$ is a family of lines which admits the envelope:

$$u = \Phi(y) - f(v).$$

We see that $y = 0$ is the only solution of (19.1), and $y = z = 0$ the only one of (19.12). □

Example 19.8. Let us set $n = 2$, $X = \{x \in \mathbb{R}^2_+ \mid x_1 \le 1\}$, $F(x) = (x_1 + x_2, -x_1 - x_2)$ and $g(x) = 1 - x_2$. Then (19.1) is fulfilled at $y = (0,0)$ and at $y = (0,1)$. In fact, (19.1) becomes:

$$\langle (y_1 + y_2, -y_1 - y_2), (x_1 - y_1, x_2 - y_2) \rangle = (y_1 + y_2)[x_1 - x_2 - (y_1 - y_2)] \ge 0,$$

for all $x \in K$, which is identically true if $y_1 + y_2 = 0$, while is equivalent to $x_2 \le x_1 - (y_1 - y_2)$ if $y_1 + y_2 > 0$.

We find:

$$\begin{aligned} \Phi(x) &= \int_0^1 \langle (x_1 - \tilde{x}_1, x_2 - \tilde{x}_2), (\tilde{x}_1 + \tilde{x}_2 + [(x_1 - \tilde{x}_1) + (x_2 - \tilde{x}_2)]t, \\ &\qquad -\{\tilde{x}_1 + \tilde{x}_2 + [(x_1 - \tilde{x}_1) + (x_2 - \tilde{x}_2)]t\}) \rangle dt \\ &= \frac{1}{2}x_1^2 + \tilde{x}_2 x_1 - \frac{1}{2}x_2^2 - \tilde{x}_1 x_2 - \tilde{x}_1^2 + \tilde{x}_2^2. \end{aligned}$$

Φ does not depend on $\tilde{x}$ in additive form; F is dissymmetric; the gradient of Φ is different from F. □

19.3 A Particular Case

Now let us consider the particular case where F is continuous and monotone, $m = n$, $X = \mathbb{R}^n$ and $g(x) = Ax - b$, with A an $n \times n$ matrix with real entries.

Proposition 19.9. *The map $\Phi(x)$ is convex and F is its subgradient, if F is continuous, monotone and*

(i) *symmetric and $C^1(\mathbb{R}^n)$; or*
(ii) *such that:*

$$\Phi(\tilde{x},x') + \Phi(x',x'') + \Phi(x'',\tilde{x}) = 0. \tag{19.16}$$

Proof. The continuity of F gives the existence of Φ. **(i)** is trivial, since $F(x) = \Phi'(x)$. **(ii)** Since $\Phi(a,b) = -\Phi(b,a)$ for all $a,b \in \mathbb{R}^n$, from (19.16) and the mean value theorem we find:

$$\Phi(\tilde{x},x') - \Phi(\tilde{x},x'') = \Phi(x'',x') = \langle x' - x'', F(x(t^*))\rangle, \tag{19.17}$$

where $x(t) := (1-t)x'' + tx'$ and $t^* \in]0,1[$ is suitable. From the assumption of monotonicity, we have:

$$\langle x' - x'', F(x(t^*))\rangle \geq \langle x' - x'', F(x'')\rangle. \tag{19.18}$$

From (19.17) and (19.18), we deduce the inequality:

$$\Phi(\tilde{x},x') \geq \Phi(\tilde{x},x'') + \langle F(x''), x' - x''\rangle,$$

which, being $\tilde{x}$ fixed, shows $F(x'')$ as subgradient of $\Phi(x')$ at x''. Since x' and x'' are arbitrary, the claim is proved. □

The convexity of (19.10) can be proved by the same arguments as for the above (ii), but without the assumption (19.16).

In the current case, we have:

$$\mathscr{K}(y;z) = \{(u,v) \in \mathbb{R} \times \mathbb{R}^n : u = \Phi(y) - \Phi(z) + \langle F(z), z - A^{-1}(v+b)\rangle\}.$$

Consider the set:

$$\mathscr{K}^e(y) := \{(u,v) \in \mathbb{R}^{1+n} : u = \varphi(y;v) := \min_{z \in \mathbb{R}^n}[\Phi(y) - \Phi(z) + \langle F(z), z - A^{-1}(v+b)\rangle]\}.$$

The map $\varphi(y;\cdot)$ is concave, since it is the minimum of a family of linear functions, namely the family $\mathscr{K} = \{(y;z),\ z \in \mathbb{R}^n\}$.

Proposition 19.10. *If F is continuous and monotone, and if (19.16) is fulfilled, then for all $y \in \mathbb{R}^n$ the map $\varphi(y,\cdot)$ is the envelope of the family $\{\mathscr{K}(y;z),\ z \in \mathbb{R}^n\}$ of hyperplanes; namely $\varphi(y,\cdot)$ is such that:*

$$\forall z \in \mathbb{R}^n,\ \exists v_z \in \mathbb{R}^n,\ \textit{with}\ \varphi(y;v_z) = \Phi(y) - \Phi(z) + \langle F(z), z - A^{-1}(v_z+b)\rangle. \tag{19.19}$$

Proof. Ab absurdo, suppose that (19.19) does not hold, so that there exists $\hat{z} \in \mathbb{R}^n$, such that, for all $v \in \mathbb{R}^n$, we have:

$$\varphi(y;v) < \Phi(y) - \Phi(\hat{z}) + \langle F(\hat{z}), \hat{z} - A^{-1}(v+b)\rangle. \tag{19.20}$$

Since the right-hand side of (19.20) is linear in v, then there exist $a>0$ and $\hat{v}\in\mathbb{R}^n$, such that:

$$\varphi(y;v) \leq \Phi(y)-\Phi(\hat{z})+\langle F(\hat{z}),\hat{z}-A^{-1}(v+b)\rangle - a, \quad \forall v\in\mathbb{R}^n, \tag{19.21a}$$
$$\varphi(y;\hat{v}) = \Phi(y)-\Phi(\hat{z})+\langle F(\hat{z}),\hat{z}-A^{-1}(\hat{v}+b)\rangle - a. \tag{19.21b}$$

It follows that, within the family $\{\mathscr{K}(y;z),\ z\in\mathbb{R}^n\}$, there must exist a hyperplane, say $\mathscr{K}(y;z^*)$, having the same gradient $F(\hat{z})$ as $\mathscr{K}(y;\hat{z})$, and such that, for every $v\in\mathbb{R}^n$, we have:

$$\Phi(y)-\Phi(z^*)+\langle F(\hat{z}),z^*-A^{-1}(v+b)\rangle = \Phi(y)-\Phi(\hat{z})+\langle F(\hat{z}),\hat{z}-A^{-1}(v+b)\rangle - a. \tag{19.22}$$

Due to the assumption (19.16), by applying Proposition 19.9, we find:

$$\Phi(\hat{z})-\Phi(z^*)-\langle F(z^*),\hat{z}-z^*\rangle \geq 0. \tag{19.23}$$

Taking into account that $\mathscr{K}(y;\hat{z})$ and $\mathscr{K}(y;z^*)$ have the same gradient or $F(\hat{z}) = F(z^*)$, from (19.22), we draw:

$$\Phi(\hat{z})-\Phi(z^*)-\langle F(z^*),\hat{z}-z^*\rangle = -a<0,$$

which contradicts (19.23). □

Proposition 19.11. *If F is continuous and monotone, and if (19.16) is fulfilled, then:*

$$\mathscr{K}(y;z)\cap\mathscr{H}=\emptyset \quad \Rightarrow \quad \mathscr{K}^e(y;z)\cap\mathscr{H}=\emptyset. \tag{19.24}$$

Proof. ($\Rightarrow$) Due to the definition of φ and its concavity, for all $z\in\mathbb{R}^n$, we have:

$$\varphi(y;v) \leq \Phi(y)-\Phi(z)+\langle F(z),z-A^{-1}(\hat{v}+b)\rangle, \quad \forall v\in\mathbb{R}^n.$$

As $z=y$, the above inequality becomes:

$$\varphi(y;v) \leq \langle F(z),z-A^{-1}(\hat{v}+b)\rangle, \quad \forall v\in\mathbb{R}^n,$$

and shows the claim.
($\Leftarrow$) *Ab absurdo*, suppose that there exists $\hat{v}\in\mathbb{R}^n$, such that

$$\langle F(y),y-A^{-1}(\hat{v}+b)\rangle > 0.$$

Then, we have

$$\Phi(y)-\Phi(y)+\langle F(y),y-A^{-1}(\hat{v}+b)\rangle > 0,$$

which shows that $\mathscr{K}(y;y)\cap\mathscr{H}\neq\emptyset$. The proof is complete. □

References

1. Dien, P. H.; Mastroeni, G.; Pappalardo, M.; Quang, P. H., *Regularity conditions for constrained extremum problems via image space.* J. Optim. Theory Appl. 80 (1994), no. 1, 19–37.
2. B. Djafari Rouhani and A.A. Khan, On the embedding of variational inequalities, *Proc. Amer. Math. Soc.,* **131** (2003), 3861–3871.
3. Giannessi, F., *Separation of sets and gap functions for quasi-variational inequalities.* Variational inequalities and network equilibrium problems (Erice, 1994), 101–121, Plenum, New York, 1995.
4. Giannessi, F. *On connections among separation, penalization and regularization for variational inequalities with point-to-set operators,* Equilibrium problems with side constraints. Lagrangean theory and duality, II. Rend. Circ. Mat. Palermo, 48 (1997), 137–145.
5. Giannessi, F., *Embedding variational inequalities and their generalizations into a separation scheme,* J. Inequal. Appl. 1 (1997), no. 2, 139–147.
6. Giannessi, F., *Constrained optimization and image space analysis. Vol. 1. Separation of sets and optimality conditions.* Springer, New York, 2005.
7. Giannessi, F. and Rapcsk, T., *Images, separation of sets and extremum problems.* Recent trends in optimization theory and applications, 79–106, World Sci. Publ., River Edge, NJ, 1995.
8. Hestenes, M. R., *Optimization theory,* Wiley-Interscience, New York, 1975.
9. Isac, G., *Tihonov's regularization and the complementarity problem in Hilbert spaces,* J. Math. Anal. Appl. 174 (1993), 53–66.
10. Mastroeni, G. and Pellegrini, L., *On the image space analysis for vector variational inequalities,* J. Ind. Manag. Optim. 1 (2005), 123–132.
11. Pellegrini, L. Coercivity and image of constrained extremum problems. J. Optim. Theory Appl. 89 (1996), no. 1, 175–188.
12. Pellegrini, L. An extension of the Hestenes necessary condition. J. Optim. Theory Appl. 69 (1991), no. 2, 297–309.
13. Quang, P. H., *Lagrangian multiplier rules via image space analysis.* Nonsmooth optimization: methods and applications (Erice, 1991), 354–365, Gordon and Breach, Montreux, 1992.

Chapter 20
On the Nonlinear Generalized Ordered Complementarity Problem

D. Goeleven

Dedicated to the memory of Professor George Isac

Abstract In this paper, we develop a new approach to study a class of nonlinear generalized ordered complementarity problems.

20.1 Introduction

The nonlinear generalized ordered complementarity problem is a mathematical model that has been especially studied by Isac and Kostreva in [2] so as to formulate complementarity conditions for several functions. More precisely, let $\Phi_1, \Phi_2 : \mathbf{R}^n \to \mathbf{R}^n$ be two given functions; the generalized ordered complementarity problem $\mathbf{GOCP}(\Phi_1, \Phi_2)$ consists of finding $x \in \mathbb{R}^n$ such that

$$\wedge\{\Phi_1(x), \Phi_2(x)\} = 0. \tag{20.1}$$

Here, for every pair $(X,Y) \in \mathbf{R}^n \times \mathbf{R}^n$, the notation $\wedge\{X,Y\}$ is used to denote the infimum of the set $\{X,Y\} \subset \mathbf{R}^n$ with respect to the ordering $\leq$ defined by the closed pointed convex cone $(\mathbf{R}_+)^n$, i.e.,

$$x \leq y \Leftrightarrow y - x \in (\mathbf{R}_+)^n.$$

In other words:

$$(\forall\, \alpha \in \{1,2,...,n\}) : \wedge\{X,Y\}_\alpha = \min\{X_\alpha, Y_\alpha\}.$$

D. Goeleven
IREMIA, University of La Reunion, 97400 Saint-Denis, France, e-mail: goeleven@univ-reunion.fr

P.M. Pardalos et al. (eds.), *Nonlinear Analysis and Variational Problems*, Springer Optimization and Its Applications 35, DOI 10.1007/978-1-4419-0158-3_20,

Here $\wedge\{X,Y\}_\alpha$ denotes the α-th component of the vector $\wedge\{X,Y\}$. Note that for our purpose, we have restricted the formulation of the generalized ordered complementarity problem to a model in $\mathbf{R}^n$ involving two functions. A more general formulation in locally convex space ordered by a general closed pointed convex cone and involving a finite number of operators may be found in the original article of Isac and Kostreva [2].

It is clear that $x \in \mathbf{R}^n$ is a solution of problem (20.1) if and only if for all $i \in \{1,...,n\}$,

$$\begin{cases} (\Phi_1(x))_i \geq 0, \\ (\Phi_2(x))_i \geq 0, \\ (\Phi_1(x))_i(\Phi_2(x))_i = 0. \end{cases} \tag{20.2}$$

Here $(\Phi_1(x))_i$ (resp. $(\Phi_2(x))_i$) denotes the i-th component of the vector $\Phi_1(x)$ (resp. $\Phi_2(x)$). The system in (20.2) can also be written equivalently as

$$\begin{cases} \Phi_1(x) \geq 0, \\ \Phi_2(x) \geq 0, \\ \langle \Phi_1(x), \Phi_2(x) \rangle = 0, \end{cases} \tag{20.3}$$

where $\langle .,. \rangle$ denotes the Euclidean scalar product in $\mathbf{R}^n$. It results in particular that the following complementarity relations hold:

$$\begin{cases} (\Phi_1(x))_i > 0 \implies (\Phi_2(x))_i = 0, \\ (\Phi_2(x))_i > 0 \implies (\Phi_1(x))_i = 0. \end{cases} \tag{20.4}$$

The model in (20.1) is a generalization of the classical complementarity problem

$$\begin{cases} x \geq 0, \\ \Phi_1(x) \geq 0, \\ \langle x, \Phi_1(x) \rangle = 0 \end{cases} \tag{20.5}$$

which can be written equivalently as

$$\wedge\{x, \Phi_1(x)\} = 0, \tag{20.6}$$

and which is thus recovered by setting $(\forall x \in \mathbf{R}^n): \; \Phi_2(x) = x$ in (20.1).

A complementarity relationship between two (or more) functions may reflect a mathematical complexity that can be found in the mathematical formulation of several important problems like mixed lubrication problems [6], singular stochastic optimal control problems [7] and global reproduction problems of economical systems involving several technologies [1].

In using the ordering, one gives a clear and natural mathematical approach to study complementarity relationships between a finite number of functions. Further theoretical and iterative results have then been obtained in [2], [3] and [5]. The approaches discussed in these papers allow the study of generalized ordered complementarity problems involving isotone functions. In this paper, we develop an approach that can be used to study another class of functions, i.e., asymptotically linear functions.

20.2 A Spectral Condition for the Generalized Ordered Complementarity Problem

We assume that **(H)**

$$(\forall x \in \mathbf{R}^n): \ \Phi_1(x) := M_1 x + F_1(x) \tag{20.7}$$

and

$$(\forall x \in \mathbf{R}^n): \ \Phi_2(x) := M_2 x + F_2(x), \tag{20.8}$$

where $M_1, M_2 \in \mathbb{R}^{n\times n}$ are two given matrices and $F_1, F_2 : \mathbf{R}^n \to \mathbf{R}^n$ are two given continuous functions satisfying the conditions:

$$\lim_{\|x\| \to +\infty} \frac{F_1(x)}{\|x\|} = 0 \tag{20.9}$$

and

$$\lim_{\|x\| \to +\infty} \frac{F_2(x)}{\|x\|} = 0. \tag{20.10}$$

Let us now introduce a concept of generalized eigenvalue for the nonlinear mapping

$$\wedge\{M_1 ., M_2 .\} : \mathbf{R}^n \to \mathbf{R}^n; x \mapsto \wedge\{M_1 x, M_2 x\}.$$

Definition 20.1. We say that $\mu \in \mathbb{R}$ is an eigenvalue of the mapping $\wedge\{M_1 ., M_2 .\}$ provided that there exists a vector $z \in \mathbb{R}^n$, $z \neq 0$ such that:

$$\wedge\{M_1 z, M_2 z\} = \mu z. \tag{20.11}$$

The system in (20.11) can also be written as

$$\wedge\{(M_1 - \mu I)z, (M_2 - \mu I)z\} = 0$$

which is equivalent to

$$\begin{cases} M_1z - \mu z \geq 0, \\ M_2z - \mu z \geq 0, \\ \langle M_1z - \mu z, M_2z - \mu z \rangle = 0. \end{cases} \tag{20.12}$$

We denote by $\sigma_+(M_1,M_2)$ the set of eigenvalues of $\wedge\{M_1.,M_2.\}$, i.e.,

$$\sigma_+(M_1,M_2) := \{\mu \in \mathbb{R} : \exists\, z \in \mathbb{R}^n \backslash \{0\} : \wedge\{M_1z, M_2z\} = \mu z\}.$$

Let us also set

$$\mathrm{SOL}((1-t)I + tM_1, (1-t)I + tM_2)) :=$$
$$= \{z \in \mathbf{R}^n : \wedge\{(1-t)z + tM_1z, (1-t)z + tM_2z = 0\}.$$

We first prove a technical result which ensures that all eigenvalues of $\wedge\{M_1.,M_2.\}$ are strictly positive if and only if for all $t \in [0,1]$, the trivial solution 0 is the unique solution of problem $\mathbf{GOCP}((\mathbf{1}-\mathbf{t})\mathbf{I} + \mathbf{tM_1}, (\mathbf{1}-\mathbf{t})\mathbf{I} + \mathbf{tM_2})$.

Lemma 20.2. *We have*

$$\sigma_+(M_1,M_2) \;\subset\,]0,+\infty[\tag{20.13}$$

if and only

$$(\forall t \in [0,1]) :\; \mathrm{SOL}((1-t)I + tM_1, (1-t)I + tM_2)) := 0.$$

Proof. a) Let us first assume that $\sigma_+(M_1,M_2) \;\subset\,]0,+\infty[$. The case $t = 0$ is trivial. Indeed, if $z \in \mathrm{SOL}(I,I)$ then $\wedge\{z,z\} = 0$ and thus $z = 0$. Let us now suppose that $t \in]0,1]$. Let $z \in \mathrm{SOL}((1-t)I + tM_1, (1-t)I + tM_2)$ be given. We have

$$\wedge\{(1-t)z + tM_1z, (1-t)z + tM_2z\} = 0.$$

Thus

$$-(1-t)z = \wedge\{tM_1z, tM_2z\}$$

and then

$$-\frac{(1-t)}{t}z = \wedge\{M_1z, M_2z\}.$$

It results that $z = 0$. Indeed, if we suppose on the contrary that $z \neq 0$, then we see that $\mu := -\frac{(1-t)}{t} \in \sigma_+(M_1,M_2)$ and we get a contradiction to (20.13) since $\mu \leq 0$.

Let us now assume that for all $t \in [0,1]$: $\mathrm{SOL}((1-t)I + tM_1, (1-t)I + tM_2) = \{0\}$. We claim that (20.13) holds. Indeed, suppose on the contrary then there exists $\mu \leq 0$ and $z \neq 0$ such that $\wedge\{(M_1 - \mu I)z, (M_2 - \mu I)z\} = 0$. We set $t := \frac{1}{1-\mu} = \frac{1}{1+|\mu|}$.

It is clear that $t \in]0,1]$ and $\wedge\{(M_1 + \frac{(1-t)}{t}I)z, (M_2 + \frac{(1-t)}{t}I)z\} = 0$. It results that $z \in \mathrm{SOL}((1-t)I + tM_1, (1-t)I + tM_2))$, which is a contradiction. □

We are now in position to prove the main result of this section:

Theorem 20.3. *Suppose that condition* **(H)** *holds. If*

$$\sigma_+(M_1, M_2) \subset]0, +\infty[$$

then problem **GOCP**(Φ_1, Φ_2) *has at least one solution.*

Proof. We have

$$\wedge\{\Phi_1(x), \Phi_2(x)\} = 0 \Leftrightarrow x = x - \wedge\{\Phi_1(x), \Phi_2(x)\}.$$

Let us now denote by $H : [0,1] \times \mathbb{R}^n \to \mathbb{R}^n$ the continuous homotopy defined as

$$H(t,x) = t(x - \wedge\{\Phi_1(x), \Phi_2(x)\}).$$

We have

$$\begin{aligned} H(t,x) &= x - (1-t)x - \wedge\{t\Phi_1(x), t\Phi_2(x)\} \\ &= x - \wedge\{(1-t)x + t\Phi_1(x), (1-t)x + t\Phi_2(x)\}. \end{aligned}$$

We claim that there exists $R_0 > 0$ such that for all $R \geq R_0$ and for all $t \in [0,1]$,

$$H(t,x) \neq x, \ \forall x \in \mathbb{R}^n, ||x|| = R. \tag{20.14}$$

Indeed, if we suppose the contrary then we may find sequences $\{t_i\}_{i\in\mathbb{N}} \subset [0,1]$ and $\{x_i\}_{i\in\mathbb{N}} \subset \mathbb{R}^n$ satisfying $||x_i|| \to +\infty$ and $x_i = H(t_i, x_i)$, i.e.,

$$\wedge\{(1-t_i)x_i + t_i\Phi_1(x_i), (1-t_i)x_i + t_i\Phi_2(x_i)\} = 0. \tag{20.15}$$

For i large enough, $||x_i|| \neq 0$ and we may set

$$z_i := \frac{x_i}{||x_i||}.$$

There exist subsequences, again denoted by $\{t_i\}$ and $\{z_i\}$, such that $\lim_{i\to+\infty} t_i = t \in [0,1]$ and $\lim_{i\to+\infty} z_i = z$ with $||z|| = 1$.

We have

$$\frac{1}{||x_i||} \wedge \{(1-t_i)x_i + t_i\Phi_1(x_i), (1-t_i)x_i + t_i\Phi_2(x_i)\} = 0$$

and thus

$$\wedge\{(1-t_i)z_i + t_i(M_1 z_i + \frac{F_1(x_i)}{||x_i||}), (1-t_i)z_i + t_i(M_2 z_i + \frac{F_2(x_i)}{||x_i||})\} = 0.$$

Taking the limit as $i \longrightarrow +\infty$, we get

$$\wedge\{(1-t)z+tM_1z,(1-t)z+tM_2z\}=0.$$

Thus

$$z \in \mathrm{SOL}((1-t)I+tM_1,(1-t)I+tM_2).$$

However $z \neq 0$, and we obtain a contradiction to Lemma 20.2.

Thus, for $R \geq R_0$, (20.14) holds and the Brouwer degree with respect to the set $D_R = \{x \in \mathbb{R}^n : ||x|| < R\}$ and 0 of the map $x \mapsto x - H(t,x)$ is well defined for all $t \in [0,1]$. Using the homotopy invariance property as well as the normalized property of Brouwer degree, we obtain

$$\deg(\wedge\{\Phi_1(.),\Phi_2(.)\},D_R,0) = \deg(id_{\mathbb{R}^n} - H(1,.),D_R,0)$$

$$= \deg(id_{\mathbb{R}^n} - H(0,.),D_R,0) = \deg(id_{\mathbb{R}^n},D_R,0) = 1.$$

The result follows from the existence property of Brouwer degree. □

20.3 Existence and Uniqueness Results

Corollary 20.4. *Let $M_1, M_2 \in \mathbf{R}^{n\times n}$ be two matrices satisfying the conditions:*

$$(\forall x \in \mathbf{R}^n) : \langle M_1x,x\rangle \geq 0$$

$$(\forall x \in \mathbf{R}^n) : \langle M_2x,x\rangle \geq 0$$

$$(\forall x \in \mathbf{R}^n, x \neq 0) : \langle M_2^T M_1x,x\rangle > 0.$$

Then for all $q_1, q_2 \in \mathbf{R}^n$, problem $\mathbf{GOCP}(\mathbf{M_1}.+\mathbf{q_1},\mathbf{M_2}.+\mathbf{q_2})$ *has a unique solution.*

Proof. Here condition **(H)** holds with

$$(\forall x \in \mathbf{R}^n) : \; F_1(x) = q_1$$

and

$$(\forall x \in \mathbf{R}^n) : \; F_2(x) = q_2.$$

Let $t \in [0,1]$ be given and $x \in \mathbf{R}^n$ such that

$$\wedge\{(1-t)x+tM_1x,(1-t)x+tM_2x\}=0.$$

Then

$$\langle(1-t)x+tM_1x,(1-t)x+tM_2x\rangle=0$$

and thus

$$(1-t)^2||x||^2+t^2\langle M_1x,M_2x\rangle+(1-t)t\langle M_1x,x\rangle+(1-t)t\langle M_2x,x\rangle=0.$$

If $t=0$, then necessarily $||x||=0$. If $t\in]0,1]$, then necessarily $\langle M_1x,M_2x\rangle=0$ and thus

$$\langle M_2^TM_1x,x\rangle=0$$

and thus necessarily $x=0$. The existence of at least one solution of $\mathbf{GOCP}(\mathbf{M_1}.+\mathbf{q_1},\mathbf{M_2}.+\mathbf{q_2})$ follows from Lemma 20.2, which ensures that

$$\sigma_+(M_1,M_2)\subset\,]0,+\infty[,$$

and Theorem 20.3.

Let us now prove the uniqueness of the solution. Let X,x be two solutions of problem $\mathbf{GOCP}(\mathbf{M_1}.+\mathbf{q_1},\mathbf{M_2}.+\mathbf{q_2})$. Then $M_1X+q_1\geq 0$, $M_1x+q_1\geq 0$ and thus

$$\langle M_1X+q_1,v\rangle\geq 0,\forall v\in\mathbf{R}^n_+$$

and

$$\langle M_1x+q_1,v\rangle\geq 0,\forall v\in\mathbf{R}^n_+.$$

We have also $\langle M_1X+q_1,M_2X+q_2\rangle=0$, $\langle M_1x+q_1,M_2x+q_2\rangle=0$ and thus

$$\langle M_1X+q_1,v-(M_2X+q_2)\rangle\geq 0,\forall v\in\mathbf{R}^n_+$$

and

$$\langle M_1x+q_1,v-(M_2x+q_2)\rangle\geq 0,\forall v\in\mathbf{R}^n_+.$$

We have $(M_2x+q_2)\geq 0$ and thus

$$0\leq\langle M_1X+q_1,(M_2x+q_2)-(M_2X+q_2)\rangle=\langle M_1X+q_1,M_2(x-X)\rangle.$$

We have also $(M_2X+q_2)\geq 0$ and thus

$$0\leq\langle M_1x+q_1,(M_2X+q_2)-(M_2x+q_2)\rangle=\langle M_1x+q_1,M_2(X-x)\rangle.$$

It results that

$$\langle -M_1X-q_1,M_2(x-X)\rangle\leq 0$$

and

$$\langle M_1x+q_1,M_2(x-X)\rangle\leq 0.$$

Thus

$$\langle M_2^TM_1(x-X),x-X\rangle=\langle M_1(x-X),M_2(x-X)\rangle\leq 0$$

and necessarily

$$x=X$$

since the matrix $M_2^TM_1$ is assumed positive definite. □

Corollary 20.5. *Let $\Phi_1, \Phi_2 : \mathbf{R}^n \to \mathbf{R}^n$ be two functions satisfying condition* **(H)** *with*

$$(\forall x \in \mathbf{R}^n) : \langle M_1 x, x\rangle \geq 0$$

$$(\forall x \in \mathbf{R}^n) : \langle M_2 x, x\rangle \geq 0$$

$$(\forall x \in \mathbf{R}^n, x \neq 0) : \langle M_2^T M_1 x, x\rangle > 0.$$

Then problem **GOCP**(Φ_1, Φ_2) *has at leat one solution.*

Proof. As in the proof of Corollary 20.4, we check that

$$\sigma_+(M_1, M_2) \subset]0, +\infty[$$

and conclude by applying Theorem 20.3. □

Corollary 20.6. *Let $\Phi_1, \Phi_2 : \mathbf{R}^n \to \mathbf{R}^n$ be two functions satisfying condition* **(H)** *with*

$$(\forall x \in \mathbf{R}^n) : \langle M_1 x, x\rangle \geq 0$$

$$(\forall x \in \mathbf{R}^n) : \langle M_2 x, x\rangle \geq 0$$

$$(\forall x \in \mathbf{R}^n, x \neq 0) : \langle M_2^T M_1 x, x\rangle > 0.$$

$$(\forall x \in \mathbf{R}^n, x \neq 0) : \langle \Phi_1(x) - \Phi_1(X), \Phi_2(x) - \Phi_2(X)\rangle > 0.$$

Then problem **GOCP**(Φ_1, Φ_2) *has a unique solution.*

Proof. The existence result is given in Corollary 20.5.

Let X, x be two solutions of problem **GOCP**(Φ_1, Φ_2). Then $\Phi_1(X) \geq 0$,$\Phi_1(x) \geq 0$ and thus

$$\langle \Phi_1(X), v\rangle \geq 0, \forall v \in \mathbf{R}^n_+$$

and

$$\langle \Phi_1(x), v\rangle \geq 0, \forall v \in \mathbf{R}^n_+.$$

We have also $\langle \Phi_1(X), \Phi_2(X)\rangle = 0$, $\langle \Phi_1(x), \Phi_2(x)\rangle = 0$ and thus

$$\langle \Phi_1(X), v - \Phi_2(X)\rangle \geq 0, \forall v \in \mathbf{R}^n_+$$

and

$$\langle \Phi_1(x), v - \Phi_2(x)\rangle \geq 0, \forall v \in \mathbf{R}^n_+.$$

We have $\Phi_2(x) \geq 0$ and thus

$$0 \leq \langle \Phi_1(X), \Phi_2(x) - \Phi_2(X)\rangle.$$

We have also $\Phi_2(X) \geq 0$ and thus

$$0 \leq \langle \Phi_1(x), \Phi_2(X) - \Phi_2(x)\rangle.$$

It results that

$$\langle -\Phi_1(X), \Phi_2(x) - \Phi_2(X)\rangle \le 0$$

and

$$\langle \Phi_1(x), \Phi_2(x) - \Phi_2(X)\rangle \le 0.$$

Thus

$$\langle \Phi_1(x) - \Phi_1(X), \Phi_2(x) - \Phi_2(X)\rangle \le 0$$

and thus necessarily

$$x = X.$$

□

References

1. R.W. Cottle and G.B. Dantzig, *A Generalization of the Linear Complementarity Problem*, Journal of Combinatorial Theory, **8**, 79–90, 1979.
2. G. Isac and M. Kostreva, *The Generalized Order Complementarity Problem*, J.O.T.A., **71**, 517–534, 1991.
3. G. Isac and D. Goeleven, *The General Order Complementarity Problem, Models and Iterative Methods*, Annals of Operation Research, **44**, 63–92, 1993.
4. G. Isac, *Complementarity Problems*, Lecture Notes in Mathematics 1528, Springer-Verlag, Heidelberg, 1993.
5. D. Goeleven, *A Uniqueness Theorem for the Generalized Order Linear Complementarity Problem*, Linear Algebra and its Applications, **235**, 221–227, 1996.
6. K. P. Oh, *The Formulation of the Mixed Lubrication Problem as a Generalized Nonlinear Complementarity Problem*, Trans. AMSE, J. of Tribology, **106**, 598–604, 1986.
7. M. Sun, *Monotonocity of Mangasarian's Iterative Algorithm for Generalized Linear Complementarity Problems*, Journal of Mathematical Analysis and Applications, **144**, 474–485, 1989.

Chapter 21
Optimality Conditions for Several Types of Efficient Solutions of Set-Valued Optimization Problems

T.X.D. Ha

Dedicated to the memory of Professor George Isac

Abstract A simple unified framework is presented for the study of strong efficient solutions, weak efficient solutions, positive proper efficient solutions, Henig global proper efficient solutions, Henig proper efficient solutions, super efficient solutions, Benson proper efficient solutions, Hartley proper efficient solutions, Hurwicz proper efficient solutions and Borwein proper efficient solutions of set-valued optimization problem with/or without constraints. Some versions of the Lagrange claim, the Fermat rule and the Lagrange multiplier rule are formulated in terms of the first- and second-order radial derivatives, the Ioffe approximate coderivative and the Clarke coderivative.

21.1 Introduction

In a pioneering paper, Pareto presented the concept of Pareto efficient point for a set and began a new branch in optimization: vector optimization. As some efficient points exhibit certain abnormal properties, various concepts of proper efficient points have been introduced in order to eliminate these points. The original concept was introduced by Kuhn–Tucker and modified by Geoffrion and later it was formulated in a more general framework by Hurwicz (Ref. 24), Benson (Ref. 3), Hartley (Ref. 20), Borwein (Ref. 4), Henig (Ref. 21), and Borwein–Zhuang (Ref. 5); also see Ref. 18. The definitions of proper efficiency emphasize different aspects but, as mentioned by Zaffaroni in Ref. 41, they can be seen as an extension of the primitive idea in that they can be geometrically described in terms of separations between

T.X.D. Ha
Researcher, Hanoi Institute of Mathematics, Hanoi, Vietnam.

P.M. Pardalos et al. (eds.), *Nonlinear Analysis and Variational Problems*,
Springer Optimization and Its Applications 35, DOI 10.1007/978-1-4419-0158-3_21,

the ordering cone and the considered set by means of an open convex cone or open convex sets.

A brief inquiry into the matter reveals that definitions of weak efficient point, strong efficient point, positive proper efficient point, Henig global proper efficient point, Henig proper efficient point, super efficient point, Benson proper efficient point, Hartley proper efficient point, Hurwicz proper efficient point and Borwein proper efficient point can be described as disjointness between some set and some nonempty open (not necessarily convex) cone. Inspired by this fact, we present the notion of Q-minimal point of a set, where Q is some nonempty open (not necessarily convex) cone, and show that the above definitions can be reduced in a unified form by mean of this notion. We then present the notion of Q-minimal solution of vector optimization problems. With these auxiliary notions in hand we are able to obtain optimality conditions for various kinds of efficient solutions and proper efficient solutions in a unified scheme: firstly, one establishes results for Q-minimal solutions and secondly derives from them the similar ones for other kinds of solutions. As the reader will see, some arguments developed for weak efficient solutions are still valid for Q-minimal solutions without assuming the ordering cone to have nonempty interior. Moreover, it turns out that Q-minimal points can be characterized by a special function introduced in optimization by Hiriart–Urruty which has a nice subdifferentiality property due to the openness of Q.

For illustration, we apply this scheme to the study of the unconstrained set-valued optimization problem (P)

$$\text{Minimize } \ F(x) \ \text{ subject to } \ x \in X$$

and the constrained set-valued optimization problem (CP) with set-inclusion together with abstract constraints

$$\text{Minimize } \ F(x) \ \text{ subject to } \ x \in \Omega \text{ and } G(x) \cap \mathscr{C} \neq \emptyset,$$

where F and G are set-valued maps from a Banach space X respectively into Banach spaces Y and Z, $\Omega \subset X$ and $\mathscr{C} \subset Z$ are nonempty sets. We note that recently the interest toward these problems has grown. Along with the Pareto efficient solutions, other types of solutions such as strong efficient solutions, weak efficient solutions, positive proper efficient solutions, Henig global proper efficient solutions, Henig proper efficient solutions, super efficient solutions and Benson proper efficient solutions are the object of many investigations (see Refs. 1, 2, 7, 9–16, 19, 23, 26, 28–30, 33, 34, 36–40, 44, 45). Our aim is to obtain optimality conditions for the above types of solutions (except efficient solutions) and also for Hartley proper efficient solutions, Hurwicz proper efficient solutions and Borwein proper efficient solutions; the latter three types of solutions are considered for the first time. We formulate versions of the Lagrange claim expressed in terms of first- and second-order radial derivatives, the Fermat rule and the Lagrange multiplier rule expressed in terms of the Ioffe approximate coderivative and the Clarke coderivative. We also compare our results to some in literature.

The paper is organized as follows. In Section 21.2 we recall the notions of subdifferentials of a function, derivatives and coderivatives of a set-valued map to be used in next sections. In Section 21.3 we first recall concepts of efficient points and proper efficient points, then we introduce the concept of Q-minimal points and establish relations among them. Afterward, we characterize Q-minimal points by Hiriart–Urruty's function and establish properties of subdifferentials of this function for several special cases. The last section is devoted to optimality conditions for set-valued optimization problems.

21.2 Subdifferentials, Derivatives and Coderivatives

In this section, we present preliminary materials on the basic generalized differentiability concepts which will be used in the following sections. We refer the reader to Refs. 1, 8, 11, 22, 25, 38 for more references and discussions.

Throughout the paper, unless otherwise specified, let $R =]-\infty, +\infty[$, $\overline{R} =]-\infty, +\infty]$, $R_+ = [0, +\infty[$ and let X, Y and Z be Banach spaces with the duals X^*, Y^* and Z^* resp. For a nonempty set S in any space, int$S$, cl$S$, bd$S$, conv$S$ and cone$S$ denote its interior, closure, boundary, convex hull and the set $\{ts : t \in \text{int}R_+, s \in S\}$; $d(.;S)$ is the distance function associated with S and $\delta(.;S)$ is the indicator function associated with S, i.e., $\delta_S(x) = 0$ if $x \in S$ and $\delta_S(x) = \infty$ otherwise. An open (closed) unit ball in a space, say X, is denoted by $\mathring{B}_X$ (resp. B_X) and we omit the subscript denoting the space when no confusion occurs.

Let F be a set-valued map from X to Y. We denote the domain, image and graph of F respectively by

$$\begin{aligned} \text{Dom } F &= \{x \in X \ : \ F(x) \neq \emptyset\}, \\ F(X) &= \cup_{x \in X} F(x), \\ \text{gr } F &= \{(x,y) \in X \times Y \ : \ x \in \text{ Dom } F,\ y \in F(x)\}. \end{aligned}$$

We say that F is closed (convex) if its graph is closed (convex).

We begin the section with the notions of first- and second-order radial derivatives which allow us to derive simple optimality conditions. Through this section, let S be a nonempty subset of X. Recall (Ref. 6) that the *radial cone* to S at $x \in S$ is given by

$$T_R(x;S) = \{u \in X : \exists (\lambda_n) > 0,\ \exists\ (u_n) \to u,\ \forall\ n,\ x + \lambda_n u_n \in S\}.$$

Here, $(\lambda_n) > 0$ means a sequence of positive scalars, $(x_n) \xrightarrow{S} x$ means that a sequence (x_n) converges to x with $x_n \in S$ for all n. Recall (Ref. 38) that the *first-order radial derivative* of F at $(x,y) \in \text{gr}F$ is the set-valued map $D_R F(x,y)$ from X to Y defined by

$$\text{gr} D_R F(x,y) = T_R((x,y); \text{gr}F).$$

Motivated by the concepts of (extended) second-order Bouligand derivative and Dini derivative presented in Ref. 11, we introduce the concept of second-order radial derivative. Given $(x',y') \in X \times Y$, the *second-order radial derivative* of F at (x,y) w.r.t. (x',y') is the set-valued map $D_R^2 F(x,y),(x',y'))$ from X to Y defined by

$$\begin{aligned} D_R^2 F((x,y),(x',y'))(u) = \{\ v \in Y :\ & \exists (t_n) > 0, \exists (u_n) \to u, \exists (v_n) \to v, \\ & \forall n, y + t_n y' + t_n^2 v_n \in F(x + t_n x' + t_n^2 u_n)\}. \end{aligned}$$

The following result will be used in Section 21.4.

Proposition 21.1. *(i)* $F(x) - y \subseteq D_R F(x,y)(x' - x)$ *for all* $x' \in X$.
(ii) $D_R F(x,y)(u) \subseteq D_R^2 F((x,y),(0,0))(u)$ *for all* $u \in X$.

Proof. (i) See Ref. 38.
(ii) Let $v \in D_R F(x,y)(u)$. By the definition, there exist sequences $(t_n) > 0$, $(u_n) \to u$ and $(v_n) \to v$ such that for all n we have $y + t_n v_n \in F(x + t_n u_n)$. Setting $r_n = \sqrt{t_n}$ for each n we have $y + r_n.0 + r_n^2 v_n \in F(x + r_n.0 + r_n^2 u_n)$. This means that $v \in D_R^2 F((x,y),(0,0))(u)$. □

Next, we recall some concepts of subdifferentials and normal cones. Let f be a function from X to $\overline{R}$. We denote its domain and epigraph respectively by

$$\begin{aligned} \text{dom}\, f &= \{x \in X : f(x) \neq +\infty\}, \\ \text{epi}\, f &= \{(x,t) \in X \times R : f(x) \leq t\}. \end{aligned}$$

Assume that f is lower semicontinuous (l.s.c.) on X and $x \in \text{dom} f$. The *Ioffe approximate subdifferential* of f at x (Ref. 25) is the set

$$\partial_A f(x) = \cap_{L \in \mathscr{F}} \limsup_{(\varepsilon,y) \to (0^+,x)} \partial_\varepsilon^- f_{y+L}(y),$$

where $\mathscr{F}$ is the collection of all finite-dimensional subspaces of X, $f_{y+L}(u) = f(u)$ if $u \in y + L$ and $f_{y+L}(u) = +\infty$ otherwise, for $\varepsilon \geq 0$

$$\partial_\varepsilon^- f_{y+L}(y) = \{x^* \in X^* : \langle x^*, v\rangle \leq \varepsilon \|v\| + \liminf_{t \to 0^+} t^{-1}[f_{y+L}(y + tv) - f_{y+L}(y)], \forall v \in X\}.$$

The *Ioffe approximate normal cone* to S at $x \in S$ (Ref. 25) is given by

$$N_A(x;S) = \cup_{\lambda > 0} \lambda \partial_A d(x;S).$$

Now we recall the concept of the Clarke subdifferential (Ref. 8). First suppose that f is Lipschitz near x. The *Clarke generalized subdifferential* of f at x is the set

$$\partial_C f(x) = \{x^* \in X^* : \langle x^*, v\rangle \leq f^0(;v),\ \forall v \in X\},$$

where $f^0(x;v)$ is the generalized directional derivative of f at x in the direction v

$$f^0(x;v) = \limsup_{y\to x,\ t\to 0+} \frac{f(y+tv)-f(y)}{t}.$$

The *Clarke normal cone* to S at $x \in S$ is given by

$$N_C(x;S) = \mathrm{cl} \cup_{\lambda>0} \lambda \partial_C d(x;S).$$

Now, assume that f is l.s.c. on X. The Clarke subdifferential of f at x is the set

$$\partial_C f(x) = \{x^* \in X^* : (x^*,-1) \in N_C((x,f(x));\mathrm{epi} f)\}.$$

When f is convex, the *subdifferential* in the sense of convex analysis of f at x is the set

$$\partial f(x) = \{x^* \in X^* : \langle x^*, v\rangle \leq f(x+v) - f(x), \forall v \in X\}.$$

The above normal cones to a set can also be equivalently defined using indicator functions associated with this set as follows

$$N_A(x;S) = \partial_A \delta(x;S),\ N_C(x;S) = \partial_C \delta(x;S).$$

Let $(x,y) \in \mathrm{gr}F$. Assuming that F is closed, the *Ioffe approximate coderivative* $D_A^*F(x,y)$ and the *Clarke coderivative* $D_C^*F(x,y)$ of F at (x,y) are set-valued maps from Y^* to X^* respectively defined by

$$D_A^*F(x,y)(y^*) = \{x^* \in X^* : (x^*,-y^*) \in N_A((x,y);\mathrm{gr}F)\}$$

and

$$D_C^*F(x,y)(y^*) = \{x^* \in X^* : (x^*,-y^*) \in N_C((x,y);\mathrm{gr}F)\}.$$

Among all the properties of the above subdifferentials and coderivative, let us recall those ones which will be used in the sequel. In the proposition below, let f and g be functions from X to $\overline{R}$ which are l.s.c. on their domains, let F be a set-valued closed map from X to Y. We will assume that $x \in \mathrm{dom} f$ or $x \in \mathrm{dom} F$.

Proposition 21.2. *(i)* $\partial_A f(x) \subset \partial_C f(x)$ *and these subdifferentials coincide with the subdifferential of convex analysis when f is convex and Lipschitz near x.*
(ii) If f attains a local minimum at x, then $0 \in \partial_A f(x)$.
(iii) $\partial_C(-f)(x) = -\partial_C f x)$ *if f is Lipschitz near x.*
(iv) $\partial_A(f+g)(x) \subset \partial_A f(x) + \partial_A g(x)$ *for* $x \in \mathrm{dom} f \cap \mathrm{dom} g$.
(iv) $D_A^*F(x,y)(y^*) \subseteq D_C^*F(x,y)(y^*)$ *for* $(x,y) \in \mathrm{gr}F$ *and* $y^* \in Y^*$.

21.3 Some Concepts of Efficient Points

In this section, we first recall concepts of efficient points and proper efficient points, then we introduce the concept of Q-minimal points and establish relations among them. Afterward, we characterize Q-minimal points by Hiriart–Urruty's function and establish properties of subdifferentials of this function in some special cases.

Throughout the paper, let $K \subset Y$ be a nonempty closed pointed convex cone with apex at zero (pointedness means $K \cap (-K) = \{0\}$). The cone K introduces a partial order in Y: for $y_1, y_2 \in Y$ we write $y_1 \geq_K y_2$ if $y_1 - y_2 \in K$ and $y_1 >_K y_2$ if $y_1 \geq y_2$, $y_1 \neq y_2$. In the sequel, we omit the subscript K as no confusion occurs.

A convex set $\Theta \subset Y$ is called a base for K if $0 \notin \mathrm{cl}\Theta$ and $K = \{t\theta : t \in R_+, \theta \in \Theta\}$. When Θ is bounded, we say that K has a bounded base. Denote

$$K^+ = \{\varphi \in Y^* : \ \varphi(k) \geq 0, \ \forall k \in K\}$$

and

$$K^{+i} = \{\varphi \in Y^* : \varphi(k) > 0, \ \forall k \in K \setminus \{0\}\}.$$

It is known that K has a base iff $K^{+i} \neq \emptyset$ and K has a bounded base iff $\mathrm{int}K^+ \neq \emptyset$ (Ref. 31). Non-negative orthants in the Banach spaces R^n, $C_{[0,1]}$, $L^p_{[0,1]}$ and l^p $(1 \leq p < \infty)$ have bases and the non-negative orthants in $L^1_{[0,1]}$, l^1 have bounded bases (Ref.31).

Throughout this section, let A be a nonempty subset of Y and $\overline{a} \in A$. The main concept in vector optimization is Pareto efficiency. Recall that $\overline{a}$ is an *efficient* (or *Pareto minimal) point* of A w.r.t. K ($\overline{a} \in Min(A,K)$) if

$$(A - \overline{a}) \cap (-K \setminus \{0\}) = \emptyset.$$

In this paper, we are concerned with the following concepts of efficiency.

Definition 21.3. We say that

(i) $\overline{a}$ is a strong (or ideal) efficient point of A w.r.t. K ($\overline{a} \in StrMin(A,K)$) if $A - \overline{a} \subseteq K$.
(ii) Supposing that int $K \neq \emptyset$, $\overline{a}$ is a weak efficient point of A w.r.t. K ($\overline{a} \in WMin(A,K)$) if $(A - \overline{a}) \cap (-\mathrm{int}K) = \emptyset$.
(iii) Supposing that $K^{+i} \neq \emptyset$, $\overline{a}$ is a positive proper efficient point of A w.r.t. K ($\overline{a} \in Pos(A,K)$) if there exists $\varphi \in K^{+i}$ such that $\varphi(a) \geq \varphi(\overline{a})$ for all $a \in A$.
(iv) $\overline{a}$ is a Hurwicz proper efficient point of A w.r.t. K ($\overline{a} \in Hu(A,K)$) if

$$\mathrm{cl\ conv\ cone}[(A - \overline{a}) \cup K] \cap (-K) = \{0\}.$$

(v) $\overline{a}$ is a Hartley proper efficient point of A w.r.t. K ($\overline{a} \in Ha(A,K)$) if $\overline{a} \in Min(A,K)$ and there exists a constant $M > 0$ such that, whenever there is $\lambda \in K^+$ with $\lambda(a - \overline{a}) > 0$ for some $a \in A$, one can find $\mu \in K^+$ with

$$\lambda(a - \overline{a})/\|\lambda\| \leq -M(\mu(a - \overline{a})/\|\mu\|).$$

(vi) $\overline{a}$ is a Benson proper efficient point of A w.r.t. K ($\overline{a} \in Be(A,K)$) if

$$\mathrm{cl\ cone}[(A - \overline{a}) + K] \cap (-K) = \{0\}.$$

(vii) $\bar{a}$ is a Borwein proper efficient point of A w.r.t. K ($\bar{a} \in Bo(A,K)$) if

$$\text{cl cone}(A-\bar{a}) \cap (-K) = \{0\}.$$

(viii) $\bar{a}$ is a Henig global proper efficient point of A w.r.t. K ($\bar{a} \in GHe(A,K)$) if there exists a convex cone C with apex at zero with $K \setminus \{0\} \subset \text{int}C$ such that $(A-\bar{a}) \cap (-\text{int}C) = \emptyset$.

(ix) Supposing that K has a base Θ, $\bar{a}$ is a Henig proper efficient point of A w.r.t. to Θ ($\bar{a} \in He(A,\Theta)$) if there is a scalar $\varepsilon > 0$ such that

$$\text{cl cone}(A-\bar{a}) \cap (-\text{cl cone}(\Theta + \varepsilon B)) = \{0\}.$$

(x) $\bar{a}$ is a super efficient point of A w.r.t. K ($\bar{a} \in SE(A,K)$) if there is a scalar $\rho > 0$ such that

$$\text{cl cone}(A-\bar{a}) \cap (B-K) \subseteq \rho B.$$

For the notions of efficient points in Definition 21.3, we refer the reader to Refs. 3–5,18, 20, 24, 27, 28, 34, 42, 43, 46. The above definition of Henig proper efficient points can be found in Refs. 5, 46 , see also Refs. 42 and 43. For an equivalent definition of Henig proper efficient point by means of a functional from K^{+i}, the reader is referred to Ref. 35. We note that positive proper efficiency has been introduced by Hurwicz, and super efficiency has been introduced by Borwein and Zhuang. We refer the reader to Ref. 18 for a survey and materials on proper efficiency.

In the sequel, when speaking of weak efficient points (resp. properly positive efficient points) we mean that int$K$ (resp. K^{+i}) is nonempty, when speaking of Henig efficient points we mean that K has a base Θ, and when speaking that K has a bounded base we mean that Θ is bounded. When no confusion arises we write simply $Min(A)$, $StrMin(A)$, $WMin(A)$, $Pos(A)$, $Hu(A)$, $Ha(A)$, $Bo(A)$, $Be(A)$, $GHe(A)$, $He(A)$ and $SE(A)$.

Many authors have investigated the relationships among various kinds of efficiency and proper efficiency. We recall several known results in the proposition below.

Proposition 21.4. *(i)* $StrMin(A) \subseteq Min(A) \subseteq WMin(A)$.
(ii) $Pos(A) \subseteq Hu(A)$*; if* Y *is separable, then* $Hu(A) \subseteq Pos(A)$.
(iii) $Pos(A) \subseteq GHe(A)$.
(iv) $SE(A) \subseteq Ha(A) \subseteq Be(A) \subseteq Bo(A) \subseteq Min(A)$.
(v) $SE(A) \subseteq GHe(A) \subseteq Be(A)$.
(vi) $SE(A) \subseteq He(A)$ *and if* K *has a bounded base then* $SE(A) = He(A)$.

Next, we present the notion of Q-minimal points. From now on unless otherwise specified let $Q \subset Y$ be an arbitrary nonempty open cone (with apex at zero) and different from Y.

Definition 21.5. We say that $\bar{a}$ is a Q-minimal point of A ($\bar{a} \in Qmin(A)$) if

$$A \cap (\bar{a} - Q) = \emptyset$$

or, equivalently,

$$(A-\overline{a})\cap(-Q)=\emptyset.$$

Remark 21.6. Makarov and Rachkovski (Ref. 35) studied in more details some concepts of proper efficiency and introduced the notion of D-efficiency, i.e., efficiency w.r.t. a family of dilating cones. Namely, given $D\in\mathscr{F}(K)$, where $\mathscr{F}(K)$ is the class of families of cones dilating K, $\overline{a}$ is said to be a *D-minimal point* of A ($\overline{a}\in DMin(A)$) if there exists $C\in D$ such that

$$(A-\overline{a})\cap(-C)=\emptyset.$$

Recall that an open cone in Y is said to be a *dilating cone* (or a *dilation*) of K, or *dilating* K if it contains $K\setminus\{0\}$. It has been established that Borwein proper efficiency, Henig global proper efficiency, Henig proper efficiency, super efficiency and Hartley proper efficiency are D-efficiency with D being appropriately chosen family of dilating cones. The reader will see that in contrast with D-efficiency, our concept includes not only some concepts of proper efficiency among which are these ones considered in Ref. 35 but also the concepts of strong efficiency and weak efficiency.

Let θ be as before a base of K. Setting

$$\delta=\inf\{\|\theta\|:\theta\in\Theta\}>0,$$

for each $0<\eta<\delta$, we can associate to K with another convex, pointed and open cone V_η, defined by

$$V_\eta=\text{cone}(\Theta+\eta\mathring{B}_Y).$$

For each scalar $\varepsilon>0$, we also associate to K with another open cone $K(\varepsilon)$

$$K(\varepsilon)=\{y\in Y:d_K(y)<\varepsilon d_{-K}(y)\}.$$

We are going to show that the efficient points of Definition 21.3 are in fact Q-minimal points with Q being appropriately chosen cones. Our main result in this section is the following.

Theorem 21.7. *(i)* $\overline{a}\in StrMin(A)$ *iff* $\overline{a}\in Qmin(A)$ *with* $Q=Y\setminus(-K)$.
(ii) $\overline{a}\in WMin(A)$ *iff* $\overline{a}\in Qmin(A)$ *with* $Q=\text{int}K$.
(iii) $\overline{a}\in Pos(A)$ *iff* $\overline{a}\in Qmin(A)$ *with* $Q=\{y\in Y\mid\varphi(y)>0\}$ *and* φ *is some functional in* K^{+i}.
(iv) $\overline{a}\in Hu(A)$ *iff* $\overline{a}\in Qmin(A)$, *with* $Q=Y\setminus-\text{cl conv cone}[(A-\overline{a})\cup K]$.
(v) $\overline{a}\in Be(A)$ *iff* $\overline{a}\in Qmin(A)$, *with* $Q=Y\setminus-\text{cl cone}[(A-\overline{a})+K]$.
(vi) $\overline{a}\in Ha(A)$ *iff* $\overline{a}\in Qmin(A)$ *with* $Q=K(\varepsilon)$ *for some* $\varepsilon>0$.
(vii) $\overline{a}\in Bo(A)$ *iff* $\overline{a}\in Qmin(A)$, *with* Q *being some open cone dilating* K.
(viii) $\overline{a}\in GHe(A)$ *iff* $\overline{a}\in Qmin(A)$, *with* Q *being some open pointed convex cone dilating* K.
(ix) $\overline{a}\in He(A)$ *iff* $\overline{a}\in Qmin(A)$ *with* $Q=V_\eta$ *and* η *is some scalar satisfying* $0<\eta<\delta$.
(x) (supposing that K *has a bounded base)* $\overline{a}\in SE(A)$ *iff* $\overline{a}\in Qmin(A)$ *with* $Q=V_\eta$ *and* η *is some scalar satisfying* $0<\eta<\delta$.

Proof. Using Definitions 21.3 and 21.5 one can easily prove the assertions (i)–(v) and (viii). The assertions (vi)–(vii) formulated in a slightly different form are established in Ref. 35. We prove now the assertion (ix), namely, we show that $\overline{a} \in He(A)$ iff there is a scalar η with $0 < \eta < \delta$ such that

$$(A - \overline{a}) \cap (-V_\eta) = \emptyset. \tag{21.1}$$

Recall that by the definition, $\overline{a} \in He(A)$ iff

$$\text{cl cone}(A - \overline{a}) \cap (-\text{cl cone}(\Theta + \varepsilon B)) = \{0\}. \tag{21.2}$$

It is also known (Ref. 42) that $\overline{a} \in He(A)$ iff

$$(A - \overline{a}) \cap (-\overline{S}_n) = \{0\} \tag{21.3}$$

for some integer $n \in N$, where $\overline{S}_n = \text{cl cone}(\Theta + \delta/(2n)B_Y)$. Now, suppose that $\overline{a} \in He(A)$. Then (21.2) holds. Without lost of generality we can assume that $0 < \varepsilon < \delta$. We show that (21.2) holds with $\eta = \varepsilon$. Suppose to the contrary that there is $a' \in A - \overline{a}$ such that $a' \in -V_\varepsilon$. Clearly, $a' \in \text{cl cone}(A - \overline{a}) \cap (-\text{cl cone}(\Theta + \varepsilon B_Y))$. On the other hand, as $0 < \eta = \varepsilon < \delta$ and by the definition of δ, $0 \notin V_\varepsilon$. Hence $a' \neq 0$. This is a contradiction to (21.2). Next, suppose that (21.1) holds for some η. Let n be an integer satisfying $n - 1 > \delta/(2\eta)$ or $\delta/(2n-2) < \eta$. By (21.1) we have

$$(A - \overline{a}) \cap (-V_{\delta/(2n-2)}) \subset (A - \overline{a}) \cap (-V_\eta) = \emptyset.$$

Then $(A - \overline{a}) \cap (-V_{\delta/(2n-2)} \cup \{0\}) = \{0\}$. On the other hand, Lemma 2.1 in Ref. 42 states that if $(A - \overline{a}) \cap (-V_{\delta/(2n-2)} \cup \{0\}) = \{0\}$, then $(A - \overline{a}) \cap (-\overline{S}_n) = \{0\}$. Thus, (21.3) holds and therefore, $\overline{a} \in He(A)$, as it was to be shown. To complete the proof note that the last assertion follows from the just proved one and the assertion (vi) in Proposition 21.4. □

Remark 21.8. The assertion (ix) in the above theorem is inspired by the definition of Henig proper efficient point for sets in locally convex spaces given by Gong in Ref. 14. One can deduce that any Henig proper efficient point is global Henig proper efficient point.

Our next aim is to provide necessary and sufficient conditions for a Q-minimal point in the form of being a solution for a scalar optimization problem. Here the scalar function is the signed distance function introduced by Hiriart–Urruty (Ref. 22). Recall that for a subset U of Y, this function is defined by

$$\Delta_U(y) = d(y;U) - d(y;Y \setminus U).$$

We collect some known properties of the function Δ_U in the proposition below.

Proposition 21.9. (Ref. 22)

(i) Δ_U is Lipschitz of rank 1.
(ii) $\Delta_{Y\setminus U} = -\Delta_U$.

(iii) Δ_U *is convex if* U *is convex and* Δ_U *is concave if* U *is reverse convex, i.e.,* $U = Y \setminus V$ *with* V *being convex.*
(iv) $\Delta_U(y) < 0$ *iff* $y \in \mathrm{int}\, U$, $\Delta_U(y) = 0$ *iff* $y \in \mathrm{bd}\, U$ *and* $\Delta_U(y) > 0$ *iff* $y \in Y \setminus \mathrm{int}\, U$.
(v) Suppose that U *is convex and has a nonempty interior, and* $0 \in \mathrm{bd}U$. *Then*

$$\partial \Delta_U(0) \subset N(0;U) \setminus \{0\}$$

where $N(0;U)$ *is the normal cone in the sense of convex analysis of* U *at zero.*

With this function in hand, we can characterize a Q-minimal point as follows.

Proposition 21.10. $\overline{a} \in Qmin(A)$ *iff the function* $\Delta_{-Q}(. - \overline{a})$ *attains its minimum at* $\overline{a}$*, that is*

$$\Delta_{-Q}(a - \overline{a}) \geq \Delta_{-Q}(0) = 0, \ \ \forall a \in A. \tag{21.4}$$

Proof. Observe first that $0 \in \mathrm{bd}(-Q)$ and $\Delta_{-Q}(0) = 0$. Now, let $\overline{a} \in Qmin(A)$. By the definition, $(A - \overline{a}) \subseteq Y \setminus (-Q)$. Consequently, $\Delta_{-Q}(a - \overline{a}) \geq 0$ for all $a \in A$, i.e., (21.4) holds. Next, suppose that (21.4) holds. If $\overline{a} \notin Qmin(A)$, then there is $a' \in A$ such that $a' - \overline{a} \in -Q$. As the set $-Q$ is open, $d(a' - \overline{a}; Y \setminus (-Q)) > 0$ and therefore, $\Delta_{-Q}(a' - \overline{a}) < 0$, a contradiction to (21.4). Thus, $\overline{a} \in Qmin(A)$. □

It is of interest to know more about the subdifferential (in some sense) of $\Delta_{-Q}(. - \overline{a})$ at the minimizer $\overline{a}$. Below we consider several cases with Q being some open cones defined in Theorem 21.7. The result to be obtained will play an important role in formulating Fermat rule and Lagrange multiplier rule.

Proposition 21.11. *(i)* $\partial \Delta_{-\mathrm{int}K}(0) \subseteq K^+ \setminus \{0\}$.
(ii) Let C *be an open convex cone dilating* K. *Then*

$$\partial \Delta_{-C}(0) \subseteq K^{+i}.$$

(iii) Assuming that K *has a bounded base, we have* $\partial \Delta_{-V_\eta}(0) \subseteq \mathrm{int}K^+$.
(iv) Assuming that A *has a nonempty interior, we have*

$$\partial_A \Delta_{Y \setminus \mathrm{cl\ conv\ cone}[(A - \overline{a}) \cup K]}(0) \subseteq K^+ \setminus \{0\}.$$

Proof. (i) Apply Proposition 21.9 (v) to $U = -\mathrm{int}K$ and take account of $N(0; -\mathrm{int}K) = K^+$.
(ii) Applying Proposition 21.9 (v) to C we get $\partial \Delta_{-C}(0) \subset N(0; -C) \setminus \{0\}$. Now take $y^* \in \partial \Delta_{-C}(0)$ and $k \in K \setminus \{0\} \subset C$. We have to show that $y^*(k) > 0$. As $y^* \neq 0$, there is $y \in Y$ such that $y^*(y) > 0$. On the other hand, as C is open, and $k \in C$, there exist a scalar $t > 0$ such that $-k + ty \in -C$. Hence, $y^*(-k + ty) \leq 0$ and $y^*(k) \geq ty^*(y) > 0$. Thus, $y^* \in K^{+i}$.
(iii) Suppose that Θ is bounded and denote

$$\overline{\delta} = \sup\{\|\theta\| : \theta \in \Theta\} < \infty.$$

Let $y^* \in \partial \Delta_{-V_\eta}(0)$. It is known that $y^* \in \mathrm{int}K^+$ iff y^* is uniformly positive on K in the sense that there exists a scalar $\alpha > 0$ such that $y^*(k) \geq \alpha \|k\|$ for all $k > 0$. Let

$k \in K \setminus \{0\}$ be an arbitrary vector. As Θ is a base of K, there exist a scalar $t > 0$ and $\theta \in \Theta$ such that $k = t\theta$. Since $-\theta + \eta \mathring{B} \subset -V_\eta$, it follows that

$$-k + t\eta \mathring{B} = t(-\theta + \eta \mathring{B}) \subset -V_\eta,$$

i.e., the open ball centered at $-k$ with the radius $t\eta$ is contained in $-V_\eta$. Therefore,

$$d(-k; Y \setminus (-V_\eta)) \geq t\eta.$$

On the other hand, $t\eta = (\|k\|/\|\theta\|)\eta \geq (\eta/\overline{\delta})\|k\|$. Hence,

$$d(-k; Y \setminus (-V_\eta)) \geq (\eta/\overline{\delta})\|k\|.$$

Clearly, $\Delta_{-V_\eta}(0) = 0$. As $y^* \in \partial \Delta_{-V_\eta}(0)$, the definition of the convex subdifferential yields

$$\begin{aligned}\langle y^*, -k \rangle &\leq \Delta_{-V_\eta}(-k) - \Delta_{-V_\eta}(0) = d(-k; -V_\eta) - d(-k; Y \setminus (-V_\eta)) \\ &= -d(-k; Y \setminus (-V_\eta)) \leq -(\eta/\overline{\delta})\|k\|\end{aligned}$$

or $\langle y^*, k \rangle \geq (\eta/\overline{\delta})\|k\|$. This means that y^* is uniformly positive on K, or $y^* \in \text{int} K^+$ as it was to be shown.

(iv) By the relation between approximate subdifferential and the Clarke subdifferential (see Proposition 21.2 (i)), it suffices to show that

$$\partial_C \Delta_{Y \setminus \text{cl conv cone}[(A - \overline{a}) \cup K]}(0) \subseteq K^+ \setminus \{0\}.$$

For simplicity, we denote $Q = Y \setminus -V$, where $V = \text{cl conv cone}[(A - \overline{a}) \cup K$. Note that V is a closed convex cone with a nonempty interior and $K \subset V$. We have to show that

$$\partial_C \Delta_{-Q}(0) \subseteq K^+ \setminus \{0\}.$$

By Proposition 21.9 (ii), we have $\Delta_{-Q}(0) = -\Delta_{Y \setminus -V}(0) = -\Delta_V(0)$. The properties of the Clarke subdifferential and the subdifferential of convex analysis yield

$$\partial_C(-\Delta_V)(0) = -\partial_C \Delta_V(0) = -\partial \Delta_V(0).$$

Applying Proposition 21.9 (v) to the closed convex cone V which has a nonempty interior gives $-\partial \Delta_V(0) \subset -N(0; V) \setminus \{0\}$. Further, since $K \subset V$ we get $-N(0, V) \subseteq K^+$. Therefore, we get $-\partial \Delta_V(0) \subseteq K^+ \setminus \{0\}$, which yields $\partial_C \Delta_{-Q}(0) \subseteq K^+ \setminus \{0\}$.

□

21.4 Optimality Conditions for Set-Valued Optimization Problem

In this section, we formulate optimality conditions for the unconstrained set-valued optimization problem (P) and the constrained set-valued optimization problem (CP) presented in the introduction.

The notions of efficient points for sets naturally induce corresponding notions of *efficient solutions* to these vector optimization problems. In this paper, we restrict ourselves to global solutions. Denote by $\mathscr{A}$ the *feasible set* of (CP), i.e.,

$$\mathscr{A} = \{x \in \Omega : \ G(x) \cap \mathscr{C} \neq \emptyset\}$$

and by $F(\mathscr{A})$ the image of $\mathscr{A}$, i.e., $F(\mathscr{A}) = \cup_{x \in \mathscr{A}} F(x)$. Through the section, let $Q \subset Y$ be as before an open cone (with apex at zero) different from Y.

Definition 21.12. Let $\overline{x} \in X$ (resp., $\overline{x} \in \mathscr{A}$) and $(\overline{x}, \overline{y}) \in \mathrm{gr}F$. We say that $(\overline{x}, \overline{y})$ is an efficient (resp., weak efficient, strong efficient, positive proper efficient, Hurwicz proper efficient, Hartley proper efficient, Benson proper efficient, Borwein proper efficient, Henig global proper efficient, Henig proper efficient, super efficient and Q-minimal) solution of (P) (resp., of (CP)) if $\overline{y}$ is an efficient (resp., weak efficient, strong efficient, positive proper efficient, Hurwicz proper efficient, Hartley proper efficient, Benson proper efficient, Borwein proper efficient, Henig global proper efficient, Henig proper efficient, super efficient and Q-minimal) point of $F(X)$ (resp., $F(\mathscr{A})$).

From Theorem 21.7, it is clear that one can obtain optimality conditions for some kinds of solutions in Definition 21.12 except efficient solutions in a unified scheme: firstly, one establishes results for Q-minimal solutions and secondly derives from them the similar ones for other solutions. We will apply this approach to obtain the Lagrange claim, the Fermat rule and the Lagrange multiplier rule. Our arguments are motivated by that used in Ref. 11 (partially) and by that of Ref. 12.

First we establish the Lagrange claim and the Fermat rule for the unconstrained problem (P). Recall that Lagrange once formulated a claim (Ref. 32) that a smooth function f from R^n into R has a local minimum at x if all the directional derivatives of f at x are non-negative. However, this claim is wrong as shown by Peano (Ref. 17). Various new versions of the Lagrange claim with sufficient and/or necessary optimal conditions have been obtained for single- or set-valued maps (for the set-valued case, see Ref. 11). We begin with establishing the Lagrange claim for Q-minimal solutions. Let $\overline{x} \in X$ and $(\overline{x}, \overline{y}) \in \mathrm{gr}F$.

Proposition 21.13. *$(\overline{x}, \overline{y})$ is a Q-minimal solution of (P) iff*

$$D_R F(\overline{x}, \overline{y})(u) \cap (-Q) = \emptyset, \quad \forall u \in X \tag{21.5}$$

and

$$D_R^2 F((\overline{x}, \overline{y}), (0,0))(u) \cap (-Q) = \emptyset, \quad \forall u \in X. \tag{21.6}$$

Proof. By Proposition 21.1 (ii), the implication (21.6) $\Longrightarrow$ (21.5) holds. Therefore, we need only to show that (21.6) holds if $(\bar{x},\bar{y})$ is a Q-minimal solution of (P) and that $(\bar{x},\bar{y})$ is a Q-minimal solution of (P) if (21.5) holds. Observe that by Definition 21.12, $(\bar{x},\bar{y})$ is a Q-minimal solution of (P) iff

$$(F(X)-\bar{y})\cap(-Q)=\emptyset. \tag{21.7}$$

First, suppose to the contrary that $(\bar{x},\bar{y})$ is a Q-minimal solution of (P) but (21.6) is false, i.e., there exists $w\in Y$ such that $w\in -Q$ and $w\in D^2_RF((\bar{x},\bar{y}),(0,0))(u)$. By the definition of the second-order radial cone, there exist sequences $(t_n)>0$, $(u_n)\to u$ and $(w_n)\to w$ s.t.

$$\bar{y}+t_n.0+t_n^2w_n\in F(\bar{x}+t_n.0+t_n^2u_n)$$

for all n. Since $(w_n)\to w$, $w\in -Q$ and the cone Q is open, we have $t_n^2w_n\in -Q$ for n large enough. Therefore,

$$t_n^2w_n\in(F(\bar{x}+t_n^2u_n)-\bar{y})\cap(-Q)$$

for n large enough. This is a contradiction to (21.7).

Next, suppose that (21.5) holds. From Proposition 21.1 (i) we get

$$(F(x)-\bar{y})\cap(-Q)\subset D_RF(\bar{x},\bar{y})(x-\bar{x})\cap(-Q)=\emptyset$$

for any $x\in X$. Therefore, (21.7) holds, which means that $(\bar{x},\bar{y})$ is a Q-minimal solution of (P), as it was to be shown. □

We can now derive the Lagrange claim for other solutions.

Theorem 21.14. *Let* $(\bar{x},\bar{y})\in gr\,F$.

(i) $(\bar{x},\bar{y})$ *is a strong efficient solution of* (P) *iff (21.5)–(21.6) hold with* $Q=Y\setminus -K$.
(ii) $(\bar{x},\bar{y})$ *is a weak efficient solution of* (P) *iff (21.5)–(21.6) hold with* $Q=-\text{int}K$.
(iii) $(\bar{x},\bar{y})$ *is a positive proper efficient solution of* (P) *iff (21.5)–(21.6) hold with* $Q=\{y:\varphi(y)<0\}$ *for some functional* $\varphi\in K^{+i}$.
(iv) $(\bar{x},\bar{y})$ *is a Hartley proper efficient solution of* (P) *iff (21.5)–(21.6) hold with* $Q=K(\varepsilon)$ *for some scalar* $\varepsilon>0$.
(v) $(\bar{x},\bar{y})$ *is a Borwein proper efficient solution of* (P) *iff (21.5)–(21.6) hold with* $Q=C$ *for some open cone* C *dilating* K.
(vi) $(\bar{x},\bar{y})$ *is a global Henig proper efficient solution of* (P) *iff (21.5)–(21.6) hold with* $Q=C$ *for some open convex cone* C *dilating* K.
(vii) $(\bar{x},\bar{y})$ *is a Henig proper efficient solution of* (P) *iff (21.5)–(21.6) hold with* $Q=V_\eta$ *for some scalar* η *satisfying* $0<\eta<\delta$.
(viii) (supposing that K *has a bounded base)* $(\bar{x},\bar{y})$ *is a super proper efficient solution of* (P) *iff (21.5)–(21.6) hold with* $Q=V_\eta$ *for some scalar* η *satisfying* $0<\eta<\delta$.

Proof. It is easy to see that from Definition 21.12 and Theorem 21.7, $(\bar{x},\bar{y})$ is an efficient solution of (P) in some sense iff it is Q-minimal solution of (P) with Q

being some corresponding open cone. The assertion follows from this and Proposition 21.13. □

Let us note that in Theorem 21.14, we include only the cases when either Q is expressed directly in terms of K ($Q = Y \setminus -K$, $Q = -\text{int}K$) or Q is a dilation of K.

Remark 21.15. (i) Some necessary or sufficient conditions (in terms of first- and second-order Dini derivatives or Bouligand contingent derivatives) have been obtained in Ref. 11 for efficient soltions but under the assumption that the ordering cone has nonempty interior or the spaces under consideration are finite dimensional. In case of weak efficient solutions, the first-order conditions of Theorem 21.14 agree with Theorem 3.4 of Ref. 13, but the second-order conditions are formulated here for the first time. In case of strong efficient solutions, Theorem 21.14 extends a result due to Aubin–Ekeland (Ref.1) which states first-order necessary and sufficient conditions in terms of contingent derivative under the assumption that F is convex. To our knowledge, a result similar to the assertions (iii)–(viii) of Theorem 21.14 does not appear in the literature. For other first-order conditions (in terms of various derivatives, contingent epiderivative or generalized contingent epiderivative) the reader is referred to Refs. 28, 34 and the references herein.
(ii) In Ref. 26, Isac and Khan explored the ideas around the so-called Dubovitskii-Milyutin approach to study necessary conditions for several types of solutions to set-valued optimization problems with or without constraints. Using the concept of contingent derivative and introducing several concepts of subgradient and scalarized subgradients of a set-valued map, they expressed optimality at a point as an empty intersection of constraint sets in the domain space, which is in contrast with the results obtained so far in set-valued optimization (in particular, with Theorem 21.14 above), where the optimality conditions have been given in the image space.

Next, we formulate the Fermat rule for Q-minimal solutions of (P). Our techniques are motivated by the ones developed in Ref. 12 for weak efficient solutions.

Proposition 21.16. *Assume that F has a closed graph. If $(\bar{x},\bar{y})$ is a Q-minimal solution of (P) then there exists $y^* \in \partial_A \Delta_{-Q}(0)$ such that the following inclusions hold:*

$$0 \in D_A^* F(\bar{x},\bar{y})(y^*) \tag{21.8}$$

and

$$0 \in D_C^* F(\bar{x},\bar{y})(y^*). \tag{21.9}$$

Proof. Using Proposition 21.10, one can easily check that $(\bar{x},\bar{y})$ is a solution of (P) iff $(\bar{x},\bar{y})$ is a solution to the following problem

$$\text{Minimize } \Delta_{-Q}(y-\bar{y}) + \delta((x,y);\text{gr } F).$$

Therefore, by Proposition 21.2, we have

$$\begin{aligned}(0,0) &\in \partial_A(\Delta_{-Q}(0) + \delta_{\mathrm{gr}\,F}(\bar{x},\bar{y})) \\ &\subset \{0\} \times \partial_A \Delta_{-Q}(0) + \partial_A \delta_{\mathrm{gr}\,F}(\bar{x},\bar{y}) \\ &= \{0\} \times \partial_A \Delta_{-Q}(0) + N_A((\bar{x},\bar{y}); \mathrm{gr}\,F)\end{aligned} .$$

Hence, there exists $y^* \in \partial_A \Delta_{-Q}(0)$ with $(0,-y^*) \in N_A((\bar{x},\bar{y}); \mathrm{gr}\,F)$. This implies $0 \in D_A^* F(\bar{x},\bar{y})(y^*)$ and (21.8) holds. Finally, (21.9) follows from (21.8) and Proposition 21.2 (iv). □

We derive now the Fermat rule for other solutions.

Theorem 21.17. *Assume that F has a closed graph.*

(i) If $(\bar{x},\bar{y})$ is a weak efficient solution of (P) or $(\bar{x},\bar{y})$ is a Hurwicz proper efficient solution of (P) and the image $F(X)$ has a nonempty interior, then there exists $y^* \in K^+ \setminus \{0\}$ such that (21.8)–(21.9) hold.

(ii) If $(\bar{x},\bar{y})$ is a global Henig proper efficient solution (in particular, if $(\bar{x},\bar{y})$ is either a positively proper efficient solution, or a Henig proper efficient solution, or a super efficient solution) of (P), then there exists $y^* \in K^{+i}$ such that (21.8)–(21.9) hold.

(iii) If $(\bar{x},\bar{y})$ is a super efficient solution of (P) and K has a bounded base, then there exists $y^* \in intK^+$ such that (21.8)–(21.9) hold.

Proof. It is easy to see that from Definition 21.12 and Theorem 21.7, $(\bar{x},\bar{y})$ is an efficient solution of (P) in some sense iff it is Q-minimal solution of (P) with Q being some open conebeing chosen in correspondence with Theorem 21.7. Theorem 21.17 then follows from Propositions 21.11 and 21.16. □

Remark 21.18. (i) One can also formulate a version of Theorem 21.17 in terms of the Mordukhovich coderivative in Asplund spaces setting.
(ii) Unfortunately, Theorem 21.17 does not hold for arbitrary Pareto efficient solutions even when F is convex. Zheng–Ng in Ref. 44 provided an example showing the existence of a set-valued convex closed map F such that (0,0) is a (global) efficient solution of (P) but $0 \notin D_C^* F(0,0)(y^*)$ for all $y^* \in Y^* \setminus \{0\}$. For Pareto efficient solutions, one can establish only a fuzzy version of the Fermat rule expressed in terms of the Clarke coderivative (Ref. 44).
(iii) The Fermat rule for weak efficient solutions has already been established in Ref. 12, and for Henig proper efficient solutions and super efficient solutions in Refs. 2, 23. For completeness we include these cases in Theorem 21.17. The Fermat rule for Hurwicz proper efficient solution, global Henig proper efficient solution and positive proper efficient solution has been formulated here for the first time. Note that the Fermat rule in terms of coderivative of convex analysis has been established for strong efficient solutions under the additional assumption that F is convex in Ref. 1.
(iv) We would also like to mention other versions of the Fermat rule which are expressed in terms of subdifferentials or generalized subdifferentials of a set-valued map (Refs. 26, 28).

Our next purpose is to establish the Lagrange multiplier rule for the constrained set-valued optimization problem (CP). Let $\bar{x} \in \mathscr{A}$ and $(\bar{x}, \bar{y}) \in \mathrm{gr}F$. Recall that F is pseudo-Lipschitz around $(\bar{x}, \bar{y})$ (Ref.1) if there exist scalars $r > 0$ and $\gamma > 0$ such that for all $x, x' \in \bar{x} + rB_X$

$$(\bar{y} + rB_Y) \cap F(x) \subset F(x') + \gamma \|x - x'\| B_Y.$$

Let $\bar{z} \in G(\bar{x}) \cap \mathscr{C}$. Recall further that G is metrically regular around $(\bar{x}, \bar{z})$ relatively to $\Omega \times \mathscr{C}$ (Ref.12) if there exist scalars $r > 0$ and $\gamma > 0$ such that for all $(x, z) \in [(\bar{x} + rB_X) \times (\bar{z} + rB_Z)] \cap (\Omega \times \mathscr{C})$

$$d((x,z); (\Omega \times \mathscr{C}) \cap Gr\, G) \leq \gamma d(z; G(x)).$$

We will need the following assumption.
Assumption (A):

(i) The sets Ω and $\mathscr{C}$ are closed;
(ii) F and G are closed and pseudo-Lipschitz around $(\bar{x}, \bar{y})$ and $(\bar{x}, \bar{z})$ resp., where $\bar{z} \in G(\bar{x}) \cap \mathscr{C}$;
(iii) G is metrically regular around $(\bar{x}, \bar{z})$ relatively to $\Omega \times \mathscr{C}$.

We will use the arguments similar to those in Ref. 12 to establish the Lagrange multiplier rule for Q-minimal solutions.

Proposition 21.19. *Let Assumption (A) be satisfied. If $(\bar{x}, \bar{y})$ is a Q-minimal solution of (CP) then there exist $y^* \in \partial_A \Delta_{-Q}(0)$ such that*

$$0 \in D_A^* F(\bar{x}, \bar{y})(y^*) + D_A^* G(\bar{x}, \bar{z})(z^*) + N_A(\bar{x}; \Omega) \tag{21.10}$$

and

$$0 \in D_C^* F(\bar{x}, \bar{y})(y^*) + D_C^* G(\bar{x}, \bar{z})(z^*) + N_C(\bar{x}; \Omega), \tag{21.11}$$

where $z^* \in N_A(\bar{z}; \mathscr{C})$ and $z^* \in N_C(\bar{z}; \mathscr{C})$, respectively.

Proof. The proof is similar to that of Theorem 3.7 in Ref. 12. Observe that by Definition 21.12, $(\bar{x}, \bar{y})$ is a Q-minimal solution of (CP) iff

$$(F(\mathscr{A}) - \bar{y}) \cap (-Q) = \emptyset.$$

Proposition 21.10 yields that $(\bar{x}, \bar{y}, \bar{z})$ is a minimizer of the problem

$$\text{Minimize } q(x,y,z) \text{ subject to } (x,y) \in \mathrm{gr}\, F \text{ and } (x,z) \in (\Omega \times \mathscr{C}) \cap \mathrm{gr}\, G.$$

Put $q(x,y,z) = \Delta_{-Q}(y - \bar{y})$. Then by the Clarke penalization (Proposition 2.4.3 in Ref. 8), the metric regularity assumption and Proposition 2.6 in Ref. 12, for some integer $l > 0$ large enough, $(\bar{x}, \bar{y}, \bar{z})$ is an unconstrained local minimizer of

$$(x,y,z) \longrightarrow q(x,y,z) + ld((x,y); \mathrm{gr}\, F) + ld((x,z); \mathrm{gr}\, G) + ld((x,z); \Omega \times \mathscr{C}).$$

By the assertions (ii) and (iv) of Proposition 21.2, 0 is in the sum of the subdifferentials, that is, there exist

$$y_1^* \in \partial_A q(\bar{x},\bar{y},\bar{z}) = \partial_A \Delta_{-Q}(0) \subset \partial_C \Delta_{-Q}(0),$$

$$(x_2^*, y_2^*) \in l\partial_A d((\bar{x},\bar{y}); \text{gr } F) \subset N_A((\bar{x},\bar{y}); \text{gr } F)$$

$$(x_3^*, z_3^*) \in l\partial_A d((\bar{x},\bar{z}); \text{gr } G) \subset N_A((\bar{x},\bar{z}); \text{gr } G)$$

and

$$(x_4^*, -z_4^*) \in l\partial_A d((\bar{x},\bar{z}); \Omega \times \mathscr{C}) \subset N_A((\bar{x},\bar{z}); \Omega \times \mathscr{C})$$

such that

$$0 = x_2^* + x_3^* + x_4^*, 0 = y_1^* + y_2^* \text{ and } 0 = z_3^* + z_4^*.$$

Putting $y^* = y_1^* = -y_2^*$ and $z^* = z_4^* = -z_3^*$, we obtain

$$0 \in D_A^* F(\bar{x},\bar{y})(y^*) + D_A^* G(\bar{x},\bar{z})(z^*) + N_A(\bar{x};\Omega)$$

and (21.10) holds. Finally, (21.11) follows from (21.10) and Proposition 21.2 (iv). □

The following Lagrange multiplier rule for several types of efficient solutions of (CP) reads as follows.

Theorem 21.20. *Let Assumption (A) be satisfied and let the notations be as in Proposition 21.19.*

(i) If $(\bar{x},\bar{y})$ is a weak efficient solution of (CP) or $(\bar{x},\bar{y})$ is a Hurwicz proper efficient solution of (CP) and the image $F(\mathscr{A})$ has a nonempty interior, then there exists $y^* \in K^+ \setminus \{0\}$ such that (21.10)–(21.11) hold.
(ii) If $(\bar{x},\bar{y})$ is a global Henig proper efficient solution (in particular, a positive proper efficient solution, or a Henig proper efficient solution) of (CP), then there exists $y^* \in K^{+i}$ such that (21.10)–(21.11) hold.
(iii) If $(\bar{x},\bar{y})$ is a super efficient solution of (CP) and K has a bounded base, then there exists $y^* \in \text{int}K^+$ such that (21.10)–(21.11) hold.

This theorem can be proved in the same way as for Theorem 21.17 (but apply Proposition 21.19 in place of Proposition 21.16) and the proof is then omitted.

Remark 21.21. (i) The above version of Lagrange multiplier rule are known for weak efficient solutions in Ref. 12, for strong efficient solutions or positive proper efficient solutions in Ref. 19 (for completeness we include these cases in Theorem 21.20) and are new in cases of Hurwicz proper efficient solution, global Henig proper efficient solution, Henig proper efficient solution and super efficient solutions. We note that for weak efficient solutions to (CP), there have also been obtained Lagrange–Kuhn–Tucker multipliers in terms of subdifferential of set-valued maps introduced by Sawaragi and Tanino (Ref. 39). Recently, the Lagrange multiplier rule for super efficient solutions to a constrained set-valued optimization problem has been obtained by Bao and Mordukhovich in Asplund spaces (Ref. 2) and

by Huang in Banach spaces (Ref. 23). The Bao–Mordukhovich technique depends on the use of extremal principle of variational analysis while Huang's technique is based on exploiting a property of the normal cone due to Clarke. In contrast to our work, they considered super efficient solutions of the set-valued optimization problem the constraints set of which contains only the geometric constraint $x \in \Omega$.
(ii) Recently, using the ideas surrounding the Dubovitskii–Milyutin approach, Isac and Khan obtained Lagrange multiplier rule for Pareto efficient solutions, proper efficient solutions, weak efficient solutions and strong efficient solutions associated with contingent derivative (Ref. 26). Note that along with (CP), Isac and Khan considered also the case when the constraint set contains the condition $x \in H(x)$ with H being some set-valued map.
(iii) Other versions of the Lagrange multiplier rule in terms of various derivatives or coderivatives can be found in Refs. 7, 10, 15, 16, 28–30, 34, 39.
(iv) One can use the techniques of Ref. 12 further to obtain other optimality conditions for (CP) for instance in terms of coderivatives of the map (F,G), where $F,G)(x) = F(x) \times G(x)$.

Using a characterization of a Benson proper efficient point in a separable or reflexive Banach space recently obtained in Ref. 36, we derive from Theorems 21.14, 21.17 and 21.20 the Lagrange claim, the Fermat rule and the Lagrange multiplier rule for Benson proper efficient solutions.

Corollary 21.22. *Assume that Y is a separable Banach space, or that Y is a reflexive Banach space and K has a base.*

(i) Let $\overline{x} \in X$ and $(\overline{x},\overline{y}) \in grF$. Assume that F has a closed graph and $F - \overline{y}$ is nearly K-subconvexlike on X, i.e., cl cone $(F(X) - \overline{y} + K)$ is convex. Then $(\overline{x},\overline{y})$ is a Benson proper efficient solution of (P) iff there exists $y^* \in K^{+i}$ such that (21.5)–(21.6) hold with $Q = \{y : \varphi(y) < 0\}$ and only if there exists $y^* \in K^{+i}$ such that (21.8)–(21.9) hold.
(ii) Let $\overline{x} \in \mathscr{A}$ and $(\overline{x},\overline{y}) \in grF$. Let the assumption (A) be satisfied and let the notations be as in Proposition 9. Assume in addition that $F - \overline{y}$ is nearly K-subconvexlike on $\mathscr{A}$, i.e., cl cone $(F(\mathscr{A}) - \overline{y} + K)$ is convex. If $(\overline{x},\overline{y})$ is a Benson efficient solution of (CP), then there exists $y^* \in K^{+i}$ such that (21.10)–(21.11) hold.

Proof. It has been established (Corollaries 4.2 and 4.5, Ref. 36) that if Y is a separable Banach space, or Y is a reflexive Banach space and K has a base and if $F - \overline{y}$ is nearly K-subconvexlike on X (respectively, on $\mathscr{A}$), then $(\overline{x},\overline{y})$ being Benson proper efficient solution of (P) (respectively, of (CP)) also is a positive proper efficient solution. The assertion follows from this fact and Theorems 21.14, 21.17 and 21.20.

□

Remark 21.23. Recently, Benson proper efficient solutions to set-valued optimization problems have received more attention, and several optimality conditions for them were obtained under assumptions on generalized convexity of set-valued data such as convexlikeness, subconvexlikeness, near convexlikeness and near subconvexlikeness (see Refs. 36, 37, 40 for references). Near subconvexlikeness firstly presented in Ref. 40 and used also in Ref. 37 is the weakest convexity among the above four kinds of generalized convexity. Under the assumptions on the near subconvexlikeness of the objective and constraints, it has been established in Ref. 37 that a Benson proper efficient solution of (CP) can be expressed in terms of saddle points defined in a suitable sense. To our knowledge, the Lagrange claim, the Fermat rule and the Lagrange multiplier rule for Benson proper efficient solutions do not appear yet in the literature.

Acknowledgment This research was initiated during the author's stay at the Institute of Applied Mathematics of the University of Erlangen-Nürnberg under the Georg Forster grant of the Alexander von Humboldt Foundation. The author thanks Professor J. Jahn for hospitality, advice, and help in her work.

References

1. J.P. Aubin, I. Ekeland, *Applied Nonlinear Analysis*, Wiley, New York, 1984.
2. T.Q. Bao, B.S. Mordukhovich, *Necessary conditions for super minimizers in constrained multiobjective optimization*, manuscript (2008).
3. H.P. Benson, *An improved definition of proper efficiency for vector minimization with respect to cones,* J. Math. Anal. Appl., 71 (1978) 232–241.
4. J.M. Borwein, *The geometry of Pareto efficiency over cones,* Mathematische Operationforschung and Statistik, Serie Optimization, 11(1980), 235–248.
5. J.M. Borwein, D. Zhuang, *Super efficiency in vector optimization,* Trans. Amer. Math. Society, 338(1993), 105–122.
6. G. Bouligand, *Sur l'existence des demi-tangentes* à une courbe de Jordan, Fundamenta Mathematicae 15 (1030), 215–218.
7. G.Y. Chen, J. Jahn, *Optimality conditions for set-valued optimization problems*, Math. Methods Oper. Res. 48 (1998) 1187–1200.
8. F.H. Clarke, *Optimization and Nonsmooth Analysis*, Wiley, New York, 1983.
9. H.W. Corley, *Optimality conditions for maximization of set-valued functions,* J. Optim. Theory Appl. 58 (1988) 1–10.
10. G.P. Crepsi, I. Ginchev, M. Rocca, *First-order optimality conditions in set-valued optimization*, Math. Meth. Oper. Res. 63, (2006), 87–106.
11. M. Durea, *First and second-order Lagrange claims for set-valued maps*, J. Optim. Theory Appl., 133 (2007), 111–116.
12. B. El Abdouni, L. Thibault, *Optimality conditions for problems with set-valued objectives*, J. Applied Analysis 2 (1996) 183–201.
13. F. Flores-Bazán, *Optimality conditions in non-convex set-valued optimization*, Math. Meth. Oper. Res. 53 (2001), 403–417.
14. X.H. Gong, *Optimality conditions for Henig and globally proper efficient solutions with ordering cone has empty interior*, J. Math. Anal. Appl. 307 (2005) 12–37.
15. X.H. Gong, H.B. Dong, S.Y. Wang, *Optimality conditions for proper efficient solutions of vector set-valued optimization,* J. Math. Anal. Appl. 284 (2003) 332–350.
16. A. Gotz, J. Jahn, *The Lagrange multiplier rule in set-valued optimization*, SIAM J. Optim. 10 (1999), 331–344.
17. E. Goursat, *Courses d'Analyse Mathématiques,* Courcier, Paris (1813).
18. A. Guerraggio. E. Molno, A. Zaffaroni, *On the notion of proper efficiency in vector optimization*, J. Optim. Theory Appl., 82 (1994), 1–21.

19. T. X. D. Ha, *Lagrange multipliers for set-valued optimization problems associated with coderivatives*, J. Math. Anal. Appl. 311 (2005) 647–663.
20. R. Hartley, *On cone efficiency, cone convexity, and cone compactness*, SIAM Journal on Applied Mathematics 34 (1978), 211–222.
21. M.I. Henig, *Proper efficiency with respect to the cones*, J. Optim. Theory Appl. 36 (1982), 387–407.
22. J. B. Hiriart-Urruty, *New concepts in nondifferentiable programming*, Bull. Soc. Math. France 60 (1979) 57–85.
23. H. Huang, *The Lagrange multiplier rule for super efficiency in vector optimization*, J. Math. Anal. Appl. 342 (2008), no. 1, 503–513.
24. L. Hurwicz, *Programming in linear spaces*, Edited by K.J. Arrow, L. Hurwicz, H. Uzawa, Standford University Press, Stanford, CA, 1958.
25. A.D. Ioffe, *Approximate subdifferentials and applications 3: Metric theory*, Mathematica 36 (1989), 1–38.
26. G. Isac, A.A. Khan, *Dubovitskii-Milyutin approach in set-valued optimization*, SIAM J. Control Optim. 47 (2008), no. 1, 144–162.
27. J. Jahn, *Mathematical vector optimization in partially ordered linear spaces*, Peter Lang, Frankfurt, 1986.
28. J. Jahn, *Vector Optimization*, Springer-Verlag, Berlin, 2004.
29. J. Jahn, A. A. Khan, *Generalized contingent epiderivatives in set-valued optimization: optimality conditions*, Numer. Funct. Anal. Optim. 23 (2002) 807–831.
30. J. Jahn, R. Rauh, *Contingent epiderivatives and set-valued optimization*, Math. Methods Oper. Res. 46 (1997) 193–211.
31. M.A. Krasnoselski, *Positive Solutions of Operator Equations*, Nordhoff, Groningen, 1964.
32. J.L. Lagrange, *Théorie des Fonctions Analytiques*, Impr.de la République, Paris, 1797.
33. S. J. Li, X. Q. Yang, G.Y. Chen, *Nonconvex vector optimization of set-valued mappings*, J. Math. Anal. Appl. 283 (2003) 337–350.
34. D.T. Luc, *Theory of Vector Optimization*, Springer-Verlag, Berlin, 1989.
35. E.K. Makarov, N.N. Rachkovski, *Unified representation of proper efficiency by means of dilating cones*, J. Optim. Theory Appl., 101 (1999),141–165.
36. J.-H. Qiu, *Dual characterization and scalarization for Benson proper efficiency*, SIAM J. Optim., 19 (2008), 144–162.
37. P.H. Sach, *Nearly subconvexlike set-valued maps and vector optimization problems*, J. Optim. Theory Appl., 119 (2003), 335–356.
38. A. Taa, *Set-valued derivatives of multifunctions and optimality conditions*, Numer. Func. Anal. Optimiz. 19 (1998), 121–140.
39. A. Taa, *Subdifferentials of multifunctions and Lagrange multipliers for multiobjective optimization*, J. Math. Anal. Appl. 283 (2003) 398–415.
40. X.M. Yang, D. Li, S.Y. Wang, *Nearly subconvexlikeness in vector optimization with set- valued functions*, J. Optim. Theory Appl., 110 (2001), 413–427.
41. A. Zaffaroni, *Degrees of efficiency and degrees of minimality*, SIAM J. Control Optim. 42 (2003), 1071–1086.
42. X.Y. Zheng, *Scalarization of Henig proper efficient points in a normed space*, J. Optim. Theory Appl., 105 (2000), 233–247.
43. X.Y. Zheng, *Proper efficiency in local convex topological vector spaces*, J. Optim. Theory Appl., 94 (1997), 469–486.
44. X.Y. Zheng, K.F. Ng, *Fermat rule for multifunctions in Banach spaces*, Math. Program, 104 (2005) 69–90.
45. X.Y. Zheng, K.F. Ng, *The Lagrange multiplier rule for multifunctions in Banach spaces*, SIAM J. Optim., 17 (2006), 1154–1175.
46. D. Zhuang, *Density results for proper efficiency*, SIAM J. Control Optim., 32 (1994), 51–58.

Chapter 22
Mean Value Theorems for the Scalar Derivative and Applications

G. Isac and S.Z. Németh

Abstract In this paper, we present some mean value theorems for the scalar derivatives. This mathematical tool is used to develop a new method applicable to the study of existence of nontrivial solutions of complementarity problems.

22.1 Introduction

This paper is related to the solvability of complementarity problems. The main goal of complementarity theory is the study of complementarity problems from several points of view such as: the existence of solutions; the existence of a nontrivial solution; the properties of the solution set, if several solutions exist; and the numerical approximation of some solutions.

The complementarity problems are mathematical models for the study of some practical problems considered in optimization, economics and engineering [4, 16, 17, 18, 19]. It is also well known that there exist deep relations between complementarity problems and variational inequalities [4, 16, 17]. In several practical problems, generally considered in engineering, such as robotics, we have complementarity problems which have the trivial solution. Because of this fact, it is very important to know if such a complementarity problem has nontrivial solutions. Mathematically speaking, it is not so easy to know under what conditions a complementarity problem which has the trivial solution has some nontrivial solutions too. Generally speaking, such a problem can be studied (if some assumptions are satisfied) by the topological degree, but the topological degree is a heavy mathematical tool. In this

G. Isac
Department of Mathematics and Computer Science, Royal Military College of Canada, P.O. Box 17000, STN Forces Kingston, Ontario K7K 7B4, Canada.

S.Z. Németh
The University of Birmingham, School of Mathematics, The Watson Building, Edgbaston, B15 2TT Birmingham, United Kingdom, e-mail: nemeths@for.mat.bham.ac.uk

P.M. Pardalos et al. (eds.), *Nonlinear Analysis and Variational Problems*, Springer Optimization and Its Applications 35, DOI 10.1007/978-1-4419-0158-3_22,

paper, we try to develop a new mathematical tool applicable to the study of the existence of multiple solutions of a complementarity problem.

The method presented in this paper can be considered as an alternative of the method based on the topological degree. Certainly, our method must be improved and adapted to other problems too. We note that our method is based on the notion of scalar derivative. Regarding the scalar derivative, the reader is referred to [33, 20, 21, 22, 25, 26, 27]. Precisely, our method is based on some mean value theorems for the scalar derivatives. It seems that no similar ideas were used and developed in complementarity theory before. Some of the examples are related to homogeneous mappings. The homogeneous complementarity is a very important special class of complementarity, and our examples could create a link with this topic. The scalar derivatives are related to almost all major derivative notions, have several applications, and constitute the topic of our book [33]. The topic of mean value theorems for scalar derivatives is considered for the first time here.

Finally, we note that our mean value theorems are based on a Rolle type theorem for scalar derivatives. Therefore, this paper can also be considered as a new direction to extend the classical Rolle's theorem to infinite-dimensional vector spaces. Generalizations of Rolle's theorem to infinite-dimensional vector spaces have been considered in several papers (see [7, 24] and the references therein).

22.2 Preliminaries

We use the following notations: $\mathbb{R}$ is the field of real numbers, $\overline{\mathbb{R}} = \mathbb{R} \cup \{-\infty, +\infty\}$ and $\mathbb{R}_+ = \{a \in \mathbb{R} \mid a \geq 0\}$.

We denote by $(H, \langle\cdot,\cdot\rangle)$ a real Hilbert space and by $(\mathbb{R}^n, \langle\cdot,\cdot\rangle)$ the n-dimensional Euclidean space.

A closed pointed convex cone in H will be a closed non-empty subset $K \subset H$ satisfying the following properties:

1. $K + K \subseteq K$
2. $\lambda K \subseteq K$, for any $\lambda \in \mathbb{R}_+$,
3. $K \cap (-K) = \{0\}$.

For example $\mathbb{R}^n_+ = \{(x_1, \cdots, x_n) \in \mathbb{R}^n \mid x_1 \geq 0, \cdots, x_n \geq 0\}$ is a closed pointed convex cone called the non-negative orthant of $\mathbb{R}^n$.

If K is a given closed pointed convex cone in H, the dual cone of K is

$$K^* = \{y \in H \mid \langle x, y\rangle \geq 0 \text{ for any } x \in K\}.$$

We recall that a linear operator $T : H \to H$ is *skew-symmetric* if for any $x, y \in H$ we have

$$\langle Tx, y\rangle + \langle Ty, x\rangle = 0.$$

It is known that if $T : H \to H$ is a linear mapping, then T is skew-symmetric if and only if $\langle Tx, x\rangle = 0$ for any $x \in H$.

Let $(H, \langle\cdot,\cdot\rangle)$ be a Hilbert space, $K \subset H$ a closed convex cone and $f : H \to H$ a mapping. The nonlinear complementarity problem defined by f and K is

$$NCP(f,K) : \begin{cases} \text{find } x_0 \in K \text{ such that} \\ f(x_0) \in K^* \text{ and } \langle x_0, f(x_0)\rangle = 0. \end{cases}$$

If $\Omega \subset H$ is a closed convex set, the variational inequality defined by f and Ω is

$$VI(f,\Omega) : \begin{cases} \text{find } x_0 \in \Omega \text{ such that} \\ \langle f(x_0), y - x_0\rangle \geq 0 \text{ for any } y \in \Omega. \end{cases}$$

Finally, we suppose that the classical concepts of Fréchet and Gâteaux derivatives are known.

22.3 Scalar Derivatives and Scalar Differentiability

The results of this section are proved in [27].

Definition 22.1. Let $C_1, C_2 \subseteq H$ such that 0 is a non-isolated point of C_1 and x_0 a non-isolated point of C_2. Let $f : C_2 \to H$. The limit

$$\underline{f}^{\#}(x_0, C_1) = \liminf_{\substack{x \to x_0 \\ x - x_0 \in C_1}} \frac{\langle f(x) - f(x_0), x - x_0\rangle}{\|x - x_0\|^2}$$

is called the *lower scalar derivative of the mapping f in x_0 in the direction of C_1*. Taking lim sup in place of lim inf, we can define the *upper scalar derivative of the mapping f at x_0 in the direction of C_1* similarly. It will be denoted by $\overline{f}^{\#}(x_0, C_1)$. If $C_1 = H$ or $C_1 = C_2$ and $x_0 = 0$, then without confusion, we can omit the phrase "in the direction of C_1" from the definitions. In this case, we omit C_1 from the corresponding notations.

Definition 22.2. Let $C_1, C_2 \subseteq H$ such that 0 is a non-isolated point of C_1 and x_0 a non-isolated point of C_2. Consider the mapping $f : C_2 \to H$. If the limit

$$\lim_{\substack{x \to x_0 \\ x - x_0 \in C_1}} \frac{\langle f(x) - f(x_0), x - x_0\rangle}{\|x - x_0\|^2} =: f^{\#}(x_0, C_1)$$

exists (here $\|x - x_0\|^2 = \langle x - x_0, x - x_0\rangle$), then it is called the *scalar derivative of the mapping f in x_0 in the direction of C_1*. If $f^{\#}(x_0, C_1) \in \mathbb{R}$, then f is said to be *scalarly differentiable in x_0 in the direction of C_1*. If C_2 is open and $f^{\#}(x, C_1) \in \mathbb{R}$ exists for every $x \in C_2$, then f is said to be *scalarly differentiable on C_2 in the direction of C_1*, with the *scalar derivative* $f^{\#}(\cdot, C_1) : C_2 \to \mathbb{R}$. If $C_1 = H$ or $C_1 = C_2$ and $x_0 = 0$, then without confusion, we can omit the phrase "in the direction of C_1" from the definitions. In this case, we omit C_1 from the corresponding notations.

It follows from this definition that both the set of mappings scalarly differentiable in x_0 and the set of mappings scalarly differentiable on H form linear spaces.

Theorem 22.3. *The linear mapping $A : H \to H$ is scalarly differentiable on H if and only if it is of the form $A = B + cI$ with B a skew-symmetric linear mapping, I the identity of H, and c a real number.*

Theorem 22.4. *Suppose that $f : H \to H$ is both Gâteaux differentiable and scalarly differentiable in x_0. Then we have for the Gateaux differential $\delta f(x_0)$ of f at x_0 the relation*

$$\delta f(x_0) = B + f^{\#}(x_0)I,$$

with $B : H \to H$ linear and skew-symmetric.

22.3.1 Computational Formulae for the Scalar Derivatives

We recall that a subset of a Hilbert space is called a cone if it is invariant under multiplication by positive scalars, and a cone is called a convex cone if it is invariant under addition. The first two theorems of this section show that the lower and upper scalar derivatives of a Frechét differentiable mapping in a point x are equal to the lower and upper scalar derivatives in 0 of its differential in x, respectively.

Theorem 22.5. *Let $x \in H$ and $K \subseteq H$ a closed cone. If $f : H \to H$ is Frechét differentiable in x, with the differential $df(x)$, then*

$$\underline{f}^{\#}(x,K) = \underline{df(x)}^{\#}(0,K),$$

$$\overline{f}^{\#}(x,K) = \overline{df(x)}^{\#}(0,K).$$

Theorem 22.6. *Let $K \subseteq H$ be a closed convex cone with non-empty interior and x an interior point of K. If $f : K \to H$ is Frechét differentiable in x, with the differential $df(x)$, then*

$$\underline{f}^{\#}(x,K) = \underline{df(x)}^{\#}(0,K),$$

$$\overline{f}^{\#}(x,K) = \overline{df(x)}^{\#}(0,K).$$

The next two theorems give the computational formulae for the lower and upper scalar derivatives of a positively homogeneous mapping in 0.

Theorem 22.7. *Let $K \subseteq H$ be a closed cone. If $\Phi : H \to H$ is positively homogeneous, then*

$$\underline{\Phi}^{\#}(0,K) = \inf_{\substack{\|u\|=1 \\ u\in K}} \langle \Phi(u), u\rangle,$$

$$\overline{\Phi}^{\#}(0,K) = \sup_{\substack{\|u\|=1 \\ u\in K}} \langle \Phi(u), u\rangle.$$

Theorem 22.8. *Let $K \subseteq H$ be a closed convex cone. If $\Phi : K \to H$ is positively homogeneous, then*

$$\underline{\Phi}^{\#}(0) = \inf_{\substack{\|u\|=1 \\ u\in K}} \langle \Phi(u), u\rangle,$$

$$\overline{\Phi}^{\#}(0) = \sup_{\substack{\|u\|=1 \\ u\in K}} \langle \Phi(u), u\rangle.$$

Theorems 22.5 and 22.7 imply:

Theorem 22.9. *Let $x \in H$ and $K \subseteq H$ a closed cone. If $f : H \to H$ is Frechét differentiable in x, with the differential $df(x)$, then*

$$\underline{f}^{\#}(x,K) = \inf_{\substack{\|u\|=1 \\ u\in K}} \langle df(x)(u), u\rangle,$$

$$\overline{f}^{\#}(x,K) = \sup_{\substack{\|u\|=1 \\ u\in K}} \langle df(x)(u), u\rangle.$$

Theorems 22.6 and 22.8 imply:

Theorem 22.10. *Let $K \subseteq H$ be a closed convex cone with non-empty interior and x an interior point of K. If $f : K \to H$ is Frechét differentiable in x, with the differential $df(x)$, then*

$$\underline{f}^{\#}(x,K) = \inf_{\substack{\|u\|=1 \\ u\in K}} \langle df(x)(u), u\rangle,$$

$$\overline{f}^{\#}(x,K) = \sup_{\substack{\|u\|=1 \\ u\in K}} \langle df(x)(u), u\rangle.$$

22.4 Mean Value Theorems

From now on, let $(H, \langle\cdot,\cdot\rangle)$ be a Hilbert space and $\|\cdot\|$ the norm generated by $\langle\cdot,\cdot\rangle$. For $A \subseteq H$ denote by $\operatorname{conv} A$, $\operatorname{cone} A$, and $\operatorname{ri} A$ the convex hull, the conic hull (smallest closed convex cone containing A), and the relative interior of A, respectively. For $a,b,c \in H$ denote $[a,b] = \operatorname{conv}\{a,b\}$, $]a,b] = \operatorname{conv}\{a,b\} \setminus \{a\}$, $]a,b[= \operatorname{ri}[a,b]$, $abc_{\bar{\triangle}} = \operatorname{conv}\{a,b,c\} \setminus \{a\}$, and $abc_{\triangle} = \operatorname{ri}\operatorname{conv}\{a,b,c\}$.

Theorem 22.11 (Rolle). *Let $K \subseteq H$ be a closed cone, $a,b \in H$ distinct points such that $b-a \in K$, and $f : H \to H$ a mapping continuous on $[a,b]$ and differentiable on $]a,b[$ with $\langle f(b)-f(a), b-a\rangle = 0$. Then, there exists a $\xi \in]a,b[$ such that*

$$\underline{f}^{\#}(\xi,K) \le 0 \le \overline{f}^{\#}(\xi,K). \tag{22.1}$$

Proof. Let $\phi : [0,1] \to \mathbb{R}$ defined by

$$\phi(t) = \frac{\langle f(a+t(b-a)), b-a\rangle}{\|b-a\|^2}.$$

It is easy to see that $\phi(0) = \phi(1)$. Therefore, Rolle's theorem for real functions implies that $\phi'(\tau) = 0$ for some $\tau \in]0,1[$. Let $\xi = a + \tau(b-a)$. Then, $\xi \in]a,b[$ and

$$0 = \phi'(\tau) = \frac{\langle df(\xi)(b-a), b-a\rangle}{\|b-a\|^2} \tag{22.2}$$

On the other hand, by Theorem 17 of [27], we have

$$\underline{f}^{\#}(\xi, K) \le \langle df(\xi)(h), h\rangle \le \overline{f}^{\#}(\xi, K), \tag{22.3}$$

where

$$h = \frac{b-a}{\|b-a\|} \in K.$$

Relations (22.3), (22.2) and the linearity of the Frechét differential imply (22.1). □

Theorem 22.12 (Cauchy's mean value theorem). *Let $K \subseteq H$ be a closed cone, $a, b \in H$ distinct points such that $b - a \in K$, $f, g : H \to H$ mappings continuous on $[a,b]$ and differentiable on $]a,b[$ such that $\langle g(b) - g(a), b-a\rangle \neq 0$. Then, there exists a $\xi \in]a,b[$ such that*

$$\underline{f}^{\#}(\xi, K) \le \frac{\langle f(b) - f(a), b-a\rangle}{\langle g(b) - g(a), b-a\rangle} \overline{g}^{\#}(\xi, K) \tag{22.4}$$

and

$$\frac{\langle f(b) - f(a), b-a\rangle}{\langle g(b) - g(a), b-a\rangle} \underline{g}^{\#}(\xi, K) \le \overline{f}^{\#}(\xi, K). \tag{22.5}$$

Proof. Let $h(x) = f(x) - \lambda g(x)$, where

$$\lambda = \frac{\langle f(b) - f(a), b-a\rangle}{\langle g(b) - g(a), b-a\rangle}. \tag{22.6}$$

Then, it is easy to verify that $\langle h(b) - h(a), b-a\rangle = 0$. By using Theorem 22.11 with h replacing f, we have that

$$\underline{h}^{\#}(\xi, K) \le 0 \le \overline{h}^{\#}(\xi, K). \tag{22.7}$$

for some $\xi \in]a,b[$. But, from the definition of the lower and upper scalar derivatives it follows that

$$\underline{h}^{\#}(\xi, K) \ge \underline{f}^{\#}(\xi, K) - \lambda \overline{g}^{\#}(\xi, K) \tag{22.8}$$

and

$$\overline{h}^{\#}(\xi, K) \le \overline{f}^{\#}(\xi, K) - \lambda \underline{g}^{\#}(\xi, K). \tag{22.9}$$

Relations (22.7), (22.8) and (22.9), with λ given by (22.6), imply (22.4) and (22.5).

□

Theorem 22.13 (Mean value theorem). *Let $K \subseteq H$ be a closed cone, $a, b \in H$ distinct points with $b - a \in K$, and $f : H \to H$ a mapping continuous on $[a,b]$ and differentiable on $]a,b[$. Then, there exists a $\xi \in]a,b[$ such that*

$$\underline{f}^{\#}(\xi, K) \leq \frac{\langle f(b) - f(a), b - a \rangle}{\|b - a\|^2} \leq \overline{f}^{\#}(\xi, K).$$

Proof. Follows from Theorem 22.12 with $g = I$, where $I : H \to H$ is the identity mapping of H, i.e., $I(x) = x$ for all $x \in H$.

□

Remark 22.14. Let H be a Hilbert space, $f : H \to H$ and $p \in H \setminus \{0\}$. Then, $f^{\#}(\xi, \mathrm{cone}\{p\})$ exists for all $x \in H$.

Corollary 22.15. *Let $K \subseteq H$ be a closed convex cone, $a, b \in H$ distinct points, and $f : H \to H$ a mapping continuous on $[a,b]$ and differentiable on $]a,b[$. Then, there exists a $\xi \in]a,b[$ such that*

$$f^{\#}(\xi, \mathrm{cone}\{b - a\}) = \frac{\langle f(b) - f(a), b - a \rangle}{\|b - a\|^2}.$$

22.5 Applications to Complementarity Problems

In this section, we present some applications to complementarity problems. Let $(H, \langle \cdot, \cdot \rangle)$ be a Hilbert space, $K \subset H$ a closed convex cone and $f : H \to H$ a mapping. We consider the general nonlinear complementarity problem defined by f and K

$$NCP(f,K) : \begin{cases} \text{find } x_0 \in K \text{ such that} \\ f(x_0) \in K^* \text{ and } \langle x_0, f(x_0) \rangle = 0. \end{cases}$$

Related to this problem, we have the following result.

Theorem 22.16. *Let $K \subseteq H$ be a closed convex cone, K^* its dual cone, and $f : H \to H$. If there are $p_0, q_0 \in K \setminus \{0\}$ such that f is continuous on $0p_0q_{0\bar{\triangle}}$, differentiable on $0p_0q_{0\triangle}$, $f(0p_0q_{0\bar{\triangle}}) \subseteq K^*$, and*

$$\langle f(0), p \rangle + \|p\|^2 f^{\#}(p, \mathrm{cone}\{p_0\})) \geq 0,$$

$$\langle f(0), q \rangle + \|q\|^2 f^{\#}(q, \mathrm{cone}\{q_0\}) \leq 0,$$

for all $p \in]0, p_0[$, $q \in]0, q_0[$, respectively, then for all $p \in]0, p_0[$ and $q \in]0, q_0[$ the nonlinear complementarity problem $NCP(f,K)$ has a solution in $[p,q]$.

Proof. Let $p \in]0, p_0[$ and $q \in]0, q_0[$. Then, by the convexity of K, $p, q \in K$. Hence, by Corollary 22.15, there exists a $\xi \in]0, p[$ and $\mu \in]0, q[$ such that

$$\langle f(p), p\rangle = \langle f(0), p\rangle + \|p\|^2 f^{\#}(\xi, \mathrm{cone}\{p\}) = \langle f(0), p\rangle + \|p\|^2 f^{\#}(\xi, \mathrm{cone}\{p_0\})$$

and

$$\langle f(q), q\rangle = \langle f(0), q\rangle + \|q\|^2 f^{\#}(\mu, \mathrm{cone}\{q\}) = \langle f(0), q\rangle + \|q\|^2 f^{\#}(\mu, \mathrm{cone}\{q_0\}).$$

By the continuity of the function $x \mapsto \langle f(x), x\rangle$ on $0p_0q_{0_{\bar{\Delta}}}$, there exists an $x^* \in [p, q]$ such that $\langle f(x^*), x^*\rangle = 0$. By the convexity of K, $x^* \in K$. Hence, we have $f(x^*) \in f(0p_0q_{0_{\bar{\Delta}}}) \subseteq K^*$. Therefore, x^* is a solution of the nonlinear complementarity problem $NCP(f, K)$.

□

Corollary 22.17. *Let $K \subseteq H$ be a closed convex cone, K^* its dual cone, and $f : H \to H$ a mapping with $f(0) = 0$. If there are $p_0, q_0 \in K \setminus \{0\}$ such that f is continuous on $0p_0q_{0_{\bar{\Delta}}}$, differentiable on $0p_0q_{0_{\Delta}}$, $f(0p_0q_{0_{\bar{\Delta}}}) \subseteq K^*$, and $f^{\#}(p, \mathrm{cone}\{p_0\})) \geq 0$, $f^{\#}(q, \mathrm{cone}\{q_0\}) \leq 0$ for all $p \in]0, p_0[$, $q \in]0, q_0[$, respectively, then for all $p \in]0, p_0[$ and $q \in]0, q_0[$ the nonlinear complementarity problem $NCP(f, K)$ has a solution in $[p, q]$.*

The proof of the following lemma is a straightforward computation and is left to the reader.

Lemma 22.18. *Denote by $d(\cdot, L)^2$ the distance function from a straight line L in H. If $L = \{x_0 + \lambda v \mid \lambda \in \mathbb{R}\}$, where $x_0 \in H$ and $v \in H \setminus \{0\}$ are constant vectors, then*

$$d(x, L)^2 = \|x - x_0\|^2 - \frac{\langle v, x - x_0\rangle^2}{\|v\|^2},$$

and consequently $d(\cdot, L)^2$ is a differentiable function.

Example 22.19. Let $K \subseteq H$ be a closed convex cone and K^* its dual cone. Denote by $\leq$ the order induced by the cone K^*. Let $a, b \in H$ with $a \neq b$ and $a \leq b$, and $p_0, q_0 \in K \setminus \{0\}$ such that $a \leq p_0 \leq b$ and $a \leq q_0 \leq b$. Denote $P_0 = \mathrm{cone}\{p_0\}$ and $Q_0 = \mathrm{cone}\{q_0\}$. Consider the mapping $f : H \to H$ defined by

$$f(x) = \begin{cases} \dfrac{d(x, Q_0)^2 (x - a) - d(x, P_0)^2 (x - b)}{d(x, Q_0)^2 + d(x, P_0)^2} & \text{if } x \in H \setminus \{0\}, \\ 0 & \text{if } x = 0. \end{cases}$$

Then, by using Lemma 22.18, it is easy to verify that f satisfies all conditions of Corollary 22.17. Therefore, for all $p \in]0, p_0[$ and $q \in]0, q_0[$ the nonlinear complementarity problem $NCP(f, K)$ has a solution in $[p, q]$. The details are left to the reader.

Example 22.20. Let $K \subseteq H$ be a closed convex cone. Denote by $\leq$ the order induced by the cone K^*. Let $c \in H$ and $A : H \to H$ a continuous linear mapping. Let $a, b \in H$ with $a \neq b$ and $a \leq c \leq b$, and $p_0, q_0 \in K \setminus \{0\}$ such that $a - c \leq A(p_0) \leq b - c$ and $a - c \leq A(q_0) \leq b - c$. Denote $P_0 = \text{cone}\{p_0\}$ and $Q_0 = \text{cone}\{q_0\}$. Consider the mapping $f : H \to H$ defined by

$$f(x) = \begin{cases} \dfrac{d(x,Q_0)^2(A(x)+c-a) - d(x,P_0)^2(A(x)+c-b)}{d(x,Q_0)^2 + d(x,P_0)^2} & \text{if } x \in H \setminus \{0\}, \\ 0 & \text{if } x = 0. \end{cases} \tag{22.10}$$

Then, for all $p \in]0, p_0[$ and $q \in]0, q_0[$, the nonlinear complementarity problem $NCP(f, K)$ has a solution in $[p, q]$.

Proof. By using Lemma 22.18, we will prove that f satisfies all conditions of Corollary 22.17. It is enough to prove that for all $p \in]0, p_0[$ and $q \in]0, q_0[$ and $x \in [p, q]$ we have $0 \leq f(x)$. All other conditions of Corollary 22.17 can be verified easily and the details are left to the reader. Consider the affine mapping $h = A + c : H \to H$ defined by $h(x) = (A+c)(x) = A(x) + c$. Let $t \in]0, 1[$ such that $p = tp_0$. We have

$$a \leq h(0) \leq b \tag{22.11}$$

and

$$a \leq h(p_0) \leq b. \tag{22.12}$$

Multiplying equation (22.11) by $1 - t$ and equation (22.12) by t and summing up, we get

$$a \leq (1-t)h(0) + th(p_0) \leq b,$$

or equivalently

$$a \leq h(p) = h(tp_0) = h((1-t)0 + tp_0) = (1-t)h(0) + th(p_0) \leq b.$$

Hence,

$$a \leq h(p) \leq b. \tag{22.13}$$

It can be similarly showed that

$$a \leq h(q) \leq b. \tag{22.14}$$

Let $\tau \in [0, 1]$ such that $x = (1-\tau)p + \tau q$. Multiplying equation (22.13) by $1 - \tau$ and equation (22.14) by τ and summing up, we get

$$a \leq h(x) = h((1-\tau)p + \tau q) = (1-\tau)h(p) + \tau h(q) \leq b,$$

or equivalently $a \leq A(x) + c \leq b$. Hence, by using (22.10), we have $0 \leq f(x)$.

□

Corollary 22.21. *Let $K \subseteq H$ be a closed convex cone, K^* its dual cone, and $f : H \to H$ a mapping with $f(0) = 0$. If there is a $p_0 \in K$ such that f is continuous on $[0, p_0]$, differentiable on $]0, p_0[$, $f(]0, p_0]) \subseteq K^*$, and $f^\#(p, \text{cone}\{p_0\}) = 0$ for all $p \in]0, p_0[$, then every $p \in]0, p_0[$ is a solution of the nonlinear complementarity problem $NCP(f, K)$.*

Proof. It follows from Corollary 22.17 with $q_0 = p_0$ and $q = p$. □

Let $(\mathbb{R}^n, \langle \cdot, \cdot \rangle)$ be the n-dimensional Euclidean space and $\mathbb{R}^n_+$ its non-negative orthant. Let $f = (f_1, \cdots, f_n) : \mathbb{R}^n \to \mathbb{R}^n$ be a mapping. Consider the nonlinear complementarity problem $NCP(f, \mathbb{R}^n_+)$. In several papers such as [2, 5, 6, 8, 9, 10, 11, 12, 13, 14, 23, 28] the problem $NCP(f, \mathbb{R}^n_+)$ has been studied by numerical methods based on the concept of *NCP*-functions, introduced for the first time by A. Fischer [8]. In the following we will give a different kind of application of the notion of *NCP*-functions to the study of complementarity. First we recall some fact related to *NCP*-functions.

Definition 22.22. A function $\phi : \mathbb{R}^2 \to \mathbb{R}$ is called an *NCP*-function if satisfies the following condition: $\phi(a, b) = 0$ if and only if $a \geq 0$, $b \geq 0$ and $ab = 0$.

Examples

1. $\phi_{FB}(a, b) = \sqrt{a^2 + b^2} - (a + b)$, for any $a, b \in \mathbb{R}$ (Fischer–Burmeister).
2. $\phi_{NR}(a, b) = \min\{a, b\}$, for any $a, b \in \mathbb{R}$ (Natural–Residual).
3. $\phi_1(a, b) = -ab + \frac{1}{2}[\max\{0, a + b\}]^2$, for all $a, b \in \mathbb{R}$.
4. $\phi_2(a, b) = \sqrt{(a - b)^2 + \lambda ab} - a - b$, for all $a, b \in \mathbb{R}$ and $\lambda \in]0, 1]$.
5. $\phi_{MS}(a, b) = ab + \frac{1}{2\alpha}[(\max\{0, a - \alpha b\})^2 - a^2 + (\max\{0, b - \alpha a\})^2 - b^2$, for all $a, b \in \mathbb{R}$ and $\alpha > 1$.

The reader can find other information about *NCP*-functions in [2, 5, 6, 8, 9, 10, 11, 12, 13, 14, 23, 28].

The following lemma follows easily from Definition 22.22.

Lemma 22.23. *Let $f = (f_1, \cdots f_n) : \mathbb{R}^n \to \mathbb{R}^n$, $\phi : \mathbb{R}^2 \to \mathbb{R}$ an NCP-function, and $\Phi = (\Phi_1, \cdots, \Phi_n) : \mathbb{R}^n \to \mathbb{R}^n$ defined by $\Phi_i(x) = \phi(x_i, f_i(x))$. Then, x^* is a solution of the nonlinear complementarity problem $NCP(f, \mathbb{R}^n_+)$ if and only if $\Phi(x^*) = 0$.*

Proposition 22.24. *Let $\beta : \mathbb{R}^n \to \mathbb{R}$ a function with $\beta(x) > 0$ for all $x \in \mathbb{R}^n_+$, $f = (f_1, \cdots f_n) : \mathbb{R}^n \to \mathbb{R}^n$, $\phi : \mathbb{R}^2 \to \mathbb{R}$ an NCP-function, $\Phi = (\Phi_1, \cdots, \Phi_n) : \mathbb{R}^n \to \mathbb{R}^n$ defined by $\Phi_i(x) = \phi(x_i, f_i(x))$, $\Psi : \mathbb{R}^n \to \mathbb{R}$ defined by $\Psi(x) = \beta(x) \|\Phi(x)\|^2/2$, and $g : \mathbb{R}^n \to \mathbb{R}^n$ defined by $g(x) = \Psi(x)x$. Then, x^* is a nonzero solution of the nonlinear complementarity problem $NCP(f, \mathbb{R}^n_+)$ if and only if it is a nonzero solution of the nonlinear complementarity problem $NCP(g, \mathbb{R}^n_+)$.*

Proof. The vector x^* is a nonzero solution of the nonlinear complementarity problem $NCP(g, \mathbb{R}^n_+)$ if and only if $x^* \geq 0$, $g(x^*) \geq 0$, and $\langle x^*, g(x^*) \rangle = 0$. By the form of g, relation $g(x^*) \geq 0$ follows from $x^* \geq 0$. Relation $\langle x^*, g(x^*) \rangle = 0$ is equivalent to $\beta(x^*) \|x^*\|^2 \Psi(x^*) = 0$. But, since $\beta(x^*) > 0$ and $\|x^*\| \neq 0$, this is equivalent to $\Psi(x^*) = 0$. Therefore, by $\Psi(x^*) = \beta(x^*) \|\Phi(x^*)\|^2/2$ and Lemma 22.23, it follows that x^* is a nonzero solution of the nonlinear complementarity problem $NCP(f,\mathbb{R}^n_+)$ if and only if it is a nonzero solution of the nonlinear complementarity problem $NCP(g,\mathbb{R}^n_+)$.

Proposition 22.25. *Let H be a Hilbert space $K \subseteq H$ be a closed convex cone, $\rho : H \to \mathbb{R}$ be a function differentiable in $x \in H$, and $f : H \to H$ defined by $f(x) = \rho(x)x$. Then, we have*

$$\underline{f}^\#(x,K) = \rho(x) + \inf_{\substack{\|u\|=1 \\ u\in K}} [\langle \nabla\rho(x), u\rangle \langle x, u\rangle] \tag{22.15}$$

and

$$\overline{f}^\#(x,K) = \rho(x) + \sup_{\substack{\|u\|=1 \\ u\in K}} [\langle \nabla\rho(x), u\rangle \langle x, u\rangle]. \tag{22.16}$$

Proof. We only prove formula (22.15). Formula (22.16) can be proved similarly. By Theorem 22.9 we have

$$\begin{aligned}
f^\#(x,K) &= \inf_{\substack{\|u\|=1 \\ u\in K}} \left\langle \lim_{t\downarrow 0} \frac{\rho(x+tu)(x+tu)-\rho(x)x}{t}, u \right\rangle \\
&= \inf_{\substack{\|u\|=1 \\ u\in K}} \left\langle \lim_{t\downarrow 0} \frac{\rho(x+tu)(x+tu)-\rho(x)(x+tu)+\rho(x)(x+tu)-\rho(x)x}{t}, u \right\rangle \\
&= \inf_{\substack{\|u\|=1 \\ u\in K}} \left\langle \rho(x)u + \lim_{t\downarrow 0} \frac{\rho(x+tu)-\rho(x)}{t} \lim_{t\downarrow 0}(x+tu), u \right\rangle \\
&= \inf_{\substack{\|u\|=1 \\ u\in K}} \langle \rho(x)u + [d\rho(x)u]x, u\rangle = \rho(x) + \inf_{\substack{\|u\|=1 \\ u\in K}} [d\rho(x)(u)\langle x,u\rangle]. \qquad \square
\end{aligned}$$

Corollary 22.26. *Let $u \in H$ with $\|u\| = 1$, $\rho : H \to \mathbb{R}$ be a function differentiable in $x \in H$, and $f : H \to H$ defined by $f(x) = \rho(x)x$. Then,*

$$f^\#(x, \text{cone}\{u\}) = \rho(x) + \langle \nabla\rho(x), u\rangle\langle x, u\rangle.$$

Definition 22.27. A closed convex cone $K \subseteq H$ is called *subdual* if $K \subseteq K^*$, where K^* is the dual cone of the cone K.

Theorem 22.28. *Let $K \subseteq H$ be a subdual closed convex cone and $\rho : H \to \mathbb{R}$ a mapping with $\rho(x) \geq 0$ for all $x \in K \setminus \{0\}$. Define $f : H \to H$ by $f(x) = \rho(x)x$. If there is $p_0 \in K \setminus \{0\}$ such that f is continuous on $0p_0q_{0\bar{\triangle}}$, differentiable on $0p_0q_{0\triangle}$ (for example this is satisfied if ρ is continuous on $0p_0q_{0\bar{\triangle}}$ and differentiable on $0p_0q_{0\triangle}$),*

$$\rho(p) + \langle \nabla\rho(p), p\rangle \geq 0,$$

and

$$\rho(q) + \langle \nabla\rho(q), q\rangle \leq 0,$$

for all $p \in]0, p_0[$, $q \in]0, q_0[$, respectively, then for all $p \in]0, p_0[$ and $q \in]0, q_0[$ the nonlinear complementarity problem $NCP(f,K)$ has a solution in $[p,q]$.

Proof. It follows from Corollaries 22.17 and 22.26. □

Let $q \in H$ and $\rho : H \to \mathbb{R}$ a mapping differentiable on $]0,q[$ with bounded differential such that $\rho(x) \geq 0$ and

$$\rho(x) + \langle \nabla\rho(x), x \rangle \leq 0,$$

for all $x \in]0,q[$. Then, $\langle \nabla\rho(x), x \rangle \to 0$ if $x \in]0,q[$ and $x \to 0$. Thus,

$$\rho(0) = \lim_{\substack{x \to 0 \\ x \in]0,q[}} (\rho(x) + \langle \nabla\rho(x), x \rangle \leq 0,$$

so $\rho(0) = 0$. Since $\rho(x) \geq 0$ for all $x \in]0,q[$, it follows that $\langle \nabla\rho(x), x \rangle \leq 0$ for all $x \in]0,q]$. But, $\rho(tq) - \rho(0) = \rho(tq) = \langle \nabla\rho(\tau q), \tau q \rangle$ for some $\tau \in]0,t[$. If there were a $t \in]0,1]$ with $\rho(tq) > 0$, then it would follow that $\langle \nabla\rho(\tau q), \tau q \rangle > 0$ which is a contradiction. Therefore, $\rho(x) + \langle \nabla\rho(x), x \rangle = 0$ for all $x \in]0,q]$. This observation should be taken into account when one tries to construct examples. For example, if the second inequality of Theorem 22.28 is strict, then the differential of f is unbounded in some point of $]0,q_0[$.

Corollary 22.29. *Let $K \subseteq H$ be a subdual closed convex cone and $\rho : H \to \mathbb{R}$ a mapping with $\rho(x) \geq 0$ for all $x \in K \setminus \{0\}$. Define $f : H \to H$ by $f(x) = \rho(x)x$. If there is $p_0 \in K \setminus \{0\}$ such that f is continuous on $[0,p_0]$, differentiable on $]0,p_0[$ (for example this is satisfied if ρ is continuous on $[0,p_0]$ and differentiable on $]0,p_0[$) and*

$$\rho(p) + \langle \nabla\rho(p), p \rangle = 0$$

for all $p \in]0,p_0[$, then each $p \in]0,p_0[$ is a solution of the nonlinear complementarity problem $NCP(f,K)$.

Recall the following definition

Definition 22.30. Let k be a real constant. The function $v : H \to \mathbb{R}$ is called *homogeneous of order* k if $v(tx) = t^k v(x)$ for all $x \in H$ and $t \in \mathbb{R}^n$ with $t \neq 0$.

In the following definition, we introduce the notion of superhomogeneous (subhomogeneous) functions.

Definition 22.31. Let k be a real constant. The function $v : H \to \mathbb{R}$ is called *superhomogeneous (subhomogeneous) of order* k if $v(tx) \geq t^k v(x)$ $(\leq)$ for all $x \in H$ and $t \in \mathbb{R}^n$ with $t \geq 1$.

Obviously, every homogeneous function of order k is both subhomogeneous and superhomogeneous of order k. Moreover, every non-negative homogeneous function of order k is superhomogeneous of order $k - l$ and is subhomogeneous of order $k + l$ for each positive real number l.

Let k be a real constant. By Euler's theorem for homogeneous functions in $\mathbb{R}^n$, the general solution of the equation

$$\langle \nabla v(x), x \rangle = kv(x) \tag{22.17}$$

is a homogeneous function of order k, that is, a function v with the property

$$v(tx) = t^k v(x), \ \forall t \neq 0.$$

A similar result is true if the equation (22.17) is considered in an arbitrary Hilbert space H.

Indeed, assume that v is a solution. Fix x and let be t nonzero real number. Then we have for

$$\varphi(t) = t^{-k} v(tx)$$

that

$$\varphi'(t) = t^{-k} dv(tx)(x) - kt^{-k-1} v(tx) = t^{-k} \langle \nabla v(tx), x \rangle - kt^{-k-1} v(tx)$$
$$= t^{-k-1} \langle \nabla v(tx), tx \rangle - kt^{-k-1} v(tx) = t^{-k-1}(kv(tx)) - kt^{-k-1} v(tx) = 0,$$

that is, φ is a constant function. Hence, $\phi(t) = \phi(1)$, or equivalently

$$v(tx) = t^k v(x), \ \forall t \neq 0.$$

The same relation shows that if φ is a constant function, then v is a solution.

Hence, the general solution of the equation

$$u(x) + \langle \nabla u(x), x \rangle = 0$$

is a homogeneous function of order -1. Hence, in Corollary 22.29 the function ρ can be chosen to be a differentiable function such that $\rho|_{K \setminus \{0\}}$ is an arbitrary homogeneous function of order -1 and $\rho(x) \geq 0$ for all $x \in K \setminus \{0\}$, and $p_0 \in K \setminus \{0\}$ can be arbitrary.

Next, consider the inequation

$$\langle \nabla v(x), x \rangle \geq kv(x).$$

The general solution of this inequation is a superhomogeneous function of order k. Indeed, assume that v is a solution. Fix x and let $t > 0$ be a real number. Then we have for

$$\varphi(t) = t^{-k} v(tx)$$

that

$$\varphi'(t) = t^{-k} dv(tx)(x) - kt^{-k-1} v(tx) = t^{-k} \langle \nabla v(tx), x \rangle - kt^{-k-1} v(tx)$$
$$= t^{-k-1} \langle \nabla v(tx), tx \rangle - kt^{-k-1} v(tx) \geq t^{-k-1}(kv(tx)) - kt^{-k-1} v(tx) = 0.$$

Hence, φ is monotone on $]0, +\infty[$. Therefore, $\varphi(t) \geq \varphi(1)$ for all $t \geq 1$, or equivalently v is superhomogeneous of order k. Conversely, suppose that v is superhomogeneous of order k. We want to prove that

$$\langle \nabla v(x), x \rangle \geq kv(x).$$

Since v is superhomogeneous of order k, we have $v(tx) \geq t^k v(x)$ for all $t > 1$ and $x \in H$. Hence,

$$\frac{v(tx)-v(x)}{t-1} \geq \frac{t^k v(x)-v(x)}{t-1},$$

or equivalently

$$\frac{v(tx)-v(x)}{t-1} \geq \frac{t^k-1}{t-1}v(x).$$

Tending with t to 1 in the above inequality and using the chain rule we obtain

$$dv(x)(x) = dv(tx)(x)|_{t=1} = \frac{d}{dt}v(tx)|_{t=1} \geq \frac{d}{dt}t^k|_{t=1}v(x) = kv(x),$$

or equivalently

$$\langle \nabla v(x), x\rangle \geq kv(x).$$

It can be similarly shown that the general solution of the inequation

$$\langle \nabla v(x), x\rangle \leq kv(x).$$

is a subhomogeneous function v of order k.

Example 22.32. Let $K \subseteq H$ be a subdual closed convex cone. Let $p_0, q_0 \in K \setminus \{0\}$. Denote $P_0 = \text{cone}\{p_0\}$ and $Q_0 = \text{cone}\{q_0\}$. Consider the mapping $f : H \to H$ defined by

$$f(x) = \begin{cases} \dfrac{d(x,Q_0)^2\rho_1(x) + d(x,P_0)^2\rho_2(x)}{d(x,Q_0)^2 + d(x,P_0)^2}x & \text{if } x \in H \setminus \{0\}, \\ 0 & \text{if } x = 0, \end{cases}$$

where $\rho_1 : H \to \mathbb{R}$ and $\rho_2 : H \to \mathbb{R}$ such that $\rho_1|_{K\setminus\{0\}}$ is superhomogeneous of order -1, $\rho_2|_{K\setminus\{0\}}$ is subhomogeneous of order -1, and $\rho_1(x) \geq 0$ and $\rho_2(x) \geq 0$ for all $x \in K \setminus \{0\}$. Then, by using Lemma 22.18 and the above results, it is easy to verify that f satisfies all conditions of Theorem 22.28. Therefore, for all $p \in]0, p_0[$ and $q \in]0, q_0[$ the nonlinear complementarity problem $NCP(f,K)$ has a solution in $[p,q]$. The details are left to the reader.

Theorem 22.33. *Let $\beta : \mathbb{R}^n \to \mathbb{R}$ a function with $\beta(x) > 0$ for all $x \in \mathbb{R}^n_+ \setminus \{0\}$, $f = (f_1, \cdots f_n) : \mathbb{R}^n \to \mathbb{R}^n$, $\phi : \mathbb{R}^2 \to \mathbb{R}$ an NCP-function, $\Phi = (\Phi_1, \cdots, \Phi_n) : \mathbb{R}^n \to \mathbb{R}^n$ defined by $\Phi_i(x) = \phi(x_i, f_i(x))$, $\Psi : \mathbb{R}^n \to \mathbb{R}$ defined by $\Psi(x) = \beta(x)\|\Phi(x)\|^2/2$, and $g : \mathbb{R}^n \to \mathbb{R}^n$ defined by $g(x) = \Psi(x)x$. If there are $p_0, q_0 \in \mathbb{R}^n_+ \setminus \{0\}$ such that g is continuous on $0p_0q_{0_{\bar{\triangle}}}$, differentiable on $0p_0q_{0_\triangle}$ (for example, this is satisfied if ϕ, β and f are continuous on $0p_0q_{0_{\bar{\triangle}}}$ and differentiable on $0p_0q_{0_\triangle}$),*

$$\Psi(p) + \langle \nabla\Psi(p), p\rangle \geq 0,$$

and

$$\Psi(q) + \langle \nabla\Psi(q), q \rangle \le 0,$$

for all $p \in]0, p_0[$, $q \in]0, q_0[$, *then for all* $p \in]0, p_0[$ *and* $q \in]0, q_0[$ *the nonlinear complementarity problem* $NCP(f,K)$ *has a solution in* $[p,q]$.

Proof. The assertion of the theorem follows from Proposition 22.24 and Theorem 22.28. □

Corollary 22.34. *Let* $\beta : \mathbb{R}^n \to \mathbb{R}$ *be a function with* $\beta(x) > 0$ *for all* $x \in \mathbb{R}^n_+ \setminus \{0\}$, $f = (f_1, \cdots f_n) : \mathbb{R}^n \to \mathbb{R}^n$, $\phi : \mathbb{R}^2 \to \mathbb{R}$ *an NCP-function,* $\Phi = (\Phi_1, \cdots, \Phi_n) : \mathbb{R}^n \to \mathbb{R}^n$ *defined by* $\Phi_i(x) = \phi(x_i, f_i(x))$, $\Psi : \mathbb{R}^n \to \mathbb{R}$ *defined by* $\Psi(x) = \beta(x)\|\Phi(x)\|^2/2$, *and* $g : \mathbb{R}^n \to \mathbb{R}^n$ *defined by* $g(x) = \Psi(x)x$. *If there is a* $p_0 \in \mathbb{R}^n_+ \setminus \{0\}$ *such that* g *is continuous on* $[0, p_0]$, *differentiable on* $]0, p_0[$ *(for example, this is satisfied if* ϕ, β *and* f *are continuous on* $[0, p_0]$ *and differentiable on* $]0, p_0[$*),*

$$\Psi(p) + \langle \nabla\Psi(p), p \rangle = 0$$

for all $p \in]0, p_0[$, *then each* $p \in]0, p_0[$ *is a solution of the nonlinear complementarity problem* $NCP(f,K)$.

We have seen that in Corollary 22.29 ρ can be chosen to be a differentiable function such that $\rho|_{K\setminus\{0\}}$ to be an arbitrary homogeneous function of order -1 and $\rho(x) \ge 0$ for all $x \in K \setminus \{0\}$, and $p_0 \in K \setminus \{0\}$ can be arbitrary. Consider Corollary 22.34. In this corollary $\rho(x) = \beta(x)\|\Phi(x)\|^2$. Let $\beta(x) = 1$ for all $x \in \mathbb{R}^n$. We want $\mathbb{R}^n_+ \setminus \{0\} \ni x \mapsto \rho(x) = \beta(x)\|\Phi(x)\|^2$ to be homogeneous of order -1. Suppose that ϕ is the Fischer–Burmeister function defined right after Definition 22.22. It is enough to have the function

$$\mathbb{R}^n_+ \setminus \{0\} \ni x \mapsto \sqrt{x_i^2 + f_i^2(x)} - x_i - f_i(x)$$

to be homogeneous of order $-1/2$ for all i. This can be achieved by expressing f_i from the relation

$$\sqrt{x_i^2 + f_i^2(x)} - x_i - f_i(x) = h_i(x)$$

in terms of $h_i(x)$, where h_i is homogeneous of order $-1/2$ such that $h_i(x) + x_i \neq 0$ for all $x \in \mathbb{R}^n_+ \setminus \{0\}$. We obtain

$$f_i(x) = -\frac{h_i(x)}{2}\left(1 + \frac{x_i}{h_i(x) + x_i}\right). \tag{22.18}$$

In this way, any homogeneous h_i of order $-1/2$, with $h_i(x) + x_i \neq 0$ for all $x \in \mathbb{R}^n \setminus \{0\}$, generates an example for Corollary 22.34. It is enough to choose f_i to be of the form (22.18) on $\mathbb{R}^n_+ \setminus \{0\}$. The condition "$h_i$ is homogeneous of order $-1/2$ such that $h_i(x) + x_i \neq 0$ for all $x \in \mathbb{R}^n_+ \setminus \{0\}$" holds for example when $h_i : \mathbb{R}^n_+ \setminus \{0\} \to \mathbb{R}$ is homogeneous of order $-1/2$ and positive, that is, $h_i(x) > 0$ for all $x \in \mathbb{R}^n_+ \setminus \{0\}$.

22.6 Comments

We presented in this paper some mean value theorems and a new method applicable to the study of multiple solutions of a complementarity problem. Our method can be considered as an alternative to the classical method based on the topological degree. Certainly, our method can be improved and developed to be applicable to variational inequalities or to other kind of nonlinear equations.

Acknowledgment The authors express their gratitude to A. B. Németh for many helpful conversations. S. Z. Németh was supported by the Hungarian Research Grant OTKA 60480.

References

1. R. Andreani, J. M. Martinez, Reformulations of variational inequalities on a simplex and compactification of complementarity problems, SIAM J. Opt. 10(3) (2000) 878–895.
2. B. Chen, X. Chen, C. Kanzow, A penalized Fischer-Burmeister NCP-function: Theoretical investigations and numerical results, Math. Programming 88(1) (2000) 211–216.
3. F. H. Clarke, Optimization and Nonsmooth Analysis, Wiley, New York, 1983.
4. R. Cottle, J. Pang, R. Stone, The Linear Complementarity Problem, Academic Press, Boston, 1992.
5. T. De Luca, F. Facchinei, C. Kanzow, A semismooth equation approach to the solution of nonlinear complementarity problems, Math. Programming 75(3) (1996) 407–439.
6. F. Facchinei, C. Kanzow, A nonsmooth inexact Newton method for the large-scale nonlinear complementarity problems, Math. Programming 76(3) (1997) 493–512.
7. J. Ferrer, On Rolle's theorem in spaces of infinite dimension, Indian J. Math. 42(1) (2000) 21–36.
8. A. Fischer, Solving linear complementarity problems by embedding, (Preprint) Dresden University of Technology (1992).
9. A. Fischer, A special Newton-type optimization method, Optimization 24(3–4) (1992) 269–284.
10. A. Fischer, An NCP-function and its use for the solution of complementarity problems, Recent advances in nonsmooth optimization, World Sci. Publ., River Edge, NJ, 88–105, 1995.
11. A. Fischer, A Newton-type method for positive semidefinite linear complementarity problems, J. Optim. Theory Appl. 86(3) (1995) 585–608.
12. A. Fischer, On the local superlinear convergence of a Newton-type method for LCP under weak conditions, Optimization Methods and Software, 6 (1995) 83–107.
13. A. Fischer, Solution of monotone complementarity problems with locally Lipschitzian functions, *Math. Programming*, 76(3), 513–532 (1997).
14. M. Fukushima, Merit functions for variational inequality and complementarity problems, Nonlinear Optimization and Applications (Eds. G. Di Billo and F. Gianessi), Plenum Press, New York, 155–170, 1996.
15. G. Isac, The numerical range theory and boundedness of solutions of the complementarity problem, J. Math. Anal. Appl. 143(1) (1989) 235–251.
16. G. Isac, Complementarity Problems, Lecture Notes in Mathematics, vol. 1528, Springer-Verlag, Berlin, 1992.
17. G. Isac, Topological Methods in Complementarity Theory, Kluwer Academic Publishers, Dordrecht, 2000.
18. G. Isac, Leray-Schauder Type Alternatives, Complementarity Problems and Variational Inequalities, Springer, Berlin, 2006.

19. G. Isac, Bulavsky, V.A. and Kalashnikov, V.V., Complementarity, Equilibrium, Efficiency and Economics, Kluwer Academic Publishers, Dordrecht, 2002.
20. G. Isac and S.Z. Németh, Scalar derivatives and scalar asymptotic derivatives: properties and some applications. J. Math. Anal. Appl. 278(1) (2003) 149–170.
21. G. Isac and S.Z. Németh, Scalar derivatives and scalar asymptotic derivatives. An Altman type fixed point theorem on convex cones and some applications. J. Math. Anal. Appl. 290(2) (2004) 452–468.
22. G. Isac and S.Z. Németh, Scalar and Asymptotic Scalar Derivatives. Theory and Applications, Springer (forthcoming).
23. C. Kanzow, N. Yamashita, M. Fukushima, New NCP-functions and their properties, J. Optim. Theory Appl. 94(1) (1997) 115–135.
24. Kobayashi, M.H., On an extension of Rolle's theorem to locally convex spaces, J. Math. Anal. Appl. 323(2) (2006) 1225–1230.
25. S.Z. Németh, A scalar derivative for vector functions, Riv. Mat. Pura Appl., 10(1992) 7–24.
26. S.Z. Németh, Scalar derivatives and spectral theory, Mathematica, Thome 35(58), N.1, (1993) 49–58.
27. S.Z. Németh, Scalar derivatives in Hilbert spaces, Positivity 10(2) (2006) 299–314.
28. D. Sun and L. Qi, On NCP-functions, Comput. Optim. Appl., 13(1–3) (1999) 201–220.
29. G. Isac and S.Z. Németh, Browder-Hartman-Stampacchia Theorem and the existence of a bounded interior band of epsilon solutions for nonlinear complementarity problems, Rocky Mountain J. Math, 6(37) (2007) 1917–1940
30. G. Isac and S.Z. Németh, Duality in multivalued complementarity theory by using inversions and scalar derivatives, J Global Optim., 33(2) (2005) 197–213.
31. G. Isac and S.Z. Németh, Duality in nonlinear complementarity theory by using inversions and scalar derivatives, Math. Inequal. Appl., 9(4) (2006) 781–795.
32. G. Isac and S.Z. Németh, Duality of implicit complementarity problems by using inversions and scalar derivatives, J Optim. Theory Appl., 128(3) (2006) 621–633.
33. G. Isac and S.Z. Németh, Scalar and Asymptotic Scalar Derivatives. Theory and Applications, Springer Optimization and Application, 13. Springer, New York, 2008.

Chapter 23
Application of a Vector-Valued Ekeland-Type Variational Principle for Deriving Optimality Conditions

G. Isac and C. Tammer

Dedicated to the memory of Professor George Isac

Abstract In order to show necessary conditions for approximate solutions of vector-valued optimization problems in general spaces, we introduce an axiomatic approach for a scalarization scheme. Several examples illustrate this scalarization scheme. Using an Ekeland-type variational principle by Isac [12] and suitable scalarization techniques, we prove the optimality conditions under different assumptions concerning the ordering cone and under certain differentiability assumptions for the objective function.

23.1 Introduction

The aim of our paper is to present necessary conditions for approximate solutions of vector-valued optimization problems in Banach spaces using an Ekeland-type variational principle by Isac [12] under different differentiability properties of the objective function. In the proofs of the assertions, a nonlinear scalarization technique plays an important role. We will use an axiomatic approach for the scalarization scheme. In order to apply the variational principle in partially ordered spaces, one needs additional assumptions for the ordering cone. Furthermore, the differentiability properties require certain assumptions concerning the ordering cone and the objective function. So a discussion of corresponding ordering and topological assumptions is important for our assertions.

G. Isac
Department of Mathematics and Computer Science, Royal Military College of Canada, P.O. Box 17000, STN Forces Kingston, Ontario K7K 7B4, Canada.

C. Tammer
Institute of Mathematics, Martin-Luther-University Halle-Wittenberg, D-06099 Halle, Germany.

P.M. Pardalos et al. (eds.), *Nonlinear Analysis and Variational Problems*,
Springer Optimization and Its Applications 35, DOI 10.1007/978-1-4419-0158-3_23,

In this paper, we will be mainly concerned with the following vector minimization problem (VP) given as

$$V - \min f(x), \quad \text{subject to} \quad x \in S,$$

where (X,d) is a complete metric space and Y is a locally convex space, $S \subseteq X$, $K \subset Y$ is a proper (i.e., $\{0\} \neq K$, $K \neq Y$) pointed closed convex cone which induces a partial order on Y (i.e., $y^1 \leq_K y^2 \Longleftrightarrow y^2 \in y^1 + K \quad (y^1, y^2 \in Y)$), $f : S \to Y$. We describe the solution concepts for the vector optimization problem (VP) with respect to the ordering cone K in Section 23.3.

In order to show necessary optimality conditions for the problem (VP) using an Ekeland-type variational principle and differential calculus, one needs certain assumptions concerning the spaces, the ordering cone and the objective function. In the assertions of our paper, we suppose some of the following assumptions with respect to the spaces:

($A_{space}1$) (X,d) is a complete metric space and Y is a locally convex space.
($A_{space}2$) X and Y are Banach spaces.
($A_{space}3$) X is an Asplund space and $Y = \mathbb{R}^n$.

Remark 23.1. A Banach space X is said to be an Asplund space (cf. Phelps [23, Def. 1.22]) if every continuous convex function defined on a non-empty open convex subset D of X is Fréchet differentiable at each point of some dense G_δ subset of D. If the dual space X^* of the Banach space X is separable, then X is an Asplund space. Every reflexive Banach space is an Asplund space. The sequence space c_0, and furthermore, the spaces l^p, $L^p[0,1]$ for $1 < p < \infty$ are examples for Asplund spaces. The space l^1 is not an Asplund space.

Concerning the objective function, we have different assumptions with respect to the derivatives (see Section 23.5) that we will use:

($A_{map}1$) The vector-valued directional derivative $Df(x,h)$ of $f : X \to Y$ at $x \in X$ in direction $h \in X$ exists for all $x, h \in X$ (cf. Definition 23.31).
($A_{map}2$) $f : X \to Y$ is strictly differentiable at $x \in X$ (cf. Definition 23.29).
($A_{map}3$) $f : X \to Y$ is locally Lipschitz at $x \in X$ (cf. Definition 23.26).

Furthermore, in order to apply an Ekeland-type variational principle (see Theorem 23.18) we suppose that (X,d) and Y fulfill ($A_{space}1$), consider $f : X \to Y$ and formulate the following assumption ($A_{map}4$) with respect to a closed normal (cf. Definition 23.2) cone $K \subset Y$ and $k^0 \in K \setminus \{0\}$:

($A_{map}4$) For every $u \in X$ and for every real number $\alpha > 0$ the set

$$\left\{x \in X \mid f(x) - f(u) + k^0 \alpha d(u,x) \in -K\right\}$$

is closed.

Most of our results are related to the non-convex case, however under certain convexity assumptions we get stronger results. Consider the proper pointed closed convex cone K in Y and a non-empty convex subset S of X. The function $f : S \to Y$ is called convex if for all $x^1, x^2 \in X$ and for all $\lambda \in [0,1]$ holds

$$f(\lambda x^1 + (1-\lambda)x^2) \in \lambda f(x^1) + (1-\lambda) f(x^2) - K.$$

$(A_{map}5)$ The function $f : S \to Y$ is convex.

Our paper is organized as follows: In Section 23.2 we give an overview on cone properties that are important for deriving existence results in infinite-dimensional spaces. Especially, we give several examples in general spaces for cones having the Daniell property and for cones with non-empty interior. Solution concepts for vector optimization problems and an Ekeland-type variational principle by Isac [12] under the assumptions $(A_{space}1)$ and $(A_{map}4)$ are presented in Section 23.3. An axiomatic scalarization scheme that is important for deriving optimality conditions is introduced in Section 23.4. We present several examples for scalarizing functionals having some of the properties supposed in the scalarization scheme. In Section 23.5 we recall differentiability properties of vector-valued functions. We show necessary optimality conditions for vector optimization problems under assumptions $(A_{space}2)$ and $(A_{map}4)$ in Section 23.6. For the case of vector optimization problems where a Lipschitz objective function takes its values in a finite-dimensional space, we prove necessary conditions for approximate solutions in Section 23.7 under assumptions $(A_{space}3)$ and $(A_{map}3)$ using the subdifferential calculus by Mordukhovich [18].

23.2 Properties of Cones

In the following, we give a survey of some properties of cones in ordered topological spaces; they are compiled in this way in order to make the choice of Y as made in Section 23.5 plausible.

In order to prove existence results for solutions of optimization problems in infinite-dimensional spaces where the solution concept is given by a partial order induced by a closed pointed and convex cone, one needs additional assumptions concerning the connections between topology and order (cf. Isac [11]).

First, we recall some corresponding cone properties (that the cone is normal, well-based, nuclear, Daniell property), compare Peressini [22], Isac [11], Isac, Bulavsky, Kalashnikov [14], Jahn [17], Hyers, Isac, Rassias [16], Göpfert, Riahi, Tammer, Zalinescu [9]. In many important cases the ordering cone does not have such a property, for instance the usual ordering cone in the space of continuous functions does not have a bounded base and the Daniell property is not given. In Figure 23.1 we give an overview on such additional cone properties and corresponding relations for the case that Y is a Banach space, C and K are proper convex cones in Y. As usual, we denote by

$$K^* := \{y^* \in Y^* \mid y^*(y) \geq 0 \; \forall y \in K\}$$

the continuous dual cone of K, and by

$$K^\# := \{y^* \in K^* \mid y^*(y) > 0 \; \forall y \in K \setminus \{0\}\}$$

the quasi-interior of K^*.

In order to study connections between *topology* and *order* we say that a nonempty subset A of the linear space Y is **full** with respect to the convex cone $K \subset Y$ if

$$A = (A + K) \cap (A - K).$$

Definition 23.2. Let (Y, τ) be a topological linear space and let $K \subset Y$ be a convex cone. Then K is called **normal** (relative to τ) if the origin $0 \in Y$ has a neighborhood base formed by **full** sets w.r.t. K.

Definition 23.3. Let Y be a Hausdorff topological vector space and $K \subset Y$ a proper convex cone.

(i) K is called **based** if there exists a convex set B, such that $K = \mathbb{R}_+ B$ and $0 \notin cl\, B$.
(ii) K is called **well-based** if there exists a bounded convex set B, such that $K = \mathbb{R}_+ B$ and $0 \notin cl\, B$.
(iii) Let the topology of Y be defined by a family $\mathscr{P}$ of seminorms. K is called **supernormal** or **nuclear** if for each $p \in \mathscr{P}$ there exists $y^* \in Y^*$, such that $p(y) \leq \langle y, y^* \rangle$ for all $y \in K$; it holds $y^* \in K^*$ in this case.
(iv) K is said to be **Daniell** if any nonincreasing net having a lower bound converges to its infimum.

Now, we give a few examples of Daniell cones.

Example 23.4. First, we recall the following result (cf. Peressini [22], Proposition 3.1, p. 90, 91): If $\{x_\alpha\}_{\alpha \in A}$ is a net which is increasing (decreasing) in a topological vector space (Y, τ) ordered by a closed convex cone K and if x_0 is a cluster point of $\{x_\alpha\}$, then $x_0 = \sup_{\alpha \in A} x_\alpha$ ($x_0 = \inf_{\alpha \in A} x_\alpha$). We recall that a convex cone is regular if any decreasing (increasing) net which has a lower bound (upper bound) is convergent. By the result, cited above we have that any regular cone is Daniell.

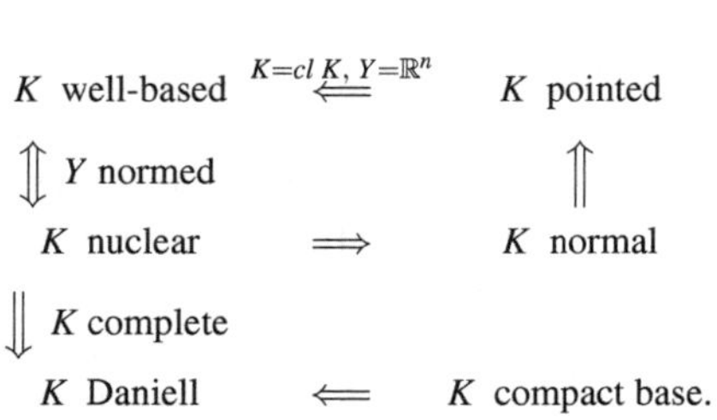

Fig. 23.1 Cone properties.

Example 23.5. If $(Y, ||\cdot||)$ is a Banach lattice, that is Y is a Banach space, vector lattice and the norm is absolute, i.e., $||x|| = ||\,|x|\,||$ for any $x \in Y$, then the cone $Y_+ = \{y \in Y \mid y \geq 0\}$ is Daniell if Y has weakly compact intervals.

Example 23.6. Finally, a convex cone with a weakly compact base is a Daniell cone.

Proposition 23.7. *(Isac [11]):*
Let $(Y, \mathscr{P})$ be an Hausdorff locally convex space and $K \subset Y$ a proper convex cone. Then

$$K \text{ well-based} \Longrightarrow K \text{ nuclear} \Longrightarrow K \text{ normal.}$$

If Y is a normed space, then

$$K \text{ nuclear} \Longrightarrow K \text{ well-based.}$$

Remark 23.8. Among the classical Banach spaces their usual positive cones are well-based only in l^1 and $L^1(\Omega)$, but l^1 is not an Asplund space.

Relations between supernormal (nuclear) cones, Pareto efficiency and geometrical aspects of Ekeland's principle are derived by Isac, Bulavsky and Kalashnikov [14].

Let Y be a topological vector space over $\mathbb{R}$. Assume (Y, K) is at the same time a vector lattice with the lattice operations $x \mapsto x^+$, $x \mapsto x^-$, $x \mapsto |x|$, $(x,y) \mapsto \sup\{x,y\}$ and $(x,y) \mapsto \inf\{x,y\}$.

Definition 23.9. A set $A \subset Y$ is called *solid,* if $x \in A$ and $|y| \leq |x|$ implies $y \in A$. The space Y is called locally solid, if it possesses a neighborhood of 0 consisting of solid sets.

Lemma 23.10 ([26]). *The following properties are equivalent:*

(i) Y is locally solid.
(ii) K is normal, and the lattice operations are continuous.

In order to derive optimality conditions in general spaces (cf. Section 23.6), there is often the assumption that the (natural) ordering cone has a non-empty interior. Now, we give some examples of convex cones with non-empty interior.

Example 23.11. Any closed convex cone K in the Euclidean space $(\mathbb{R}^n, \langle\cdot,\cdot\rangle)$ such that K is self-adjoint (i.e., $K = K^{**}$) has a non-empty interior.

Example 23.12. We consider the space of continuous functions $C[a,b]$ with the norm $||x|| = \sup\{|x(t)| \mid t \in [a,b]\}$. The cone of positive functions in $C[a,b]$

$$K_{C[a,b]} := \{x \in C[a,b] \mid x(t) \geq 0\, \forall t \in [a,b]\}$$

has a non-empty interior.

Example 23.13. Let $Y = l^2(\mathbb{N}, \mathbb{R})$ with the well-known structure of a Hilbert space. The convex cone

$$K_{l^2} := \{x = \{x_i\}_{i \geq 0} \mid x_0 \geq 0 \text{ and } \sum_{i=1}^{\infty} x_i^2 \leq x_0^2\}$$

has a non-empty interior

$$\text{int } K_{l^2} := \{x = \{x_i\}_{i \geq 0} \mid x_0 > 0 \text{ and } \sum_{i=1}^{\infty} x_i^2 < x_0^2\}.$$

Example 23.14. Let l^∞ be the space of bounded sequences of real numbers, equipped with the norm $||x|| = \sup_{n \in \mathbb{N}}\{|x_n|\}$. The cone

$$K_{l^\infty} := \{x = \{x_n\}_{n \in \mathbb{N}} \mid x_n \geq 0 \text{ for any } n \in \mathbb{N}\}$$

has a non-empty interior (cf. Peressini [22], p. 186).

Example 23.15. Let $C^1[a,b]$ be the real vector space formed by all real continuously differentiable functions defined on $[a,b]$ $(a, b \in \mathbb{R}, a < b)$, equipped with the norm

$$||f||_1 := \{\int_a^b (f(t))^2 \, \mathrm{d}t + \int_a^b (f'(t))^2 \, \mathrm{d}t\}^{1/2}$$

for any $f \in C^1[a,b]$. Using a Sobolev's embedding theorem, we can show that the natural ordering cone

$$K_{C^1} := \{f \in C^1[a,b] \mid f \geq 0\}$$

has a non-empty interior. The proof is based on some technical details (cf. Da Silva [4]).

Example 23.16. About the locally convex spaces, we put in evidence the following result. If (Y, τ) is a real locally convex space, then for every closed convex cone $K \subset Y$, with non-empty interior, there exists a continuous norm $||\cdot||$ on Y such that K has a non-empty interior in the normed space $(Y, ||\cdot||)$.

Furthermore, in order to show optimality conditions one has sometimes both assumptions: that the ordering cone has a non-empty interior and has the Daniell property. So it is important to ask for examples in infinite-dimensional spaces, where the ordering cone has both properties.

Example 23.17. (see Jahn [17]) Consider the real linear space $L_\infty(\Omega)$ of all (equivalence classes of) essentially bounded functions $f : \Omega \to \mathbb{R}$ $(\emptyset \neq \Omega \subset \mathbb{R}^n)$ equipped with the norm $||\cdot||_{L_\infty(\Omega)}$ given by

$$||f||_{L_\infty(\Omega)} := \operatorname*{ess\,sup}_{x \in \Omega}\{|f(x)|\} \text{ for all } f \in L_\infty(\Omega).$$

The ordering cone

$$K_{L_\infty(\Omega)} := \{f \in L_\infty(\Omega) \mid f(x) \geq 0 \text{ almost everywhere on } \Omega\}$$

has a non-empty interior and is weak* Daniell.

23.3 An Ekeland-Type Variational Principle for Vector Optimization Problems

Concerning the vector optimization problem (VP), we use the following (approximate) solution concepts: Assume $(A_{space}1)$. Let us consider $A \subset Y$, a pointed closed convex cone $K \subset Y$, $\varepsilon \geq 0$ and $k^0 \in K \setminus \{0\}$.

- A point $y_0 \in A$ is said to be an *εk^0-minimal point of A with respect to K*, if there exists no other point $y \in A$ such that $y - y_0 \in -\varepsilon k^0 - (K \setminus \{0\})$. We denote this by $y_0 \in \varepsilon k^0 - Eff(A,K)$, where $\varepsilon k^0 - Eff(A,K)$ is the set of εk^0 -minimal points of A with respect to the ordering cone K. A point $x_0 \in S$ is called an *εk^0 -efficient point* of (VP), if $f(x_0) \in \varepsilon k^0 - Eff(f(S),K)$. Is $x_0 \in S$ an εk^0 -efficient point of (VP) with $\varepsilon = 0$ we say that x_0 is an efficient point of (VP) and we write $f(x_0) \in Eff(f(S),K)$.
- A point $y_0 \in A$ is said to be an *εk^0 -properly minimal element of A with respect to K*, if there is a closed normal cone $B \subset Y$ with $K \setminus \{0\} \subset \operatorname{int} B$ such that $y_0 \in \varepsilon k^0 - Eff(A,B)$. The set of εk^0 -properly minimal elements of A with respect to K is denoted by $\varepsilon k^0 - pEff(A,K)$. A point $x_0 \in S$ is called an *εk^0 -properly efficient* point for (VP), if $f(x_0) \in \varepsilon k^0 - Eff(f(S),B)$ where B is a closed normal cone with $K \setminus \{0\} \subset \operatorname{int} B$. Is $x_0 \in S$ an εk^0 -properly efficient point of (VP) with $\varepsilon = 0$ we say that x_0 is a properly efficient point of (VP) and we write $f(x_0) \in pEff(f(S),K)$.

We will apply a vector-valued variational principle of Ekeland's type in order to show necessary conditions for approximately efficient solutions of the vector optimization problem (VP). There are many vector-valued variants of Ekeland's variational principle (and equivalent assertions) with different assumptions concerning the ordering cone in Y and concerning the properties of the objective function $f : X \to Y$ (cf. [12], [13], [19], [27], [28]). Here we recall the variational principle by Isac [12, Theorems 4 and 7], [16, Theorem 8.4] that is shown for the case that the ordering cone K in Y is normal without assuming that the interior of K is non-empty.

Theorem 23.18. *(Isac [12]) Assume $(A_{space}1)$, K is a closed normal cone, $k^0 \in K \setminus \{0\}$. Furthermore, suppose that for $f : X \to Y$ the assumption $(A_{map}4)$ with respect to K and k^0 is fulfilled. If $\varepsilon > 0$ is an arbitrary real number and $x_0 \in X$ is an element with $f(x_0) \leq_K f(x) + \varepsilon k^0$ for all $x \in X$ then there exists $x_\varepsilon \in X$ such that $f(x_\varepsilon) \leq_K f(x_0)$, $d(x_\varepsilon, x_0) \leq \sqrt{\varepsilon}$ and, moreover,*

$$f(x) + k^0 \sqrt{\varepsilon} d(x, x_\varepsilon) \leq_K f(x_\varepsilon) \quad \Longrightarrow \quad x = x_\varepsilon. \tag{23.1}$$

Remark 23.19. The assertion (23.1) in Theorem 23.18 means that x_ε is an efficient element of the perturbed objective function $f_{\sqrt{\varepsilon}k^0}(x) := f(x) + k^0\sqrt{\varepsilon}d(x, x_\varepsilon)$ with respect to K, i.e., $f_{\sqrt{\varepsilon}k^0}(x_\varepsilon) \in Eff(f_{\sqrt{\varepsilon}k^0}(X), K)$.

23.4 Nonlinear Scalarization Scheme

In order to prove optimality conditions, we will introduce an axiomatic approach for scalarization by means of (in general nonlinear) functionals. We consider a linear topological space Y, a proper set $K \subset Y$ and a scalarizing functional $\varphi : Y \to \mathbb{R} \cup \{\pm\infty\}$ having some of the following properties:

$(A_\varphi 1)$ The functional φ is K-monotone, i.e., $y, w \in Y$, $y \in w - K$ implies $\varphi(y) \leq \varphi(w)$.
$(A_\varphi 1')$ The functional φ is strictly K-monotone, i.e., $y, w \in Y$, $y \in w - (K \setminus \{0\})$ implies $\varphi(y) < \varphi(w)$.
$(A_\varphi 2)$ The functional φ is convex.
$(A_\varphi 2')$ The functional φ is sublinear.
$(A_\varphi 2'')$ The functional φ is linear.
$(A_\varphi 3)$ The functional φ enjoys the translation property

$$\forall s \in \mathbb{R},\ \forall y \in Y:\ \varphi\left(y + sk^0\right) = \varphi(y) + s. \tag{23.2}$$

$(A_\varphi 4)$ The functional φ is lower continuous.
$(A_\varphi 4')$ The functional φ is continuous.

Examples for functionals satisfying the axioms given above are listed in the following:

Example 23.20. Assume that B is a closed proper subset of Y and $K \subset Y$ is a proper set with $B + K \subset B$. Let $k^0 \in Y \setminus \{0\}$ be such that $B + [0, +\infty)k^0 \subset B$. Consider the functional $\varphi := \varphi_{B,k^0} : Y \to \mathbb{R} \cup \{\pm\infty\}$, defined by

$$\varphi(y) := \inf\left\{t \in \mathbb{R} \mid y \in tk^0 - B\right\}. \tag{23.3}$$

We use the convention $\inf\emptyset = +\infty$. Then it holds $\operatorname{dom}\varphi = \mathbb{R}k^0 - B$.

If $B = K$ is a proper closed convex cone and $k^0 \in int\ K$, the functional (23.3) fulfills $(A_\varphi 1)$, $(A_\varphi 2')$, $(A_\varphi 3)$ and $(A_\varphi 4')$.
Moreover, if B is a proper closed convex subset of Y with $B + (K \setminus \{0\}) \subset \operatorname{int} B$, B does not contain lines parallel to k^0 (i.e., $\forall y \in Y$, $\exists t \in \mathbb{R}\ :\ y + tk^0 \notin D$) and $\mathbb{R}k^0 - B = Y$ the functional (23.3) is finite-valued and fulfills $(A_\varphi 1')$, $(A_\varphi 2)$, $(A_\varphi 3)$ and $(A_\varphi 4)$. These properties of the functional (23.3) are shown in [9, Theorem 2.3.1].

Example 23.21. The scalarizing functional by Pascoletti and Serafini [21] for a vector optimization problem (VP)

$$\mathrm{V}-\min_{x\in S\subset \mathbb{R}^n} f(x) = (f_1(x),...,f_m(x))^T$$

(where $Y=\mathbb{R}^m$, $K=\mathbb{R}^m_+$, S convex and $f_i: S\to\mathbb{R}$ convex for all $i=1,...,m$) given by

$$\min\ t \tag{23.4}$$

$$\text{subject to the constraints}$$

$$f(x)\in a+tr-K,$$

$$x\in S,\ t\in\mathbb{R},$$

(with parameters $a\in\mathbb{R}^m$ and $r\in \operatorname{int}\mathbb{R}^m_+$) satisfies the axioms $(A_\varphi 1)$, $(A_\varphi 2)$, $(A_\varphi 3)$, $(A_\varphi 4')$ (cf. [9, Theorem 2.3.1]).

Example 23.22. The following functional was introduced by Hiriart–Urruty [15]: Assume that Y is a normed space. For a non-empty set $A\subset Y$, $A\neq Y$, the oriented distance function $\Delta_A: Y\to\mathbb{R}$ is given as $\Delta_A(y)=d_A(y)-d_{Y\setminus A}(y)$ (where $d_A(y)=\inf\{||a-y|| \mid a\in A\}$ is the distance function to a set A). It is well known that this function has the following properties (see [29, Proposition 3.2]):
(i) Δ_A is Lipschitzian of rank 1.
(ii) If A is convex, then Δ_A is convex and if A is a cone, then Δ_A is positively homogeneous.
(iii) Assume that A is a closed convex cone. If $y_1,y_2\in Y$ with $y_1-y_2\in A$, then $\Delta_A(y_1)\leq\Delta_A(y_2)$.

The functional Δ_A satisfies the axioms $(A_\varphi 1)$ with respect to $k=-A$, $(A_\varphi 2)$, $(A_\varphi 4')$ if A is a closed convex cone.

Example 23.23. Certain nonlinear functionals are used in financial mathematics in order to express a risk measure (for example a valuation of risky investments) with respect to an acceptance set $B\subset Y$. Artzner, Delbean, Eber and Heath [1] (compare Heyde [10]) introduced coherent risk measures. Risk measures are functionals $\mu: Y\to\mathbb{R}\cup\{\pm\infty\}$, where Y is a vector space of random variables. In the papers by Artzner, Delbean, Eber and Heath [1] and Rockafellar, Uryasev and Zabarankin [24] the following properties of coherent risk measures μ are supposed:

$(P1)$ $\mu(y+tk^0)=\mu(y)-t$,

$(P2)$ $\mu(0)=0$ and $\mu(\lambda y)=\lambda\mu(y)$ for all $y\in Y$ and $\lambda>0$,

$(P3)$ $\mu(y^1+y^2)\leq\mu(y^1)+\mu(y^2)$ for all $y^1,y^2\in Y$,

$(P4)$ $\mu(y^1) \le \mu(y^2)$ if $y^1 \ge y^2$.

The sublevel set $L_\mu(0) =: B$ of μ to the level 0 is a convex cone and corresponds to the acceptance set. It can be shown that a coherent risk measure admits a representation as

$$\mu(y) = \inf\{t \in \mathbb{R} \mid y + tk^0 \in B\}. \tag{23.5}$$

It can be seen that a coherent risk measure can be identified with the functional $\varphi_{B,k^0}(-y)$ (see (23.3)) by

$$\varphi_{B,k^0}(y) = \mu(-y).$$

We get corresponding properties $(A_\varphi 1)$, $(A_\varphi 2')$, $(A_\varphi 3)$ for the functional $\mu(-y)$ like in Example 23.20 for the functional φ_{B,k^0} depending from the properties of the set B, i.e., of the acceptance set B in mathematical finance.

Examples for coherent risk measures are the **conditional value at risk** (cf. [7], Section 4.4, Definition 4.43) and the **worst-case risk measure** (cf. Example 23.25).

Example 23.24. (**value at risk**) Let Ω be a fixed set of scenarios. A financial position is described by a mapping $x : \Omega \to \mathbb{R}$ and x belongs to a given class $\mathscr{X}$ of financial positions. Assume that $\mathscr{X}$ is the linear space of bounded measurable functions containing the constants on some measurable space (Ω, A). Furthermore, let P be a probability measure on (Ω, A). A position x is considered to be acceptable if the probability of a loss is bounded by a given level $\lambda \in (0,1)$, i.e., if $P[x < 0] \le \lambda$. The corresponding monetary risk measure $V@R_\lambda$, defined by

$$V@R_\lambda(x) := \inf\{m \in \mathbb{R} \mid P(m + x < 0) \le \lambda\}$$

is called *value at risk*. $V@R_\lambda$ is the smallest amount of capital which, if added to x and invested in the risk-free asset, keeps the probability of a negative outcome below the level λ.

$V@R_\lambda$ is positively homogeneous but in general it is not convex (cf. Föllmer and Schied [7], Example 4.11), this means that $(A_\varphi 2)$ and $(A_\varphi 2')$ are not fulfilled.

Example 23.25. (**worst-case risk measure**) Consider the *worst-case risk measure* ρ_{max} defined by

$$\rho_{max}(x) := -\inf_{w \in \Omega} x(w) \text{ for all } x \in \mathscr{X},$$

where Ω is a fixed set of scenarios, $x : \Omega \to \mathbb{R}$ and x belongs to a given class $\mathscr{X}$ of financial positions. Assume that $\mathscr{X}$ is the linear space of bounded measurable functions containing the constants on some measurable space (Ω, A). The value $\rho_{max}(x)$ is the least upper bound for the potential loss which occur in any scenario. ρ_{max} is a coherent risk measure (cf. Föllmer and Schied [7], Example 4.8) such that we get the properties mentioned in Example 23.23.

23.5 Differentiability Properties of Vector-Valued Functions

In this section, we suppose that assumption $(A_{space}2)$ is fulfilled and consider $f : X \to Y$. Furthermore, assume that $K \subset Y$ is a proper pointed closed convex cone.

First of all, we introduce a concept of a vector-valued local Lipschitz property for $f : X \to Y$.

Definition 23.26 ([26]). $f : X \to Y$ is called **locally Lipschitz** at $x \in X$, if there is a function $P : X \times \mathbb{R} \to K$ such that

$$\left| \frac{f(u+th) - f(u)}{t} \right| \leq P(h,\varepsilon) \qquad \forall u \in U(x,\varepsilon),\ t \in (0,\varepsilon) \tag{23.6}$$

for all sufficiently small $\varepsilon > 0$. Therein, P is supposed to be continuous in h, and $\lim_{h\to 0} P(h,\varepsilon) = 0$ for each $\varepsilon > 0$.

This property is a basis for the definition of a directional derivative, which follows the idea of the Clarke directional derivative of real-valued functions. First, we recall Clarke's generalized directional derivative:

Definition 23.27 ([3]). Let X be a Banach space and let f be Lipschitz near a given point x and let v be any other vector in X. A mapping $f^\circ : X \to Y$ defined by

$$f^\circ(x,v) := \limsup_{t\downarrow 0\ y\to x} \frac{f(y+tv) - f(y)}{t}$$

is called Clarke's generalized directional derivative of f at x in direction v.

Definition 23.28. Clarke's tangent cone (contingent cone) is defined by

$$\mathscr{T}(S,x) := \{h \in X \mid d_S^\circ(x,h) = 0\},$$

where $d_S(x) := \inf\{||y-x|| \mid y \in S\}$ is the distance function to a non-empty set $S \subset X$, X is a Banach space and $x \in X$.

This cone can be described also in the following way:

$$\mathscr{T}(S,x) := \{h \in X \mid \forall \{x_n\}_{n\in N} \subseteq S,\ x_n \to x, \forall \{t_n\}_{n\in N} \in (0,+\infty),\ t_n \to 0$$

$$\exists \{h_n\}_{n\in N} \subseteq X :\ h_n \to h,\ x_n + t_n h_n \in S\, \forall n \in N\}.$$

Furthermore, we study strictly differentiable mappings:

Definition 23.29 ([3]). $f : X \to Y$ is called strictly differentiable at $x \in X$ if there is a linear continuous mapping $D_S f(x) : X \to Y$ such that for each $h \in X$, for each sequence $\{t_n\}_{n\in\mathbb{N}} \subseteq \mathbb{R}_+$ and for each sequence $\{x_n\}_{n\in\mathbb{N}} \in X$ with $x_n \to x$ and $t_n \to 0$ the following holds

$$D_S f(x)(h) = \lim_{n\to\infty} \frac{f(x_n + t_n h) - f(x_n)}{t_n},$$

provided the convergence is uniform for h in compact sets.

Remark 23.30. If f is Lipschitz near x, the convergence is uniform for h in compact sets. Definition 23.29 is a certain "Hadamard type strict derivative."

Definition 23.31 ([26]). We define the vector-valued directional derivative $Df(x,h)$ of f at $x \in X$ in direction $h \in X$ by $Df(x,\cdot) : X \to Y$,

$$Df(x,h) := \lim_{\varepsilon \downarrow 0} \sup_{u \in U(x,\varepsilon),\ t \in (0,\varepsilon)} \frac{f(u+th) - f(u)}{t}.$$

Remark 23.32. In the following, we assume that certain directional derivatives exist. In order to have sufficient conditions for the existence of the directional derivative, one can suppose that Y is a Daniell locally convex vector lattice and f is locally Lipschitz.

Using the vector-valued directional derivative, we introduce the subdifferential of $f : X \to Y$:

Definition 23.33. The subdifferential of $f : X \to Y$ at the point $x \in X$ is defined by

$$\partial f(x) := \{L \in \mathscr{L}(X,Y) \mid L(h) \leq_K Df(x,h)\ \forall h \in X\},$$

where $\mathscr{L}(X,Y)$ denotes the space of linear continuous operators from X to Y.

Under certain conditions on a set $D \subset Y$, we can conclude from the derivatives being an element of D that certain differential quotients are elements of D as well:

Lemma 23.34 ([26]). *Let $D \subset Y$ be such that*

(i) *$\mathrm{int}\, D \neq \emptyset$*
(ii) *$\mathrm{int}\, D - K \subset \mathrm{int}\, D$.*

Assume $Df(x,h) \in \mathrm{int}\, D$ for $x,h \in X$. Then there is a real number $\varepsilon(h) > 0$ such that

$$\frac{f(u+th) - f(u)}{t} \in \mathrm{int}\, D \qquad \forall u \in U(x,\varepsilon(h)),\ t \in (0,\varepsilon(h)). \tag{23.7}$$

What is more, also with small perturbations of the direction h an estimation for the differential quotient can be given.

Lemma 23.35. *Assume that $D \subset Y$ satisfies the conditions (i) and (ii) from Lemma 23.34, $x,h \in X$. Moreover, suppose that f is locally Lipschitz at $x \in X$, the vector-valued directional derivative $Df(x,h)$ exists and $Df(x,h) \in \mathrm{int}\, D$. Then, for each neighborhood V of 0 in Y satisfying $Df(x,h) + V \subset \mathrm{int}\, D$ there is a real number $\varepsilon(h) > 0$ and a neighborhood U' of h such that*

$$\frac{f(u+th') - f(u)}{t} \in Df(x,h) + V - K \qquad \forall u \in U(x,\varepsilon(h)),\ h' \in U',\ t \in (0,\varepsilon(h)). \tag{23.8}$$

In particular this implies

$$\frac{f(u+th')-f(u)}{t} \in int\, D \qquad \forall u \in U(x,\varepsilon(h)),\ h' \in U',\ t \in (0,\varepsilon(h)). \tag{23.9}$$

Proof. There is a neighborhood $V \subset Y$ of 0 such that $Df(x,h)+V \subset int\, D$. Without loss of generality, we assume V to be solid.

Choose a solid neighborhood V' of 0 such that $V'+V' \subset V$; furthermore, choose $\varepsilon_0 > 0$ such that

$$\sup_{u\in U(x,\varepsilon_0),\ t\in(0,\varepsilon_0)} \frac{f(u+th)-f(u)}{t} \in Df(x,h)+V', \text{ hence} \tag{23.10}$$

$$\frac{f(u+th)-f(u)}{t} \in Df(x,h)+V'-K \tag{23.11}$$

$$\forall u \in U(x,\varepsilon_0),\ t \in (0,\varepsilon_0).$$

Finally, fix $U' \in \mathscr{U}(h)$ such that for each $h' \in U'$ holds

$$P(h'-h,\varepsilon_0) \in V',$$

with P being the function corresponding to (23.6). Now, define $\varepsilon(h) := \min\{\frac{\varepsilon_0}{2},\ \frac{\varepsilon_0}{2\|h\|}\}$. For each $u \in U(x,\varepsilon(h))$ and each $t \in (0,\varepsilon)$ we have $u+th \in u(x,\varepsilon_0)$. Hence, the vector-valued local Lipschitz property of f yields for these u and t

$$\left|\frac{f(u+th')-f(u+th)}{t}\right| \leq P(h'-h,\varepsilon_0) \in V'. \tag{23.12}$$

The solidity of V' now leads to

$$\frac{f(u+th')-f(u+th)}{t} \in V'. \tag{23.13}$$

For each $u \in U(x,\varepsilon(h))$, $t \in (0,\varepsilon(h))$ and $h' \in U'$ we thus have derived

$$\frac{f(u+th')-f(u)}{t} = \underbrace{\frac{f(u+th)-f(u)}{t}}_{\in Df(x,h)+V'-K} + \underbrace{\frac{f(u+th')-f(u+th)}{t}}_{\in V'}$$

$$\subset Df(x,h)+V-K \subset int\, D.$$

□

Lemma 23.36. *Assume that $g: X \to Y$ is sublinear and $f: X \to Y$ convex. Then it holds:*

(i) $Dg(x,h) \leq g(h)$ *for all* $x \in X$ *and all* $h \in X$.
(ii) $Dg(0,h) = g(h)$ *for all* $h \in X$.
(iii) $\partial f(x) = \partial Df(x,\cdot)(0)$.

Proof. (i). By the sublinearity of g, we have for all $x,h \in X$ and $t \in (0,1)$

$$g(u+th) \leq g(u)+g(th), \qquad \text{also}$$
$$\frac{g(u+th)-g(u)}{t} \leq \frac{g(th)}{t} = g(h);$$

the last equality holds because g is positively homogeneous. Consequently, it follows that $\sup_{u\in U(x,\varepsilon),\, t\in(0,\varepsilon)} \frac{g(u+th)-g(u)}{t} \leq g(h)$ for $\varepsilon < 1$. For the limes (which is guaranteed to exist) this implies

$$\lim_{\varepsilon\downarrow 0} \sup_{u\in U(x,\varepsilon),\, t\in(0,\varepsilon)} \frac{g(u+th)-g(u)}{t} \leq g(h).$$

(ii) For $x = 0$ the supremum is attained at $u = 0$.

(iii) Set $\hat{g} = Df(x,\cdot)$. Then $\hat{g}$ is subadditive and positively homogeneous (cf. Staib [26, Lemma 1.2.6]. Hence $D\hat{g}(0,h) = \hat{g}(h)$ by (ii). Thus, $D\hat{g}(0,h) = Df(x,h)$ holds according to the definition of $\hat{g}$, and the assertion follows. □

In Section 23.7, we will show necessary conditions for approximately efficient elements of a vector optimization problem using the Mordukhovich subdifferential. Here we recall the corresponding definition.

Definition 23.37. [18] Let S be a non-empty subset of X and let $\alpha \geq 0$. Given $x \in \text{cl } S$ the non-empty set

$$N^F_\alpha(S,x) = \left\{ x^* \in X^* : \limsup_{y\to x, y\in S} \frac{\langle x^*, y-x\rangle}{\|y-x\|} \leq \alpha \right\}$$

is called the set of Fréchet α-normals to S at x. When $\alpha = 0$, then the above set is a cone, called the set of Fréchet normals and denoted by $N^F(S,x)$.
Let $x_0 \in \text{cl } S$. The non-empty cone

$$N_L(S,x_0) = \limsup_{x\to x_0, \alpha\downarrow 0} N^F_\alpha(S,x)$$

is called the limiting normal cone or the Mordukhovich normal cone to S at x_0.

Definition 23.38. [18] Let $f : X \to \mathbb{R}\cup\{\pm\infty\}$ be a given proper function and $x_0 \in \text{dom} f$. The set

$$\partial_L f(x_0) = \{x^* \in X^* : (x^*,-1) \in N_L(\text{epi} f, (x_0, f(x_0)))\}$$

is called the limiting subdifferential or the Mordukhovich subdifferential of f at x_0. If $x_0 \notin \text{dom} f$, then we set $\partial_L f(x_0) = \emptyset$.

Remark 23.39. If $f : X \to \mathbb{R}\cup\{\pm\infty\}$ is convex, then $\partial_L f(x)$ coincides with the Fenchel subdifferential $\partial f(x)$.

23.6 Necessary Optimality Conditions for Vector Optimization Problems in General Spaces Based on Directional Derivatives

In this section, we derive necessary conditions for approximate solutions of the vector optimization problem (VP). Under the assumption that the ordering cone K has a non-empty interior, we show in Theorem 23.41 necessary conditions for approximately efficient elements of (VP). Furthermore, in Theorem 23.43 we derive necessary conditions for approximately efficient points of (VP) without the assumption that $\operatorname{int} K \neq \emptyset$. Here $\varepsilon > 0$ and $k^0 \in K \setminus \{0\}$ are fixed arbitrarily and represent an admissible error of the approximate solutions.

Lemma 23.40. *Suppose that $K \subset Y$ is a pointed closed convex cone with* $\operatorname{int} K \neq \emptyset$. *Fix an arbitrary $c > 0$, $k^0 \in \operatorname{int} K$ and set $D = -ck^0 - K$. Then it holds* $\operatorname{int} D - K \subset \operatorname{int} D$.

Proof. Fix an arbitrary $y \in \operatorname{int} D - K$. This means $y = y_1 - y_2$ with certain $y_1 \in \operatorname{int} D$ and $y_2 \in K$, where again $y_1 = -ck^0 - y_3$ with an $y_3 \in \operatorname{int} K$. Hence we have $y = -ck^0 - (y_2 + y_3)$. Now, by the convexity of K we conclude $(y_2 + y_3) \in \operatorname{int} K$ and consequently $y \in \operatorname{int} D$. □

Theorem 23.41. *Consider the vector optimization problem (VP). Suppose that $K \subset Y$ is a closed normal cone, $\operatorname{int} K \neq \emptyset$ and $k^0 \in K \setminus \{0\}$. Assume $(A_{space}2)$, $(A_{map}1)$, and $(A_{map}4)$ with respect to K and k^0. Furthermore, suppose that $S \subseteq X$ is closed. If $\varepsilon > 0$ is an arbitrary real number and $x_0 \in S$ is an element with $f(x_0) \leq_K f(x) + \varepsilon k^0$ for all $x \in S$ then there exists an element $x_\varepsilon \in S$ with $f(x_\varepsilon) \leq_K f(x_0)$ and*

(i) $\|x_0 - x_\varepsilon\| \leq \sqrt{\varepsilon}$;

(ii) $Df(x_\varepsilon, h) \notin -\sqrt{\varepsilon} k^0 \|h\| - \operatorname{int} K$ *for all* $h \in \mathscr{T}(S, x_\varepsilon)$.

(iii) $C \subset \mathscr{T}(S, x_\varepsilon)$, *$C$ is a convex cone, implies the existence of an $y^* \in K^*$, $y^* \neq 0$, satisfying*

$$y^* \circ Df(x_\varepsilon, h) \geq -\sqrt{\varepsilon}\, y^*(k^0) \text{ for all } h \in C \text{ with } \|h\| = 1.$$

(iv) *Assume $(A_{map}5)$ and $S = X$. For C as above there is an element $y^* \in K^* \setminus \{0\}$ such that*

$$0 \in \overline{y^* \circ \partial f(x_\varepsilon)}^{w^*} - C^* + \sqrt{\varepsilon}\, y^*(k^0) B^0_{X^*}$$

(where ∂ is the usual convex subdifferential and $B^0_{X^*}$ is the unit ball in X^*).
If the order intervals in Y are weakly compact, there holds even

$$0 \in y^* \circ \partial f(x_\varepsilon) - C^* + \sqrt{\varepsilon}\, y^*(k^0) B^0_{X^*}.$$

Proof. The assumptions $(A_{space}2)$, $(A_{map}4)$ and that K is a closed normal cone are fulfilled. Furthermore, since S is a closed set in a Banach space, it is a complete metric space endowed with the distance given by the norm such that the assumptions of Theorem 23.18 are fulfilled. Choose $x_\varepsilon \in S$ according to Theorem 23.18; this directly implies (i).

(ii) Furthermore, for the element $x_\varepsilon \in S$, the following holds:

$$f_{\sqrt{\varepsilon}k^0}(x_\varepsilon) \in Eff(f_{\sqrt{\varepsilon}k^0}[S], K), \quad \text{where} \quad f_{\sqrt{\varepsilon}k^0}(x) := f(x) + \sqrt{\varepsilon}k^0\|x - x_\varepsilon\|$$

taking into account Theorem 23.18. This means

$$f(x) + \sqrt{\varepsilon}k^0\|x - x_\varepsilon\| \notin f(x_\varepsilon) - K \setminus \{0\} \qquad \forall x \in S.$$

Fix an $h \in \mathscr{T}(S, x_\varepsilon)$. Then there are sequences $h_n \to h$, $t_n \downarrow 0$ such that $x_\varepsilon + t_n h_n \in S$. For these we have

$$f(x_\varepsilon + t_n h_n) + \sqrt{\varepsilon}k^0 t_n \|h_n\| - f(x_\varepsilon) \notin -K \setminus \{0\},$$

hence

$$\frac{f(x_\varepsilon + t_n h_n) - f(x_\varepsilon)}{t_n} \notin -\sqrt{\varepsilon}k^0\|h_n\| - K \setminus \{0\}. \tag{23.14}$$

Assume now $Df(x_\varepsilon, h) \in -\sqrt{\varepsilon}k^0\|h\| - \text{int } K$; this means

$$Df(x_\varepsilon, h) = -\sqrt{\varepsilon}k^0\|h\| - y_1$$

with an $y_1 \in \text{int } K$.

Choose a neighborhood V of 0 in Y in such a way that $y_1 + 2V \subset \text{int } K$. According to Lemma 23.35, with $D := -\sqrt{\varepsilon}k^0\|h\| - K$ there is a number $\varepsilon(h) > 0$ and a neighborhood U' of h, such that (in particular, with $u = x_\varepsilon$)

$$\frac{f(x_\varepsilon + th') - f(x_\varepsilon)}{t} \in Df(x_\varepsilon, h) + V - K = -\sqrt{\varepsilon}k^0\|h\| - y_1 + V - K$$

holds for all $t \in (0, \varepsilon(h))$ and $h' \in U'$. For sufficiently large indices n this implies

$$\frac{f(x_\varepsilon + t_n h_n) - f(x_\varepsilon)}{t_n} \in -\sqrt{\varepsilon}k^0\|h\| - y_1 + V - K.$$

Finally, choose n large enough to satisfy

$$-\sqrt{\varepsilon}k^0\|h\| = -\sqrt{\varepsilon}k^0\|h_n\| + v_n \quad \text{with a } v_n \in V,$$

which is possible because of $h_n \to h$. This, however, means

$$\frac{f(x_\varepsilon + t_n h_n) - f(x_\varepsilon)}{t_n} \in -\underbrace{\sqrt{\varepsilon}k^0\|h_n\| - y_1 + 2V}_{\subset \text{int } D} - K \subset \text{int } D,$$

contradicting (23.14).

(iii) Let B^0 denote the unit ball in X. The set $Df(x_\varepsilon, C \cap B^0) + K$ is convex. Since $-\sqrt{\varepsilon}k^0\|h\| - \text{int } K \supset -\sqrt{\varepsilon}k^0 - \text{int } K$ for elements h with $\|h\| \leq 1$, we have

$(Df(x_\varepsilon, C\cap B^0)+K)\cap(-\sqrt{\varepsilon}k^0-\text{int } K)=\emptyset$ by (ii); this means

$$\left[(Df(x_\varepsilon, C\cap B^0)+\sqrt{\varepsilon}k^0+K\right]\cap -\text{int } K=\emptyset.$$

By a separation argument, we find an element $y^*\in Y^*$ with $y^*\neq 0$ and an $\beta\in\mathbb{R}$ satisfying

$$y^*(y)\geq\beta \qquad \forall y\in Df(x_\varepsilon, C\cap B^0)+\sqrt{\varepsilon}k^0+K \tag{23.15}$$
$$y^*(y)<\beta \qquad \forall y\in -\text{int } K. \tag{23.16}$$

Since $0\in \text{cl int }(-K)=-K$, from (23.16) follows that $\beta\geq 0$.

Now assume that $y^*(y)>0$ for an element $y\in -\text{int } K$. For a certain positive multiple $cy\in -\text{int } K$ of y this implies $y^*(cy)>\beta$, contradicting (23.16). Hence, $y^*(y)\leq 0$ for each $y\in -\text{int } K$; this inequality even holds for each $y\in -\text{cl int } K$ because of the continuity of y^*. This means $y^*\in K^*\setminus\{0\}$.

In the following we exploit (23.15):
Let $h\in C$, $\|h\|=1$. With $y\in Df(x_\varepsilon,h)+\sqrt{\varepsilon}k^0+v$ ($v\in K$ arbitrary) we also have $y^*(y)\geq 0$. Hence

$$y^*(Df(x_\varepsilon,h)+\sqrt{\varepsilon}k^0+v)\geq 0;$$

in particular, with $v=0$ we get

$$y^*\circ Df(x_\varepsilon,h)\geq -\sqrt{\varepsilon}\,y^*(k^0).$$

(iv). For C as in (iii) choose $y^*\in K^*$ according to (iii); this is,

$$y^*\circ Df(x_\varepsilon,h)\geq -\sqrt{\varepsilon}\,y^*(k^0)\|h\|$$

for all $h\in C$. Define $p(h):=y^*\circ Df(x_\varepsilon,h)+\sqrt{\varepsilon}\,y^*(k^0)\|h\|$ for $h\in C$ and the sets S_1 and S_2 in $X\times\mathbb{R}$ by

$$S_1 := \text{epi }(p),$$
$$S_2 := \{(h,\alpha)\in X\times\mathbb{R}:\quad h\in C,\ \alpha\leq 0\}.$$

Both S_1 as well as S_2 is convex. Furthermore, we have int $S_1\neq\emptyset$ and int $S_1\cap S_2=\emptyset$. By a separation argument we conclude the existence of an $(x^*,\alpha^*)\in(X\times\mathbb{R})^*=X^*\times\mathbb{R}$, $(x^*,\alpha^*)\neq 0$ and $\beta\in\mathbb{R}$ satisfying

$$(x^*,\alpha^*)(h,\alpha)\geq\beta \qquad \forall(h,\alpha)\in S_1, \tag{23.17}$$
$$(x^*,\alpha^*)(h,\alpha)\leq\beta \qquad \forall(h,\alpha)\in S_2. \tag{23.18}$$

With $(0,0)\in S_1\cap S_2$ we deduce $\beta=0$, and $(0,\alpha)\in \text{int } S_1$ for $\alpha>0$ yields $\alpha^*>0$. Setting $\alpha=0$ in (23.17) leads to $\frac{x^*}{-\alpha^*}\in C^*$. Using (23.18) this yields

$$\frac{x^*}{-\alpha^*}(h) \leq y^* \circ Df(x_\varepsilon, h) + \sqrt{\varepsilon}\|h\|\, y^*(k^0).$$

Since $y^* \in K^*$, $y^* \circ Df(x_\varepsilon, h)$ is a convex function in h; this is passed on to the whole right side of the above inequality. Hence, we have

$$\frac{x^*}{-\alpha^*} \in \partial(y^* \circ Df(x_\varepsilon, \cdot) + \sqrt{\varepsilon}\|\cdot\|\, y^*(k^0))(0)$$

with the usual convex subdifferential ∂. Subdifferential calculus further yields

$$\begin{aligned}\frac{x^*}{-\alpha^*} &\in \partial(y^* \circ Df(x_\varepsilon, \cdot))(0) + \partial(\sqrt{\varepsilon}\|\cdot\|\, y^*(k^0))(0)\\ &\subset \overline{y^* \circ \partial(Df(x_\varepsilon, \cdot))(0)}^{w^*} + \sqrt{\varepsilon}\, y^*(k^0) B^0_{X^*}.\end{aligned}$$

By Lemma 23.36 (iii), under convexity assumptions concerning f, this implies

$$\frac{x^*}{-\alpha^*} \in \overline{y^* \circ \partial f(x_\varepsilon)}^{w^*} + \sqrt{\varepsilon}\, y^*(k^0) B^0_{X^*}.$$

This yields

$$0 \in \overline{y^* \circ \partial f(x_\varepsilon)}^{w^*} + \sqrt{\varepsilon}\, y^*(k^0) B^0_{X^*} - C^*.$$

Regarding the formula when the order intervals are weakly compact: We have to show that $y^* \circ \partial f(x_\varepsilon)$ is weakly closed. This follows by an argumentation along the lines of [2, Theorem 6.3] for convex operators.

□

Remark 23.42. The assertions in Theorem 23.41 are corrections of corresponding results in [26, Theorem 2.2.1] and an extension of the results in [26, Theorem 2.2.1] to approximate solutions.

Using a closed normal cone $B \subset Y$ with $K \setminus \{0\} \subset \text{int}\, B$ like in the concept of proper efficiency, we can drop the strong assumption $\text{int}\, K \neq \emptyset$ for the ordering cone K in Y. We will show a necessary condition under the assumption that f is strictly differentiable using the abstract nonlinear scalarizing scheme and Clarke's strict derivative $D_S f(x)$ of f at $x \in X$.

Theorem 23.43. *Consider the vector optimization problem (VP) with $S = X$ assuming $(A_{space}2)$, $(A_{map}2)$. Let $K \subset Y$ be a pointed closed convex cone, $k^0 \in K \setminus \{0\}$ and $B \subset Y$ a closed normal cone with $K \setminus \{0\} \subset \text{int}\, B$. We suppose that $(A_{map}4)$ with respect to B and k^0 is fulfilled. If $\varepsilon > 0$ is an arbitrary real number and there exists an element $x_0 \in X$ such that $f(x_0) \leq_B f(x) + \varepsilon k^0$ for all $x \in X$ then there is an element $x_\varepsilon \in X$ with $f(x_\varepsilon) \leq_B f(x_0)$ such that*

(i) $\|x_0 - x_\varepsilon\| \leq \sqrt{\varepsilon}$.
(ii) *There exists $y^* \in K^\#$ such that*

$$\|y^* \circ D_S f(x_\varepsilon)\|_* \leq \sqrt{\varepsilon}.$$

Proof. Consider $x_0 \in X$ such that $f(x_0) \leq_B f(x) + \varepsilon k^0$ for all $x \in X$, where B is a closed normal cone with $K \setminus \{0\} \subset \text{int } B$. Because of $(A_{space}2)$ (this implies $(A_{space}1)$) and $(A_{map}4)$ with respect to B and k^0 the assumptions of Theorem 23.18 are fulfilled.

According to Theorem 23.18, we get the existence of an element $x_\varepsilon \in X$ such that (i) holds. Furthermore, $f(x_\varepsilon) \in Eff(f_{\sqrt{\varepsilon}k^0}(X), B)$ holds for x_ε, where $f_{\sqrt{\varepsilon}k^0}(x) := f(x) + \sqrt{\varepsilon}k^0 \|x - x_\varepsilon\|$.
This means

$$f(x) + \sqrt{\varepsilon}k^0 \|x - x_\varepsilon\| \notin f(x_\varepsilon) - (B \setminus \{0\}) \qquad \forall x \in X,$$

i.e.,

$$f(x) \notin f(x_\varepsilon) - \sqrt{\varepsilon}k^0 \|x - x_\varepsilon\| - (B \setminus \{0\}) \qquad \forall x \in X. \tag{23.19}$$

Consider the functional (23.3) and take $\varphi(y) := \varphi_{B,k^0}(y - f(x_\varepsilon))$. For B, K and k^0 the assumptions in Example 23.20 are fulfilled and so we get the properties $(A_\varphi 1')$ with respect to K, $(A_\varphi 2)$ and $(A_\varphi 4')$ for φ. Assume that there exists $x \in X$ such that

$$\varphi(f(x)) + \sqrt{\varepsilon}\|x - x_\varepsilon\| < \varphi(f(x_\varepsilon)) = 0.$$

Then there exists $t < -\sqrt{\varepsilon}\|x - x_\varepsilon\|$ with $f(x) - f(x_\varepsilon) \in tk^0 - B$ and so

$$\begin{aligned} f(x) &\in f(x_\varepsilon) - \sqrt{\varepsilon}\|x - x_\varepsilon\|k^0 - (B + (-\sqrt{\varepsilon}\|x - x_\varepsilon\| - t)k^0) \\ &\subset f(x_\varepsilon) - \sqrt{\varepsilon}\|x - x_\varepsilon\|k^0 - \text{int } B \\ &\subset f(x_\varepsilon) - \sqrt{\varepsilon}\|x - x_\varepsilon\|k^0 - (B \setminus \{0\}), \end{aligned}$$

a contradiction to (23.19). So we get

$$\varphi(f(x)) \geq \varphi(f(x_\varepsilon)) - \sqrt{\varepsilon}\|x - x_\varepsilon\| \qquad \forall x \in X.$$

Because of $(A_\varphi 2)$ and $(A_\varphi 4')$ we get that the scalarizing functional φ is locally Lipschitz. Furthermore, f is supposed to be a strictly differentiable mapping and so locally Lipschitz. Hence the composition $\varphi \circ f$ is locally Lipschitz such that we can use Clarke's generalized directional derivative $(\varphi \circ f)^\circ$.

Consider now for $n \in \mathbb{N}$, $t_n > 0$, $x := x_\varepsilon + t_n h_n$ with $h_n \in U$ (U is a neighborhood of $h \in X$) and $\|h\| = 1$. For these we have

$$\frac{\varphi(f(x_\varepsilon + t_n h_n)) - \varphi(f(x_\varepsilon))}{t_n} \geq -\sqrt{\varepsilon}\|h_n\|. \tag{23.20}$$

Taking the limits for $t_n \to 0$ and $h_n \to h$ we get for Clarke's generalized directional derivative

$$(\varphi \circ f)^\circ(x, h) \geq -\sqrt{\varepsilon} \quad \forall h \in X \text{ with } \|h\| = 1.$$

Using the chain rule given by [3] (Theorem 2.3.10 and Proposition 2.1.2) we get that there is an element $y^* \in \partial\varphi(f(x_\varepsilon))$ such that for all $h \in X$ with $\|h\| = 1$

$$y^* \circ D_S f(x_\varepsilon)(h) \geq -\sqrt{\varepsilon}.$$

Taking into account the linearity of $D_S f(x_\varepsilon)$, we get (if we replace h by $-h$)

$$y^* \circ D_S f(x_\varepsilon)(h) \leq \sqrt{\varepsilon}$$

such that

$$||y^* \circ D_S f(x_\varepsilon)||_* \leq \sqrt{\varepsilon}.$$

Finally, we will show $y^* \in K^\#$ using the property $(A_\varphi 1')$ with respect to K of φ. Let $k \in K \setminus \{0\}$. Thus we have $\varphi(y) > \varphi(y-k)$. Since φ is a continuous convex function on the Banach space Y one has $\partial\varphi(y) \neq \emptyset$ for each $y \in Y$. Thus we have

$$\varphi(y) > \varphi(y-k) \geq \varphi(y) + y^*(-k) \quad \forall y^* \in \partial\varphi(y).$$

This shows that $y^*(k) > 0$ for any $k \in K \setminus \{0\}$. This immediately yields that $y^* \in K^\#$. This completes the proof. □

For problems with restrictions we get the following result:

Theorem 23.44. *Consider the vector optimization problem (VP) under the assumptions* $(A_{space}2)$ *and* $(A_{map}2)$*. Suppose that* $S \subseteq X$ *is closed. Let* $K \subset Y$ *be a pointed closed convex cone,* $k^0 \in K \setminus \{0\}$ *and* $B \subset Y$ *a closed normal cone with* $K \setminus \{0\} \subset$ *int* B*. We suppose that* $(A_{map}4)$ *with respect to* B *and* k^0 *is fulfilled. If* $\varepsilon > 0$ *is an arbitrary real number and there exists an element* $x_0 \in S$ *such that* $f(x_0) \leq_B f(x) + \varepsilon k^0$ *for all* $x \in S$ *then there is an element* $x_\varepsilon \in S$ *with* $f(x_\varepsilon) \leq_B f(x_0)$ *such that*

(i) $\|x_0 - x_\varepsilon\| \leq \sqrt{\varepsilon}$*.*
(ii) There exists $y^* \in K^\#$ *such that*

$$y^* \circ D_S f(x_\varepsilon)(h) \geq -\sqrt{\varepsilon} \quad \forall h \in \mathscr{T}(S, x_\varepsilon) \text{ with } ||h|| = 1.$$

Proof. We follow the line of the proof of Theorem 23.43. □

Remark 23.45. The assertions in Theorems 23.43 and 23.44 are related to the proper efficiency of the element x_ε. Especially, (23.19) says that x_ε is a properly efficient point of $f_{\sqrt{\varepsilon}k^0}$ over X with respect to K because B is a closed normal cone with $K \setminus \{0\} \subset \text{int}\, B$, i.e., $f_{\sqrt{\varepsilon}k^0}(x_\varepsilon) \in pEff(f_{\sqrt{\varepsilon}k^0}(X), K)$. The property $K \setminus \{0\} \subset \text{int}\, B$ implies in both theorems the strong assertion $y^* \in K^\#$ for the multiplier y^*.

Remark 23.46. In order to derive necessary conditions for εk^0 -efficient points of (VP) (with $\varepsilon > 0$), i.e., for $x_0 \in S$ with $f(x_0) \in \varepsilon k^0 - Eff(f(S), K)$, it would be possible to use the same procedures like in the proofs of Theorems 23.41, 23.43, and 23.44 using corresponding variational principles (for instance [9, Corollary 3.10.14]). The same holds for εk^0 -properly efficient points of (VP) (with $\varepsilon > 0$), i.e., for $x_0 \in S$ with $f(x_0) \in \varepsilon k^0 - pEff(f(S), K)$.

23.7 Vector Optimization Problems with Finite-Dimensional Image Spaces

As seen in Theorem 23.41, the assumptions concerning the ordering cone K for deriving optimality conditions in general spaces are strong. Now, we will show necessary optimality conditions for vector optimization problems where the objective function f takes its values in a finite-dimensional space $\mathbb{R}^n$ under weaker assumptions. Corresponding results are shown in [5], [6] and [20].

For a locally Lipschitz function f we derive Lagrangian multiplier rules for approximately efficient elements of (VP) using Mordukhovich's subdifferential calculus (see Definition 23.38).

Theorem 23.47. *Consider the vector optimization problem (VP). Assume that* $(A_{map}3)$ *and* $(A_{space}3)$ *are satisfied. Suppose that* $S \subseteq X$ *is closed. Let* $K \subset Y$ *be a pointed closed convex cone,* $k^0 \in K \setminus \{0\}$ *and* $B \subset Y$ *a closed normal cone with* $K \setminus \{0\} \subset \operatorname{int} B$*. If* $\varepsilon > 0$ *is an arbitrary real number and there exists an element* $x_0 \in S$ *such that* $f(x_0) \leq_B f(x) + \varepsilon k^0$ *for all* $x \in S$*, then there are elements* $x_\varepsilon \in S$ *with* $f(x_\varepsilon) \leq_B f(x_0)$*,* $u^* \in K^{\#}$ *with* $u^*(k^0) = 1$ *and* $x^* \in X^*$ *with* $\|x^*\| \leq 1$ *such that*

$$0 \in \partial_L(u^* \circ f)(x_\varepsilon) + u^*(k^0)\sqrt{\varepsilon}x^*(x_\varepsilon) + N_{\partial_L}(S, x_\varepsilon).$$

Proof. Consider an element $x_0 \in S$ such that $f(x_0) \leq_B f(x) + \varepsilon k^0$ for all $x \in S$, where $B \subset Y$ is a closed normal cone with $K \setminus \{0\} \subset \operatorname{int} B$. Taking into account that $(A_{map}3)$ is fulfilled for the function f, it is continuous as well and since S is a closed set in a Asplund space, it is a complete metric space endowed with the distance given by the norm such that the assumptions of Theorem 23.18 are fulfilled. From this variational principle we get the existence of an element $x_\varepsilon \in S$ such that $f(x_\varepsilon) \leq_B f(x_0)$. Moreover, $f(x_\varepsilon) \in Eff(f_{\sqrt{\varepsilon}k^0}(S), B)$ holds for x_ε, where $f_{\sqrt{\varepsilon}k^0}$ the perturbed objective function is given by

$$f_{\sqrt{\varepsilon}k^0}(x) := f(x) + k^0\sqrt{\varepsilon}\,\|x - x_\varepsilon\|\,.$$

Now, applying Theorem 3.1 in [5] we can find $u^* \in \partial\varphi(v)$ with $u^* \in K^*$, $u^*(k^0) = 1$ (where the scalarizing function φ is given by $\varphi(y) := \varphi_{B,k^0}(y - f_{\sqrt{\varepsilon}k^0}(x_\varepsilon))$ (cf. (23.3)) and has the properties $(A_\varphi 1')$ with respect to K, $(A_\varphi 2)$ and $(A_\varphi 4')$) such that by the calculation rules for the Mordukhovich subdifferential

$$0 \in \partial_L(u^* \circ f_{\sqrt{\varepsilon}k^0})(x_\varepsilon) + N_{\partial_L}(S, x_\varepsilon). \tag{23.21}$$

Consider an element $x_\varepsilon^* \in \partial_L(u^* \circ f_{\sqrt{\varepsilon}k^0})(x_\varepsilon)$ involved in (23.21). Because of

$$\partial_L(u^* \circ f_{\sqrt{\varepsilon}k^0})(x) = \partial_L(u^* \circ (f(\cdot) + k^0\sqrt{\varepsilon}\,\|\cdot - x_\varepsilon\|))(x),$$

by use of the rule for sums and the property that the Mordukhovich subdifferential coincides in the convex case with the Fenchel subdifferential, we get

$$x_\varepsilon^* \in \partial_L(u^* \circ f)(x_\varepsilon) + u^*(k^0)\sqrt{\varepsilon}\partial\,\|\cdot - x_\varepsilon\|\,(x_\varepsilon). \tag{23.22}$$

From (23.21) and (23.22) and taking into account the calculation rule for the subdifferential of the norm, it follows that there is $x^* \in X^*$ with $\|x^*\| \leq 1$ such that

$$0 \in \partial_L(u^* \circ f)(x_\varepsilon) + u^*(k^0)\sqrt{\varepsilon}x^*(x_\varepsilon) + N_{\partial_L}(S, x_\varepsilon)).$$

The property $u^* \in K^\#$ follows analogously like in the proof of Theorem 23.43. □

References

1. Artzner, P., Delbean, F., Eber, J.-M. and Heath, D., 1999, *Coherent measures of risk*. Math. Finance, 9 (3) , 203–228.
2. Borwein, J.M., 1982, *Continuity and Differentiability Properties of Convex Operators*, Proc. London Math. Soc., 44, 420–444.
3. Clarke, F. H., 1983 *Optimization and Nonsmooth Analysis*, Wiley Interscience, New York.
4. Da Silva, A.R., 1987, *Evaluation functionals are the extreme points of a basis for the dual of* $C_1^+[a,b]$. In: Jahn, J. and Krabs, W.: *Recent Advances and Historical Development of Vector Optimization*, Lecture Notes in Economics and Mathematical Systems, Springer, Berlin, 294, 86–95.
5. Durea, M., Tammer, C., 2009, *Fuzzy necessary optimality conditions for vector optimization problems*, Reports of the Institute for Mathematics, Martin-Luther-University Halle-Wittenberg, Report 08. To appear in Optimization 58, 449–467.
6. Dutta, J. and Tammer, Chr., 2006, *Lagrangian conditions for vector optimization in Banach spaces*. Mathematical Methods of Operations Research, 64, 521–541.
7. Föllmer, H., Schied, A., 2004, *Stochastic Finance*. Walter de Gruyter, Berlin.
8. Gerth, C., and Weidner, P., 1990, *Nonconvex Separation Theorems and Some Applications in Vector Optimization*. Journal of Optimization Theory and Applications, 67, 2, 297–320.
9. Göpfert, A., Riahi, H., Tammer, Chr. and Zălinescu, C., 2003, *Variational Methods in Partially Ordered Spaces*, Springer, New York.
10. Heyde, F., 2006, *Coherent risk measures and vector optimization*. In: Küfer, K.-H., Rommelfanger, H., Tammer, Chr., Winkler, K. *Multicriteria Decision Making and Fuzzy Systems*, SHAKER Verlag, 3–12; *Reports of the Institute of Optimization and Stochastics, Martin-Luther-University Halle-Wittenberg*, No. 19 (2004), 23–29.
11. Isac, G., 1983, *Sur l'existence de l'optimum de Pareto*, Riv. Mat. Univ. Parma (4) 9, 303–325.
12. Isac, G., 1996, *The Ekeland's principle and the Pareto ε-efficiency*, Multiobjective Programming and Goal Programming: Theories and applications (M. Tamiez ed.) Lecture Notes in Economics and Mathematical Systems, Springer, Berlin, 432, 148–163.
13. Isac, G., 2004, *Nuclear cones in product spaces, Pareto efficiency and Ekeland-type variational principles in locally convex spaces*, Optimization, 53, 3, 253–268.
14. Isac, G., Bulavsky, A.V., Kalashnikov, V.V., 2002, *Complementarity, Equilibrium, Efficiency and Economics*. Kluwer Academic Publishers, Dordrecht.
15. J.-B. Hiriart-Urruty, 1979, *Tangent cones, generalized gradients and mathematical programmming in Banach spaces*, Math. Oper. Res., 4, 79–97.
16. Hyers, D. H., and Isac, G., and Rassias, T. M., 1997, *Topics in Nonlinear Analysis and Applications*, World Scientific, Singapore.
17. Jahn, J., 2004, *Vector Optimization. Theory, Applications, and Extensions*. Springer-Verlag, Berlin.
18. B. S. Mordukhovich and Y. Shao, 1996, Nonsmooth sequential analysis in Asplund spaces, *Transactions of American Mathematical Society*, 348, 1235–1280.
19. Nemeth, A.B., 2004, *Ordered uniform spaces and variational problems*, Italian Journal of Pure and Applied Mathematics, 16, 183–192.

20. Ng, K. F. and Zheng, X. Y., 2005, *The Fermat rule for multifunctions on Banach spaces*. Mathematical Programming, Ser. A, 104, 69–90.
21. Pascoletti, A. and Serafini, P., 1984, *Scalarizing Vector Optimization Problems*. J. Opt. Theory Appl., 42, 499–524.
22. Peressini, A.L., 1967, *Ordered Topological Vector Space*, Harper and Row, New York.
23. Phelps, R.R., 1993, *Convex Functions, Monotone Operators and Differentiability* (2nd ed.). Lect. Notes Math., 1364, Springer, Berlin.
24. Rockafellar, R. T., Uryasev, S. and Zabarankin, M., 2006, *Deviation measures in risk analysis and optimization*, Finance Stochastics, 10, 51–74.
25. Schaefer, H.H., 1986, *Topological Vector Spaces*, Springer-Verlag, Berlin.
26. Staib, T., 1989, *Notwendige Optimalitätsbedingungen in der mehrkriteriellen Optimierung mit Anwendung auf Steuerungsprobleme*, Friedrich-Alexander-Universität Erlangen-Nürnberg.
27. Tammer, Chr., 1994, *A Variational Principle and a Fixed Point Theorem*, System Modelling and Optimization, Henry, J., and Yven, J.-P., 197, Lecture Notes in Control and Informatics Sciences, Springer, Berlin, 248–257.
28. Turinici, M., 1981, *Maximality Principles and Mean-Value Theorems*, Anais Acad. Brasil. Ciencias 53, 653–655.
29. A. Zaffaroni, 2003, *Degrees of efficiency and degrees of minimality*, SIAM Journal on Control and Optimization, 42, 1071–1086.

Chapter 24
Nonlinear Variational Methods for Estimating Effective Properties of Multiscale Materials

Dag Lukkassen, Annette Meidell, and Lars-Erik Persson

Dedicated to the memory of Professor George Isac

Abstract We consider homogenization of sequences of integral functionals with natural growth conditions. Some means are analyzed and used to discuss some fairly new bounds for the homogenized integrand corresponding to integrands which are periodic in the spatial variable. These bounds, which are obtained by variational methods, are compared with the nonlinear bounds of Wiener and Hashin–Shtrikman type. We also point out conditions that make our bounds sharp. Several applications are presented. Moreover, we also discuss bounds for some linear reiterated two-phase problems with m different micro-levels in the spatial variable. In particular, the results imply that the homogenized integrand becomes optimal as m turns to infinity. Both the scalar case (the conductivity problem) and the vector-valued case (the elasticity problem) are considered. In addition, we discuss bounds for estimating the effective behavior described by homogenizing a problem which is a generalization of the Reynold equation.

24.1 Introduction

Variational methods are important for many problems in analysis. For example, the importance of variational methods for the development of modern complementary theory has been pointed out in several papers and books by G, Isac, see, e.g., [18] and

Dag Lukkassen
Narvik University College, and Norut Narvik, P.O.B. 385 N-8505 Narvik, Norway.

Annette Meidell
Narvik University College, P.O.B. 385 N-8505 Narvik, Norway.

Lars-Erik Persson
Department of Mathematics, Lulea University, S-97187, Lulea, Sweden.

P.M. Pardalos et al. (eds.), *Nonlinear Analysis and Variational Problems*,
Springer Optimization and Its Applications 35, DOI 10.1007/978-1-4419-0158-3_24,

the references therein. In this work we use variational methods and homogenization theory to estimate effective properties of multiscale materials.

Many problems in homogenization theory are devoted to the study of the asymptotic behavior as h goes to ∞ of integral functionals of the form

$$\mathscr{F}_h(u) = \int_\Omega f_h(x, Du(x))\,dx,\ \Omega \subset \mathbf{R}^n$$

defined on some (subset of a) Sobolev space $\mathrm{H}^{1,p}(\Omega, \mathbf{R}^N)$ (in general vector-valued functions) where the functions f_h are increasingly oscillating in the first variable, as $h \to +\infty$. Such functionals appear naturally in connection with boundary value problems described by some minimum energy principle of the form

$$E_h = \min\left\{\mathscr{F}_h(u) + \int_\Omega gu dx : u \in \mathrm{H}^{1,p}(\Omega, \mathbf{R}^N),\ u = \phi \text{ on } \gamma_0 \subset \partial\Omega\right\}.$$

In several important cases it happens that the "energy" E_h converges (as h goes to $+\infty$) to

$$E_{\mathrm{hom}} = \min\left\{\mathscr{F}_{\mathrm{hom}}(u) + \int_\Omega gu dx : u \in \mathrm{H}^{1,p}(\Omega, \mathbf{R}^N),\ u = \phi \text{ on } \gamma_0\right\},$$

where $\mathscr{F}_{\mathrm{hom}}$ is of the form

$$\mathscr{F}_{\mathrm{hom}}(u) = \int_\Omega f_{\mathrm{hom}}(Du(x))\,dx.$$

Hence, E_{hom} approximates the actual energy E_h for cases where the characteristic length of the oscillations is small compared with the size of Ω. The problem of proving the convergence $E_h \to E_{\mathrm{hom}}$ is often treated by using Γ-convergence, introduced by Ennio De Georgie in the early 1970s. Concerning this fact and other basic information in homogenization theory we refer to the literature, e.g., the books [6, 2, 4, 9, 10, 11, 19, 47] and the papers [5, 12, 14, 43, 46].

There are many examples of the close connections and interactions between the theory of means and theory of bounds for the homogenization integrand f_{hom}. For instance, it is well known that if $N = 1$ and the functions f_h are of the form $f_h(x, \xi) = \lambda(hx)\,|\xi|^2$ where λ is periodic and positively upper and lower bounded, then

$$q_h\,|\xi|^2 \le f_{\mathrm{hom}}(\xi) \le q_a\,|\xi|^2,$$

where q_h and q_a denote the harmonic mean of λ and the arithmetic mean of λ, respectively. These inequalities have been improved in many different ways (see [19], [23]–[40] and the references given there).

In this chapter we discuss some fairly new results of this kind obtained by the authors of this paper. By assuming periodicity in the first variable and certain growth-conditions in the second variable of the functions f_h, we present upper and lower bounds for the homogenized integrand $f_{\mathrm{hom}}(\xi)$. These bounds turn out to be some kind of mixture of various types of means. Some of the keys for proving these bounds are to prove several results of independent interest connected to some useful means of power type.

The bounds are analyzed and compared with other types of bounds. In particular, we establish conditions where these bounds are much sharper than the bounds of Wiener and Hashin–Shtrikman type. We also point out conditions that make our bounds sharp. Several applications are presented. For instance, we give some concrete nonlinear examples for which our bounds are very close to each other and where it is almost impossible to give good estimates on the homogenized integrand by more direct numerical treatments.

Motivated by the main ideas in the proof of these results and by applying the iterated homogenization theorem for elliptic operators of Bensoussan, Lions and Papanicolaou [4] (later on generalized in [21], [22] and [35]), we obtain similar estimates for some interesting elliptic cases where the functions f_h are of the form

$$f_h(x,\xi) = f(hx, h^2x, ..., h^mx, \xi).$$

In particular, we discuss a two-phase structure, *the reiterated cell structure,* described by characteristic functions χ_h, and find upper and lower bounds for the homogenized integrand corresponding to functions f_h of the form

$$f_h(x,\xi) = \chi_h(x)g_1(\xi) + (1-\chi_h(x))g_2(\xi).$$

These bounds turn out to be very close to each other for large values of the reiteration number m. In fact, our results show that they converge to the same limit as $m \longrightarrow +\infty$. Moreover, we prove that this limit is optimal within the class of two-phase structures with predescribed volume fractions. Both the scalar case $N = 1$ (the conductivity problem) and the vector-valued case $N = n$ (the elasticity problem) are considered.

We also discuss upper and lower bounds for estimating the effective behavior described by homogenizing a problem which is a generalization of the Reynold equation.

The paper is organized as follows. In Section 24.2 we have collected some necessary preliminaries and notations. In Section 24.3 we state and prove a few results on the nonlinear Wiener bounds and some nonlinear bounds of Hashin–Shtrikman type. Definitions and results connected to means of power type are given in Section 24.4. The nonlinear bounds are stated and discussed in Section 24.5. In Section 24.6 we consider in more detail the special case when $f_h(x,\cdot)$ grows quadratically, and we present some numerical experiments which illustrate our theoretical results in Section 24.5. The results on the bounds connected to the reiterated cell structure are given in Section 24.7. In Section 24.8 we discuss the bounds related to a Reynold-type equation. Some final comments can be found in Section 24.9.

24.2 Preliminaries

Let us start with some notations. If $\square$ is the unit-cube $\square =]0,1[^n$ we shall consider the usual $L^p(\square)$, $[L^p(\square)]^n$ spaces of measurable functions defined on $\square$ with values in $\mathbf{R}$ and $\mathbf{R}^n$, respectively, and the usual Sobolev spaces $H^{1,p}(\square)$, $H_0^{1,p}(\square)$ of measurable functions defined on $\square$ with values in $\mathbf{R}^N$. The real number $p \in]1,+\infty[$ is fixed and we let p' be the number such that $1/p + 1/p' = 1$. Moreover, we shall denote with $\mathscr{C}_{\text{per}}^\infty(\square)$ the space of $\mathscr{C}^\infty$ $\square$-periodic functions, and with $H_{\text{per}}^{1,p}(\square)$ the closure of $\mathscr{C}_{\text{per}}^\infty(\square)$ in $H^{1,p}(\square)$. With $|E|$ we shall denote the Lebesgue measure $m_n(E)$ of a measurable set E, where m_s denotes the Lebesgue measure on $\mathbf{R}^s$. With an abuse of notation we shall identify $\mathbf{R}^{nN}$ with the space of real $N \times n$ matrices.

Let $f : \mathbf{R}^n \times \mathbf{R}^{nN} \to [0,+\infty[$ have the following properties:

$$f(\cdot,z) \text{ is a measurable function} \tag{24.1}$$

for all $z \in \mathbf{R}^{nN}$;

$$f(x,\cdot) \text{ is a convex function} \tag{24.2}$$

for all $x \in \mathbf{R}^n$; the function $f(\cdot,z)$ is $\square$-periodic, i.e., we have that

$$f(x+e_h,z) = f(x,z) \tag{24.3}$$

for all $x \in \mathbf{R}^n, z \in \mathbf{R}^{nN}$ and $h = 1,\ldots,n$ ($e_1,\ldots,e_n$ is the canonical basis of $\mathbf{R}^n$);

$$|z|^p \le f(x,z) \le C(1+|z|^p). \tag{24.4}$$

We recall the minimal energy principle

$$f_{\text{hom}}(\xi) = \min_{u \in H_{\text{per}}^{1,p}(\square)} \int_\square f(x, \xi + Du(x))\,dx, \ \xi \in \mathbf{R}^{nN},$$

where f_{hom} is the homogenized integrand corresponding to the functions f_h, $h = 1,2,3,\ldots$, defined by $f_h(x,z) = f(hx,z)$. We also recall the minimal principle of the complementary energy for $N = 1$:

$$f_{\text{hom}}^*(\eta) = \min_{v \in V_{\text{sol}}^{p'}(\square)} \int_\square f^*(x, \eta + v)\,dx, \ \xi \in \mathbf{R}^n.$$

Here,

$$V_{\text{sol}}^{p'}(\square) = \left\{ v \in \left[L^{p'}(\square)\right]^n : \int_\square v\,dx = 0 \text{ and } \int_\square v \cdot D\varphi\,dx = 0\ \forall \varphi \in H_{\text{per}}^{1,p}(\square) \right\},$$

and f^* is the Legendre–Fenchel dual (convex polar) of f defined by

$$f^*(x,\eta) = \sup_{E \in \mathbf{R}^n} \{E \cdot \eta - f(x,E)\}$$

(the minimal principle of the complementary energy for $N = n$ will be formulated later).

Finally, let us recall the following version of the iterated homogenization theorem for elliptic operators (c.f. [4]):

Theorem 24.1. *Let* $f_h : \mathbf{R}^n \times \mathbf{R}^{nN} \to [0, +\infty[$, $h = 1, 2, 3, \ldots$, *be functions of the form*

$$f_h(x, \xi) = f(hx, h^2 x, \ldots, h^m x, \xi),$$

where f *is* $\square$*-periodic and piecewise constant in the* m *first variables and a quadratic form in the last variable. Then, the corresponding homogenized integrand* f_{hom} *is given by*

$$f_{hom}(\xi) = \psi_m(\xi),$$

where ψ_m *is computed according to the following iterative scheme:*

$$\psi_i(y_1, \ldots, y_{m-i}, \xi) = \min_{u \in \mathrm{H}^{1,p}_{\mathrm{per}}(\square)} \int_{\square} \psi_{i-1}(y_1, \ldots, y_{m-i+1}, \xi + Du(y_{m-i+1}))\, dy_{m-i+1},$$

$$\psi_0(y_1, \ldots, y_m, \xi) = f(y_1, \ldots, y_m, \xi).$$

We also refer to [7] for a nonlinear generalization of Theorem 24.1.

While the minimal energy principle and the minimal principle of the complementary energy frequently will be used in this paper, Theorem 24.1 will only be used in Section 24.6.

24.3 Some Nonlinear Bounds of Classical Type

In this section, we present and discuss a few results connected to some nonlinear bounds of classical type. These bounds will later be compared with the nonlinear bounds (see Section 24.4).

Let a be a m-tuple of the form $a = (a_1, \ldots, a_m)$ such that

$$0 < a_i < 1 \text{ and } \sum_{i=1}^{m} a_i = 1.$$

Similarly, let $g = (g_1, \ldots, g_m)$, where g_i is a convex function $g_i : \mathbf{R}^n \to [0, +\infty[$

$$\lambda_i^- |\xi|^p \le g_i(\xi) \le \lambda_i^+ (|\xi|^p + l),\ l \ge 0,\ g_i(0) = 0,$$

and $\lambda_i^-, \lambda_i^+ > 0$. We let $\mathfrak{I}_{a,g}$ denote the family of functions f of the form

$$f(x, \xi) = \sum_{i=1}^{m} g_i(\xi)\, \chi_{A_i}(x),$$

where $\cup A_i = \mathbf{R}^n$, $A_i \cap A_j = \emptyset$ for $i \neq j$, and such that the characteristic function of the set A_i, χ_{A_i} is $\square$-periodic and $a_i = \int_\square \chi_{A_i} dx$. We recall the well known nonlinear Wiener bounds :

$$f_{W-}(\xi) \leq f_{\text{hom}}(\xi) \leq f_{W+}(\xi),$$

where

$$f^*_{W-}(\xi) = \int_\square f^*(x,\xi)\, dx \text{ and } f_{W+}(\xi) = \int_\square f(x,\xi)\, dx.$$

In the case when f_{hom} is isotropic, i.e., only dependent of $|\xi|$, we have the following sharper nonlinear bounds of Hashin–Shtrikman type (denoted the *HS*-bounds):

$$f_{HS-}(\xi) \leq f_{\text{hom}}(\xi) \leq f_{HS+}(\xi),$$

where

$$f_{\text{HS}-}(\xi) = \sup\left\{c(\xi) : c(\eta) \leq h_{\text{hom}}(\eta)\ \forall\, \eta \in \mathbf{R}^n \text{ and } \forall h \in \mathfrak{I}^{\text{iso}}_{a,g}\right\},$$
$$f_{\text{HS}+}(\xi) = \inf\left\{c(\xi) : c(\eta) \geq h_{\text{hom}}(\eta) \forall\, \eta \in \mathbf{R}^n \text{ and } \forall h \in \mathfrak{I}^{\text{iso}}_{a,g}\right\},$$
$$\mathfrak{I}^{\text{iso}}_{a,g} = \left\{h \in \mathfrak{I}_{a,g} : h_{\text{hom}} \text{ is isotropic}\right\}.$$

In general, f_{hom} is anisotropic for nonlinear problems with periodic structures. No simple condition that would guarantee its isotropy has yet been found, in contrast to the linear case. The above nonlinear bounds of Hashin–Shtrikman type are therefore relevant only in cases where $f(\cdot,\xi)$ has a random nature. Since this type of problem is not within the scope of this paper, we introduce some modified bounds $f_{\text{HSC}-}$ and f_{HSC+} that can be used when $f(\cdot,\xi)$ satisfies the property of cubic symmetry:

$$f_{\text{HSC}-}(\xi) \leq f_{\text{hom}}(\xi) \leq f_{\text{HSC}+}(\xi),$$

where

$$f_{\text{HSC}-}(\xi) = \sup\left\{c(\xi) : c(\eta) \leq h_{\text{hom}}(\eta)\ \forall\, \eta \in \mathbf{R}^n \text{ and } \forall\, h \in \mathfrak{I}^{\text{cub}}_{a,g}\right\},$$
$$f_{\text{HSC}+}(\xi) = \inf\left\{c(\xi) : c(\eta) \geq h_{\text{hom}}(\eta) \forall\, \eta \in \mathbf{R}^n \text{ and } \forall\, h \in \mathfrak{I}^{\text{cub}}_{a,g}\right\},$$
$$\mathfrak{I}^{\text{cub}}_{a,g} = \left\{h \in \mathfrak{I}_{a,g} : h(\cdot,\xi) \text{ satisfies the property of cubic symmetry}\right\}.$$

We recall that $f(\cdot,\xi)$ satisfies the property of cubic symmetry iff

$$f(x,\xi) = f(\sigma x,\xi),$$

where σ is the rotation by $\pi/2$ in the plane of coordinates x_i, x_j, $i \neq j$, $i,j = 1,...,n$.

Remark 24.2. By the definition, the nonlinear bounds of Hashin–Shtrikman type f_{HSC-} and $f_{\text{HSC}+}$ are the best possible bounds among the bounds that can be obtained for the class $\mathfrak{I}^{\text{cub}}_{a,g}$ without taking into account other geometrical properties than the volume-fractions $\{a_i\}$. For the simple case of linear two-phase problems it turns out that both the *HS*-bounds and the *HSC*-bounds coincide with the well known Hashin–Shtrikman bounds. Generally, explicit formulae for these bounds, in terms of $\{a_i\}$ and $\{g_i\}$, are not available. However, there exist some interesting "estimates" on the *HS*-bounds (see [48, 49, 53] and the references given there).

Definition 24.3. We say that $V \subset \mathbf{R}^n$ is $\square$-periodic iff the characteristic function of V is $\square$-periodic.

Definition 24.4. Let $V \subset \mathbf{R}^n$ be a $\square$-periodic set. We say that V is a disperse set if is the union $\cup_{i=1}^{\infty}\overline{O}_i$ of mutually disjoint components $\overline{O}_i$, each being the closure of a smooth bounded domain O_i, and such that at most a finite collection of these components intersects $\square$.

In order to be able to make further study of the nonlinear Wiener bounds and the nonlinear bounds of Hashin–Shtrikman type, we now present the following statement of independent interest:

Theorem 24.5. *Suppose that $f \in \mathfrak{I}_{a,g}$ is of the form*

$$f(x,\xi) = \sum_{i=1}^{m} g_i(\xi)\, \chi_{A_i}(x),$$

such that for a fixed $j \in \{1,...,m\}$ we have that the closure of $\mathbf{R}^n \backslash A_j$ is a disperse set. Then,

$$\lambda_j^- c^- |\xi|^p \le f_{hom}(\xi) \le \lambda_j^+ c^+ \left(|\xi|^p + l\right),$$

where $0 < c^-, c^+ < +\infty$ are only dependent on A_j and p.

Proof. $f_{\text{hom}}(\xi) \le \lambda_j^+ c^+ (|\xi|^p + l)$: Let $\mathbf{R}^n \backslash A_j = \cup_{i=1}^{\infty}\overline{O}_i$ and let $\cup_{i=1}^{\infty}\overline{O}_i'$ be a $\square$-periodic union of mutually disjoint components $\overline{O}_i'$, each being the closure of a smooth bounded domain O_i' such that $\overline{O}_i \subset O_i'$. Let $u_i \in \mathscr{C}^{\infty}_{\text{per}}(\square)$ be such that $u_i = 0$ on $\mathbf{R}^n \backslash \cup_{i=1}^{\infty}\overline{O}_i'$ and $Du_i = 1$ on $\cup_{i=1}^{\infty}\overline{O}_i$. It follows that the function $u = -\sum_{i=1}^{n}\xi_i u_i$ has the property $Du + \xi = 0$ on $\mathbf{R}^n \backslash A_j$, and hence $f(\cdot, Du+\xi) = 0$ on $\mathbf{R}^n \backslash A_j$. Moreover, on A_j we have

$$\begin{aligned} |Du+\xi| &= \left|\xi - \sum_{i=1}^{n}\xi_i Du_i\right| \le \\ &\le |\xi| + \sum_{i=1}^{n}|\xi_i|\,|Du_i| \le |\xi|\left(1 + \sum_{i=1}^{n}|Du_i|\right). \end{aligned}$$

Therefore,

$$f(\cdot, Du+\xi) \le \lambda_j^+ \left(|Du+\xi|^p + l\right) \le \lambda_j^+ \left(|\xi|^p \left(1 + \sum_{i=1}^{n}|Du_i|\right)^p + l\right)$$

on A_j. These facts give

$$f_{\text{hom}}(\xi) \le \int_{\Box} f(x, Du+\xi)\, dx = \int_{A_j} f(x, \xi + Du)\, dx$$
$$\le \lambda_j^+ \int_{A_j} (|Du+\xi|^p + l)\, dx \le \lambda_j^+ c^+ (|\xi|^p + l),$$

where $c^+ = \int_{A_j} (1+\sum_{i=1}^n |Du_i|)^p \, dx$, and the upper bound for f_{hom} is proved.

We turn to prove the lower bound for f_{hom}. First, we note that if $k|\xi|^q \le h(\xi)$, where $k>0$ and $q>1$, then

$$h^*(\xi) \le \frac{1}{q'} (qk)^{\frac{1}{1-q}} |\xi|^{q'}. \tag{24.5}$$

Indeed,

$$h^*(\xi) = \sup_{E\in\mathbf{R}^n} \{E\cdot\xi - h(E)\} \le \sup_{E\in\mathbf{R}^n} \{E\cdot\xi - k|E|^q\}$$
$$= \sup_{|E|\in\mathbf{R}^n} \{|E|\cdot|\xi| - k|E|^q\} = \frac{1}{q'}(qk)^{\frac{1}{1-q}} |\xi|^{q'}.$$

Hence, for each $x \in A_j$ we have that

$$f^*(x,\xi) \le \frac{1}{p'} (p\lambda_k^-)^{\frac{1}{1-p}} |\xi|^{p'}.$$

For each $i \in \{1,\dots,n\}$ we choose $k \ne i$ and define

$$v_i = \frac{\partial u_k}{\partial x_k} e_i - \frac{\partial u_k}{\partial x_i} e_k.$$

Hence $v_i \in \mathrm{V}_{\text{sol}}^{p'}(\Box)$ and $v_i = e_i$ on $\Box\backslash A_j$. For any vector $\xi \in \mathbf{R}^n$ we let $v = -\sum_{i=1}^n \xi_i v_i$. By continuing similarly as above we find that

$$f_{\text{hom}}^*(\xi) \le k_0 |\xi|^{p'},$$

where

$$k_0 = \frac{1}{p'}\left(p\lambda_j^-\right)^{\frac{1}{1-p}} \int_{A_j} \left(1+\sum_{i=1}^n |Du_i|\right)^{p'} dx.$$

Thus, using (24.5) once more we deduce that

$$f_{\text{hom}}(\xi) = f_{\text{hom}}^{**}(\xi) \ge \frac{1}{p}(p'k_0)^{\frac{1}{1-p'}} |\xi|^p = \lambda_k^- c^- |\xi|^p,$$

where

$$c^- = \left(\int_{A_j} \left(1+\sum_{i=1}^n |Du_i|\right)^{\frac{p}{p-1}} dx\right)^{\frac{1}{1-p}}$$

and the lower bound is also proved. The proof is complete. □

Corollary 24.6. *For every $k \in \{1,...,m\}$ it holds that*

$$f_{W-}(\xi) \leq f_{HSC-}(\xi) \leq \lambda_k^+ c^+ (|\xi|^p + l)$$

and

$$\lambda_k^- c^- |\xi|^p \leq f_{HSC+}(\xi) \leq f_{W+}(\xi)$$

for all $\xi \in \mathbf{R}^n$, where $0 < c^-, c^+ < +\infty$ are only dependent on p, n and a_k.

Proof. By letting $f \in \mathfrak{I}_{a,g}$, defined by

$$f(x,\xi) = \sum_{i=1}^{m} g_i(\xi)\chi_{A_i}(x),$$

be such that for a fixed $k \in \{1,...,m\}$ $\square \backslash A_k$ is a cube, and the other sets $\{A_i\}_{i \neq k}$ are constructed such that $f(\cdot,\xi)$ satisfies the property of cubic symmetry, we get that

$$f_{\mathrm{W}-}(\xi) \leq f_{\mathrm{HSC}-}(\xi) \leq f_{\mathrm{hom}}(\xi) \leq f_{\mathrm{HSC}+}(\xi) \leq f_{\mathrm{W}+}(\xi).$$

Hence, the result follows directly from Theorem 24.5. $\square$

Remark 24.7. According to Corollary 24.6 we have that

$$\lim_{\lambda_k^+ \to 0} f_{\mathrm{W}-}(\xi) = \lim_{\lambda_k^+ \to 0} f_{\mathrm{HSC}-}(\xi) = 0$$

and

$$\lim_{\lambda_k^- \to +\infty} f_{\mathrm{W}+}(\xi) = \lim_{\lambda_k^- \to +\infty} f_{\mathrm{HSC}+}(\xi) = +\infty.$$

Hence, from Theorem 24.5 we conclude that the nonlinear bounds of Wiener and Hashin–Shtrikman type, $f_{\mathrm{W}\pm}$ and $f_{\mathrm{HSC}\pm}$, are not well fitted for estimating f_{hom} in the case when $\mathbf{R}^n \backslash A_j$ is a disperse set and λ_k^- is relatively large or λ_k^+ is relatively small for a fixed $k \neq j$. In such cases the bounds defined in Section 24.5 turn out to be much more suitable.

24.4 Some Useful Means of Power Type

In this section, we present some results on means of independent interest. These results will be an important tool in connection with the definition and proof of the bounds presented in the next section.

Definition 24.8. For every $\xi = \sum_{i=1}^{m} k_i e_{r_i}$ where $n \geq m \geq 1$, $k_i \neq 0$ and $\{r_i\}$ are distinct integers in $\{1,...,n\}$, let $w(\xi)$ be the n-tuple with components $(w(\xi))_i = \xi_i^2/|\xi|^2$. Furthermore, let a be any positive n-tuple. Then, the r-th power-mean of a with weights $w(\xi)$, $P^r(a,w(\xi))$, is defined by

$$P^r(a,w(\xi)) = \left(\sum_{i=1}^n a_i^r (w(\xi))_i\right)^{\frac{1}{r}}, \; r \neq 0,\; r \neq \pm\infty$$
$$= \prod_{i=1}^n a_i^{(w(\xi))_i}, \quad r = 0$$
$$= \min_{i=1}^m \{a_{r_i}\}, \quad r = -\infty$$
$$= \max_{i=1}^m \{a_{r_i}\}, \quad r = +\infty.$$

We now recall some well known results on power means (for the proof and some further information, see [8] and [44]):

Lemma 24.9. *Let $r \leq s$, $a_i > 0$ and let ξ be of the form $\xi = \sum_{i=1}^m k_i e_{r_i}$ as in Definition 24.8. Then,*

1. *$P^{[r]}(a,w(\xi)) \leq P^{[s]}(a,w(\xi))$ and equality occurs if and only if $a_{r_1} = a_{r_2} = \cdots = a_{r_m}$ or $r = s$.*
2. $\min_{i=1}^m \{a_{r_i}\} \leq P^{[r]}(a,w(\xi)) \leq \max_{i=1}^m \{a_{r_i}\}$.
3. $\lim_{r\to+\infty} P^{[r]}(a,w(\xi)) = \max_{i=1}^m \{a_{r_i}\}$.
4. $\lim_{r\to-\infty} P^{[r]}(a,w(\xi)) = \min_{i=1}^m \{a_{r_i}\}$.
5. *$\lim_{a_{r_i}\to+\infty} P^{[r]}(a,w(\xi)) \leq +\infty$, $i \in \{1,\ldots,m\}$ and equality occurs if and only if $r \geq 0$.*
6. *$\lim_{a_{r_i}\to 0} P^{[r]}(a,w(\xi)) \geq 0$, $i \in \{1,\ldots,m\}$ and equality occurs if and only if $r < 0$.*
7. *$P^{[r]}(a,w(\xi))$ is continuous in r and $\xi(\neq 0)$ and continuous and non-decreasing in each a_i, $i = 1,2,\ldots,n$.*

24.4.1 A Particular Power Type Mean

For the main result of the next section, it will be convenient to discuss a mean which was introduced in [28] and [31]. It is denoted $L^{[p]}$ and defined by

$$L^{[p]}(a,w(\xi)) = \max_{|\eta|=1} \left(\frac{\eta\cdot\xi}{|\xi|}\right)^p P^{[s]}(a,w(\eta)),\; p > 1,$$

where $s = 1/(1-p)$ if $p \leq 2$ and $s = -2/p$ if $p \geq 2$.

Theorem 24.10. *It yields that*

1. *$L^{[p]}(a,w(\xi))$ is continuous in p and $\xi(\neq 0)$ and continuous and non-decreasing in each $a_i > 0$, $i = 1,2,\ldots,n$.*
2. *$P^{[1/(1-p)]} \leq L^{[p]} \leq P^{[2/p]}$. In particular, $P^{[-\infty]} \leq \lim_{p\to 1_+} L^{[p]} \leq P^{[2]}$ and it yields that $P^{[-\infty]}$ and $P^{[2]}$ are the sharpest possible bounds for the limit $\lim_{p\to 1_+} L^{[p]}$ within the class of power means.*

3. $L^{[p]} = P^{[2/p]}$ *for every* $p \geq 2$.

4. $K^{[p]} \leq L^{[p]}$ *where* $K^{[p]} = \left(P^{[k]}(P^{[2k]})^{(p-2)}\right)^k$ *and* $k = (p-1)^{-1}$.

Remark 24.11. Since

$$P^{[1/(1-p)]} \to P^{[-\infty]} \text{ and } P^{[2/p]} \to P^{[2]}$$

as $p \to 1_+$, Theorem 24.10(2) gives that $P^{[1/(1-p)]}$ and $P^{[2/p]}$ are almost optimal upper and lower bounds for $L^{[p]}$ within the class of power means when p is close to 1. Moreover, since

$$K^{[p]}, P^{[2/p]} \to P^{[1]} \text{ as } p \to 2,$$

we find that the lower bound $K^{[p]}$ and the upper bound $P^{[2/p]}$ are close to $L^{[p]}$ when p is close to 2.

Proof. Let

$$\vartheta(v, \xi, a, p) = \left|\frac{v \cdot \xi}{|\xi|}\right|^p P^{[s]}(a, w(v)),$$

where $s = 1/(1-p)$ if $p \leq 2$ and $s = -2/p$ if $p \geq 2$ and let

$$F(v) = \left|\frac{v \cdot \xi}{|\xi|}\right|^p P^{[1/(1-p)]}(a, w(v)).$$

1: Consider convergent sequences $\xi_k \to \xi_0$, $a_k \to a_0$, $p_k \to p_0$ and let $\vartheta_k(v) = \vartheta(v, \xi_k, a_k, p_k)$. Since $\{\vartheta_k\}$ is a sequence of bounded continuous functions converging pointwise to ϑ_0 on the compact set $\{v \in \mathbf{R}^n : |v| = 1\}$, it follows that the convergence is uniform and

$$|\|\vartheta_k\|_{\max} - \|\vartheta_0\|_{\max}| \leq \|\vartheta_k - \vartheta_0\|_{\max} \to 0,$$

where

$$\|\vartheta_k\|_{\max} = \max\{\vartheta_k(v) : |v| = 1\}.$$

Hence,

$$L^{[p_k]}(a_k, w(\xi_k)) = \|\vartheta_k\|_{\max} \to \|\vartheta_0\|_{\max} = L^{[p_0]}(a_0, w(\xi_0))$$

and $L^{[p]}(a, w(\xi))$ is therefore continuous in p, a and ξ. The fact that $L^{[p]}(a, w(\xi))$ is non-decreasing in the components of a follows easily by the fact that $P^{[s]}(a, w(v))$ is non-decreasing in the components of a.

2 and 3: Using Lemma 24.9(1) we get that $F(v) \leq L^{[p]}(a, w(\xi))$ for every $|v| = 1$. Let η be the vector with components $\eta_i = \xi_i / |\xi|$. We easily obtain that $F(\eta) = P^{[1/(p-1)]}(a, w(\xi))$. Thus, $P^{[1/(p-1)]} \leq L^{[p]}$. We next prove the inequality $L^{[p]} \leq P^{[2/p]}$. Observe that

$$L^{[p]}(a, w(\xi)) \leq \left(\frac{1}{|\xi|^2} \max_{v \neq 0} \left\{\frac{(v \cdot \xi)^2}{\sum_{i=1}^n a_i^{-\frac{2}{p}} v_i^2}\right\}\right)^{\frac{p}{2}}. \tag{24.6}$$

Indeed, according to the definition of $L^{[p]}$ and Lemma 24.9(1) we find that

$$\begin{aligned} L^{[p]}(a,w(\xi)) &= \max_{|v|=1}\left\{\left(\frac{v\cdot\xi}{|\xi|}\right)^p P^{[s]}(a,w(v))\right\} \\ &\le \max_{v\neq 0}\left\{\left(\frac{v\cdot\xi}{|v|\,|\xi|}\right)^p P^{[-\frac{2}{p}]}(a,w(v))\right\} \\ &= \max_{v\neq 0}\left\{\left(|\xi|^{-2}\frac{(v\cdot\xi)^2}{\sum_{i=1}^n a_i^{-\frac{2}{p}} v_i^2}\right)^{\frac{p}{2}}\right\} \\ &= \left(|\xi|^{-2}\max_{v\neq 0}\left\{\frac{(v\cdot\xi)^2}{\sum_{i=1}^n a_i^{-\frac{2}{p}} v_i^2}\right\}\right)^{\frac{p}{2}}. \end{aligned}$$

Moreover, for every $v \neq 0$ it is easy to check that

$$\begin{aligned} \frac{(v\cdot\xi)^2}{\left(\sum_{i=1}^n a_i^{-\frac{2}{p}} v_i^2\right)^{-1}} &= \max_{t\in\mathbf{R}}\left\{tv\cdot\xi - \frac{1}{4}t^2\sum_{i=1}^n a_i^{-\frac{2}{p}} v_i^2\right\} \\ &= \max_{t\in\mathbf{R}}\left\{\sum_{j=1}^n\left((tv_j)\,\xi_j - \frac{1}{4}a_i^{-\frac{2}{p}}(tv_j)^2\right)\right\}, \end{aligned}$$

and therefore,

$$\max_{v\neq 0}\left\{\frac{(v\cdot\xi)^2}{\sum_{i=1}^n a_i^{-\frac{2}{p}} v_i^2}\right\} = \max_{v\neq 0}\max_{t\in\mathbf{R}}\left\{\sum_{j=1}^n\left((tv_j)\,\xi_j - \frac{1}{4}a_i^{-\frac{2}{p}}(tv_j)^2\right)\right\}$$

$$\begin{aligned} &= \max_{v\neq 0}\left\{\sum_{j=1}^n\left(v_j\xi_j - \frac{1}{4}a_i^{-\frac{2}{p}} v_j^2\right)\right\} = \sum_{j=1}^n \max_{v_j\in\mathbf{R}}\left\{v_j\xi_j - \frac{1}{4}a_i^{-\frac{2}{p}} v_j^2\right\} \\ &= \sum_{j=1}^n a_i^{\frac{2}{p}}\xi_j^2 = |\xi|^2\left(P^{[\frac{2}{p}]}(a,w(\xi))\right)^{\frac{2}{p}}. \end{aligned}$$

Thus, according to (24.6) we finally get that

$$L^{[p]}(a,w(\xi)) \le P^{[\frac{2}{p}]}(a,w(\xi)). \tag{24.7}$$

If $p \ge 2$, then all the above $\le$ signs can be replaced by $=$. Hence $L^{[p]} \le P^{[\frac{2}{p}]}$, and equality occurs iff $p \ge 2$. By letting $\xi_k = \xi_0$, $a_k = a_0$ and $p_k \to 1$ we can argue as above and obtain from Lemma 24.9(4) that

$$\lim_{p\to 1} L^{[p]}(a,w(\xi)) = \max_{|v|=1} \frac{|v\cdot\xi|}{|\xi|} P^{[-\infty]}(a,w(v)).$$

Moreover, if

$$P^{[r]} \leq \lim_{p\to 1} L^{[p]} \leq P^{[s]},$$

then

$$P^{[r]}(a,w(\xi)) \leq \lim_{p\to 1} L^{[p]}(a,w(\xi)) \leq P^{[s]}(a,w(\xi))$$

for all a and ξ. In particular, if

$$a_1 = k > 1,\ a_2 = a_3 = \cdots = a_n = 1,\ \xi = e_1\cos\theta + e_2\sin\theta,\ \theta\in\mathbf{R},$$

then

$$\max_{|v|=1} \frac{|v\cdot\xi|}{|\xi|} P^{[-\infty]}(a,w(v)) = \begin{cases} k|\cos\theta| & \text{if } |\cos\theta| \geq 1/k \\ 1 & \text{if } |\cos\theta| \leq 1/k \end{cases}.$$

Hence, if $0 < |\cos\theta| \leq 1/k$ then $P^{[r]}(a,w(\xi)) \leq 1$, and according to Lemma 24.9(1) this implies that $r = -\infty$. Consequently, $P^{[-\infty]}$ is the sharpest possible lower bound for $\lim_{p\to 1} L^{[p]}$ within the class of power means. Let us prove that the upper bound $P^{[2]}$ also is optimal within the class of power means. Assume on the contrary that $1/k < |\cos\theta| < 1$, $P^{[s]}(a,w(\xi)) \geq k|\cos\theta|$ and $s < 2$. Then we easily deduce that

$$\sin^2\theta \geq k^s\left(\left(\cos^2\theta\right)^{\frac{s}{2}} - \cos^2\theta\right).$$

Since $P^{[s]}(a,w(\xi))$ is non-decreasing in s by Lemma 24.9(1), we can in addition assume that $0 < s < 2$. But since $\sin^2\theta < 1$, this is impossible if k is large enough, and we have reached a contradiction, i.e., we can conclude that $P^{[2]}$ is the sharpest possible upper bound for $\lim_{p\to 1} L^{[p]}$ within the class of power means.

4: Let ζ be the vector with components

$$\zeta_i = a_i^{1/(p-1)} \frac{\xi_i}{|\xi|} \left(\sum_{i=1}^{n} a_i^{2/(p-1)} w_i(\xi)\right)^{-\frac{1}{2}}.$$

We easily get that $F(\zeta) = K^{[p]}(a,w(\xi))$ and since $F(v) \leq L^{[p]}(a,w(\xi))$ for every $|v| = 1$, it follows that $K^{[p]} \leq L^{[p]}$. The proof is complete. □

Proposition 24.12. *Let $\xi = \sum_{i=1}^{m} k_i e_{r_i}$ as in Lemma 24.9. It yields that*

1. $\min_{i=1}^{m}\{a_{r_i}\} \leq L^{[p]}(a,w(\xi)) \leq \max_{i=1}^{m}\{a_{r_i}\}$ *and equalities occur if and only if* $a_{r_1} = a_{r_2} = \cdots = a_{r_m}$.
2. *If* $a_{r_1} < a_{r_2} < \cdots < a_{r_m}$ *then* $\lim_{p\to 1_+} K^{[p]}(a,w(\xi)) = \sqrt{w_{r_m}(\xi)}a_{r_m}$.
3. $\lim_{a_{r_i}\to+\infty} L^{[p]}(a,w(\xi)) = \lim_{a_{r_i}\to+\infty} K^{[p]}(a,w(\xi)) = +\infty$, $i \in \{1,...,m\}$.
4. *If* $m \geq 2$ *then* $\lim_{a_{r_i}\to 0} L^{[p]}(a,w(\xi)) \geq \lim_{a_{r_i}\to 0} K^{[p]}(a,w(\xi)) > 0$, $i = 1,...,m$.

Remark 24.13. Clearly, it is Proposition 24.12(1) which justifies that $L^{[p]}$ actually is a mean (cf. [8, p. 35, Remark 5]). By Proposition 24.12(2) and Lemma 24.9(4) we see that it is generally not possible to decide which one of lower bounds $K^{[p]}$ or $P^{[1/(1-p)]}$ that is closest to $L^{[p]}$, in particular when p is close to 1. On the other hand, since

$$P^{[1/(1-p)]} \leq P^{[-1]} \text{ and } K^{[p]} \to P^{[1]} \text{ as } p \to 2,$$

we conclude that $K^{[p]}$ is a sharper lower bound for $L^{[p]}$ than $P^{[1/(1-p)]}$ if p is close to 2. Moreover, Proposition 24.12(3 and 4) shows that $L^{[p]}$ and $K^{[p]}$ are large if some a_{r_i} is large and even large compared with a_{r_i} if a_{r_i} is small. Due to the fact that

$$\lim_{a_{r_i} \to +\infty} P^{[1/(1-p)]}(a, w(\xi)) < +\infty$$

and

$$\lim_{a_{r_i} \to 0} P^{[1/(1-p)]}(a, w(\xi)) = 0,$$

we also have that $K^{[p]}$ is a sharper lower bound than $P^{[1/(1-p)]}$ in such cases.

Proof. 1: This follows directly by Theorem 24.10(2) and Lemma 24.9(2).

2, 3 and 4: It is easily seen that

$$K^{[p]}(a, w(\xi)) = \left(S^{[p]}(a, w(\xi))\right)^{\frac{1}{2}} P^{[2/(p-1)]}(a, w(\xi)),$$

where

$$S^{[p]}(a, w(\xi)) = \frac{\left(\sum_{i=1}^{m} a_{r_i}^{\frac{1}{p-1}} w_{r_i}(\xi)\right)^2}{\sum_{i=1}^{m} \left(a_{r_i}^{\frac{1}{p-1}}\right)^2 w_{r_i}(\xi)}.$$

We also easily find the limits for $S^{[p]}(a, w(\xi))$ as $p \to 1_+, a_{r_k} \to +\infty$ and $a_{r_k} \to 0$, respectively. The corresponding limits for $P^{[2/(p-1)]}(a, w(\xi))$ follow from Lemma 24.9(3, 4 and 5) and 2, 3 and 4 follow thereby directly. □

24.4.2 Composition of Power Means

We start with some notations. Let $0 < \alpha \leq \beta < +\infty$ and let $\wp_{\alpha,\beta}$ be the class of every Lebesgue measurable and □-periodic function ρ such that $\alpha \leq \rho \leq \beta$ a.e.

Let $\rho \in \wp_{\alpha,\beta}$, $r \neq 0$ and let $\tau_-^{[t]}(\rho)$ and $\tau_+^{[t]}(\rho) : \mathbf{R}^n \to \mathbf{R}^n$ be the functions with components defined almost everywhere by the following power means:

$$\left(\tau_-^{[t]}(\rho)\right)_i = \left(\int_0^1 \rho^t dx_i\right)^{\frac{1}{t}},$$

$$\left(\tau_+^{[t]}(\rho)\right)_i = \left(\int_0^1 \cdots \int_0^1 \rho^t \, dx_1 \cdots dx_{i-1} dx_{i+1} \cdots dx_n\right)^{\frac{1}{t}}.$$

Moreover, let $q_-^{r,s}(\rho)$ and $q_+^{r,s}(\rho)$ be the n-tuples with components defined by

$$(q_-^{r,s}(\rho))_i = \left(\tau_+^{[r]}(\left(\tau_-^{[s]}(\rho)\right)_i)\right)_i,$$
$$(q_+^{r,s}(\rho))_i = \left(\tau_-^{[r]}(\left(\tau_+^{[s]}(\rho)\right)_i)\right)_i.$$

Let A_i (resp. A^i) be the family of every set $E_{a,b}$, $a = (a_1, ..., a_n)$, $b = (b_1, ..., b_n)$, $a_j < b_j$, $j = 1, ..., n$, of the form

$$E_{a,b} = \left\{x \in \Box : a_j < x_j < b_j \text{ if } j \neq i,\ 0 < x_i < 1\right\} \tag{24.8}$$

(resp.

$$E_{a,b} = \left\{x \in \Box : 0 < x_j < 1 \text{ if } j \neq i,\ a_i < x_i < b_i\right\}) \tag{24.9}$$

or any set obtained by replacing any or all of the $<$ signs in (24.8) and (24.9)) by $\leq$. Moreover, let W_i (resp. W^i) be the family of all finite collections $\{E_j\}$ of members of A_i (resp. A^i) with the following property: $E_j \cap E_k = \emptyset$ whenever $j \neq k$, and $\cup E_j = \Box$. For every $A = \{E_t\}_{t=1}^m = \left\{E_{a_t,b_t}\right\}_{k=1}^m \in W_i$ (resp. W^i) we associate the numbers

$$k_A^+ = \max_{1 \leq t \leq m,\ j \neq i} \left\{(b_t)_j - (a_t)_j\right\} \text{ and } k_A^- = \min_{1 \leq t \leq m,\ j \neq i} \left\{(b_t)_j - (a_t)_j\right\}$$

(resp.

$$k_A^+ = \max_{1 \leq t \leq m} \{(b_t)_i - (a_t)_i\}) \text{ and } k_A^- = \min_{1 \leq t \leq m} \{(b_t)_i - (a_t)_i\}).$$

We define a partial order relation $\prec$ on W_i and W^i by saying that $A \prec B$ if each member of A is a union of members of B. Moreover, if $\rho \in \wp_{\alpha,\beta}$ we let $q_A^{c,d}(\rho)$ denote the number

$$q_A^{c,d}(\rho) = \left(\int_\Box \rho_A^c dx\right)^{\frac{1}{c}},$$

where ρ_A is defined by

$$\rho_A(x) = (\left|E_j\right|^{-1} \int_{E_j} \rho^d dx)^{\frac{1}{d}}$$

for all $x \in E_j$.

Theorem 24.14. *Let $r < 0 < s$ and suppose that $\rho \in \wp_{\alpha,\beta}, A_1, A_2, A_3, ..., \in W_i$ where $A_1 \prec A_2 \prec A_3 \cdots$ and $B_1, B_2, B_3, ..., \in W^i$ such that $B_1 \prec B_2 \prec B_3 \cdots$. Then,*

$$\begin{aligned} q_{A_1}^{s,r}(\rho) \leq q_{A_2}^{s,r}(\rho) \leq \cdots &\leq (q_-^{s,r}(\rho))_i \\ &\leq (q_+^{r,s}(\rho))_i \leq \cdots \leq q_{B_2}^{r,s}(\rho) \leq q_{B_1}^{r,s}(\rho). \end{aligned}$$

Moreover, if $k^+_{A_m}, k^+_{B_m} \to 0$ *as* $m \to +\infty$, *then* $q^{s,r}_{A_m}(\rho) \to \left(q^{s,r}_-(\rho)\right)_i$ *and* $q^{r,s}_{B_m}(\rho) \to \left(q^{r,s}_+(\rho)\right)_i$.

Remark 24.15. The above result is an extension of the well-known inequality:

$$\left(q^{s,r}_-(\rho)\right)_i \leq \left(q^{r,s}_+(\rho)\right)_i,$$

which has been known in the theory of inequalities for quite a long time (see, e.g., [17, p. 148], [3] or [8, p. 170]). For related subjects, we refer to the book [17, p. 148]. We note that Theorem 24.14 contains a useful algorithm for computing lower and upper approximations, $q^{s,r}_{A_m}(\rho)$ and $q^{r,s}_{B_m}(\rho)$, of $\left(q^{s,r}_-(\rho)\right)_i$ and $\left(q^{r,s}_+(\rho)\right)_i$, respectively.

Proof. For the proof of the inequality $\left(q^{s,r}_-(\rho)\right)_i \leq \left(q^{r,s}_+(\rho)\right)_i$, see Remark 24.15. Let $A = \left\{E_j\right\}$ and $B = \left\{F_j\right\}$, where $A, B \in W_i$ or $A, B \in W^i$, and assume that $A \prec B$. Then each $E_j = \cup F_{j_k}$ where $F_{j_k} \in B$. Moreover, put $a_k = \left|F_{j_k}\right|$ and $b_k = \int_{F_{j_k}} \rho^d dx$. If $c < 0 < d$ or $d < 0 < c$, then the reversed discrete Hölder inequality (see, e.g., [8, p. 136]) yields that

$$\left(\sum a_k\right)^{\frac{d-c}{d}} \left(\sum b_k\right)^{\frac{c}{d}} \leq \sum a_k^{\frac{d-c}{d}} b_k^{\frac{c}{d}}.$$

Therefore, in view of the definitions of ρ_A and ρ_B

$$\begin{aligned}\int_{E_j} \rho_A^c\, dx &= \left|E_j\right|^{\frac{d-c}{d}} \left(\int_{E_j} \rho^d dx\right)^{\frac{c}{d}} = \left(\sum \left|F_{j_k}\right|\right)^{\frac{d-c}{d}} \left(\int_{E_j} \rho^d dx\right)^{\frac{c}{d}} \\ &\leq \sum \left|F_{j_m}\right|^{\frac{d-c}{d}} \left(\int_{F_{j_k}} \rho^d dx\right)^{\frac{c}{d}} = \sum \int_{F_{j_k}} \rho_B^c\, dx = \int_{E_j} \rho_B^c\, dx.\end{aligned}$$

Hence

$$\left(q^{c,d}_A(\rho)\right)^c \leq \left(q^{c,d}_B(\rho)\right)^c,$$

which gives that

$$q^{s,r}_A(\rho) \leq q^{s,r}_B(\rho) \text{ and } q^{r,s}_B(\rho) \leq q^{r,s}_A(\rho). \tag{24.10}$$

This proves the inequalities $q^{s,r}_{A_1}(\rho) \leq q^{s,r}_{A_2}(\rho) \leq \cdots$ and $q^{r,s}_{B_1}(\rho) \geq q^{r,s}_{B_2}(\rho) \geq \cdots$.

Next, let λ be a positive function. The Hölder inequality implies that

$$\left|E_j\right| = \int_{E_j} \lambda^{\frac{r}{s-r}} \lambda^{-\frac{r}{s-r}} dx \leq \left(\int_{E_j} \lambda^{\frac{r}{s}} dx\right)^{\frac{s}{s-r}} \left(\int_{E_j} \lambda\, dx\right)^{\frac{-r}{s-r}}.$$

Hence,

$$\left|E_j\right|^{\frac{s-r}{s}} \left(\int_{E_j} \lambda\, dx\right)^{\frac{r}{s}} \leq \int_{E_j} \lambda^{\frac{r}{s}} dx$$

and

$$\left|E_j\right|^{\frac{r-s}{r}} \left(\int_{E_j} \lambda^{\frac{r}{s}} dx\right)^{\frac{s}{r}} \leq \int_{E_j} \lambda\, dx.$$

Thus, by putting $\lambda = \left(\tau_+^{[s]}(\rho)\right)^s$ and $A \in W^i$ (resp. $\lambda = \left(\tau_-^{[r]}(\rho)\right)^s$ and $A \in W_i$) we obtain

$$\int_{E_j} (\rho_A)^r dx = |E_j|^{\frac{s-r}{s}} (\int_{E_j} \rho^s dx)^{\frac{r}{s}} = |E_j|^{\frac{s-r}{s}} (\int_{E_j} \left(\tau_+^{[s]}(\rho)\right)^s dx)^{\frac{r}{s}}$$
$$= |E_j|^{\frac{s-r}{s}} (\int_{E_j} \lambda\, dx)^{\frac{r}{s}} \le \int_{E_j} \lambda^{\frac{r}{s}} dx = \int_{E_j} \left(\tau_+^{[s]}(\rho)\right)^r dx$$

(resp.

$$\int_{E_j} (\rho_A)^s dx = |E_j|^{\frac{t-s}{r}} (\int_{E_j} \rho^r dx)^{\frac{s}{r}} = |E_j|^{\frac{t-s}{r}} (\int_{E_j} \left(\tau_-^{[r]}(\rho)\right)^r dx)^{\frac{s}{r}}$$
$$= |E_j|^{\frac{t-s}{r}} (\int_{E_j} \lambda^{\frac{r}{s}} dx)^{\frac{s}{r}} \le \int_{E_j} \lambda\, dx = \int_{E_j} \left(\tau_+^{[r]}(\rho)\right)^s dx),$$

and it follows that

$$\left(q_+^{r,s}(\rho)\right)_i \le q_A^{r,s}(\rho) \;\; (\text{resp. } q_A^{s,r}(\rho) \le \left(q_-^{s,r}(\rho)\right)_i). \tag{24.11}$$

Moreover, if $C_1 \prec C_2 \prec \cdots$ and $k_{C_m}^+ \to 0$, then it is easy to see that we can construct a sequence $\{C_m'\}$, $C_m' \prec C_m$, $C_1' \prec C_2' \prec \cdots$, $k_{C_m'}^+ \to 0$ such that $k_{C_m'}^+ \le c k_{C_m'}^-$ for a constant $c > 0$. Let $\{A_m'\}$ and $\{B_m'\}$ be such sequences associated with $\{A_m\}$ and $\{B_m\}$, respectively. Let $\rho_l : \mathbf{R}^{n-1} \to \mathbf{R}$ and $\rho_t : \mathbf{R} \to \mathbf{R}$ be the functions defined by $\rho_l(t) = \rho(x)$ and $\rho_t(l) = \rho(x)$ where $l = x_i$ and $t = (x_1, ..., x_{i-1}, x_{i+1}, ..., x_n)$. According to the Fubini theorem, ρ_l and ρ_t are measurable for almost every l and t with respect to the Lebesgue measures on $\mathbf{R}^{n-1}$ and $\mathbf{R}^1$, respectively. For each $t \in [0,1]^{n-1}$ we let A_m^t be the set such that $t \in A_m^t$ and $A_m^t \times [0,1] \in A_m'$. Similarly, for each $l \in [0,1]$ we let B_m^l be the set such that $l \in B_m^l$ and $B_m^l \times [0,1]^{n-1} \in B_m'$. It is clear that the sequences $\{A_m^t\}$ and $\{B_m^l\}$ shrink to t and l nicely, respectively (we recall that a sequence $\{E_j\}$ of Borel sets in $\mathbf{R}^k$ is said to shrink to x nicely if there is a number $c > 0$ with the following property: There is a sequence of balls $B(x, r_j)$, with $\lim r_j = 0$, such that $E_j \subset B(x, r_j)$ and $m_k(E_j) \ge c m_k(B(x, r_j))$ for $j = 1, 2, 3, ...,$). Hence, according to [54, Theorem 7.10], we get that

$$\rho^r(x) = \rho_l^r(t) = \lim_{m \to +\infty} \frac{1}{m_{n-1}(A_m^t)} \int_{A_m^t} \rho_l^r \, dm_{n-1}$$

and

$$\rho_t^s(l) = \lim_{m \to +\infty} \frac{1}{m_1(B_m^l)} \int_{B_m^l} \rho_t^s \, dm_1$$

at every Lebesgue point of ρ_l^r and ρ_t^s, respectively, and, hence, for almost every l and t. Moreover, for a.e. x we have the identities:

$$\rho_{A_m'}^r(x) = \frac{1}{m_n\left(A_m^t \times [0,1]\right)} \int_{A_m^t \times [0,1]} \rho^r dm_n$$

$$= \int_{[0,1]} \left(\frac{1}{m_{n-1}(A^t_m)} \int_{A^t_m} \rho^r_l \, dm_{n-1} \right) dm_1$$

and

$$\rho^s_{B'_m}(x) = \frac{1}{m_n\left(B^l_m \times [0,1]^{n-1}\right)} \int_{B^l_m \times [0,1]^{n-1}} \rho^s dm_n$$

$$= \int_{[0,1]^{n-1}} \left(\frac{1}{m_1(B^l_m)} \int_{B^l_m} \rho^s_t \, dm_1 \right) dm_{n-1}.$$

Thus, according to Lebesgue dominated convergence theorem (LDCT) we find that

$$\lim_{m\to+\infty} \rho_{A'_m}(x) = \left(\int_{[0,1]} \rho^r_l \, dm_1 \right)^{\frac{1}{r}} = \left(\int_0^1 \rho^r(x) \, dx_i \right)^{\frac{1}{r}} = \left(\tau^{[r]}_-(\rho) \right)_i$$

and

$$\lim_{m\to+\infty} \rho_{B'_m}(x) = \left(\int_{[0,1]^{n-1}} \rho^s_t(l) \, dl \right)^{\frac{1}{s}}$$

$$= \left(\int_0^1 \cdot\cdot \int_0^1 \rho^s(x) \, dx_1 \cdot\cdot dx_{i-1} dx_{i+1} \cdot\cdot dx_n \right)^{\frac{1}{s}} = \left(\tau^{[s]}_+(\rho) \right)_i (x).$$

By using LDCT once more, we easily obtain the convergences $q^{s,r}_{A'_m}(\rho) \to \left(q^{s,r}_-(\rho)\right)_i$ and $q^{r,s}_{B'_m}(\rho) \to \left(q^{r,s}_+(\rho)\right)_i$. Moreover, according to (24.10) and (24.11)

$$q^{s,r}_{A'_m}(\rho) \le q^{s,r}_{A_m}(\rho) \le \left(q^{s,r}_-(\rho)\right)_i \text{ and } \left(q^{r,s}_+(\rho)\right)_i \le q^{r,s}_{B_m}(\rho) \le q^{r,s}_{B'_m}(\rho).$$

Therefore, we obtain that $q^{s,r}_{A_m}(\rho) \to \left(q^{s,r}_-(\rho)\right)_i$ and $q^{r,s}_{B_m}(\rho) \to \left(q^{r,s}_+(\rho)\right)_i$. The proof is complete. □

Proposition 24.16. *Let $r < 0 < s$ and assume that $C : \mathbf{R}^n \times \mathbf{R}_+ \to \mathbf{R}_+$ is such that $C(\cdot,t) \in \wp_{\alpha,\beta}$ for all t and $C(x,\cdot)$ is continuous and non-decreasing. It yields that:*

1. *$\left(q^{s,r}_-(C(\cdot,t))\right)_i$ and $\left(q^{r,s}_+(C(\cdot,t))\right)_i$ are continuous and non-decreasing in t.*
2. *If $C(x,t)$ is of the form $C(x,t) = \sum_{i=1}^m C_i(t)\chi_{A_i}(x)$, where $\{A_i\}$ is a disjoint partition of $\mathbf{R}^n$ such that for a fixed $j \in \{1,...,m\}$ we have that $\square \backslash A_j$ is a genuine subset of $\square$, then*

$$c^-\lambda^-_j \le \left(q^{s,r}_-(C(\cdot,t))\right)_i \le \left(q^{r,s}_+(C(\cdot,t))\right)_i \le c^+\lambda^+_j$$

for $i = 1,...,n$, where $\lambda^-_j = C_j(0)$, $\lambda^+_j = C_j(+\infty)$ and $0 < c^-, c^+ < +\infty$. Moreover, c^- and c^+ are only dependent of A_j, r and s.

Proof. 1: If λ and τ are positive measurable functions such that $\lambda \le \tau$ a.e., then it is obvious that

$$\left(\tau_-^{[r]}(\lambda)\right)_i \leq \left(\tau_-^{[r]}(\tau)\right)_i \text{ and } \left(\tau_+^{[s]}(\lambda)\right)_i \leq \left(\tau_+^{[s]}(\tau)\right)_i. \tag{24.12}$$

Therefore, since $C(x,t)$ is non-decreasing in t, it also follows that $\left(q_-^{s,r}(C(\cdot,t))\right)_i$ and $\left(q_+^{r,s}(C(\cdot,t))\right)_i$ are non-decreasing in t. Let us prove the continuity in t. According to the Fubini theorem, we have that for each t there exists a Lebesgue measurable set $A_t \subset \mathbf{R}$ with $m_1(A_t) = 0$ such that $C(x_1,...,x_n,t)$ is $\mathbf{R}^{n-1}$-Lebesgue measurable in $(x_1,...,x_{i-1},x_{i+1},...,x_n)$ for every $x_i \notin A_t$. Let $\{t_k\}$ be a sequence of positive numbers converging to $t_0 \in [0,+\infty]$ as $k \to +\infty$, and put $A = \cup_{n=1}^{\infty} A_{t_n}$. Then $m_1(A) \leq \sum_{n=1}^{\infty} m_1(A_{t_n}) = 0$, i.e., $m_1(A) = 0$. For every $x_i \notin A$ let $\gamma_k(x_i)$ be the value

$$\gamma_k(x_i) = (\int_0^1 \cdot\cdot \int_0^1 (C(x_1,...,x_n,t_k))^s \, dx_1 \cdot\cdot dx_{i-1}dx_{i+1} \cdot\cdot dx_n)^{\frac{1}{s}}.$$

Since $C(x,t_k) \to C(x,t_0)$ for every $x \in \mathbf{R}^n$, it follows from LDCT that $\gamma_k(x_i) \to \gamma_0(x_i)$ for every $x_i \notin A$, i.e. $\gamma_k(\cdot) \to \gamma_0(\cdot)$ almost everywhere on $\mathbf{R}$. Using LDCT once more we get that

$$\left(q_+^{r,s}(C(\cdot,t_k))\right)_i = \left(\int_0^1 (\gamma_k(x_i))^{\frac{1}{r}} \, dx_i\right)^r$$

$$\to \left(\int_0^1 (\gamma_0(x_i))^{\frac{1}{r}} \, dx_i\right)^r = \left(q_+^{r,s}(C(\cdot,t_0))\right)_i.$$

Hence, $\left(q_+^{r,s}(C(\cdot,t))\right)_i$ is continuous in t. The continuity of $\left(q_-^{s,r}(C(\cdot,t))\right)_i$ follows by similar arguments.

2: Consider a cube $V = [0,a]^n$ where $0 < a < 1$ such that $\Box \backslash A_j \subset V$. Let $k^- = \min\{\lambda_i^-\}$, $k^+ = \max\{\lambda_i^+\}$, and let $\lambda_{\min}$ and $\lambda_{\max}$ the functions $\lambda_{\min} = k^- \chi_V + \lambda_j^- \chi_{\Box\backslash V}$ and $\lambda_{\max} = k^+ \chi_V + \lambda_j^+ \chi_{\Box\backslash V}$. Due to (24.12)

$$\left(q_-^{s,r}(\lambda)\right)_i \leq \left(q_-^{s,r}(\tau)\right)_i \text{ and } \left(q_+^{r,s}(\lambda)\right)_i \leq \left(q_+^{r,s}(\tau)\right)_i$$

if $\lambda \leq \tau$. Moreover,

$$\left(q_-^{s,r}(\tau)\right)_i \leq \left(q_+^{r,s}(\tau)\right)_i$$

by Theorem 24.14. Therefore, since $\lambda_{\min} \leq C(\cdot,t) \leq \lambda_{\max}$, it follows that

$$\left(q_-^{s,r}(\lambda_{\min})\right)_i \leq \left(q_-^{s,r}(C(\cdot,t))\right)_i \leq \left(q_+^{r,s}(C(\cdot,t))\right)_i \leq \left(q_+^{r,s}(\lambda_{\max}\right)_i. \tag{24.13}$$

Furthermore, we have that

$$\left(q_-^{s,r}(\lambda_{\min})\right)_i = \left(\left(a\left(k^-\right)^r + (1-a)\left(\lambda_j^-\right)^r\right)^{\frac{s}{r}} a^{n-1} + (\lambda_j^-)^s\left(1-a^{n-1}\right)\right)^{\frac{1}{s}}$$

$$\geq \left((\lambda_j^-)^s\left(1-a^{n-1}\right)\right)^{\frac{1}{s}} = \lambda_j^-\left(1-a^{n-1}\right)^{\frac{1}{s}}$$

and

$$\left(q_+^{r,s}(\lambda_{\max}\right)_i = \left(a\left(a^{n-1}\left(k^+\right)^s + \left(1-a^{n-1}\right)\left(\lambda_j^+\right)^s\right)^{\frac{r}{s}} + (1-a)\left(\lambda_j^+\right)^r\right)^{\frac{1}{r}}$$

$$\leq \left((1-a)\left(\lambda_j^+\right)^r\right)^{\frac{1}{r}} = \lambda_j^+(1-a)^{\frac{1}{r}}.$$

Thus, by (24.13) we finally obtain that

$$\lambda_j^- c^- \leq \left(q_-^{s,r}(C(\cdot,t))\right)_i \leq \left(q_+^{r,s}(C(\cdot,t))\right)_i \leq \lambda_j^+ c^+$$

for $c^- = \left(1-a^{n-1}\right)^{\frac{1}{s}}$ and $c^+ = (1-a)^{\frac{1}{r}}$, and the proof is complete. □

24.5 Nonlinear Bounds

In this section, we discuss some bounds on the homogenized integrand f_{hom} corresponding to functions $\{f_h\}$ of the form

$$f_h(x,\xi) = f(hx,\xi) = C(hx,|\xi|)\,|\xi|^p,\ \xi \in \mathbf{R}^n.$$

The bounds appear to be some kind of "mixed averages" of the function C obtained by using the power mean $U^{[p]}$ defined by

$$U^{[p]} = \begin{cases} P^{[p/2]} & \text{if } p \leq 2 \\ P^{[1]} & \text{if } 2 \leq p \end{cases}$$

and the means $L^{[p]}$, $q_-^{r,s}$ and $q_+^{s,r}$. They were first presented in [28], [29], [30] and [32]. In the presentation below, we recall that the latter three means were discussed in the previous section.

The function $C : \mathbf{R}^n \times \mathbf{R}_+ \to \mathbf{R}_+$ has the following properties: $\alpha \leq C(x,t) \leq \beta$ for all x and t for some constants $0 < \alpha \leq \beta < +\infty$, $C(x,t)$ is □-periodic and Lebesgue measurable in x and differentiable and non-decreasing in t. Moreover, $C(x,t)t^p$ is convex in t and

$$\frac{d\left(C(x,t)t^p\right)}{dt} \leq p\beta t^{p-1} \tag{24.14}$$

for all x and t.

Furthermore, we will use the following notations. For every $t \geq 0$, let $c^+(t)$ and $c^-(t)$ be the n-tuples

$$c^+(t) = q_+^{r_1,\,r_2}\left(C\left(\cdot, t\left(\frac{\alpha}{\beta}\right)^{r_1}\right)\right),$$

$$c^-(t) = q_-^{s_1,\,s_2}\left(C\left(\cdot, t\left(\frac{\alpha}{\beta}\right)^{\frac{1+s_1}{p-1}}\right)\right),$$

where

$$[r_1,\ r_2, s_1, s_2] = \begin{cases} \left[-\frac{2}{p}, \frac{2}{p}, 1, \frac{1}{1-p}\right] & \text{if } p \leq 2 \\ \left[\frac{1}{1-p}, 1, \frac{2}{p}, -\frac{2}{p}\right] & \text{if } 2 \leq p \end{cases}.$$

Moreover, let the functions $C_p^+, C_p^- : \mathbf{R}^n \to \mathbf{R}_+$ be defined by

$$C_p^+(\xi) = U^{[p]}(c^+(|\xi|), w(\xi)),$$
$$C_p^-(\xi) = L^{[p]}(c^-(|\xi|)), w(\xi)).$$

We are now ready to formulate the main result of this section:

Theorem 24.17. *It holds that*

$$|\xi|^p C_p^-(\xi) \leq f_{hom}(\xi) \leq |\xi|^p C_p^+(\xi)$$

for all $\xi \in \mathbf{R}^n$.

Remark 24.18. According to Theorem 24.17 and the results of the previous section we note in particular that:

1. $C_p^-(\xi)$ and $C_p^+(\xi)$ are continuous in p and ξ.
2. The functions $k \mapsto C_p^+(k\xi)$ and $k \mapsto C_p^+(k\xi)$ are non-decreasing in $\mathbf{R}_+$.
3. If $p = 2$, then

$$\sum_{i=1}^n (c^-(|\xi|))_i \xi_i^2 \leq f_{\text{hom}}(\xi) \leq \sum_{i=1}^n (c^+(|\xi|))_i \xi_i^2.$$

4. If $t > 0$, then

$$C_p^-(te_i) = (c^-(t))_i \text{ and } C_p^+(te_i) = (c^+(t))_i.$$

5. If $C(\cdot, t)$ satisfies the property of cubic symmetry for every $t > 0$, then

$$C_p^-(\xi) = (c^-(|\xi|))_1 \text{ and } C_p^+(\xi) = (c^+(|\xi|))_1.$$

Remark 24.19. By applying the algorithm in Theorem 24.14, we can easily compute lower and upper approximations for the n-tuples $c^-(t)$ and $c^+(t)$, respectively. Hence, in the cases when $L^{[p]}$ and $U^{[p]}$ are power means, we can compute lower and upper approximations for $C_p^-(\xi)$ and $C_p^+(\xi)$, respectively, and thereby obtain bounds for $f_{hom}(\xi)$ arbitrarily close to those in Theorem 24.17. It is more difficult to compute $C_p^-(\xi)$ if $p < 2$. In this case $L^{[p]}$ is no longer a power-mean and can only be computed numerically. Nevertheless, we can always use the lower bounds $K^{[p]}$ or $P^{[1/(1-p)]}$ for $L^{[p]}$ (see Theorem 24.10 and Proposition 24.12) in order to obtain a lower bound for $C_p^-(\xi)$.

Remark 24.20. Assume that C is of the form

$$C(x,t) = \sum_{i=1}^m C_i(t)\, \chi_{A_i}(x),$$

where $\lambda_i^- \leq C_i(t) \leq \lambda_i^+$ for all $t > 0$ and $\{A_i\}$ is a disjoint partition of $\mathbf{R}^n$ such that for a fixed $j \in \{1, ..., m\}$, $\square \setminus A_j$ is a genuine subset of $\square$. Then, it follows from Proposition 24.16(2) that

$$c^{-}\lambda_j^{-} \leq C_p^{-}(\xi) \leq C_p^{+}(\xi) \leq c^{+}\lambda_j^{+},$$

where $0 < c^{-}, c^{+} < +\infty$. Moreover, c^{-} and c^{+} are only dependent of A_j and p. Hence, according to Corollary 24.6 (see also Remark 24.7) we can in this case conclude that $C_p^{-}(\cdot)\,|\cdot|^p$ is a much sharper lower bound for f_{hom} than the Wiener and Hashin–Shtrikman lower bounds f_{W-} and f_{HSC-} when λ_k^{-} is relatively large and $k \neq j$. Similarly, we see that $C_p^{+}(\cdot)\,|\cdot|^p$ is a much sharper upper bound for f_{hom} than the Wiener and Hashin–Shtrikman upper bounds f_{W+} and f_{HSC+} when λ_k^{+} is relatively small.

Remark 24.21. In Theorem 24.17 the upper bound for $p \geq 2$ and the lower bound for $p \leq 2$ reduce to the bounds presented in [37] for the special case when $\xi = ke_i$, $k \in \mathbf{R}$ and f is of the form $f(x,\xi) = \lambda(x)\,|\xi|^p$(see also [24, 25]). However, the upper bound and lower bound in [37] for $p < 2$ and $p > 2$, respectively, are different from those presented in Theorem 24.17. Moreover, the proof given below is independent of that presented in [37].

In order to illustrate the usefulness of Theorem 24.17, we present the following simple example.

Example 24.22. Assume that $n = 2$, $p = 2$ and

$$C(x,t) = g(x_1,t)g(x_2,t)$$

for all x, where

$$g(s,t) = \frac{1 + a(s)t}{1+t},$$

$a(\cdot)$ is $[0,1]$-periodic, $a(s) = 1$ for $0 \leq s \leq \frac{1}{2}$ and $a(s) = k \geq 1$ for $\frac{1}{2} < s < 1$. For this case we find that

$$\left(q_{-}^{1,-1}(C(\cdot,t))\right)_i = \left(q_{+}^{-1,1}(C(\cdot,t))\right)_i = \frac{1+tk}{1+t}$$

for $i = 1,2$. Moreover, $1 \leq C(x,t) \leq k$, and, hence, according to Theorem 24.17 we obtain the bounds

$$\frac{1+|\xi|\,k^{-1}}{1+|\xi|\,k^{-2}}\,|\xi|^2 \leq f_{\text{hom}}(\xi) \leq \frac{1+|\xi|\,k^3}{1+|\xi|\,k^2}\,|\xi|^2.$$

We observe that the lower bound and the upper bound are very close to each other, especially for small values of $|\xi|$ and also for large values of $|\xi|$. Hence, in these cases we obtain a very good estimate of $f_{\text{hom}}(\xi)$.

The inequalities in Theorem 24.17 are sharp and equalities occur for example when f is of the form

$$f(x,\xi) = C(x,|\xi|)\,|\xi|^2 = \lambda(x_i)\,|\xi|^2.$$

More generally, we will prove the following result:

Proposition 24.23. *Assume that C is of the form $C(x,t) = \lambda(x)$ and let $r_1,...,r_m \in \{1,...,n\}$ be fixed distinct integers. If $\xi = \sum_{i=1}^{m} k_i e_{r_i}$ where $n \geq m \geq 1$, then*

$$f_{hom}(\xi) = |\xi|^p C_p^-(\xi)$$

for $p \leq 2$ and

$$f_{hom}(\xi) = |\xi|^p C_p^+(\xi)$$

for $p \geq 2$ and all $k_i \in R$, $i \in \{1,...,m\}$ if and only if λ is of the form

$$\lambda(x) = (\lambda_1(x_{r_1})\lambda_2(x_{r_2})\cdots\lambda_m(x_{r_m}))^{\delta_{1,m}+\delta_{2,p}}\lambda_{m+1}(x_{r_{m+1}},x_{r_{m+2}},...,x_{r_n}).$$

Example 24.24. Assume that $C(x,t) = \lambda(x) = g(x_1)g(x_2)\cdots g(x_n)$ for all x, where g is non-constant, and let q be the function defined by

$$q(p) = \left(\int_0^1 g(t)\,dt\right)^{n-1}\left(\int_0^1 (g(t))^{\frac{1}{1-p}}\,dt\right)^{1-p}.$$

It is easy to see that $C_p^-(\xi) = q(p)$ if $p \leq 2$ and $C_p^+(\xi) = q(p)$ if $p \geq 2$. Hence, by Proposition 24.23 it follows that $f_{\text{hom}}(\xi) = q(p)|\xi|^p$ when $\xi = k_i e_i$ for $p > 1$. Moreover, according to Theorem 24.17

$$f_{\text{hom}}(\xi) \leq q(p)|\xi|^p \text{ if } p \geq 2, \tag{24.15}$$

$$f_{\text{hom}}(\xi) \geq q(p)|\xi|^p \text{ if } p \leq 2,$$

and by Proposition 24.23 we can find cases where these inequalities are strict.

Remark 24.25. In the above example $f(\cdot,\xi)$ possesses the property of cubic symmetry. In the case when f is of the form $f(x,\xi) = \lambda(x)|\xi|^2$ it is well known that this property always yields isotropy for f_{hom}, i.e., that f_{hom} is a function of $|\xi|$ only. Example 24.24 shows that this is not always true when $p \neq 2$.

Proof of Theorem 24.17: By putting $r = 2$ in Lemma 24.9(1), we obtain that

$$\left(\sum_{i=1}^{n}(a_i\xi_i)^2\right)^{\frac{s}{2}} \leq |\xi|^s\left(\sum_{i=1}^{n} a_i^s w_i(\xi)\right) \tag{24.16}$$

for any positive n-tuple a_i if $s \geq 2$.

$f_{\text{hom}}(\xi) \leq C_p^+(\xi)|\xi|^p$: Put

$$\rho = C(\cdot,(\frac{\alpha}{\beta})^{r_1}|\xi|).$$

We start with the case $p \geq 2$. Put

$$\lambda_i = \left(\tau_+^{[1]}(\rho)\right)_i$$

and let the function u be defined by

$$u(y_1,y_2,...,y_n)=\sum_{i=1}^{n}\xi_i(-y_i+(\int_0^{y_i}\lambda_i^{1/(1-p)}dx_i)(\int_0^1\lambda_i^{1/(1-p)}dx_i)^{-1}).$$

Thus u is absolutely continuous on almost all line segments in $\square$ parallel to the coordinate axes and the classical partial derivatives of u belong to $L^p(\square)$ (Compare with [54, Theorem 7.11]). Hence by [55, Theorem 2.1.4] it follows that $u\in H^{1,p}(\square)$. Moreover, we observe that u is $\square$-periodic. Therefore, $u\in H^{1,p}_{\mathrm{per}}(\square)$. We also have that

$$(\xi+Du)_i=a_i\xi_i, \tag{24.17}$$

where

$$a_i=\lambda_i^{1/(1-p)}(\int_0^1\lambda_i^{1/(1-p)}dx_i)^{-1}.$$

Since $\alpha\le\rho\le\beta$, we get that

$$a_i\le(\frac{\alpha}{\beta})^{1/(1-p)}=(\frac{\alpha}{\beta})^{r_1},$$

and, hence,

$$|\xi+Du|\le(\frac{\alpha}{\beta})^{r_1}|\xi|.$$

Therefore,

$$\begin{aligned}f(x,\xi+Du(x))&=C(x,|\xi+Du(x)|)\,|\xi+Du(x)|^p\\ &\le C(x,|\xi|(\frac{\alpha}{\beta})^{r_1})\,|\xi+Du(x)|^p=\rho(x)\,|\xi+Du(x)|^p.\end{aligned} \tag{24.18}$$

Moreover, we can prove that

$$\int_\square\rho\,|\xi+Du|^p\,dx\le C_p^+(\xi)\,|\xi|^p. \tag{24.19}$$

In fact, by (24.16) and (24.17) with $s=p\ge 2$ we see that

$$|\xi+Du(x)|^p\le|\xi|^p\left(\sum_{i=1}^{n}(a_i(x))^p w_i(\xi)\right).$$

Thus

$$\int_\square\rho\,|\xi+Du|^p\,dx\le|\xi|^p\int_\square\rho\left(\sum_{i=1}^{n}a_i^p w_i(\xi)\right)dx. \tag{24.20}$$

Moreover,

$$\int_\square\rho\left(\sum_{i=1}^{n}a_i^p w_i(\xi)\right)dx=\sum_{i=1}^{n}\int_\square\rho\lambda_i^{\frac{p}{1-p}}(\int_0^1\lambda_i^{\frac{1}{1-p}}dx_i)^{-p})\,w_i(\xi)\,dx$$

$$= \sum_{i=1}^{n} \int_0^1 \left(\int_0^1 \cdots \int_0^1 \rho \lambda_i^{\frac{p}{1-p}} (\int_0^1 \lambda_i^{\frac{1}{1-p}} dx_i)^{-p} w_i(\xi)\, dx_1 \cdots dx_{i-1} dx_{i+1} \cdots dx_n \right) dx_i$$

$$= \sum_{i=1}^{n} (\int_0^1 \lambda_i^{\frac{1}{1-p}} dx_i)^{-p} \int_0^1 \left(\left(\int_0^1 \cdots \int_0^1 \rho\, dx_1 \cdots dx_{i-1} dx_{i+1} \cdots dx_n \right) \lambda_i^{\frac{p}{1-p}} \right) dx_i\, w_i(\xi)$$

$$= \sum_{i=1}^{n} (\int_0^1 \lambda_i^{\frac{1}{1-p}} dx_i)^{-p} \int_0^1 (\lambda_i \lambda_i^{\frac{p}{1-p}})\, dx_i\, w_i(\xi) = \sum_{i=1}^{n} (\int_0^1 \lambda_i^{\frac{1}{1-p}} dx_i)^{1-p} w_i(\xi)$$

$$= \sum_{i=1}^{n} (q_+^{1/(1-p),1}(\rho))_i w_i(\xi) = P^{[1]}(q_+^{1/(1-p),1}(\rho), w(\xi)) = C_p^+(\xi). \tag{24.21}$$

Hence, according to (24.20) and (24.21) we get (24.19). Now, using (24.18) and (24.19) we find that

$$f_{\text{hom}}(\xi) = \min_{v \in \mathrm{H}^{1,p}_{\text{per}}(\Box)} \int_\Box f(x, \xi + Dv(x))\, dx \le \int_\Box f(x, \xi + Du(x))\, dx$$

$$\le \int_\Box \rho\, |\xi + Du(x)|^p\, dx \le C_p^+(\xi)\, |\xi|^p .$$

The upper bound is thereby proved for $p \ge 2$. If $p \le 2$, we put

$$\lambda_i = \left(\tau_+^{[2/p]}(\rho) \right)_i^{2/p}$$

and let the function u be defined by

$$u(y_1, \ldots, y_n) = \sum_{i=1}^{n} \xi_i (-y_i + (\int_0^{y_i} \lambda_i^{-1} dx_i)(\int_0^1 \lambda_i^{-1} dx_i)^{-1}).$$

By using the same arguments as above we get that $u \in \mathrm{H}^{1,p}_{\text{per}}(\Box)$ and

$$(\xi + Du)_i = \xi_i a_i,$$

where

$$a_i = \lambda_i^{-1} (\int_0^1 \lambda_i^{-1} dx_i)^{-1},$$

and it follows that

$$f(x, \xi + Du(x)) \le \rho(x)\, |\xi + Du(x)|^p = |\xi|^p \rho(x) \left(\sum_{i=1}^{n} (a_i(x))^2 w_i(\xi) \right)^{\frac{p}{2}} . \tag{24.22}$$

Thus, the Hölder inequality yields that

$$\int_\Box \rho\, |\xi + Du|^p\, dx = |\xi|^p \int_\Box \left(\sum_{i=1}^{n} \rho^{\frac{2}{p}} (a_i(x))^2 w_i(\xi) \right)^{\frac{p}{2}} dx \le$$

$$\leq |\xi|^p \left(\sum_{i=1}^{n} \int_{\square} \rho^{\frac{2}{p}} (a_i(x))^2 \, dx \, w_i(\xi) \right)^{\frac{p}{2}}. \tag{24.23}$$

Arguing similarly as we did in (24.21) we obtain

$$\int_{\square} \rho^{\frac{2}{p}} (a_i(x))^2 \, dx = \left(q_+^{-2/p,2/p}(\rho) \right)_i^{2/p}.$$

Hence, by (24.22) and (24.23) we find that

$$f_{\mathrm{hom}}(\xi) \leq \int_{\square} f(x, \xi + Du(x)) \, dx$$

$$\leq \int_{\square} \rho \, |\xi + Du|^p \, dx \leq |\xi|^p \left(\sum_{i=1}^{n} \left(q_+^{-2/p,2/p}(\rho) \right)_i^{\frac{2}{p}} w_i(\xi) \right)^{\frac{p}{2}}$$

$$= |\xi|^p \, P^{[2/p]}(q_+^{-2/p,2/p}(\rho), w(\xi)) = C_p^+(\xi) \, |\xi|^p.$$

This completes the proof of the upper bound.

$C_p^-(\xi) \, |\xi|^p \leq f_{\mathrm{hom}}(\xi)$: We have that

$$f^*(x, \eta) = \sup_{E \in \mathbf{R}^n} \{ E \cdot \eta - f(x, \eta) \}$$

$$= \max_{t \geq 0} \{ t \cdot |\eta| - C(x,t) \, |t|^p \} = t_m \, |\eta| - C(x, t_m) \, |t_m|^p,$$

where t_m solves the equation

$$|\eta| - \frac{d \left(C(x, t_m) t_m^p \right)}{dt} = 0.$$

Hence, by (24.14) we have that

$$\left(\frac{|\eta|}{p\beta} \right)^{\frac{1}{p-1}} \leq t_m.$$

Moreover, we can show that

$$f^*(x, \eta) \leq \frac{1}{p'} |\eta|^{p'} \left(pC(x, \left(\frac{|\eta|}{p\beta} \right)^{\frac{1}{p-1}}) \right)^{\frac{1}{1-p}}. \tag{24.24}$$

In fact,

$$f^*(x, \eta) = t_m \, |\eta| - C(x, t_m) \, |t_m|^p \leq t_m \, |\eta| - C(x, \left(\frac{|\eta|}{p\beta} \right)^{\frac{1}{p-1}}) \, |t_m|^p$$

$$\leq \max_{t\geq 0}\left\{t\cdot|\eta|-C(x,\left(\frac{|\eta|}{p\beta}\right)^{\frac{1}{p-1}})\,|t|^p\right\} = \frac{1}{p'}|\eta|^{p'}\left(pC(x,\left(\frac{|\eta|}{p\beta}\right)^{\frac{1}{p-1}})\right)^{\frac{1}{1-p}}.$$

Let

$$\rho = C(\cdot,\left(\frac{\alpha}{\beta}\right)^{\frac{1+s_1}{p-1}}|\xi|).$$

Assume first that $p\leq 2$. Put

$$\tau_i = (\tau_-^{1/(1-p)}(\rho))_i^{1/(1-p)} \text{ and } q_i^-(\rho) = \left(q_-^{1,1/(1-p)}(\rho)\right)_i.$$

Let $\eta\in\mathbf{R}^n$ be such that $|\eta|\geq p|\xi|^{p-1}k(\xi)$, where

$$k(E) = \left(\sum_{i=1}^n (q_i^-(\rho))^{\frac{1}{1-p}}\,w_i(E)\right)^{1-p}, E\in\mathbf{R}^n.$$

Next let $\vartheta:\mathbf{R}^n\to\mathbf{R}^n$ be the vector function with components

$$\vartheta_i = \eta_i(-1+\frac{\tau_i^{1-p}}{q_i^-(\rho)}).$$

Then, $\vartheta\in[\mathrm{L}^p(\Box)]^n$, $\int_\Box \vartheta\,dx = 0$ and ϑ_i is independent of the i-th coordinate. The latter property implies that

$$\int_\Box D\varphi\cdot\vartheta\,dx = 0$$

for all $\varphi\in\mathrm{H}^{1,p}_{\text{per}}(\Box)$. Consequently, $\vartheta\in\mathrm{V}^{p'}_{\text{sol}}(\Box)$. Moreover, we note that $k(\xi)\geq$ $\min_{j=1}^n\left\{q_j^-(\rho)\right\}$, and, hence,

$$\frac{|\vartheta(x)+\eta|}{p\beta} = \frac{1}{p\beta}\left(\sum_{i=1}^n\left(\frac{\eta_i\tau_i^{1-p}}{q_i^-(\rho)}\right)^2\right)^{\frac{1}{2}} \geq \frac{p|\xi|^{p-1}}{p\beta|\eta|}\left(\sum_{i=1}^n\left(\frac{k(\xi)\eta_i\tau_i^{1-p}}{q_i^-(\rho)}\right)^2\right)^{\frac{1}{2}}$$

$$\geq \frac{|\xi|^{p-1}}{|\eta|}\left(\sum_{i=1}^n\left(\frac{\eta_i\left(\min_{j=1}^n\left\{q_j^-(\rho)\right\}\right)\tau_i^{1-p}}{\beta q_i^-(\rho)}\right)^2\right)^{\frac{1}{2}}$$

$$\geq \frac{|\xi|^{p-1}}{|\eta|}\left(\sum_{i=1}^n\left(\eta_i\frac{\alpha^2}{\beta^2}\right)^2\right)^{\frac{1}{2}} = |\xi|^{p-1}\frac{\alpha^2}{\beta^2}.$$

According to (24.24) and the fact that $C(x,\cdot)$ is non-decreasing, this shows that

$$f^*(x,\vartheta(x)+\eta)\leq\frac{1}{p'}|\vartheta(x)+\eta|^{p'}\left(pC(x,\left(\frac{|\vartheta(x)+\eta|}{p\beta}\right)^{\frac{1}{p-1}})\right)^{\frac{1}{1-p}}$$

$$\leq \frac{1}{p'}\left|\vartheta(x)+\eta\right|^{p'}\left(pC(x,\left(\frac{\alpha}{\beta}\right)^{\frac{2}{p-1}}|\xi|)\right)^{\frac{1}{1-p}} = \frac{1}{p'}\left|\vartheta(x)+\eta\right|^{p'}\left(p\rho(x)\right)^{\frac{1}{1-p}},$$

i.e.,

$$f^*(x,\vartheta(x)+\eta) \leq \frac{1}{p'}\left|\vartheta(x)+\eta\right|^{p'}\left(p\rho(x)\right)^{\frac{1}{1-p}}. \tag{24.25}$$

Moreover, (24.16) with $s=p'$ and $a_i=\tau_i^{1-p}/q_i^-(\rho)$, we have that

$$|\vartheta+\eta|^{p'} = \left(\sum_{i=1}^n\left(\eta_i\frac{\tau_i^{1-p}}{q_i^-(\rho)}\right)^2\right)^{\frac{p'}{2}} \leq |\eta|^{p'}\sum_{i=1}^n\left(\frac{\tau_i^{1-p}}{q_i^-(\rho)}\right)^{p'} w_i(\eta).$$

Therefore,

$$\int_\Box f^*(x,\vartheta+\eta)\,dx \leq \int_\Box \frac{1}{p'}|\vartheta+\eta|^{p'}\left(p\rho(x)\right)^{\frac{1}{1-p}}dx \tag{24.26}$$

$$\leq \frac{1}{p'}(p)^{\frac{1}{1-p}}|\eta|^{p'}\sum_{i=1}^n\left(\int_\Box \left(q_i^-(\rho)\right)^{-p'}\left(\rho(x)\right)^{\frac{1}{1-p}}\tau_i^{-p}w_i(\eta)dx\right).$$

In addition,

$$\int_0^1\left(\left(q_i^-\rho\right)^{-p'}(\rho)^{\frac{1}{1-p}}\tau_i^{-p}w_i(\eta)\right)dx_i$$

$$= w_i(\eta)\left(q_i^-(\rho)\right)^{-p'}\tau_i^{-p}(p)^{\frac{1}{1-p}}\int_0^1\rho(x)^{\frac{1}{1-p}}dx_i = w_i(\eta)\left(q_i^-(\rho)\right)^{-p'}\tau_i^{1-p}.$$

Therefore,

$$\int_\Box\left(\left(q_i^-(\rho)\right)^{-p'}\rho^{\frac{1}{1-p}}\tau_i^{-p}w_i(\eta)\right)dx$$

$$= w_i(\eta)\left(q_i^-(\rho)\right)^{-p'}\int_0^1\cdot\cdot\int_0^1\tau_i^{1-p}dx_1\cdot\cdot dx_{i-1}dx_{i+1}\cdot\cdot dx_n$$

$$= w_i(\eta)\left(q_i^-(\rho)\right)^{-p'}q_i^-(\rho) = w_i(\eta)\left(q_i^-(\rho)\right)^{\frac{1}{1-p}} = w_i(\eta)\left(c_i^-(|\xi|)\right)^{\frac{1}{1-p}}, \tag{24.27}$$

and according to the complementary energy principle and (24.26) we obtain that

$$f^*_{\text{hom}}(\eta) \leq \int_\Box f^*(x,\vartheta+\eta)\,dx \leq \frac{1}{p'}|\eta|^{p'}\sum_{i=1}^n\left(pq_i^-(\rho)\right)^{\frac{1}{1-p}}w_i(\eta).$$

Hence,

$$f_{\text{hom}}(\xi) = \sup_{\eta\in R^n}\{\eta\cdot\xi - f^*_{\text{hom}}(\eta)\} \geq \sup_{|\eta|\geq pk(\xi)|\xi|^{p-1}}\{E\cdot\eta - f^*_{\text{hom}}(\eta)\} \tag{24.28}$$

$$\geq \sup_{|\eta|\geq pk(\xi)|\xi|^{p-1}}\left\{\eta\cdot\xi - \frac{1}{p'}|\eta|^{p'}\sum_{i=1}^n\left(pq_i^-(\rho)\right)^{\frac{1}{1-p}}w_i(\eta)\right\}$$

$$= \sup_{|E|=1} \left\{ \sup_{t \geq pk(\xi)|\xi|^{p-1}} \{g(t,E)\} \right\},$$

where

$$g(t,E) = tE \cdot \xi - \frac{1}{p'} t^{p'} \sum_{i=1}^{n} \left(pq_i^-(\rho)\right)^{\frac{1}{1-p}} w_i(E).$$

We note that

$$\sup_{t \geq 0} \{g(t,E)\} = |\xi|^p \frac{|E \cdot \xi|^p}{|\xi|^p} k(E). \tag{24.29}$$

Thus,

$$\sup_{|E|=1} \left\{ \sup_{t \geq 0} \{g(t,E)\} \right\} = C_p^-(\xi) |\xi|^p.$$

The maximum in (24.29) is attained for

$$t = p \left(\frac{|E \cdot \xi|}{|\xi|} \right)^{p-1} |\xi|^{p-1} k(E).$$

Moreover, the function

$$E \mapsto \frac{|E \cdot \xi|^p}{|\xi|^p} k(E)$$

is continuous and its maximum is therefore attained on the compact set $\{E : |E| = 1\}$. Hence, we can find a maximizer v such that

$$\frac{|v \cdot \xi|^p}{|\xi|^p} k(v) = \max_{|E|=1} \left\{ \frac{|E \cdot \xi|^p}{|\xi|^p} k(E) \right\}.$$

Therefore, for $E = v$ we have that

$$t = p \left(\frac{|v \cdot \xi|}{|\xi|} \right)^{p-1} k(v) |\xi|^{p-1} \geq p \left(\frac{|v \cdot \xi|}{|\xi|} \right)^{p} k(v) |\xi|^{p-1}$$

$$\geq p \left(\frac{\left| \xi |\xi|^{-1} \cdot \xi \right|}{|\xi|} \right)^{p} k(\xi |\xi|^{-1}) |\xi|^{p-1} = pk(\xi) |\xi|^{p-1},$$

and hence,

$$\sup_{|E|=1} \left\{ \sup_{t \geq pk(\xi)|\xi|^{p-1}} \{g(t,E)\} \right\} = \sup_{|E|=1} \left\{ \sup_{t \geq 0} \{g(t,E)\} \right\} = C_p^-(\xi) |\xi|^p.$$

According to (24.28) this implies that

$$C_p^-(\xi) |\xi|^p \leq f_{\text{hom}}(\xi),$$

and the lower bound is thereby proved for $p \leq 2$. The proof of the case $p \geq 2$ is quite similar and we therefore only give a sketch of it. We put $\tau_i = (\tau_-^{-2/p}(\rho))_i^{-2/p}$,

$$q_i^-(\rho) = \left(q_-^{2/p,-2/p}(\rho)\right)_i, k(E) = \left(\sum_{i=1}^n \left(q_i^-(\rho)\right)^{-\frac{2}{p}} w_i(E)\right)^{-\frac{p}{2}},$$

and let $\vartheta : \mathbf{R}^n \to \mathbf{R}^n$ be the vector function with components

$$\vartheta_i = \eta_i(-1 + \frac{\tau_i^{-1}}{(q_i^-(\rho))^{\frac{2}{p}}}).$$

As above we get that $\vartheta \in \mathrm{V}_{\mathrm{sol}}^{p'}(\square)$ and

$$f^*(x, \vartheta + \eta) \leq \frac{1}{p'} |\vartheta + \eta|^{p'} (p\rho)^{\frac{1}{1-p}} = \frac{1}{p'} \left(\sum_{i=1}^n \frac{\tau_i^{-2} (p\rho)^{-\frac{2}{p}}}{(q_i^-(\rho))^{\frac{4}{p}}} w_i(\eta)\right)^{\frac{p'}{2}}.$$

Therefore, according to the Hölder inequality,

$$\int_\square f^*(x, \vartheta + \eta)\, dx \leq \frac{1}{p'} \left(\sum_{i=1}^n \int_\square \frac{\tau_i^{-p} (p\rho)^{-\frac{2}{p}}}{(q_i^-(\rho))^2} dx\, w_i(\eta)\right)^{\frac{p'}{2}}.$$

Moreover, by arguing similarly as in (24.27), we find that

$$\int_\square \frac{\tau_i^{-p} (\rho(x))^{-\frac{2}{p}}}{(q_i^-(\rho))^2} dx = q_i^-(\rho)^{-\frac{2}{p}}.$$

Hence,

$$\int_\square f^*(x, \vartheta(x) + \eta)\, dx \leq \frac{1}{p'} \left(\sum_{i=1}^n \left(p q_i^-(\rho)\right)^{-\frac{2}{p}} w_i(\eta)\right)^{\frac{p'}{2}}.$$

Now, following the lines in the above proof from (24.28) we obtain that

$$|\xi|^p \max_{|v|=1} \left\{ \left(\frac{|v \cdot \xi|}{|\xi|}\right)^p \left(\sum_{i=1}^n \left(q_i^-(\rho)\right)^{-\frac{2}{p}} w_i(v)\right)^{-\frac{p}{2}} \right\} \leq f_{\mathrm{hom}}(\xi),$$

i.e.,

$$C_p^-(\xi) |\xi|^p \leq f_{\mathrm{hom}}(\xi).$$

This completes the proof. □

For the proof of Proposition 24.23 we need the following result (cf. [25, Theorem 1]):

Lemma 24.26. *If $t \in \mathbf{R}$, then*

$$f_{hom}(te_{r_i}) = |t|^p \left(q_-^{1,1/(1-p)}(\lambda)\right)_{r_i} \text{ for } p \leq 2$$

and

$$f_{hom}(te_{r_i}) = |t|^p \left(q_+^{1/(1-p),1}(\lambda)\right)_{r_i} \text{ for } p \geq 2,$$

$i = 1,...,m$, if and only if λ is of the form

$$\lambda(x) = (\lambda_1(x_{r_1})\lambda_2(x_{r_2})\cdots\lambda_m(x_{r_m}))\,\lambda_{m+1}(x_{r_{m+1}}, x_{r_{m+2}}, ..., x_{r_n}). \tag{24.30}$$

Proof of Proposition 24.23. If $m = 1$, the proof follows directly by Lemma 24.26. Moreover, if $p = 2$, then it follows from Theorem 24.17 and Lemma 24.26 that

$$\sum_{i=1}^{m} \left(q_-^{1,1/(1-p)}(\lambda)\right)_{r_i} k_i^2 = f_{\text{hom}}(\xi) = \sum_{i=1}^{m} \left(q_+^{1/(1-p),1}(\lambda)\right)_{r_i} k_i^2$$

for all $\xi = \sum_{i=1}^m k_i e_{r_i}$, $k_i \in \mathbf{R}$ if and only if λ is of the form (24.30). It only remains to prove that

$$f_{\text{hom}}(\xi) = |\xi|^p C_p^-(\xi) \text{ for } p < 2 \tag{24.31}$$

and

$$f_{\text{hom}}(\xi) = |\xi|^p C_p^+(\xi) \text{ for } p > 2, \tag{24.32}$$

for all $\xi = \sum_{i=1}^m k_i e_{r_i}$, $k_i \in \mathbf{R}$, $i \in \{1,...,m\}$, $m > 1$ if and only if λ is of the form

$$\lambda(x) = \lambda_{m+1}(x_{r_{m+1}}, x_{r_{m+2}}, ..., x_{r_n}). \tag{24.33}$$

In fact, by Lemma 24.26, it yields that the equalities (24.31) and (24.32) hold only if λ is of the form (24.30). On the other hand, suppose that λ_{r_i} or λ_{r_j}, $i \neq j$, $i, j \in \{1,...,m\}$ are non-constant. Then for the functions $\{a_i\}$ given in the proof of Theorem 24.17 we find that $a_{r_i} \neq a_{r_j}$ on a set of strictly positive measure. Hence, according to Lemma 24.9 this implies that the inequalities (24.20) and (24.26) are strict. But this contradicts (24.31) and (24.32), thus we have that our assumption is wrong, i.e., we can conclude that λ is of the form (24.33). Conversely, let $\xi = \sum_{i=1}^m k_i e_{r_i}$ and let λ be of the form (24.33). Then it is easy to see (by inspection) that $C_p^-(\xi) = \int_\square \lambda\, dx$ for $p < 2$ and $C_p^+(\xi) = \int_\square \lambda\, dx$ for $p > 2$. Moreover, a straightforward computation via Euler–Lagrange equation shows that

$$f_{\text{hom}}(\xi) = \left(\int_\square \lambda\, dx\right) |\xi|^p.$$

Thus, we obtain (24.31) and (24.32) and the proof is complete. □

24.6 Further Results for the Case $p = 2$

In this section, we discuss in more detail the special case when f is of the form

$$f(x,\xi) = C(x,|\xi|)\,|\xi|^2,\ \xi \in \mathbf{R}^n.$$

In addition to the assumptions of the previous section, we let C be of the form $C(\cdot,t) = \sum_{i=1}^{m} \lambda_i(t)\chi_{A_i}$ and satisfies the property of cubic symmetry, i.e., $f \in \mathfrak{I}_{a,g}^{\text{cub}}$, where

$$\int_{\square} \chi_{A_i}\,dx = a_i \text{ and } g_i(\xi) = \lambda_i(|\xi|)\,|\xi|^2.$$

Moreover, we assume that

$$\lambda_1(0) < \lambda_2(0) < \cdots < \lambda_m(0)$$

and

$$\lambda_1(\infty) < \lambda_2(\infty) < \cdots < \lambda_m(\infty).$$

We define the following constants:

$$\begin{aligned}
q_t^- &= \left(q_-^{1,1/(1-p)}(C(\cdot,t))\right)_1, \ q_t^+ = \left(q_+^{1/(1-p),1}(C(\cdot,t))\right)_1,\\
w_t^- &= \left(\int_{\square} (C(\cdot,t))^{-1}\,dx\right)^{-1}, \ w_t^+ = \int_{\square} C(\cdot,t)\,dx,\\
h_0^- &= -(n-1)\lambda_1(0) + \left(\int_{\square} \frac{1}{C(\cdot,0)+(n-1)\lambda_1(0)}dx\right)^{-1},\\
h_\infty^+ &= -(n-1)\lambda_m(\infty) + \left(\int_{\square} \frac{1}{C(\cdot,\infty)+(n-1)\lambda_m(\infty)}dx\right)^{-1},\\
t_0^+ &= \lambda_m(0)\left(1 - \frac{n(1-a_m)\,(\lambda_m(0)-\lambda_1(0))}{n\lambda_m(0)+a_m\,(\lambda_1(0)-\lambda_m(0))}\right),\\
t_\infty^- &= \lambda_1(\infty)\left(1 + \frac{n\,(1-a_1)\,(\lambda_m(\infty)-\lambda_1(\infty))}{n\lambda_1(\infty)+a_1\,(\lambda_m(\infty)-\lambda_1(\infty))}\right).
\end{aligned}$$

Remark 24.27. According to Remark 24.18(5) we have that

$$(c^-(|\xi|))_1\,|\xi|^2 \le f_{\text{hom}}(\xi) \le (c^+(|\xi|))_1\,|\xi|^2. \tag{24.34}$$

Since $(c^-(\cdot))_1$ and $(c^+(\cdot))_1$ are non-decreasing, we deduce that

$$q_0^-\,|\xi|^2 \le f_{\text{hom}}(\xi) \le q_\infty^+\,|\xi|^2.$$

In many cases, q_0^- and q_∞^+ happen to be sufficiently close to each other to give a sharp estimate of f_{hom}. The general bounds (24.34) also yield that

$$q_0^- |\xi|^2 \le f_{\mathrm{hom}}(\xi) \le q_0^+ |\xi|^2$$

for small values of $|\xi|$ and

$$q_\infty^- |\xi|^2 \le f_{\mathrm{hom}}(\xi) \le q_\infty^+ |\xi|^2$$

for large values of $|\xi|$.

The main result of this section reads:

Theorem 24.28. *There exist functions g, $h \in \mathfrak{I}_{a,g}^{cub}$ such that*

$$h_0^- |\xi|^2 \le f_{HSC\text{-}}(\xi) \le g_{hom}(\xi) \le h_\infty^- |\xi|^2,$$

$$h_0^+ |\xi|^2 \le h_{hom}(\xi) \le f_{HSC+}(\xi) \le h_\infty^+ |\xi|^2$$

for all $\xi \in \mathbf{R}^n$, where $h_0^+ = \max\{q_0^-, t_0^+\}$ and $h_\infty^- = \min\{q_\infty^+, t_\infty^-\}$. In addition, it holds that

$$w_0^+ |\xi|^2 \le f_{W+}(\xi) \le w_\infty^+ |\xi|^2,$$

$$w_0^- |\xi|^2 \le f_{W\text{-}}(\xi) \le w_\infty^- |\xi|^2$$

for all $\xi \in \mathbf{R}^n$.

Before continuing our discussion and proving the theorem, we present the following numerical experiments:

Let $g_i : \mathbf{R}^n \to \mathbf{R}_+$ be defined by

$$g_i(\xi) = \begin{cases} k_1 |\xi|^2 & \text{for } i = 3 \\ \frac{k_2 + k_3 |\xi|}{1 + |\xi|} |\xi|^2 & \text{for } i = 2 \\ k_4 |\xi|^2 & \text{for } i = 1 \end{cases}.$$

Here, $\{k_i\}$ are positive constants with $k_2 \le k_3$. Furthermore, let $a_1 = 0.970$, $a_2 = 0.015$ and $a_3 = 0.015$. Hence, $\{g_i\}$ and $\{a_i\}$ defines a family $\mathfrak{I}_{a,g}^{\mathrm{cub}}$. Now, consider the concentric cubes B_1, B_2 and B_3 with volumes $0,970$, $0,985$ and 1, respectively, such that $B_3 = \square =]0,1[^3$. Furthermore, let A_i, $i = 1,2,3$, be a $\square$-periodic set such that $A_1 \cap \square = B_1$, $A_2 \cap \square = B_2 \backslash B_1$ and $A_3 \cap \square = B_3 \backslash B_2$, and let

$$f(x,\xi) = g_1(\xi)\chi_{A_1}(x) + g_2(\xi)\chi_{A_2}(x) + g_3(\xi)\chi_{A_3}(x).$$

Thus $f \in \mathfrak{I}_{a,g}^{\mathrm{cub}}$ of the form

$$f(x,\xi) = C(x,|\xi|)\,|\xi|^2.$$

In particular, we note that $C(\cdot,0)$ (resp. $C(\cdot,\infty)$) is equal to k_1, k_2 (resp. k_3) and k_4 on A_3, on A_2 and A_1, respectively.

In the table below, we have listed the constants appearing in Remark 24.27 and Theorem 24.28 for four different combinations of $\{k_i\}$.

		Case 1		Case 2		Case 3		Case 4	
k_1		1		1		10^4		10^4	
k_2		1.1		10^2		10^2		$6{\cdot}10^3$	
k_3		1.5		10^3		10^3		$8{\cdot}10^3$	
k_4		10^4		10^4		1		1	
t_0^+	t_∞^-	9561	197	9561	197	101	1.10	101	1.10
q_0^-	q_0^+	102	104	191	194	102	102	160	162
q_∞^-	q_∞^+	116	118	194	196	111	112	180	182
h_0^-	h_∞^-	100	118	190	196	1.09	1.10	1.09	1.10
h_0^+	h_∞^+	9561	9561	9561	9591	102	115	160	187
w_0^+	w_∞^+	9703	9703	9704	9717	152	165	239	268
w_0^-	w_∞^-	35	40	67	66	1.03	1.03	1.03	1.03

Remark 24.29. First of all, we observe that for the four cases we obtain very sharp estimates of f_{hom}, $f_{\text{HSC}\pm}$ and $f_{\text{W}\pm}$. It is interesting to note that the bounds $q_0^- |\cdot|^2$ and $q_\infty^+ |\cdot|^2$ play an important role, not only for estimating f_{hom} but also for estimating the nonlinear bounds of Hashin–Shtrikman type $f_{\text{HSC}-}$ and $f_{\text{HSC}+}$. Particularly we see that for all four cases, h_0^- is close to h_∞^- and h_0^+ is close to h_∞^+. Hence, according to Theorem 24.28, we find a very good estimate of homogenized integrands g_{hom} and h_{hom} ($g,h \in \mathfrak{I}_{a,g}^{\text{cub}}$) that are close to $f_{\text{HSC}-}$ and $f_{\text{HSC}+}$, respectively. Hence our approach yields important knowledge of the extreme effective properties of the class $\mathfrak{I}_{a,g}^{\text{cub}}$.

Remark 24.30. In the definition of f we have that $\square \backslash A_3$ is a genuine subset of $\square$. Moreover, by Remark 24.27 it holds that $q_0^- \leq C_p^-(\xi) \leq C_p^+(\xi) \leq q_\infty^+$, and, thus, we observe in particular that the numerical results fit very well to the theoretical results of the previous section (see Remark 24.20).

Remark 24.31. We see that for case 1 and case 2, f_{hom} is very close to $f_{\text{HSC}-}$. This shows that f_{hom} is nearly lower optimal in the class $\mathfrak{I}_{a,g}^{\text{cub}}$. Correspondingly, we observe that f_{hom} is very close to $f_{\text{HSC}+}$ for case 3 and case 4. Hence, in these cases f_{hom} is nearly upper optimal in $\mathfrak{I}_{a,g}^{\text{cub}}$.

We note that $f(\cdot,\xi) = \lambda_2(|\xi|)\,|\xi|^2$ in A_2, where $\lambda_2\ (\cdot)$ ranges over the whole interval $]k_2,k_3[$. Therefore, the Euler equation corresponding to the minimum problem for finding $f_{\text{hom}}(\xi)$ is highly nonlinear in the region A_2, especially for case 2 and case 3. For a fixed ξ, any direct numerical treatment of this Euler equation leads to severe problems. In fact, if we for example choose the finite element method, we will need a large number of finite elements in the very small subset of the unit-cube where $f(\cdot,\xi)$ varies non-quadratically between $|\xi|^2$ and $10^4\,|\xi|^2$. Moreover, the table values show that the properties of $f(\cdot,\xi)$ on A_2 are significant. Hence, any kind of averaging in this region will be misleading. Besides, in contrast to the case when $f(x,\cdot)$ is a quadratic form, the values $\{f_{\text{hom}}(e_i)\}$, $i=1,\ldots,n$, alone will give no general information of f_{hom}. Therefore, we would have to compute $f_{\text{hom}}(\xi)$ numerically for a vast number of vectors $\xi \in \mathbf{R}^3$ in order to get a good picture of f_{hom}.

As discussed in the introduction, our ultimate goal is always to find the homogenized energy within Ω,

$$E(f_{\text{hom}}) = \min\left\{ \int_\Omega f_{\text{hom}}(Du)\,dx + \int_\Omega gu\,dx : u \in W^{1,p}(\Omega, R^N),\ u = \phi \text{ on } \gamma_0 \subset \partial\Omega \right\}, \tag{24.35}$$

which approximates the energy $E(f_h)$ for large values of h. Now, since $q_0^- |\cdot|^2 \leq f_{\text{hom}} \leq q_\infty^+ |\cdot|^2$, it follows that

$$E(q_0^- |\cdot|^2) \leq E(f_{\text{hom}}) \leq E(q_\infty^+ |\cdot|^2).$$

Remark 24.32. We note that the Euler equations associated with the upper and lower bounds

$$E(q_0^- |\cdot|^2) \text{ and } E(q_\infty^+ |\cdot|^2)$$

are linear. Hence, these bounds can be computed numerically, e.g., by standard FEM-algorithms, and since q_0^- is very close to q_∞^+, we get a good approximation of $E(f_{\text{hom}})$. Moreover, this shows that the Euler equation associated with $E_{\text{hom}}(f)$ is almost linear. Similarly, we can obtain good approximations of the optimal bounds for the homogenized energy $E_{\text{sup}} = E(f_{\text{HSC}-})$ and $E_{\text{inf}} = E(f_{\text{HSC}+})$.

The constants q_0^- and q_∞^+ are very close to each other in all four cases discussed above. This is however not a general property for all $f \in \mathfrak{I}_{a,g}^{\text{cub}}$. In fact, let the concentric cubes B_1, B_2, B_3 have volumes $0{,}015$, $0{,}985$ and 1, respectively, and let $A_3 \cap \Box = B_1$, $A_1 \cap \Box = B_2 \backslash B_1$ and $A_2 \cap \Box = B_3 \backslash B_2$. For the function $f(x,\xi) = \sum g_i(\xi)\chi_{A_i}(x)$, it turns out that $q_0^- / q_0^+ = 193 / 200$ and $q_\infty^- / q_\infty^+ = 1091 / 1095$ when the constants $\{k_i\}$ take values as in case 2. In order to estimate $E(f_{\text{hom}})$ in this case we have to apply the general bounds

$$E(f^-) \leq E(f_{\text{hom}}) \leq E(f^+),$$

where

$$f^-(\cdot) = (c^-(|\cdot|))_1 |\cdot|^2 \text{ and } f^+(\cdot) = (c^+(|\cdot|))_1 |\cdot|^2.$$

It is straightforward to express $(c^-(t))_1$ and $(c^+(t))_1$ in terms of t,$\{k_i\}$ and $\{a_i\}$. However, the Euler equations associated with the minimum problems for finding $E(f^-)$ and $E(f^+)$ are highly nonlinear and therefore more complicated to solve. Nevertheless, due to the facts that q_0^- is close to q_0^+ and q_∞^- is close to q_∞^+, we observe that the bounds $E(f^-)$ and $E(f^+)$ are close to each other, particularly when the variations on the boundary conditions ϕ and the values of the source g in (24.35) are such that the gradient of the solution, Du, is either large or small. In the latter case we can just use $E(q_0^- |\cdot|^2)$ as a lower bound for $E(f_{\text{hom}})$. As an upper bound for $E(f_{\text{hom}})$ we can use the value $E(f^+)'$ defined by

$$E(f^+)' = \int_\Omega f^+(Du')\,dx + \int_\Omega gu'\,dx,$$

where u' is the solution corresponding to the linear minimum problem for finding $E(q_\infty^- |\cdot|^2)$. Clearly, if $|Du|$ is small enough, then

$$E(q_0^+ |\cdot|^2) \simeq E(f^+)', \tag{24.36}$$

i.e., we find that

$$E(q_0^- |\cdot|^2) \leq E(f_{\text{hom}}) \leq E(f^+)' \simeq E(q_0^+ |\cdot|^2),$$

and, hence, we obtain a sharp estimate of the energy $E(f_{hom})$.

Proof of Theorem 24.28. Let $\varphi \in \mathfrak{I}_{a,g}^{\text{cub}}$ of the form

$$\varphi(x,\xi) = |\xi|^2 \sum_{i=1}^{m} \lambda_i(|\xi|)\chi_{A_i}.$$

We define the following functions associated with φ:

$$\mu^-(x) = \lambda_m(0)\chi_{A_m}(x) + \lambda_1(0)\chi_{\mathbf{R}^n\setminus A_m}(x),$$

$$\mu^+(x) = \lambda_1(\infty)\chi_{A_1}(x) + \lambda_m(\infty)\chi_{\mathbf{R}^n\setminus A_1}(x),$$

$$s^-(x,\xi) = |\xi|^2 \sum_{i=1}^{m} \lambda_i(0)\chi_{A_i},\ s^+(x,\xi) = |\xi|^2 \sum_{i=1}^{m} \lambda_i(\infty)\chi_{A_i},$$

$$g^-(x,\xi) = \mu^-(x)\,|\xi|^2 \text{ and } g^+(x,\xi) = \mu^+(x)\,|\xi|^2.$$

Since,

$$g^-(x,\xi) \leq s^-(x,\xi) \leq \varphi(x,\xi) \leq s^+(x,\xi) \leq g^+(x,\xi),$$

we find that

$$g_{\text{hom}}^-(\xi) \leq s_{\text{hom}}^-(\xi) \leq \varphi_{\text{hom}}(\xi) \leq s_{\text{hom}}^+(\xi) \leq g_{\text{hom}}^+(\xi). \tag{24.37}$$

Due to the fact that these functions satisfy the property of cubic symmetry, it holds that $g_{\text{hom}}^\pm$ and $s_{\text{hom}}^\pm$ are of the forms $g_{\text{hom}}^\pm(\xi) = k_{\text{hom}}^\pm |\xi|^2$ and $s_{\text{hom}}^\pm(\xi) = l_{\text{hom}}^\pm |\xi|^2$ for some positive constants $k_{\text{hom}}^\pm$ and $l_{\text{hom}}^\pm$ (see, e.g., [19, p. 39]). According to the Hashin–Shtrikman bounds for linear problems, we have that $h_0^- \leq l_{\text{hom}}^-$ and $l_{\text{hom}}^+ \leq h_\infty^+$ (see, e.g., [19, p. 188]). Hence,

$$h_0^- |\xi|^2 \leq \varphi_{\text{hom}}(\xi) \leq h_\infty^+ |\xi|^2,$$

and it follows by the definition that

$$h_0^- |\xi|^2 \leq f_{\text{HSC}-}(\xi) \leq f_{\text{HSC}+}(\xi) \leq h_\infty^+ |\xi|^2.$$

Moreover, if A_m is the classical Hashin–Shtrikman coated sphere assemblages, then it is well known that $k_{\text{hom}^-} = t_0^+$ (for the original proof, see [16]). Combined with (24.37) this shows that

$$t_0^+ |\xi|^2 \leq \varphi_{\text{hom}}(\xi) \leq f_{\text{HSC}+}(\xi).$$

Moreover, by Remark 24.27 and the definition of $f_{\text{HSC}+}$ we have

$$q_0^- |\xi|^2 \leq f_{\text{hom}}(\xi) \leq f_{\text{HSC}+}(\xi).$$

Thus,

$$h_0^+ |\xi|^2 \leq h_{\text{hom}}(\xi) \leq f_{\text{HSC+}}(\xi) \leq h_\infty^+ |\xi|^2,$$

where $h = f$ if $t_0^+ \leq q_0^-$, and $h = \varphi$ if $t_0^+ > q_0^-$. Similarly, if A_1 is the classical Hashin–Shtrikman coated sphere assemblages, we get that $k_{\text{hom}}^+ = t_\infty^-$, and, hence, by using (24.37) once more we obtain that

$$f_{\text{HSC}-}(\xi) \leq \varphi_{\text{hom}}(\xi) \leq t_\infty^- |\xi|^2.$$

In addition, Remark 24.27 and the definition of $f_{\text{HSC}-}$ yield that

$$f_{\text{HSC}-}(\xi) \leq f_{\text{hom}}(\xi) \leq q_\infty^+ |\xi|^2.$$

Hence,

$$h_0^- |\xi|^2 \leq f_{\text{HSC-}}(\xi) \leq g_{\text{hom}}(\xi) \leq h_\infty^- |\xi|^2,$$

where $h = f$ if $q_\infty^+ \leq t_\infty^-$, and $h = \varphi$ if $q_\infty^+ > t_\infty^-$. The final part of the theorem follows directly by integrating the inequalities

$$C(x,0) |\xi|^2 \leq f(x,\xi) \leq C(x,\infty) |\xi|^2$$

and

$$(C(x,0))^{-1} |\xi|^2 \geq f^*(x,\xi) \geq (C(x,\infty))^{-1} |\xi|^2$$

over $\square$. This completes the proof. $\square$

24.7 The Reiterated Cell Structure

In this section, we describe two-phase structures called *reiterated cell structures,* described by characteristic functions $\chi_{\text{cell},h}$ of the form

$$\chi_{\text{cell},h}(x,\xi) = \chi_{\text{cell}}(hx, h^2x, ..., h^m x).$$

These and similar structures where first introduced in [28] and [33]. We also want to refer to the well-known book of Milton [42] and the article [36] for a comparison between these and other more classical structures (see also the discussion below).

Inspired by the main ideas in the proof of Theorem 24.17, it is possible to obtain estimates for the homogenized integrand corresponding to functions of the form

$$f_h(x,\xi) = \chi_h g_1(\xi) + (1-\chi_h) g_2(\xi),$$

where g_1 and g_2 are quadratic forms. Below, both the scalar case $N = 1$ and the vector-valued case $N = n$ are considered.

We start with some notations. Let $0 \leq v \leq 1$. With $\Im_v$ we will denote the family of all sequences $\{\chi_h\}$ of characteristic functions on $\mathbf{R}^n$ such that $\chi_h \rightharpoonup v$ weakly in $L^2(\Box)$. If z is a n-tuple $(z_1,...,z_n)$ with $0 < z_i \leq 1$, $\prod_{i=1}^n z_i = v$, and m is a positive integer, then we let $\chi_{m,z}$ denote the characteristic function of the set $\prod_{i=1}^n]0, z_i^{\frac{1}{m}}[$, extended $\Box$-periodically to $\mathbf{R}^n$. Moreover, let

$$\chi_{\text{cell},h}(x) = \chi_{\text{cell}}[h,m,z](x) = \prod_{i=1}^m \chi_{m,z}(h^i x), \; h \in \mathbf{N}.$$

The sequence $\{\chi_{\text{cell},h}\}$ will hereafter be called the *reiterated cell structure*. It is easily seen that the mean value $\langle \chi_{\text{cell},h} \rangle$ of $\chi_{\text{cell},h}$ equals v. Moreover, $\chi_{\text{cell},h} \rightharpoonup \langle \chi_{\text{cell},h} \rangle$ weakly in $L^2(Y)$. This fact follows from almost the same arguments as those in the pure periodic case (compare with [50, p. 57]) and we omit the details. Consequently, $\{\chi_{\text{cell},h}\} \in \Im_v$.

24.7.1 The Scalar Case

Let $k > 0$ and suppose that

$$f_{\text{cell},h}(x,\xi) = \left(k|\xi|^2 \chi_{\text{cell},h}(x) + |\xi|^2 (1-\chi_{\text{cell},h}(x))\right), \; \xi \in \mathbf{R}^n.$$

Furthermore, let f_{cell} be the homogenized integrand corresponding to the functions $f_{cell,h}$. The proof of the following theorem can be found in [33] (see also [38] for an alternative proof of a fundamental step in that proof).

Theorem 24.33. *It holds that f_{cell} is of the form $f_{cell}(\xi) = \sum_{i=1}^n \lambda_{cell,i} |\xi_i|^2$, where*

$$\frac{1}{1-\lambda_{cell,i}} = \frac{1}{1-k}\frac{1}{v} - r_i \frac{1-z_i^{\frac{1}{m}}}{1-v^{\frac{1}{m}}} \frac{1-v}{v}, \tag{24.38}$$

$v = \prod_{i=1}^n z_i$ *and* $\left(\frac{v}{z_i}\right)^{\frac{1}{m}} \leq r_i \leq 1$. *In particular,*

$$\frac{1}{(1-\lambda_{cell,i})} \to \frac{1}{(1-k)}\frac{1}{v} - \frac{\ln z_i}{\sum_{j=1}^n \ln z_j} \frac{(1-v)}{v} \tag{24.39}$$

as $m \to +\infty$.

Before commenting on Theorem 24.33, we recall the following G-closure result: Let

$$f_h(x,\xi) = (a_1\chi_h(x) + a_2(1-\chi_h(x)))\sum_{i=1}^{n}|\xi_i|^2,$$

where $\{\chi_h\} \in \mathfrak{I}_{p_1}$, $0 < a_1 \leq a_2$ and $0 \leq p_1 = 1 - p_2 \leq 1$ and suppose that we have the existence of the corresponding homogenized integrand f_{hom} of the form $f_{\text{hom}}(\xi) = \sum_{i=1}^{n}\lambda_i|\xi_i|^2$. Then

$$\sum_{i=1}^{n}\frac{1}{\lambda_i - a_1} \leq \frac{n}{p_2(a_2-a_1)} + \frac{p_1}{a_1 p_2}, \tag{24.40}$$

$$\sum_{i=1}^{n}\frac{1}{a_2 - \lambda_i} \leq \frac{n}{p_1(a_2-a_1)} - \frac{p_2}{a_2 p_1} \tag{24.41}$$

(the generalized Hashin–Shtrikman bounds) and

$$\frac{a_1 a_2}{a_2 p_1 + a_1 p_2} \leq \lambda_i \leq a_1 p_1 + a_2 p_2 \tag{24.42}$$

(Voigt–Reuss inequality). Structures which realize equality in (24.40) or (24.41) are usually referred to as lower optimal and upper optimal, respectively. There exist two well-known types of optimal two-phase structures with completely different geometries from the reiterated cell structure: the *Hashin ellipsoidal structure* and the *stratified structure of rank n* (layered in the directions $e_1,\ldots,e_n$) (see [41, 52]).

Remark 24.34. In similar way as in Theorem 24.33 we consider (with no loss of generality) the case when $f_h(x,Du) = (k\chi_h(x) + (1-\chi_h(x)))|Du|^2$ and $\{\chi_h\} \in \mathfrak{I}_\nu$. It is obvious that if

$$\frac{1}{1-\lambda_i} = \frac{1}{\nu(1-k)} - \frac{1-\nu}{\nu}h_i, \tag{24.43}$$

where $h_i \geq 0$ and $\sum_{i=1}^{n} h_i = 1$, then for $k > 1$ the equality is attained in (24.40), and for $k < 1$ the equality is attained in (24.41). Moreover, one can verify that any n-tuple $(\lambda_1,\ldots,\lambda_n)$ can be written in the form (24.43), provided that it satisfies (24.40), (24.41) and (24.42), and turns either (24.40) or (24.41) into an equality (see [19, p. 196]). Hence, according to Theorem 24.33, we conclude that for any optimal effective property $(\lambda_1,\ldots,\lambda_n)$, we can find a reiterated cell structure with effective property $(\lambda_{cell,1},\ldots,\lambda_{cell,n})$ which is arbitrarily close to $(\lambda_1,\ldots,\lambda_n)$.

24.7.2 The Vector-Valued Case

Let $\mathbf{S}_n$ be the space of symmetric $n \times n$ matrices. As usual, we define the scalar product $(\cdot)$ between two matrices $A = \{a_{ij}\}$ and $B = \{b_{ij}\}$ in $\mathbf{S}_n$ by $A \cdot B = \sum_{ij} a_{ij}b_{ij}$ and let $|A|^2 = \sum_{ij} a_{ij}^2$. The space $\mathbf{S}_n$ can be divided into the orthogonal subspaces $\mathbf{S}_{\text{off}}$,

$\mathbf{S}_{\text{dia}}$ and $\mathbf{S}_{\text{tr}}$, of matrices with zero diagonal elements, diagonal matrices with zero trace, and matrices of the form αI, $\alpha \in \mathbf{R}$ (I is the identity matrix), respectively. Let $\mathscr{D}_{\text{off}}$, $\mathscr{D}_{\text{dia}}$ and $\mathscr{D}_{\text{tr}}$ be the orthogonal projections of $\mathbf{S}_n$ onto the spaces $\mathbf{S}_{\text{off}}$, $\mathbf{S}_{\text{dia}}$ and $\mathbf{S}_{\text{tr}}$. Any matrix $\xi \in \mathbf{R}^{n^2}$ can be written uniquely on the form $\xi = \xi_s + \xi_a$, where ξ_s and ξ_a denote the symmetric part and anti-symmetric part of ξ, respectively.

Let $\{\chi_{cube,h}\} \in \mathfrak{I}_v$ be the reiterated cell structure with reiteration number m for the case when the cells are cubes, i.e.,

$$\chi_{cube,h} = \chi_{cell}[h,m,z], \text{ where } z_1 = \cdots = z_n = v^{\frac{1}{n}}.$$

The sequence $\{\chi_{cube,h}\}$ will be referred to as the *reiterated cube structure.* Consider the functions f_h defined by

$$f_h(x,\xi) = \chi_{cube,h} g_a(\xi) + (1-\chi_{cube,h}) g_b(\xi), \quad \xi \in \mathbf{R}^{n^2},$$

where

$$g_i(\xi) = \kappa_i \left|\mathscr{D}_{\text{off}}\xi_s\right|^2 + \mu_i \left|\mathscr{D}_{\text{dia}}\xi_s\right|^2 + \frac{K_i}{2}\left|\mathscr{D}_{\text{tr}}\xi_s\right|^2, \ i = a,b,$$

for some positive real constants κ_i, μ_i and K_i.

We will now study the homogenized integrand f_{cube} corresponding to the functions f_h. For the sake of simplicity, we restrict our attention to the case $n = 2$, but the general case can be handled analogously. The proof of the following theorem can be found in [33].

Theorem 24.35. *Let $n = 2$. It holds that f_{cube} is of the form*

$$f_{cube}(\xi) = \widehat{\kappa}\left|\mathscr{D}_{off}\xi_s\right|^2 + \widehat{\mu}\left|\mathscr{D}_{dia}\xi_s\right|^2 + \frac{\widehat{K}}{2}\left|\mathscr{D}_{tr}\xi_s\right|^2 \tag{24.44}$$

for all $\xi \in \mathbf{R}^{2^2}$, where the constants $\widehat{K}, \widehat{\mu}$ and $\widehat{\kappa}$ are obtained by the following formulae:

$$\left(\widehat{K} - K_b\right)^{-1} = \frac{1}{v}(K_a - K_b)^{-1} + \frac{(1-v)v_m}{2v}\left(R_K + 2(K_b+\mu_b)^{-1}\right),$$

$$\left(\widehat{\mu} - \mu_b\right)^{-1} = \frac{1}{v}(\mu_a - \mu_b)^{-1} + \frac{(1-v)v_m}{2v}\left(R_\mu + 2(K_b+\mu_b)^{-1}\right),$$

$$\left(\widehat{\kappa} - \kappa_b\right)^{-1} = \frac{1}{v}(\kappa_a - \kappa_b)^{-1} + \frac{(1-v)}{v} v_m R_\kappa \kappa_b^{-1}.$$

Here,

$$0 \le R_K \le \frac{1 - v^{\frac{1}{2m}}}{K_b v^{\frac{1}{2m}}}, \ 0 \le R_\mu \le \frac{1 - v^{\frac{1}{2m}}}{\mu_b v^{\frac{1}{2m}}}, \ 1 \le R_\kappa \le \frac{v^{\frac{1}{2m}} + 1}{2v^{\frac{1}{2m}}}$$

and $v_m = 2v^{\frac{1}{2m}} / \left(v^{\frac{1}{2m}} + 1\right)$.

Before we comment Theorem 24.35, we recall the following results: Suppose that

$$f_h(x,\xi) = \chi_h g_1(\xi) + (1-\chi_h) g_2(\xi),$$

where $\{\chi_h\} \in \Im_{p_1}$ and

$$g_i(\xi) = \mu_i \left|\mathscr{D}_{\rm off}\xi_s\right|^2 + \mu_i \left|\mathscr{D}_{\rm dia}\xi_s\right|^2 + \frac{K_i}{2}\left|\mathscr{D}_{\rm tr}\xi_s\right|^2, \ i=1,2$$

with $\mu_1 < \mu_2$. In addition, assume that the homogenized integrand $f_{\rm hom}$ corresponding to the functions f_h exists and that it is *square symmetric*, i.e., of the form

$$f_{\rm hom}(\xi) = \widehat{\kappa}\left|\mathscr{D}_{\rm off}\xi_s\right|^2 + \widehat{\mu}\left|\mathscr{D}_{\rm dia}\xi_s\right|^2 + \frac{\widehat{K}}{2}\left|\mathscr{D}_{\rm tr}\xi_s\right|^2.$$

Then, it is possible to prove that

$$p_1K_1 + p_2K_2 - \frac{p_1p_2(K_1-K_2)^2}{p_2K_1+p_1K_2+\mu_1} \le \widehat{K} \tag{24.45}$$

$$\le p_1K_1 + p_2K_2 - \frac{p_1p_2(K_1-K_2)^2}{p_2K_1+p_1K_2+\mu_2}$$

and

$$\left(\mu_1^{-1}p_1 + \mu_2^{-1}p_2\right)^{-1} \le \widehat{\kappa},\ \widehat{\mu} \le p_1\mu_1 + p_2\mu_2 - \frac{p_1p_2(\mu_1-\mu_2)^2}{p_2\mu_1+p_1\mu_2+K_2} \tag{24.46}$$

(here $p_2 = 1-p_1$). Moreover, there exist laminate structures where $\widehat{K}$ is equal to the upper bound (resp. lower bound) in (24.45) and at the same time such that $\widehat{\mu}$ and $\widehat{\kappa}$ coincide with the upper bound and the lower bound in (24.46), respectively. Concerning these facts and some further information, see [1, 13, 15, 20].

Remark 24.36. In the case of the reiterated cube structure, suppose that $\mu_i = \kappa_i$, $K_a \le K_b$ and $\mu_a \le \mu_b$ (resp. $\mu_a \ge \mu_b$). According to Theorem 24.35, it is easy to check that $\widehat{K}$ converges to the upper bound (resp. lower bound) of (24.45) as m goes to $+\infty$, $\widehat{\mu}$ converges to the upper bound of (24.46) and $\widehat{\kappa}$ converges to the lower bound of (24.46). Thus, the reiterated cube structure yields optimal effective properties also in the vector-valued case.

24.8 Bounds Related to a Reynold-Type Equation

Let us consider an equation of the form

$$\frac{\partial}{\partial x_1}\left(a_1(x,\frac{x}{\varepsilon})\frac{\partial u_\varepsilon}{\partial x_1} - b_1(x,\frac{x}{\varepsilon})\right) + \frac{\partial}{\partial x_2}\left(a_2(x,\frac{x}{\varepsilon})\frac{\partial u_\varepsilon}{\partial x_2} - b_2(x,\frac{x}{\varepsilon})\right) = f(x), \tag{24.47}$$

$x \in \Omega \subset R^2$, $u_\varepsilon \in \mathrm{H}_0^{1,2}(\Omega)$. Here, a_i and b_i are assumed to be piecewise continuous in the first variable and measurable and periodic relative to a cell $\square = [0,1]^2$ in the second variable. In addition, we assume that there exist constants k_- and k_+ such that

$$0 < k_- \le a_i(x,y) \le k_+ < \infty \text{ and } |b_i(x,y)| \le k_+$$

for all x and y. Moreover, $\varepsilon > 0$ is a small parameter.

Many important physical problems can be described by a partial differential equation of the type (24.47). The standard example is stationary heat conduction in which $b_i = 0$. An other example concerns flow behavior between two surfaces in relative motion in the theory of lubrication for thin films for which $f(x) = 0$ and $c_i b_i^3(x,y) = a_i(x,y) = a(x,y)$ for some constant c_i which is propositional to the relative motion. For that special case (24.47) is the incompressible Reynolds equation.

For simplicity, we put $f = 0$. The corresponding weak formulation takes the form: Find $u_\varepsilon \in \mathrm{H}_0^{1,2}(\Omega)$ such that

$$\int_\Omega (a_1(x,\frac{x}{\varepsilon})\frac{\partial u_\varepsilon}{\partial x_1} - b_1(x,\frac{x}{\varepsilon}))\frac{\partial v}{\partial x_1} + (a_2(x,\frac{x}{\varepsilon})\frac{\partial u_\varepsilon}{\partial x_2} - b_2(x,\frac{x}{\varepsilon}))\frac{\partial v}{\partial x_2} dx = 0 \quad (24.48)$$

for all $v \in \mathrm{H}_0^{1,2}(\Omega)$. The equivalent variational formulation is then: Find $u_\varepsilon \in \mathrm{H}_0^{1,2}(\Omega)$ such that

$$F_\varepsilon(Du_\varepsilon) = \min_{v \in \mathrm{H}_0^{1,2}(\Omega)} F_\varepsilon(Dv),$$

where

$$F_\varepsilon(Dv) = \int_\Omega f_\varepsilon(x, Dv)\, dx,$$

and

$$f_\varepsilon(x,\xi) = f(x,\frac{x}{\varepsilon},\xi) = \sum_{i=1}^{2} \frac{1}{2} a_i(x,\frac{x}{\varepsilon})\xi_i^2 - b_i(x,\frac{x}{\varepsilon})\xi_i.$$

Putting $v = u_\varepsilon$ into the weak formulation (24.48), we observe that the "energy"

$$F_\varepsilon(Du_\varepsilon) = -\frac{1}{2}\int_\Omega b_1(x,\frac{x}{\varepsilon})\frac{\partial u_\varepsilon}{\partial x_1} + b_2(x,\frac{x}{\varepsilon})\frac{\partial u_\varepsilon}{\partial x_2} dx.$$

The right-hand side (multiplied with some known constant) often represents some important physical property. Examples of such properties are resultant force, resultant moment, total heat flux, total current, etc. Consequently, in many cases the calculation of $F_\varepsilon(u_\varepsilon)$ is the main purpose of the investigation.

Using the theory of gamma-convergence, it is possible to prove that

$$F_\varepsilon(Du_\varepsilon) \to F_0(D\overline{u}) = \min_{u \in \mathrm{H}_0^{1,2}(\Omega)} F_0(Du),$$

where

$$F_0(Du) = \int_\Omega f_0(x, Du)\, dx,$$

$$f_0(x,\xi) = \min_{u\in \mathrm{H}^{1,2}_{\mathrm{per}}(\square)} \int_\square f(x,y,Du+\xi)\,dy. \tag{24.49}$$

Putting

$$A = \begin{bmatrix} a_1 & 0 \\ 0 & a_2 \end{bmatrix}, \; B = \begin{bmatrix} b_1 \\ b_2 \end{bmatrix}$$

we have that f and f_0 can be written in the forms

$$f(x,y,\xi) = \frac{1}{2}\xi \cdot A(x,y)\xi - B(x,y)\cdot\xi,$$

$$f_0(x,\xi) = \frac{1}{2}\xi \cdot A_0\xi - B_0\cdot\xi + f_0(x,0).$$

Here,

$$A_0\xi = \int_\square A\left(Du_\xi + \xi\right)\,dy,$$

$$B_0 = \int_\square B - ADu_0\,dy,$$

where $u_\xi \in \mathrm{H}^{1,2}_{\mathrm{per}}(\square)$ and $u_0 \in \mathrm{H}^{1,2}_{\mathrm{per}}(\square)$ are the solutions of the local problems

$$\int_\square Dv\cdot A\left(Du_\xi + \xi\right)\,dy = 0, \quad \forall v \in \mathrm{H}^{1,2}_{\mathrm{per}}(\square), \tag{24.50}$$

$$\int_\square Dv\cdot (ADu_0 - B)\,dy = 0, \quad \forall v \in \mathrm{H}^{1,2}_{\mathrm{per}}(\square). \tag{24.51}$$

For the latter problem, we note that the corresponding variational problem takes the form

$$f_0(x,0) = \min_{v\in \mathrm{H}^{1,2}_{\mathrm{per}}(\square)} \int_\square \frac{1}{2}Dv\cdot ADv - B\cdot Dv\,dy.$$

Thus, by putting $v = u_0$ in (24.51), we get that

$$f_0(x,0) = \int_\square \frac{1}{2}Du_0\cdot ADu_0 - B\cdot Du_0\,dy = -\frac{1}{2}\int_\square B\cdot Du_0\,dy.$$

Below, we discuss sharp upper and lower bounds $f^-(x,\xi)$ and $f^+(x,\xi)$ for $f_0(x,\xi)$. These bounds were first presented in [34] and are described by integral averages in orthogonal directions and can easily be found explicitly or at least by performing numerical integration. We also characterize all cases when these bounds coincide. When these bounds are close to each other, we are able to find close upper and lower bounds $F_-(Du_-)$ and $F_+(Du_+)$ for the homogenized energy $F_0(D\overline{u})$ by solving the global problems

$$F_-(Du_-) = \min_{u\in \mathrm{H}^{1,2}_0(\Omega)} F_+(Du) \text{ and } F_+(Du_+) = \min_{u\in \mathrm{H}^{1,2}_0(\Omega)} F_+(Du),$$

where

$$F_{\pm}(Du) = \int_{\Omega} f^{\pm}(x, Du)\, dx.$$

As we will see, the bounds are closely connected to the ones obtained in the previous sections. The derivation is however different due to the presence of $b_i(x,y)$.

For $(i,j) = (1,2)$ or $(i,j) = (2,1)$, let us define the following constants associated with the functions $a_i = a_i(x,(y_1,y_2))$ and $b_i = b_i(x,(y_1,y_2))$:

$$a_i^+(x) = \left(\int_0^1 \left(\int_0^1 a_i\, dy_j\right)^{-1} dy_i\right)^{-1},$$

$$a_i^-(x) = \int_0^1 \left(\int_0^1 a_i^{-1} dy_i\right)^{-1} dy_j,$$

$$c_i^+(x) = \left(\int_0^1 \left(\int_0^1 a_i\, dy_j\right)^{-1} dy_i\right)^{-1} \int_0^1 \frac{\int_0^1 b_i\, dy_j}{\int_0^1 a_i\, dy_j} dy_i,$$

$$c_i^-(x) = \int_0^1 \left(\int_0^1 a_i^{-1} dy_i\right)^{-1} \int_0^1 \frac{b_i}{a_i} dy_i dy_j,$$

$$\begin{aligned} d_i^+(x) &= \frac{1}{2}\left(\int_0^1 \left(\int_0^1 a_i\, dy_j\right)^{-1} dy_i\right)^{-1} \left(\int_0^1 \frac{\int_0^1 b_i\, dy_j}{\int_0^1 a_i\, dy_j dy_i}\right)^2 \\ &\quad -\frac{1}{2}\int_0^1 \frac{\left(\int_0^1 b_i\, dy_j\right)^2}{\int_0^1 a_i\, dy_j} dy_i, \end{aligned}$$

$$d_i^-(x) = \frac{1}{2}\int_0^1 \left(\int_0^1 a_i^{-1}\, dy_i\right)^{-1} \left(\int_0^1 \frac{b_i}{a_i} dy_i\right)^2 dy_j - \frac{1}{2}\int_0^1 \int_0^1 \frac{b_i^2}{a_i} dy_i dy_j.$$

The proof of the following theorem can be found in [34].

Theorem 24.37. *It holds that*

$$f^-(x,\xi) \le f_0(x,\xi) \le f^+(x,\xi),$$

where

$$f^-(x,\xi) = \sum_{i=1}^{2}\left(\frac{1}{2}a_i^-(x)\xi_i^2 - c_i^-(x)\xi_i + d_i^-(x)\right),$$

$$f^+(x,\xi) = \sum_{i=1}^{2}\left(\frac{1}{2}a_i^+(x)\xi_i^2 - c_i^+(x)\xi_i + d_i^+(x)\right).$$

Moreover, $f^- = f^+$ *if and only if* $a_i = a_i^i a_i^j$ *and* $b_i = a_i^j f_i^i + f_i^j$ $(j \neq i)$ *where* a_i^k *and* f_i^k *are functions of the form*

$$a_i^k = a_i^k(x, y_k), \; f_i^k = f_i^k(x, y_k).$$

Remark 24.38. In the special case when $b_i = 0$, the bounds in Theorem 24.37 reduces to the ones obtained in [27] concerning the homogenized p-Laplace equation for the case $p = 2$.

Remark 24.39. As a consequence of Theorem 24.37, we obtain the inequality $f^- \leq f^+$. In particular this gives the inequality $a_i^-(x) \leq a_i^+(x)$, discussed in Remark 24.15.

Remark 24.40. The above theorem shows that $f^+(x,\xi) = f_0(x,\xi) = f^-(x,\xi)$ if $a_i = a_i^i a_i^j$ and $b_i = a_i^j f_i^i + f_i^j$. For this case

$$a_i^-(x) = a_i^+(x) = \left(\int_0^1 a_i^j \, dy_j\right)\left(\int_0^1 \left(a_i^i\right)^{-1} dy_i\right)^{-1},$$

$$c_i^-(x) = c_i^+(x)$$

$$= \left(\int_0^1 a_i^j \, dy_j\right)\left(\int_0^1 \left(a_i^i\right)^{-1} dy_i\right)^{-1}\left(\int_0^1 \frac{f_i^i}{a_i^i} dy_i + \frac{\int_0^1 f_i^j \, dy_j}{\int_0^1 a_i^j \, dy_j}\int_0^1 \left(a_i^i\right)^{-1} dy_i\right),$$

$$d^-(x) = d^+(x)$$

$$= \frac{1}{2}\left(\int_0^1 a_i^j \, dy_j\right)\left(\left(\int_0^1 \left(a_i^i\right)^{-1} dy_i\right)^{-1}\left(\int_0^1 \frac{f_i^i}{a_i^i} dy_i\right)^2 - \int_0^1 \frac{\left(f_i^i\right)^2}{a_i^i} dy_i\right).$$

This is seen directly for $a_i^{\pm}(x)$ and $b_i^{\pm}(x)$. The expression for $d_i^{\pm}(x)$ is, however, more complicated to verify (see [34]).

Example 24.41. As an example, we consider the case when the unit cell $\square$ consists of two materials, material 1 and material 2, with conductivities $a_1(x,y) = a_2(x,y) = 1$ and $a_1(x,y) = a_2(x,y) = 2$, respectively. Material 2 occupies a square with size $l \times l$ and material 2 is the surrounding material. In both materials we assume that $b_1(x,y) = b_2(x,y) = 0$. Due to symmetry, it holds that $f_0(x,\xi) = a\left(\xi_1^2 + \xi_2^2\right)$. In the table below we have listed a and $a^{\pm} = a_1^{\pm} = a_2^{\pm}$ for the case $l = 0.5$ and $l = 0.9$.

l	a	a^+	a^-
0.5	1.6903	1.7143	1.6667
0.9	1.1494	1.1518	1.1474

The values in the second column are found numerically by using the finite element method. The errors in these computations are estimated to be less than 4×10^{-5}.

For more information about numerical computation of effective properties and comparison with similar bounds, we refer to the literature, see, e.g., the thesis [39] and the article [40].

24.9 Some Final Comments

The problem of determining the homogenized integrand is often very delicate. Even in the linear case, there exist only a few structures for which f_{hom} can be found explicitly. Moreover, in the nonlinear case, the complexity of this problem increases tremendously due to the fact that $f_{\mathrm{hom}}(\xi)$ cannot be found generally by the knowledge of $f_{\mathrm{hom}}(\xi_i)$ for a finite number of vectors ξ_i (as we can in the linear case).

Nevertheless, by applying the upper and lower bounds presented in Section 24.5, Section 24.7 and Section 24.8, we have seen that it is possible to analyze several types of linear and nonlinear problems and obtain a very sharp estimate of the macroscopic behavior in cases where classical numerical treatments seem to be useless. Particularly, in the scalar case the bounds of Section 24.7 even allow us to verify that the reiterated cell structure can be used to obtain any optimal effective property achievable within the class $\mathfrak{I}_v$, and we have also illustrated that we obtain similar results in the vector-valued case. Accordingly, the results obtained in this paper seem to be useful for applications, e.g., in optimal structural design and homogenization of linear and nonlinear materials.

References

1. M. Avellaneda, *Optimal bounds and microgeometries for elastic composites*, SIAM J. Appl. Math., 47 (1987), 1216–1228.
2. N. Bakhvalov and G. Panasenko, *Homogenization: Averaging Processes in Periodic Media,* Kluwer Academic Publisher, Dordrecht, 1989.
3. E.F. Beckenbach, *Convexity properties of generalized mean value functions,* Ann. Math. Statist. 13 (1942), 88–90.
4. A. Bensoussan, J. L. Lions and G. Papanicolaou, *Asymptotic Analysis for Periodic Structures,* North–Holland, Amsterdam, 1978.
5. A. Braides, *An introduction to homogenization and Γ-convergence.* Lecture Notes, School on homogenization, ICTP, Trieste, 1993.
6. A. Braides and A. Defranceschi. *Homogenization of Multiple Integrals.* Oxford Science Publications, Oxford, 1998.
7. A. Braides and D. Lukkassen. *Reiterated homogenization of integral functionals.* Math. Mod. Meth. Appl. Sci. 10, 1 (2000), 1–25.
8. P.S. Bullen, D.S. Mitrinović and P.M. Vasić, *Means and their inequalities*, D. Reidel Publishing Company, Dordrecht, 1988.
9. G. A. Chechkin, A.L. Piatnitski and A. S. Shamaev, *Homogenization. Methods and Applications.* Translations of Mathematical Monographs, 234. American Mathematical Society, Providence, RI, 2007.
10. D. Cioranescu and P. Donato, *An Introduction to Homogenization*, Oxford Lecture Series in Mathematics and its Applications, 17, Oxford University Press, New York, 1999.
11. G. Dal Maso, *An Introduction to Γ-convergence,* Birkhäuser, Boston 1993.
12. E. De Georgie and S. Spagnolo, *Sulla congergenza degli integrali dell'energia per operatori ellittici del 2' ordine*, Boll. Un. mat. Ital. 8 (1973), 391–411.
13. G.A. Francfort and F. Murat, *Homogenization and optimal bounds in linear elasticity*, Arch. Rational Mech. Anal. 94 (1986), 307–334.
14. N. Fusco and G. Moscariello, *On the homogenization of quasilinear divergence structure operators*, Ann. Mat. Pura. Appl., 146 (1987), 1–13.

15. L.V. Gibiansky, *Bounds on effective moduli of composite materials*, Lecture Notes, School on Homogenization, ICTP, Trieste, 1993.
16. Z. Hashin and S. Shtrikman, *A variational approach to the theory of effective magnetic permeability of multiphase materials.* J. Appl. Phys. 33 (1962), 3125–3131.
17. G.H. Hardy, J.E. Littlewood and G. Polya, *Inequalities,* Cambridge University Press, Cambridge, 1934.
18. G. Isac, *Leray-Schauder Type Alternatives, Complementarity Problems and Variational Inequalities.* Nonconvex Optimization and its Applications, 87, Springer, New York, 2006.
19. V.V. Jikov, S.M. Kozlov and O.A. Oleinik, *Homogenization of Differential Operators and Integral Functionals*, Springer-Verlag, Berlin, 1994.
20. R. James, R. Lipton and A. Lutoborski, *Laminar elastic composites with crystallographic symmetry*, SIAM J. Appl. Math. 50, (1990), 683–702.
21. J.-L. Lions, D. Lukkassen, L.-E. Persson and P. Wall. *Reiterated homogenization of monotone operators,* C. R. Acad. Sci., Paris, Ser. I , Math. 330, 8 (2000), 675–680.
22. J.-L. Lions, D. Lukkassen, L.-E. Persson and P. Wall. *Reiterated homogenization of nonlinear monotone operators,* Chin. Ann. Math., Ser. B , 22, 1 (2001), 1–14.
23. D. Lukkassen, *Some sharp estimates connected to the homogenized p-Laplacian equation*, ZAMM-Z. Angew. Math. Mech. 76, (1996) S2, 603–604.
24. D. Lukkassen, *Upper and lower bounds for averaging coefficients*, Russian Math. Surveys 49, 4, (1994), 114–115.
25. D. Lukkassen, *On estimates of the effective energy for the Poisson equation with a p-Laplacian.* Russian Math. Surveys 51, 4, (1996), 739–740.
26. D. Lukkassen, *On some sharp bounds for the off-diagonal elements of the homogenized tensor,* Applications of Math. 40 (1995), 401–406.
27. D. Lukkassen. *Some sharp estimates connected to the homogenized p-Laplacian equation.* ZAMM-Z. Angew. Math. Mech. 76 (S2), (1996), 603–604.
28. D. Lukkassen, *Formulae and bounds connected to homogenization and optimal design of partial differential operators and integral functionals.* Ph.D thesis, University of Tromsø, 1996.
29. D. Lukkassen. *Bounds and homogenization of integral functionals.* Acta Sci. Math. 64, (1998), 121–141.
30. D. Lukkassen. *Homogenization of integral functionals with extreme local properties.* Math. Balk. New Series, 12, Fasc. 3–4, (1998), 339–358.
31. D. Lukkassen. *Means of power type and their inequalities.* Math. Nachr. 205, (1999), 131–147.
32. D. Lukkassen. *Sharp inequalities connected to the homogenized p-Poisson equation.* Math. Inequal. Appl. 2, 2, (1999), 243–250.
33. D. Lukkassen. *A new reiterated structure with optimal macroscopic behavior.* SIAM J. Appl. Math. 59, 5, (1999), 1825–1842.
34. D. Lukkassen, A. Meidell and P. Wall. *Bounds on the effective behavior of a homogenized Reynold-type equation.* J. Funct. Spaces Appl. 5, 2, (2007), 133–150.
35. D. Lukkassen, A. Meidell and P. Wall. *Multiscale homogenization of monotone operators.* To appear in: Discrete and Continuous Dynamical Systems - Ser. A, 22, 3, (2008), 711–727.
36. D. Lukkassen and G. W. Milton, *On hierarchical structures and reiterated homogenization,* Proceedings of the Conference on Function Spaces, Interpolation Theory and Related Topics in Honour of Jaak Peetre on his 65th Birthday, August 17-22, 2000, 311–324, Walter de Gruyter, Berlin 2002.
37. D. Lukkassen, L.E. Persson and P. Wall. *On some sharp bounds for the homogenized p-Poisson equation*, Applicable Anal. 58 (1995), 123–135.
38. D. Lukkassen, J. Peetre and L.-E. Persson. *On some iterated means arising in homogenization theory.* Applications of Math. 49, 4, (2004), 343–356.
39. A. Meidell. *Homogenization and computational methods for calculating effective properties of some cellular solids and composite structures*, Norwegian University of Science and Technology (NTNU), Dr. ing. thesis 2001:19, ISBN 82-7984-181-4, ISSN 0809-103X, Trondheim, Norway.

40. A. Meidell and P. Wall. *Homogenization and design of structures with optimal macroscopic behaviour*. In: *Computer Aided Optimum Design of Structures V* (Eds. S. Hernández, C. A. Breddia), 393–402. Computational Mechanics Publications, Southhampton, 1997.
41. G.W. Milton, *Bounds on the complex dielectric constant of a composite material*, Appl. Phys. Lett. 37 (1980), 300–320.
42. G.W. Milton, *The Theory of Composites*, Cambridge University Press, London, 2002.
43. F. Murat and L. Tartar, *H-convergence*, in: *Topics in Mathematical Modelling of Composite Materials*, R. V. Kohn, ed., Progress in Nonlinear Differential Equations and their Applications, Birkhäuser, Boston, 1994.
44. J. Peetre and L.E. Persson, *A general Beckenbach's inequality with applications*, in: *Functions Spaces Operator and Nonlinear Analysis*, Pitman Research Notes in Mathematics 211 (1989), 125–139.
45. C. Niculescu and L.E. Persson, *Convex Functions and Their Applications*. A Contemporary Approach, CMS Books in Mathematics, Springer, Berline, 2006
46. L.E. Persson, *The homogenization method and some of its applications*, Mathematica Balkanika, new series 7 (1993), 179–190.
47. L.E. Persson, L. Persson, N. Svanstedt and J. Wyller, *The Homogenization Method: An Introduction*, Studentlitteratur, Lund, 1993.
48. P. Ponte Castañeda, *A second-order theory for nonlinear composite materials*, C. R. Acad. Sci. Paris, t. 322, Série II b, (1996), 3–10.
49. P. Ponte Castañeda, *New variational principle and its application to nonlinear heterogeneous systems*, SIAM J. Appl. Math. 52 (1992), 1321–1341.
50. E. Sanches-Palencia, *Nonhomogeneous Media and Vibration Theory*, Lecture Notes in Physics 127, Springer, Berlin, 1980.
51. L. Tartar, *Estimation de coefficients homogeneises*, Lecture Notes in Mathematics 704 (1979), Springer, Berlin, 364–373.
52. L. Tartar, *Estimation fines de coefficients homogénéisés*, in: *Ennio De Giorgi's Colloquium*. P. Kree, ed., Pitman Research Notes in Math., London, 1985.
53. J.R. Willis, *BVO Fields and bounds for nonlinear composite response*, in: *Proc. of the Second Workshop on Composite Media and Homogenization Theory*, G. Dal Maso and G. Dell'Antonio, eds, World Scientific, Singapore, 1995.
54. W. Rudin, *Real and Complex Analysis*, McGraw-Hill, New York, 1987.
55. W.P. Ziemer, *Weakly Differentiable Functions*, Springer-Verlag, Berlin, 1989.

Chapter 25
On Common Linear/Quadratic Lyapunov Functions for Switched Linear Systems

Melania M. Moldovan and M. Seetharama Gowda

Dedicated to the memory of Professor George Isac

Abstract Using duality, complementarity ideas, and **Z**-transformations, in this chapter we discuss equivalent ways of describing the existence of common linear/quadratic Lyapunov functions for switched linear systems. In particular, we extend a recent result of Mason–Shorten on positive switched system with two constituent linear time-invariant systems to an arbitrary finite system.

25.1 Introduction

Given a finite set of matrices $\{A_1, A_2, \ldots, A_m\}$ in $R^{n\times n}$, the dynamical system

$$\dot{x} + A_\sigma x = 0, \quad \sigma \in \{1, 2, \ldots, m\}, \tag{25.1}$$

where the switching signal σ is a piecewise constant function from $[0, \infty)$ to $\{1, 2, \ldots, m\}$ is called a *switched (continuous) linear system*. Because of numerous applications of such systems, these and their variants have been well studied in the literature, see, e.g., the monograph [16] and recent survey article [27]. Given a signal σ, a solution to (25.1) is a continuous and piecewise continuously differentiable function $x(t)$ that satisfies $\dot{x} + A_\sigma x(t) = 0$ for all $t \in [0, \infty)$ except at the switching instances of σ. For (uniform exponential) asymptotic stability of (25.1)

Melania M. Moldovan
Department of Mathematics & Statistics, University of Maryland, Baltimore County, Baltimore, Maryland 21250, USA, e-mail: melania1@umbc.edu

M. Seetharama Gowda
Department of Mathematics & Statistics, University of Maryland, Baltimore County, Baltimore, Maryland 21250, USA, e-mail: gowda@math.umbc.edu

P.M. Pardalos et al. (eds.), *Nonlinear Analysis and Variational Problems*,
Springer Optimization and Its Applications 35, DOI 10.1007/978-1-4419-0158-3_25,

corresponding to arbitrary signals, which means that there exist numbers $M \geq 1$ and $\beta > 0$ such that

$$||x(t)|| \leq Me^{-\beta t}||x(0)||$$

for all $t \geq 0$, for all solutions $x(t)$, and all signals, a sufficient condition is the existence of a *common quadratic Lyapunov function* (CQLF) $V(x) := x^T Px$, where P is a symmetric matrix satisfying the conditions

$$P \succ 0 \quad \text{and} \quad A_i^T P + PA_i \succ 0 \text{ for all } i = 1,2,\ldots,m. \tag{25.2}$$

The existence of such a P has been studied by numerous authors under various sufficient conditions, such as commutativity [22], simultaneous triangularization [21], Lie algebraic conditions [17], etc. See also, [25], [26], [15], [24].

Similar to the continuous case, there is a *switched (discrete) linear system* given by

$$x(k+1) + A_\sigma x(k) = 0, \quad \sigma \in \{1,2,\ldots,m\},\ k = 1,2,\ldots. \tag{25.3}$$

These systems have also been well studied, see, e.g., [19]. As in the continuous case, the stability can be studied by constructing a CQLF $V(x) := x^T Px$, where P is a symmetric matrix satisfying the conditions

$$P \succ 0 \quad \text{and} \quad P - A_i^T PA_i \succ 0 \text{ for all } i = 1,2,\ldots,m. \tag{25.4}$$

If we restrict the dynamics of the system (25.1) to the non-negative orthant R^n_+ of R^n, we get a *positive switched system* [7]. Here we require any trajectory of (25.1) with a starting point in R^n_+ to remain in R^n_+. This condition requires all matrices A_i to be **Z**-matrices (which are matrices with nonpositive off-diagonal entries). In this setting, a common linear Lyapunov function $V(x) = x^T d$ can be constructed, where d is a vector in R^n satisfying the conditions

$$d > 0 \quad \text{and} \quad A_i^T d > 0 \text{ for all } i = 1,2,\ldots,m. \tag{25.5}$$

Alternatively, for the same positive switched system, a common quadratic copositive Lyapunov function $V(x) := x^T Px$ can be constructed, where P is a symmetric matrix satisfying the conditions

$$P \quad \text{and} \quad A_i^T P + PA_i \text{ are strictly copositive on } R^n_+ \text{ for all } i = 1,2,\ldots,m. \tag{25.6}$$

(Here, a matrix A is strictly copositive on R^n_+ means that $x^T Ax > 0$ for all $0 \neq x \in R^n_+$.)

By considering the cone of (symmetric) positive semidefinite matrices and transformation

$$L(X) = A^T X + XA,$$

various authors have used theorems of alternative (or duality theory) to formulate conditions equivalent to (25.2). For example, a result of Kamenetskiy and Pyatnitskiy [14] says that condition (25.2) holds if and only if there do not exist positive semidefinite matrices Y_i $(i = 1,2,\ldots,m)$ with at least one Y_i nonzero such that

$$\sum_{i=1}^{m}(A_iY_i + Y_iA_i^T) = 0.$$

It is possible to write similar equivalent conditions for (25.4), (25.5), and (25.6). Our first objective in this paper is to present a unified result that covers all these equivalent conditions. This result was motivated by the observation that the underlying transformations in (25.2), (25.4), (25.5), and (25.6), namely, the Lyapunov transformation L_A defined on the cone of positive semidefinite matrices, the Stein transformation S_A on the cone of positive semidefinite matrices, the matrix A on the cone of non-negative orthant in R^n, and the Lyapunov transformation L_A (corresponding to a **Z**-matrix) on the cone of symmetric copositive matrices have the following common property:

$$x \in K,\, y \in K^*, \text{ and, } \langle x,y\rangle = 0 \Rightarrow \langle L(x),y\rangle \leq 0,$$

where K denotes a cone with dual K^*.

Transformations satisfying the above property are called **Z**-transformations. By using the properties of **Z**-transformations [12], we formulate our unifying condition, see Theorem 25.8.

Our second objective is to demonstrate the relevance of complementarity problems in the study of common linear/quadratic Lyapunov functions. Specifically, we show how a complementarity result of Song, Gowda, and Ravindran [28] can be used to extend a recent result of Mason–Shorten on positive switched systems [20].

Here is an outline of this paper. In Section 25.2, we recall necessary matrix theory concepts and describe essential properties of **Z**-transformations. Section 25.3 deals with complementarity ideas. Section 25.4 deals with duality ideas and our unifying result covering the existence of linear/quadratic Lyapunov functions. Finally, in Section 25.5, we present our extension of the Mason–Shorten result.

A word about our notation. Throughout this paper, we will use standard matrix theory and complementarity theory notation. For example, we will use **Z**-matrices and positive stable matrices (instead of Metzler matrices and Hurwitz matrices), work with positive definite matrices (instead of negative definite matrices), etc. Also, our (dynamical) systems are written in the form $\dot{x} + Ax = 0$ instead of $\dot{x} = Ax$.

25.2 Preliminaries

25.2.1 Matrix Theory Concepts

The space R^n carries the usual inner product which is written as either $\langle x,y\rangle$ or $x^T y$. In R^n, we denote the non-negative orthant by R^n_+ and its interior by $(R^n_+)^\circ$. We write

$$d \geq 0 \text{ when } d \in R^n_+ \text{ and } d > 0 \text{ when } d \in (R^n_+)^\circ.$$

A matrix $A \in R^{n\times n}$ is said to be

- a **Z**-matrix if all its off-diagonal entries are nonpositive. The negative of a **Z**-matrix is called a Metzler matrix;
- a **P**-matrix if all its principal minors are positive;
- a *copositive* matrix (*strictly copositive* matrix) if $x^T Ax \geq 0\,(>0)$ for all $0 \neq x \geq 0$;
- an **S**-matrix if there is a $d > 0$ such that $Ad > 0$;
- *positive stable* if the real part of any eigenvalue of A is positive;
- *Schur stable* if the absolute value of any eigenvalue of A is less than one;
- *completely positive* if there is a non-negative (rectangular) matrix B with $A = BB^T$, or equivalently, A is a sum of matrices of the form xx^T with $x \geq 0$.

Consider a nonempty set $\mathscr{C}$ of matrices in $R^{n\times n}$. A matrix $A \in R^{n\times n}$ is called a *column representative* of $\mathscr{C}$ if for every $j = 1,2,...,n$, the jth column of A is the jth column of some matrix in $\mathscr{C}$. Similarly, row representatives of $\mathscr{C}$ are defined.
For $A_1, A_2, ..., A_m, D_1, ..., D_m \in R^{n\times n}$, where each D_i is a diagonal matrix, the following holds [29]:

$$\det(\sum_1^m A_i D_i) = \sum \det(K)\det(E), \tag{25.7}$$

where K is a column representative of $\{A_1, A_2, ..., A_m\}$, E is a column representative of $\{D_1, ..., D_m\}$, and the same indexed columns are selected to form K and E. Here the summation is over all column representatives of $\{A_1, ..., A_m\}$.

25.2.2 Z-Transformations

Throughout this paper, H denotes a finite-dimensional real Hilbert space. We reserve the symbol K for a *proper* cone in H, that is, K is a closed convex cone in H such that

$$K \cap (-K) = \{0\} \quad \text{and} \quad K - K = H.$$

We denote the dual of K by K^* and the interior by K°.

A linear transformation $L : H \to H$ is said to be a **Z**-*transformation* (or said to have the **Z**-property) on K if

$$x \in K,\, y \in K^*,\, \langle x, y\rangle = 0 \Rightarrow \langle L(x), y\rangle \leq 0. \tag{25.8}$$

When $H = R^n$ and $K = R^n_+$, a **Z**-transformation is nothing but a **Z**-matrix. We remark that a **Z**-transformation is the negative of a "cross-positive" transformation introduced in [23]. It is known [5] that L has the **Z**-property on K if and only if any trajectory of the dynamical system $\dot{x} + L(x) = 0$ which starts in K stays in K.

The above concept is illustrated in the following examples. Examples 25.1 and 25.3 given below appear in [12]. Example 25.2 is new.

Example 25.1. Let $H = \mathscr{S}^n$, the space of all $n \times n$ real symmetric matrices with inner product given by $\langle X, Y\rangle = trace(XY)$, where the trace of a matrix is the sum of its diagonal elements (or the sum of its eigenvalues). Let $K = \mathcal{S}^n_+$, the cone of positive semidefinite matrices in $\mathcal{S}^n$. We use the notation

$$X \succeq 0 \text{ when } X \in \mathcal{S}^n_+ \text{ and } X \succ 0 \text{ when } X \in (\mathcal{S}^n_+)^\circ.$$

For any $A \in R^{n\times n}$, let

$$L_A(X) = \frac{1}{2}(AX + XA^T) \quad \text{and} \quad S_A(X) = X - AXA^T$$

denote the *Lyapunov and Stein transformations* corresponding to A. We verify that L_A and S_A are **Z**-transformations on $\mathcal{S}^n_+$: Since $\mathcal{S}^n_+$ is self-dual and

$$X \succeq 0,\ Y \succeq 0, \text{ and } \langle X, Y\rangle = 0 \Rightarrow XY = 0,$$

$$\langle L_A(X), Y\rangle = trace(AXY) = 0 \ \& \ \langle S_A(X), Y\rangle = trace(XY) - trace(AXA^TY) \leq 0, \tag{25.9}$$

where the last inequality comes from the facts that $AXA^T \succeq 0$ (when $X \succeq 0$) and the inner product of any two elements in $\mathcal{S}^n_+$ is non-negative. Thus, both L_A and S_A are **Z**-transformations on $\mathcal{S}^n_+$.

Example 25.2. As in Example 25.1, let $H = \mathcal{S}^n$ with $\langle X, Y\rangle = trace(XY)$. Let

$$K = \{X \in \mathcal{S}^n : X \text{ is copositive on } R^n_+\}.$$

Then K is a proper cone with dual

$$K^* = \{X \in \mathcal{S}^n : X \text{ is completely positive}\},$$

see [3], Thm. 2.3. Given $X \in K$ and $Y = \sum_{i=1}^N y_i y_i^T \in K^*$ (where $y_i \in R^n_+$ for all i) and $\langle X, Y\rangle = 0$, we claim that XY is a non-negative matrix with zero diagonal. To see this, first observe that $\sum_i^N y_i^T X y_i = \sum_i^N trace(X y_i y_i^T) = \langle X, Y\rangle = 0$. As X is copositive, we get $y_i^T X y_i = 0$ for all i; because X is copositive and symmetric, we have $Xy_i \geq 0$. (This follows from $\lim_{t\downarrow 0} \frac{1}{t}(tx + y_i)^T X (tx + y_i) \geq 0$ for all $x \geq 0$.) From this, we see that $XY = \sum_i^N X y_i y_i^T$ is a non-negative matrix. As $trace(XY) = 0$, XY must have zero diagonal.
Now suppose $A \in R^{n\times n}$. We claim that

$$L_A \text{ is a } \mathbf{Z}\text{-transformation on } K \text{ if and only if } A \text{ is a } \mathbf{Z}\text{-matrix.}$$

To see this, suppose that L_A is a **Z**-transformation on K. We show that the $(1,2)$ and $(2,1)$ entries of A are nonpositive. Let $e_1, e_2, \ldots, e_n$ denote the standard unit coordinate vectors in R^n. Let

$$X = [x_{ij}], \quad Y = e_1 e_1^T, \quad \text{and} \quad W = e_2 e_2^T,$$

where all entries of X are zero except $x_{12} = x_{21} = 1$. We see that $X \in K$, $Y, W \in K^*$, and $\langle X, Y\rangle = 0 = \langle X, W\rangle$. A simple computation shows that

$$trace(L_A(X)Y) = a_{12} \quad \text{and} \quad trace(L_A(X)W) = a_{21}.$$

By the **Z**-property of L_A, we must have $a_{12} \leq 0$ and $a_{21} \leq 0$. A similar argument will prove that other off-diagonal entries are also nonpositive. Hence A is a **Z**-matrix. Now conversely, suppose that A is a **Z**-matrix. Let $X \in K$, $Y \in K^*$, with $\langle X, Y \rangle = 0$. Then XY and YX are non-negative matrices with zero diagonals. Then

$$trace(L_A(X)Y) = trace(AXY) \leq 0.$$

This proves that when A is a **Z**-matrix, L_A has the **Z**-property on K.

Example 25.3. Let H be a Euclidean Jordan algebra with inner product $\langle \cdot, \cdot \rangle$ and Jordan product $x \circ y$ [6]. Let K denote the cone of squares $\{x \circ x : x \in H\}$. Then K is a self-dual (proper) cone. Examples of Euclidean Jordan algebras include R^n with the usual inner product and componentwise product (as the Jordan product), the Jordan spin algebra $\mathscr{L}^n$ $(n > 1)$ whose underlying space is $R \times R^{(n-1)}$ with the usual inner product and $(x_0, \overline{x}) \circ (y_0, \overline{y}) = (x_0 y_0 + \langle \overline{x}, \overline{y} \rangle, x_0 \overline{y} + y_0 \overline{x})$, and matrix algebras $\mathscr{S}^n$ of all $n \times n$ real symmetric matrices, $\mathscr{H}^n$ of all $n \times n$ complex Hermitian matrices, $\mathscr{Q}^n$ of all $n \times n$ quaternion Hermitian matrices, and $\mathscr{O}^3$ of all 3×3 octonion Hermitian matrices. In the matrix algebras, the Jordan and inner product are given respectively by $X \circ Y := \frac{1}{2}(XY + YX)$ and $\langle X, Y \rangle := Retrace(XY)$.

In any Euclidean Jordan algebra, for any element a, the so-called Lyapunov transformation L_a is a **Z**-transformation on the cone of squares, where,

$$L_a(x) = a \circ x.$$

We now recall some properties of **Z**-transformations [12].

Theorem 25.4. *Suppose L is a* **Z***-transformation on a proper cone K in a Hilbert space H. Then the following are equivalent:*

(1) There exists a $d \in K^\circ$ *such that* $L(d) \in K^\circ$.
(2) L is invertible with $L^{-1}(K^\circ) \subseteq K^\circ$.
(3) L is positive stable, that is, the real part of any eigenvalue of L is positive.
(4) $L + tI$ *is invertible for all* $t \in [0, \infty)$.
(5) All real eigenvalues of L are positive.
(6) There is an $e \in (K^*)^\circ$ *such that* $L^T(e) \in (K^*)^\circ$.

Moreover, when $H = R^n$ *and* $K = R^n_+$, *the above properties (for a* **Z***-matrix) are further equivalent to*

(7) L is a **P***-matrix.*

Remark 25.5. When $A \in R^{n \times n}$ is a **Z**-matrix, the positive stability of A can be described in more than 50 equivalent ways, see [2]. In particular, we have the equivalence of the following for a **Z**-matrix:

(1) A is positive stable.
(2) A is a **P**-matrix.
(3) There exists a $d > 0$ such that $Ad > 0$ (or $A^T d > 0$).
(4) There exists a (diagonal) $D \succ 0$ in $\mathscr{S}^n$ such that $AD + DA^T \succ 0$.

(5) L_A is positive stable.

(Regarding item (5), we note that the eigenvalues of L_A are of the form $\frac{1}{2}(\lambda+\mu)$, where λ and μ are eigenvalues of A.) We now add one more condition to this list:

A **Z**-*matrix A is positive stable if and only if there exists a strictly copositive matrix C such that* $AC+CA^T$ *is strictly copositive.*

This follows from the above theorem applied to $L=L_A$ on the cone of symmetric copositive matrices.

25.3 Complementarity Ideas

Suppose K is a self-dual cone in H. For any two elements $x,y\in H$, let

$$x\sqcap y = x-\Pi_K(x-y),$$

where Π_K denotes the (orthogonal) projection of H onto K. It is known, see [11], that $'\sqcap'$ is a commutative, but (in general) non-associative binary operation on H. In the case of $H=R^n$ and $K=R^n_+$, we use the standard notation $x\wedge y$ in place of $x\sqcap y$; we observe that

$$x\wedge y=\min\{x,y\}$$

and that $'\wedge'$ is associative. Given a matrix-tuple $\mathbf{A}=(A_1,A_2,\ldots,A_m)$ of matrices A_i in $R^{n\times n}$ and vector-tuple $\mathbf{q}=(q_1,q_2,\ldots,q_m)$ of vectors q_i in R^n, the vertical linear complementarity problem [10] VLCP$(\mathbf{A},\mathbf{q})$ is to find a vector $x\in R^n$ such that

$$x\wedge(A_1x+q_1)\wedge\cdots\wedge(A_mx+q_m)=0.$$

If $-q_i\in(R^n_+)^\circ$, then a solution to the above problem satisfies $x\geq 0$ and $A_ix\geq -q_i>0$ for all i; hence a small perturbation of such an x produces a vector $d>0$ such that $A_id>0$ for all i.

Theorem 25.6. *([10], Section 6.3) The* VLCP $(\mathbf{A},\mathbf{q})$ *has a unique solution for all* $\mathbf{q}$ *if and only if every row representative of the set* $\{A_1,A_2,\ldots,A_m\}$ *is a* **P**-*matrix.*

When $m=1$, the VLCP reduces to the well-known linear complementarity problem [4] that has been extensively studied in the optimization literature.

Now coming to the general case of a self-dual K, it can be easily verified that $x\sqcap y\in K\Rightarrow x,y\in K$. Thus, for a finite set of linear transformations $L_i:H\to H$ and vectors $-e_i\in K^\circ$, if an $x\in H$ satisfies the complementarity problem

$$x\sqcap f(x)=0,$$

where $f(x)$ is a combination of $L_i(x)-e_i$, $i=1,2,\ldots,m$ under the binary operation $'\sqcap'$ in some order, then we can assert the existence of a $d\in K^\circ$ such that $L_i(d)\in K^\circ$. In particular, when $H=\mathscr{S}^n$ and $K=\mathscr{S}^n_+$ and $L_i=L_{A_i^T}$, we can relate the existence of a common quadratic Lyapunov function to a solution of a complementarity problem. We refer the reader to [9] and [12] for results highlighting this connection. See also Section 25.5.

25.4 Duality Ideas

In this section, we present our result that generalizes the result of Kamenetskiy and Pyatnitskiy [14] and at the same time unifies several similar results.

We begin with a theorem of alternative.

Theorem 25.7. *(Theorem of Alternative [2], Page 9) Let H_1 and H_2 be Hilbert spaces with proper cones $K_1 \subseteq H_1$ and $K_2 \subseteq H_2$. For a linear transformation $L : H_1 \to H_2$, consider the following systems:*

(i) $L(x) \in (K_2)^\circ, x \in (K_1)^\circ$,
(ii) $L^T(y) \in K_1^*, 0 \neq y \in -K_2^*$.
(iii) $L(x) \in K_2, 0 \neq x \in K_1$,
(iv) $L^T(y) \in (K_1^*)^\circ, y \in -(K_2^*)^\circ$.

Then, exactly one of the systems (i) and (ii) is consistent and exactly one of the systems (iii) and (iv) is consistent.

Theorem 25.8. *Let H be a finite-dimensional real Hilbert space and K be a proper cone in H. Let $L_i : H \to H$ be linear for $i = 1, 2, \ldots, m$, where L_1 is positive stable and has the* **Z***-property on K. Consider the following statements:*

(1) There exists a $d \in K^\circ$ such that $L_i(d) \in K^\circ$ for all $i = 1, ..., m$.
(2) There do not exist $x_i \in K^*, i = 1, ..., m$ with some x_i nonzero such that $\sum_1^m L_i^T(x_i) = 0$.
(3) There exists a nonzero $d \in K$ such that $L_i(d) \in K$ for all $i = 1, ..., m$.
(4) There do not exist $x_i \in (K^*)^\circ, i = 1, ..., m$ such that $\sum_1^m L_i^T(x_i) = 0$.

Then

$$(1) \Leftrightarrow (2) \quad \textit{and} \quad (3) \Leftrightarrow (4).$$

Proof. Assume that (1) holds but not (2). Then there are $x_1, x_2, \ldots, x_m$ in K^* with at least one x_i nonzero such that $\sum_1^m L_i^T(x_i) = 0$. Then

$$0 = \langle \sum_1^m L_i^T(x_i), d \rangle = \sum_1^m \langle x_i, L_i(d) \rangle > 0$$

as $\langle x_i, L_i(d) \rangle \geq 0$ for all $i = 1, 2, 3, \ldots, m$ and $\langle x_i, L_i(d) \rangle > 0$ when $x_i \neq 0$. This contradiction proves that $(1) \Rightarrow (2)$. Now assume the negation of (1). Since L_1 is positive stable and has the **Z**-property on K, it is invertible, see Theorem 25.4; hence (as $m \neq 1$) we can define $L : H \to H^{m-1}$ by

$$L(x) := (L_2 L_1^{-1}(x), ..., L_m L_1^{-1}(x)).$$

We now apply Theorem 25.7 with $K_1 := K$ in H and $K_2 := K \times \ldots \times K$ ($m-1$ times) in H^{m-1}. We claim that condition (i) of Theorem 25.7 cannot hold. Assuming the contrary, there is an $x \in K^\circ$ such that $L(x) \in (K_2)^\circ = K^\circ \times \cdots \times K^\circ$. This implies that $L_i(L_1^{-1}(x)) \in K^\circ$ for all $i = 2, 3, \ldots, m$. Let

$$d := L_1^{-1}(x).$$

Since $L_1^{-1}(K^\circ) \subseteq K^\circ$ (by Theorem 25.4), $d \in K^\circ$ and $L_1(d) = x \in K^\circ$. From $L_i(L_1^{-1}(x)) \in K^\circ$, we get $L_i(d) \in K^\circ$ for $i = 2, 3, \ldots, m$. This cannot happen as we have assumed the negation of (1). So condition (i) of Theorem 25.7 fails. Therefore, by condition (ii) of Theorem 25.7, we get a y such that

$$0 \neq y \in -(K^* \times \cdots \times K^*) \quad \text{and} \quad L^T(y) \in K^*.$$

Now let

$$-y = (x_2, \ldots, x_m) \in K^* \times \ldots \times K^*.$$

Then for any $u \in H$ we have

$$\begin{aligned} \langle u, L^T(y) \rangle &= \langle L(u), y \rangle = -\langle (L_2 L_1^{-1}(u), \ldots, L_m L_1^{-1}(u)), (x_2, \ldots, x_m) \rangle \\ &= -\sum_{i=2}^{m} \langle L_i L_1^{-1}(u), x_i \rangle = -\sum_{i=2}^{m} \langle u, L_1^{-T} L_i^T(x_i) \rangle = \left\langle u, -\sum_{i=2}^{m} L_1^{-T} L_i^T(x_i) \right\rangle. \end{aligned}$$

Hence

$$x_1 := L^T(y) = -\sum_{i=2}^{m} L_1^{-T} L_i^T(x_i).$$

Applying L_1^T to both sides of the above equation, we get

$$L_1^T(x_1) + L_2^T(x_2) + \ldots + L_m^T(x_m) = 0.$$

As $x_1 \in K^*$ and $(x_2, x_3, \ldots, x_m)$ is nonzero, we get the negation of (2). Hence $\sim (1) \Rightarrow \sim (2)$, or $(2) \Rightarrow (1)$. This completes the proof of the equivalence $(1) \Leftrightarrow (2)$. The proof of $(3) \Leftrightarrow (4)$ is similar, but one has to use conditions (*iii*) and (*iv*) of Theorem 25.7. □

We illustrate the theorem by the following examples.

Example 25.9. Let $H = \mathscr{S}^n$, $K = \mathscr{S}^n_+$, and

$$L_{A_i}(X) = \frac{1}{2}(A_i X + X A_i^T) \quad (X \in \mathscr{S}^n),$$

where $A_i \in R^{n \times n}$ ($i = 1, 2, \ldots, m$) are positive stable. As L_A and $(L_A)^T = L_{A^T}$ are both **Z**-transformations on $\mathscr{S}^n_+$ for any $A \in R^{n \times n}$, the above theorem gives the equivalence of the following statements [14]:

(i) There exists $D \succ 0$ in $\mathcal{S}^n$ such that $A_i^T D + DA_i = 2L_{A_i^T}(D) \succ 0$ for all $i = 1,2,\ldots,m$.

(ii) There do not exist $Y_i \succeq 0$ $(i = 1,2,\ldots,m)$ with at least one nonzero Y_i such that $\sum_{i=1}^m (A_i Y_i + Y_i A_i^T) = 0$.

We note that the matrix D in item (i) produces a common quadratic Lyapunov function $V(x) := x^T Dx$ for the switched linear system $\dot{x} + A_\sigma x = 0$.

Example 25.10. Let $H = \mathcal{S}^n$, $K = \mathcal{S}^n_+$, and $S_{A_i}(X) = X - A_i X A_i^T \quad (X \in \mathcal{S}^n)$, where $A_i \in R^{n\times n}$ $(i = 1,2,\ldots,m)$ are Schur stable. Now, S_A and $(S_A)^T = S_{A^T}$ are **Z**-transformations on $\mathcal{S}^n_+$ for any $A \in R^{n\times n}$, see Example 25.1. Also, the Schur stability of A_i implies that S_{A_i} is positive stable for each i, see, e.g., Theorem 11 in [8] and Theorem 25.4. Hence the above theorem gives the equivalence of the following statements:

(i) There exists $D \succ 0$ in $\mathcal{S}^n$ such that $D - A_i^T DA_i \succ 0$ for all $i = 1,2,\ldots,m$.

(ii) There do not exist $Y_i \succeq 0$ $(i = 1,2,\ldots,m)$ with at least one nonzero Y_i such that $\sum_{i=1}^m (Y_i - A_i Y_i A_i^T) = 0$.

The matrix D in item (i) produces a common quadratic Lyapunov function $V(x) := x^T Dx$ for the discrete switched system $x(k+1) + A_\sigma x(k) = 0$.

Example 25.11. Let $H = R^n$, $K = R^n_+$, and $A_i \in R^{n\times n}$ $(i = 1,2,\ldots,m)$ be positive stable **Z**-matrices. Then the following are equivalent:

(i) There exists $d > 0$ in R^n such that $A_i^T d > 0$ for all $i = 1,2,\ldots,m$.

(ii) There do not exist vectors $y_i \geq 0$ $(i = 1,2,\ldots,m)$ with at least one nonzero y_i such that $\sum_{i=1}^m A_i y_i = 0$.

In this setting, the vector d in item (i) produces a common linear Lyapunov function $V(x) := x^T d$ for the positive switched system $\dot{x} + A_\sigma x = 0$.

Example 25.12. Let $H = \mathcal{S}^n$, K be the cone of symmetric copositive matrices and $A_i \in R^{n\times n}$ for $i = 1,2,\ldots,m$ be positive stable **Z**-matrices. Then the following are equivalent:

(i) There exists a strictly copositive matrix C in $\mathcal{S}^n$ such that $A_i^T C + CA_i$ is strictly copositive for all $i = 1,2,\ldots,m$.

(ii) There do not exist completely positive matrices Y_i $(i = 1,2,\ldots,m)$ with at least one nonzero Y_i such that $\sum_{i=1}^m (A_i Y_i + Y_i A_i^T) = 0$.

Here, the matrix C in item (i) produces a common copositive quadratic Lyapunov function $V(x) := x^T Cx$ for the positive switched system $\dot{x} + A_\sigma x = 0$.

We remark that if condition (i) of Example 25.11 holds, then $C := dd^T$ satisfies condition (i) of Example 25.12. However, the converse is false, see Remark 25.20 in Section 25.5.

25.5 Positive Switched Linear Systems

In a recent paper [20], Mason and Shorten prove that for two positive stable **Z**-matrices A_1 and A_2, the following are equivalent:

(i) There exists a vector $d > 0$ in R^n such that $A_1^T d > 0$ and $A_2^T d > 0$.
(ii) Every column representative of $\{A_1, A_2\}$ is positive stable.
(iii) Every column representative of $\{A_1, A_2\}$ has positive determinant.

In an earlier work, based on complementarity ideas, Song, Gowda, and Ravindran [28] proved the equivalence of the following for any compact set $\mathscr{A}$ of positive stable **Z**-matrices in $R^{n\times n}$:

(1) There exists a $d > 0$ such that $A^T d > 0$ for all $A \in \mathscr{A}$.
(2) Every column representative of $\mathscr{A}$ is a **P**-matrix.

In this section, we extend the above Mason–Shorten result to any compact set of matrices and at the same time improve the result of Song, Gowda, and Ravindran by requiring that only the determinant be positive for every column representative of $\mathscr{A}$.

We recall that for a **Z**-matrix (see Remark 25.5), **P**, **S**, and positive stability properties are equivalent.

We begin with some lemmas.

Lemma 25.13. *Suppose* $\mathscr{A} = \{A_1, A_2, ..., A_m\} \subseteq R^{n\times n}$ *with* $\det(C) > 0$ *for any column representative* C *of* $\mathscr{A}$. *Then for any set of positive diagonal matrices* $\{D_1, ..., D_m\}$ *we have*

$$\det(A_1D_1 + ... + A_mD_m) > 0.$$

Proof. The result follows from the identity (25.7). □

Lemma 25.14. *Suppose* $A_1, A_2, ..., A_m$ *are positive stable* **Z**-*matrices and* $\det(C) > 0$ *for any column representative* C *of* $\{A_1, A_2, ..., A_m\}$. *Then there exists* $0 \neq u \geq 0$ *such that* $A_i^T u \geq 0$ *for all* $i = 1, ..., m$.

Proof. If the stated conclusion fails, then by the equivalence of items (3) and (4) in Theorem 25.8 (see Example 25.11), there exist $y_i > 0$ $(i = 1, \ldots, m)$ such that

$$A_1y_1 + A_2y_2 + \ldots + A_my_m = 0.$$

If D_i denotes the diagonal matrix with diagonal y_i, then

$$(A_1D_1 + A_2D_2 + \ldots + A_mD_m)e = 0,$$

where e is the transpose of the vector $(1, 1, \ldots, 1)$ in R^n. This means that

$$\det(A_1D_1 + A_2D_2 + \ldots + A_mD_m) = 0.$$

However, Lemma 25.13 together with the hypothesis shows that this is not possible. □

Lemma 25.15. *Suppose that $A_1, A_2, ..., A_m$ are positive stable* **Z***-matrices and* $\det(C) > 0$ *for any column representative C of $\{A_1, A_2, ..., A_m\}$. Then there exists $d > 0$ such that $A_i^T d > 0$ for all $i = 1, ..., m$.*

Proof. Let E be the matrix in $R^{n\times n}$ with every entry 1. We can choose a small positive ε such that $A_i - \varepsilon E$ is a $\mathbf{P} \cap \mathbf{Z}$-matrix for all i and any column representative of $\{A_1 - \varepsilon E, A_2 - \varepsilon E, \ldots, A_m - \varepsilon E\}$ has positive determinant. An application of the previous lemma produces a nonzero $u \geq 0$ such that $(A_i^T - \varepsilon E)u \geq 0$ for all i. This implies that $0 \neq u \geq 0$ and $A_i^T u > 0$ for all i; by continuity, there exists a $d > 0$ such that $A_i^T d > 0$ for all i. □

For a set $\mathscr{A} = \{A_1, A_2, ..., A_m\}$ in $R^{n\times n}$, let

$$\mathscr{A}^{\#} := \text{set of all column representatives of } \mathscr{A}$$

and

$$\widehat{\mathscr{A}} := \left\{ \sum_1^m A_i D_i \;:\; D_i \text{ is a nonnegative diagonal matrix and } \sum_1^m D_i = I \right\}.$$

We note that

$$\mathscr{A} \subseteq \mathscr{A}^{\#} \subseteq \widehat{\mathscr{A}}$$

and that $\widehat{\mathscr{A}}$ is convex.

Theorem 25.16. *Let $\mathscr{A} = \{A_1, A_2, ..., A_m\}$ be a set of positive stable* **Z***-matrices in $R^{n\times n}$. Then the following are equivalent:*

(1) There exists a $d > 0$ such that $A_i^T d > 0$ for all $i = 1, 2, \ldots, m$.
(2) There exists a $d > 0$ such that $C^T d > 0$ for all $C \in \widehat{\mathscr{A}}$.
(3) Every matrix in $\widehat{\mathscr{A}}$ is positive stable (equivalently, a **P***-matrix).*
(4) Every matrix in $\mathscr{A}^{\#}$ is positive stable (equivalently, a **P***-matrix).*
(5) Every matrix in $\mathscr{A}^{\#}$ has positive determinant.
(6) Every matrix in $\widehat{\mathscr{A}}$ has positive determinant.

Proof. If (1) holds, then for any matrix $C = \sum_1^m A_i D_i \in \widehat{\mathscr{A}}$, $C^T d = \sum_1^m D_i A_i^T d > 0$ by the imposed conditions on D_i. Thus condition (2) holds. Now each matrix C in $\widehat{\mathscr{A}}$ is a **Z**-matrix. The implication (2) $\Rightarrow$ (3) follows from Remark 25.5.
The implications (3) $\Rightarrow$ (4) $\Rightarrow$ (5) are obvious.
The implication (5) $\Rightarrow$ (6) follows from the identity (25.7). Finally, the implication (6) $\Rightarrow$ (5) is obvious and the implication (5) $\Rightarrow$ (1) follows from Lemma 25.15. This completes the proof of the theorem. □

Several remarks are in order.

Remark 25.17. Thanks to Theorem 25.6, we can relate item (4) above to the uniqueness of solution in the vertical linear complementarity problem VLCP($\mathbf{A}^T, \mathbf{q}$) for any vector-tuple $\mathbf{q}$. Here, $\mathbf{A}^T = (A_1^T, A_2^T, \ldots, A_m^T)$. Item (5) is related to the so-called column $\mathscr{W}$-property and horizontal linear complementarity problems, see [29].

Remark 25.18. A conjecture due to Mason and Shorten [18] and D. Angeli says that for a set $\mathscr{A} = \{A_1, A_2, \ldots, A_m\}$ of positive stable $\mathbf{Z}$-matrices, the following are equivalent:

(a) Every matrix in the convex hull of $\mathscr{A}$ is positive stable.
(b) The switched system $\dot{x} + A_\sigma x = 0$ is (uniformly) asymptotically stable for arbitrary signals.

In [13], Gurvits, Shorten, and Mason disprove this conjecture by constructing a set $\{A_1, A_2\}$ of positive stable $\mathbf{Z}$-matrices in $R^{2\times 2}$ for which item (a) holds but not (b). We note here that if we replace (a) by

(a') Every matrix in $\mathscr{A}^\#$ is positive stable (or equivalently, every matrix in the convex set $\widehat{\mathscr{A}}$ is positive stable),

then $(a') \Rightarrow (b)$. This raises the question whether $(b) \Rightarrow (a')$. Based on the work of Akar et al. [1], we show that this is false.

Example 25.19. Consider the following positive stable $\mathbf{Z}$-matrices

$$A_1 = \begin{bmatrix} 1 & -1 \\ -\frac{1}{2} & 1 \end{bmatrix} \quad \text{and} \quad A_2 = \begin{bmatrix} 1 & -\frac{1}{2} \\ -1 & 1 \end{bmatrix}.$$

It is easily verified that every convex combination of A_1 and A_2 is positive stable. In this context (because every diagonal entry of A_1 and A_2 is one), by a result of Akar et al. [1], the positive system $\dot{x} + A_\sigma x = 0$ is (uniformly) asymptotically stable for arbitrary signals. Thus condition (b) above holds. However, the column representative formed by the first column of A_2 and the second column of A_1 is not a $\mathbf{P}$-matrix (that is, it is not positive stable). Hence (a') fails to hold.

Remark 25.20. It is well known that a $\mathbf{Z}$-matrix A is positive stable if and only if there is a diagonal matrix $D \succ 0$ such that $AD + DA^T \succ 0$. Now for a finite set $\mathscr{A} = \{A_1, A_2, \ldots, A_m\}$ of $\mathbf{Z}$-matrices, consider the following statements:

(i) Every row and column representative of $\mathscr{A}$ is positive stable.
(ii) There exists a diagonal matrix $D \succ 0$ such $A_i D + DA_i^T \succ 0$ for all $i = 1, 2 \ldots, m$.

We claim that (i)$\Rightarrow$ (ii) but not conversely. To see this, assume that (i) holds, in which case, by the above theorem, there exist vectors $u > 0$ and $v > 0$ in R^n such that

$$A_i u > 0 \quad \text{and} \quad A_i^T v > 0 \quad (i = 1, 2, \ldots, m).$$

Let D be a diagonal matrix with $Dv = u$. Then D has positive diagonal entries and

$$(A_i D + DA_i^T)v = A_i u + D(A_i^T v) > 0$$

for all i. This means that for each i, the symmetric **Z**-matrix $A_iD + DA_i^T$ is a **P**-matrix, hence $A_iD + DA_i^T \succ 0$. Thus item (ii) holds. To see that (ii) need not imply (i), we merely construct two 2×2 symmetric positive definite **Z**-matrices $\{A_1, A_2\}$ such that a column representative of $\{A_1, A_2\}$ is not a **P**-matrix. Then with $D = I$, item (ii) holds but not (i).

Theorem 25.21. *Let $\mathscr{A}$ be a compact set of positive stable* **Z***-matrices in $R^{n\times n}$. Then the following are equivalent:*

(1) There exists a $d > 0$ such that $A^T d > 0$ for all $A \in \mathscr{A}$.
(2) Every column representative of $\mathscr{A}$ is a **P***-matrix.*
(3) Every column representative of $\mathscr{A}$ has positive determinant.

Proof. The equivalence between (1) and (2) was proved by Song, Gowda, and Ravindran in [28]. The implication (2)⇒(3) is obvious and (3)⇒(2) follows from the previous theorem applied to a finite set of matrices. □

References

1. M. Akar, A. Paul, M.G. Safonov, and U. Mitra, Conditions on the stability of a class of second-order switched systems, IEEE Transactions on Automatic Control, 51 (2006) 338–340.
2. A. Berman and R.J. Plemmons, Nonnegative matrices in the mathematical sciences, SIAM, Philadelphia, PA, 1994.
3. A. Berman and N. Shaked-Monderer, Completely positive matrices, World Scientific, Singapore, 2003.
4. R.W. Cottle, J.S. Pang, and R.E. Stone, The linear complementarity problem, Academic Press, Boston, 1992.
5. L. Elsner, Quasimonotonie und Ungleischungen in halbgeordneten Räumen, Linear Algebra and its Applications, 8 (1974) 249–261.
6. J. Faraut and A. Koranyi, Analysis on symmetric cones, Oxford University Press, Oxford 1994.
7. L. Farina and S. Rinaldi, Positive linear systems: Theory and applications, Wiley, New York 2000.
8. M.S. Gowda and T. Parthasarathy, Complementarity forms of theorems of Lyapunov and Stein, and related results, Linear Algebra and its Applications, 320 (2000) 131–144.
9. M.S. Gowda and Y. Song, Some new results for the semidefinite linear complementarity problems, SIAM Journal on Matrix Analysis, 24 (2002) 25–39.
10. M.S. Gowda and R. Sznajder, The generalized order linear complementarity problem, SIAM Journal on Matrix Analysis, 15 (1994) 779–795.
11. M.S. Gowda, R. Sznajder, and J. Tao, Some P-properties for linear transformations on Euclidean Jordan algebras, Linear Algebra and its Applications, 393 (2004) 203–232.
12. M.S. Gowda and J. Tao, **Z**-transformations on proper and symmetric cones, Mathematical Programming, Series B, 117 (2009) 195–221.
13. L Gurvits, R. Shorten, and O. Mason, On the stability of switched positive linear systems, IEEE Transactions on Automatic Control, 52 (2007) 1099–1103.
14. V.A. Kamenetskiy and Ye.S. Pyatnitskiy, An iterative method of Lyapunov function construction for differential inclusions, Systems and Control Letters, 8 (1987) 445–451.
15. C. King and M. Nathanson, On the existence of a common quadratic Lyapunov function for a rank one difference, Linear Algebra and its Applications, 419 (2006) 400–416.

16. D. Liberzon, Switching in systems and control, Birkhäuser, Boston, 2003.
17. D. Liberzon, J.P. Hespanha, and A.S. Morse, Stability of switched systems: a Lie-algebraic condition, Systems and Control Letters, 37 (1999) 117–122.
18. M. Mason and R. Shorten, A conjecture on the existence of common quadratic Lyapunov functions for positive linear systems, Proceedings of American Control Conference, 2003.
19. M. Mason and R. Shorten, On common quadratic Lyapunov functions for stable discrete-time LTI systems, IMA Journal of Applied Mathematics, 69 (2004) 271–283.
20. M. Mason and R. Shorten, On linear copositive Lyapunov functions and the stability of positive switched linear systems, IEEE Transactions on Automatic Control, 52 (2007) 1346-1349.
21. Y. Mori, T. Mori, and Y. Kuroe, A solution to the common Lyapunov function problem for continuous-time systems, Decision and Control, 4 (1997) 3530–3531.
22. K.S. Narendra and J. Balakrishnan, A common Lyapunov function for stable LTI systems with commuting A-matrices, Automatic Control, IEEE Transactions, 39 (1994) 2469–2471.
23. H. Schneider and M. Vidyasagar, Cross-positive matrices, SIAM Journal on Numerical Analysis, 7 (1970) 508–519.
24. R. Shorten, O. Mason, and C. King, An alternative proof of the Barker, Berman, Plemmons (BBP) result on diagonal stability and extensions, Linear Algebra and its Applications, 430 (2009) 34–40.
25. R.N. Shorten and K.S. Narendra, On the stability and existence of common Lyapunov functions for stable linear switching systems, Decision and Control, 4 (1998) 3723–3724.
26. R.N. Shorten and K.S. Narendra, Necessary and sufficient conditions for the existence of a common quadratic Lyapunov function for M stable second order linear time-invariant systems, American Control Conference, Chicago, 2000.
27. R. Shorten, F. Wirth, O. Mason, K. Wulff, and C. King, Stability criteria for switched and hybrid systems, SIAM Review, 49 (2007) 545–592.
28. Y. Song, M.S. Gowda, and G. Ravindran, On some properties of P-matrix sets, Linear Algebra and its Applications, 290 (1999) 237–246.
29. R. Sznajder and M.S. Gowda, Generalizations of P_0- and P-properties; extended vertical and horizontal linear complementarity problems, Linear Algebra and its Applications, 223/224 (1995) 695–715.

Chapter 26
Nonlinear Problems in Mathematical Programming and Optimal Control

Dumitru Motreanu

Dedicated to the memory of Professor George Isac

Abstract Necessary conditions of optimality are obtained for general mathematical programming problems on a product space. The cost functional is locally Lipschitz and the constraints are expressed as inclusion relations with unbounded linear operators and multivalued term. The abstract result is applied to an optimal control problem governed by an elliptic differential inclusion.

26.1 Introduction

In this work, we study the nonlinear mathematical programming problem

$$(P) \qquad \begin{array}{l} \text{Minimize (locally) } \Phi(y,u) \\ \text{subject to } Ay + Cu \in B(y,u). \end{array}$$

The data entering (P) have the following meaning: given real Banach spaces X, Y, E, $\Phi : X \times E \to \mathbb{R}$ is a locally Lipschitz function, $A : D(A) \subset X \to Y$ and $C : D(C) \subset E \to Y$ are (possibly unbounded) closed linear operators with dense domains $D(A)$, $D(C)$ in X, E, respectively, and $B : X \times E \to 2^Y$ is a multivalued mapping.

The nonconvex programming problem (P) encompasses important models as for instance general optimal control problems. The study of problem (P) was started in [8] with the case where $X = Y = E$ is a Hilbert space, A is a closed range self-adjoint linear operator, C is a bounded linear operator, B is a constant mapping $B(y,u) = \{f\}$ and Φ is a function of the form $\Phi(y,u) = G(y) + H(u)$ with G and H everywhere

Dumitru Motreanu
University of Perpignan, Department of Mathematics, 66860 Perpignan, France, e-mail: motreanu@univ-perp.fr

P.M. Pardalos et al. (eds.), *Nonlinear Analysis and Variational Problems*, Springer Optimization and Its Applications 35, DOI 10.1007/978-1-4419-0158-3_26,

defined, convex, continuous functions (so, locally Lipschitz). In [1], [2], and [3], the data X, Y, E, A, C are like in (P), but with B a (single-valued) Gâteaux differentiable mapping. Some particular situations in (P) with a multivalued mapping B are treated in [13].

Here we extend the approach of [1]–[3] to the case of a set-valued mapping B in problem (P). Considering the set of constraints

$$M := \{(y,u) \in D(A) \times D(C) : Ay + Cu \in B(y,u)\} \tag{26.1}$$

supposed to be nonempty; the key idea is to associate to every element $(y,u) \in M$ a pair of bounded linear operators $A_{y,u} \in L(X,Y)$ and $C_{y,u} \in L(E,Y)$ acting in a compatible way with the tangent cone of M at the point (y,u). The pattern for such operators $A_{y,u}$ and $C_{y,u}$ is supplied by the partial differentials of the mapping B whenever they exist at least in a generalized sense. We derive in Theorem 26.1 below necessary conditions of optimality for problem (P) in terms of these operators. Specifically, we obtain a system of relations that involve an optimal pair $(\overline{y},\overline{u}) \in M$ and a co-state variable $\overline{p} \in Y^*$ as follows

$$\begin{cases} A\overline{y} + C\overline{u} \in B(\overline{y},\overline{u}) \\ ((A - A_{\overline{y},\overline{u}})^*\overline{p}, (C - C_{\overline{y},\overline{u}})^*\overline{p}) \in \partial\Phi(\overline{y},\overline{u}). \end{cases} \tag{26.2}$$

The notation $\partial\Phi$ stands for the generalized gradient of $\Phi(y,u)$ (in the sense of Clarke [7], p. 39), and the superscript $*$ denotes the adjoint of a densely defined linear operator. If Φ satisfies Clarke's regularity condition, the inclusion $\partial\Phi(y,u) \subset \partial_y\Phi(y,u) \times \partial_u\Phi(y,u)$ holds for all $(y,u) \in X \times E$ (cf. [7], p. 48), where $\partial_y\Phi$ and $\partial_u\Phi$ represent the partial generalized gradients of Φ. Accordingly, (26.2) becomes a system of three relations with three unknowns (the optimal pair $(\overline{y},\overline{u})$ and the co-state $\overline{p}$):

$$\begin{cases} A\overline{y} + C\overline{u} \in B(\overline{y},\overline{u}) \\ (A - A_{\overline{y},\overline{u}})^*\overline{p} \in \partial_y\Phi(\overline{y},\overline{u}) \\ (C - C_{\overline{y},\overline{u}})^*\overline{p} \in \partial_u\Phi(\overline{y},\overline{u}). \end{cases} \tag{26.3}$$

A possible way to handle efficiently the system (26.3) is to eliminate, when possible, the co-state $\overline{p}$. An application to an optimal control problem illustrating this procedure is presented in Section 26.4. In order to make effective (26.2) and (26.3), the question of existence of optimal pairs $(\overline{y},\overline{u})$ for problem (P) is also addressed.

The rest of the paper is organized as follows. Section 26.2 contains our main abstract result. Its proof is given in Section 26.3. An application to an optimal control problem is discussed in Section 26.4.

26.2 Main Result

Given $(y,u) \in M$, the tangent cone $T_{(y,u)}M$ of the set M in (26.1) at the point $(y,u) \in M$ is defined by

$$T_{(y,u)}M = \{(z,w) \in X \times E : \exists\, t_n \to 0^+ \text{ in } \mathbb{R},\ p_n \to 0 \text{ in } X \text{ and } q_n \to 0 \text{ in } E \text{ as } n \to \infty \text{ such that } (y + t_n(z+p_n), u + t_n(w+q_n)) \in M \text{ for all } n\}. \tag{26.4}$$

Relation (26.4) describes the contingent cone (see, e.g., [10]). Noticing that

$$(z,w) \in T_{(y,u)}M \iff \liminf_{t \to 0^+} \frac{1}{t} d((y,u) + t(z,w), M) = 0,$$

where the notation $d(\cdot, M)$ stands for the distance to the set M in $X \times E$, it slightly extends the definition of tangent cone as used in [12].

Assume that to any $(y,u) \in M$ there are associated bounded linear operators $A_{y,u} \in L(X,Y)$ and $C_{y,u} \in L(E,Y)$ such that the following hypotheses hold:

(H_1) If $(y,u) \in M$, $z \in D(A)$ and $w \in D(C)$ satisfy

$$Az + Cw = A_{y,u}z + C_{y,u}w, \tag{26.5}$$

then $(z,w) \in T_{(y,u)}M$.

(H_2) Either the range $R(C - C_{y,u})$ of the linear operator $C - C_{y,u}$ is closed in Y and

$$R(A - A_{y,u}) \subseteq R(C - C_{y,u}), \tag{26.6}$$

or the range $R(A - A_{y,u})$ of the linear operator $A - A_{y,u}$ is closed in Y and

$$R(C - C_{y,u}) \subseteq R(A - A_{y,u}). \tag{26.7}$$

We now state our main abstract result.

Theorem 26.1. *Let $(\overline{y}, \overline{u})$ be a (locally) optimal solution of problem (P). Then, under assumptions (H_1) and (H_2), there exists*

$$\overline{p} \in D(A^*) \cap D(C^*) \tag{26.8}$$

such that

$$((A - A_{\overline{y},\overline{u}})^* \overline{p}, (C - C_{\overline{y},\overline{u}})^* \overline{p}) \in \partial \Phi(\overline{y}, \overline{u}). \tag{26.9}$$

If Φ is regular at $(\overline{y}, \overline{u})$ (in the sense of Clarke [7]), then (26.9) implies

$$((A - A_{\overline{y},\overline{u}})^* \overline{p}, (C - C_{\overline{y},\overline{u}})^* \overline{p}) \in \partial_y \Phi(y,u) \times \partial_u \Phi(y,u). \tag{26.10}$$

Remark 26.2. The conclusion of Theorem 26.1 remains valid if assumptions (H_1) and (H_2) are satisfied only for $(y,u) = (\overline{y}, \overline{u})$, as seen from the proof of Theorem 26.1.

The proof of Theorem 26.1 is given in Section 26.3.

For making consistent the applicability of Theorem 26.1, we indicate a verifiable sufficient condition ensuring the existence of optimal solutions $(\overline{y}, \overline{u})$.

Proposition 26.3. *Assume the Banach spaces X, E are reflexive and*

(i) the set M in (1) is bounded or Φ is coercive on M, i.e., $\Phi(y,u) \to +\infty$ as $\|(y,u)\| \to \infty$ with $(y,u) \in M$;
(ii) the set M is sequentially weakly closed in $X \times E$, i.e., if $(y_n,u_n) \to (y,u)$ weakly in $X \times E$ with $(y_n,u_n) \in M$ then $(y,u) \in M$;
(iii) Φ is sequentially weakly l.s.c. on M, i.e., if $(y_n,u_n) \to (y,u)$ weakly in $X \times E$ and $(y_n,u_n) \in M$, one has $\Phi(y,u) \le \liminf_{n\to\infty} \Phi(y_n,u_n)$.

Then problem (P) admits an optimal solution.

Proof. Let $\{(y_n,u_n)\} \subset M$ be a minimizing sequence for (P). Then condition (i) ensures that $\{(y_n,u_n)\}$ is bounded in $X \times E$. The reflexivity of X and E guarantees that along a relabeled subsequence we have $(y_n,u_n) \to (\overline{y},\overline{u})$ weakly in $X \times E$, with some $(\overline{y},\overline{u}) \in X \times E$. By (ii) it is known that $(\overline{y},\overline{u}) \in M$. According to (iii), one gets $\Phi(\overline{y},\overline{u}) \le \liminf\limits_{n\to\infty} \Phi(y_n,u_n)$, from which we conclude that $(\overline{y},\overline{u})$ is an optimal solution. □

Remark 26.4. Necessary optimality conditions for a general constrained minimization problem which is based on an idea close to the method developed here can be found in [11]. In the current paper, working on a product space, we derive necessary conditions of optimality that take into account in a distinguished way each one of the involved variables. This permits us to study problems where the variables play different parts, for instance to describe the state and the control of a system with optimum criteria. Section 26.4 presents an application in this direction. The approach can be extended to establish necessary optimality conditions for set-valued optimization problems on a product space (see [5], [10]) by means of contingent epiderivative of the vector-valued objective function which offers a kind of substitute for the generalized gradient in Theorem 26.1. We refer to [9] for various optimization models and applications.

26.3 Proof of Theorem 26.1

Let $(z,w) \in D(A) \times D(C)$ satisfy (26.5) with $(y,u) = (\overline{y},\overline{u})$. Assumption (H_1) ensures that $(z,w) \in T_{(\overline{y},\overline{u})}M$. By (26.4), we know there exist sequences $\{t_n\} \subset \mathbb{R}^+$ and $\{(p_n,q_n)\} \subset X \times E$ such that

$$t_n \to 0^+ \text{ in } \mathbb{R},\ p_n \to 0 \text{ strongly in } X,\ q_n \to 0 \text{ strongly in } E \quad \text{as } n \to \infty$$

and

$$(\overline{y}+t_n(z+p_n), \overline{u}+t_n(w+q_n)) \in M \quad \forall n \ge 1.$$

The optimality of $(\overline{y},\overline{u})$ implies

$$\Phi(\overline{y}+t_n(z+p_n), \overline{u}+t_n(w+q_n)) \ge \Phi(\overline{y},\overline{u}) \quad \forall n \ge 1.$$

Letting $n \to \infty$ yields

$$\begin{aligned}
0 &\le \limsup_{n\to\infty} \frac{1}{t_n}[\Phi(\bar{y}+t_n(z+p_n),\bar{u}+t_n(w+q_n)) - \Phi(\bar{y},\bar{u})] \\
&\le \limsup_{n\to\infty} \frac{1}{t_n}[\Phi(\bar{y}+t_n(z+p_n),\bar{u}+t_n(w+q_n)) - \Phi(\bar{y}+t_n p_n,\bar{u}+t_n q_n)] \\
&+ \limsup_{n\to\infty} \frac{1}{t_n}[\Phi(\bar{y}+t_n p_n,\bar{u}+t_n q_n) - \Phi(\bar{y},\bar{u})].
\end{aligned}$$

Denoting by L the Lipschitz constant of Φ near $(\bar{y},\bar{u})$ and using the definition of generalized directional derivative Φ^0 (see [7], p. 25), one obtains

$$0 \le \Phi^0(\bar{y},\bar{u};z,w) + L \lim_{n\to\infty} \|(p_n,q_n)\| = \Phi^0(\bar{y},\bar{u};z,w).$$

Since this holds for all $(z,w) \in D(A) \times D(C)$ satisfying (26.5) with $(y,u) = (\bar{y},\bar{u})$, applying Lemma 1 in [2] there exists $(\xi,\eta) \in \partial\Phi(\bar{y},\bar{u}) \subset X^* \times E^*$ such that

$$\langle \xi, z\rangle_{X^*,X} + \langle \eta, w\rangle_{E^*,E} = 0 \tag{26.11}$$

whenever $(z,w) \in D(A) \times D(C)$ verifies (26.5) with $(y,u) = (\bar{y},\bar{u})$.

Let us make a choice in (H_2) supposing that the range $R(C - C_{\bar{y},\bar{u}})$ is closed in Y and inclusion (26.6) is fulfilled. The case of (26.7) can be handled in a similar manner. Setting $z = 0$ in (26.11) gives $\langle \eta, w\rangle_{E^*,E} = 0$ for all $w \in D(C)$ with $Cw = C_{\bar{y},\bar{u}}w$ because, for such an element w, the pair $(0,w)$ verifies (26.5). This is equivalent to

$$\eta \in (N(C - C_{\bar{y},\bar{u}}))^{\perp}, \tag{26.12}$$

where the notation $N(T)$ stands for the null-space of the linear operator T. Since, by hypothesis, $R(C - C_{\bar{y},\bar{u}})$ is closed in Y, we have the orthogonality relation

$$(N(C - C_{\bar{y},\bar{u}}))^{\perp} = R((C - C_{\bar{y},\bar{u}})^*)$$

(see, e.g., [6, Theorem II.18]). Then we deduce from (26.12) that

$$\eta \in R((C - C_{\bar{y},\bar{u}})^*).$$

So there exists an element $\bar{p} \in D(C^*) \subset Y^*$ such that $\eta = (C - C_{\bar{y},\bar{u}})^*\bar{p}$. Then (26.11) leads to

$$\langle \xi, z\rangle_{X^*,X} + \langle \bar{p}, Cw - C_{\bar{y},\bar{u}}w\rangle_{Y^*,Y} = 0 \tag{26.13}$$

for all $(z,w) \in D(A) \times D(C)$ verifying (26.5) with $(y,u) = (\bar{y},\bar{u})$.

In view of (26.6), for every $z \in D(A)$ there exists $w \in D(C)$ satisfying

$$(A - A_{\bar{y},\bar{u}})z = (C - C_{\bar{y},\bar{u}})(-w).$$

Then (26.13) allows to write

$$\langle \xi, z \rangle_{X^*,X} = \langle \overline{p}, Az - A_{\overline{y},\overline{u}} z \rangle_{Y^*,Y}$$

for all $z \in D(A)$. Using the density of $D(A)$ in X, we derive that $\overline{p} \in D(A^*) \subset X^*$ and $\xi = (A - A_{\overline{y},\overline{u}})^* \overline{p}$. Consequently, the assertions expressed in (26.8) and (26.9) are proved. Finally, assuming the regularity for Φ at $(\overline{y},\overline{u})$, property (26.10) follows from (26.9). The proof is thus complete.

26.4 An Application

Consider a bounded domain Ω in $\mathbb{R}^N$ with a C^1 boundary $\partial\Omega$. For a later use we denote by $|\Omega|$ the Lebesgue measure of Ω. Let $f_i : \Omega \times \mathbb{R} \to \mathbb{R}$, $i = 1,...,m$, be Carathéodory functions (i.e., $f_i(\cdot,t) : \Omega \to \mathbb{R}$ is measurable for all $t \in \mathbb{R}$ and $f_i(x,\cdot) : \mathbb{R} \to \mathbb{R}$ is continuous for a.a. $x \in \Omega$). Assume that, for every $i = 1,...,m$, the partial derivative $\frac{\partial f_i}{\partial t}(x,t)$ of f_i with respect to the second variable $t \in \mathbb{R}$ exists, is a Carathéodory function and verifies

$$|\frac{\partial f_i}{\partial t}(x,t)| \le c_0 \quad \forall (x,t) \in \Omega \times \mathbb{R}, \tag{26.14}$$

with a constant $c_0 > 0$. In particular, we have that for a.a. $x \in \Omega$ and for each $i = 1,...,m$, the function $f(x,\cdot)$ is Lipschitz continuous. Admit further that $f_i(\cdot,0) \in L^1(\Omega)$ for all $i = 1,...,m$.

Let $g : \Omega \times \mathbb{R} \to \mathbb{R}$ be a function satisfying the requirements: $g(\cdot,t)$ is measurable on Ω for all $t \in \mathbb{R}$, $g(\cdot,0) \in L^\infty(\Omega)$, $g(x,\cdot)$ is locally Lipschitz on $\mathbb{R}$ for a.a. $x \in \Omega$ and its generalized gradient $\partial g(x,\cdot)$ verifies the growth condition

$$|\zeta| \le c_1(1+|t|) \quad \forall \zeta \in \partial g(x,t) \tag{26.15}$$

for almost all $x \in \Omega$ and for all $t \in \mathbb{R}$, where c_1 is a positive constant.

Given the numbers $a > 0$ and $b > 0$, we state the optimal control problem:

$$(\tilde{P}) \text{ Minimize (locally) } \left[\int_\Omega g(x,y(x))dx + \tfrac{b}{2}\|y\|^2_{L^2(\Omega)} + \tfrac{1}{2}\|u\|^2_{L^2(\Omega)} \right]$$

$$\text{subject to} \begin{cases} (y,u) \in (H^1_0(\Omega) \cap H^2(\Omega)) \times L^2(\Omega), \\ \text{there exist numbers } \varepsilon_i \in [0,a],\ i = 1,...,m, \text{ such that} \\ -\Delta y(x) + u(x) = \sum_{i=1}^m \varepsilon_i f_i(x,y(x)) \quad \text{for a.e. } x \in \Omega. \end{cases}$$

Here the Laplacian Δ is regarded as an unbounded linear operator on the space $L^2(\Omega)$ with domain $H^1_0(\Omega) \cap H^2(\Omega)$.

We point out that problem $(\tilde{P})$ is a particular case of the statement in (P) for the following choices:

$$X = Y = E = L^2(\Omega),$$

$$A = -\Delta \quad \text{with} \quad D(A) = H^1_0(\Omega) \cap H^2(\Omega), \quad C = id_{L^2(\Omega)},$$

$$B(y,u) := \{v \in L^2(\Omega) : \text{ there are numbers } \varepsilon_1, ..., \varepsilon_m \in [0,a] \text{ such that } v = \sum_{i=1}^{m} \varepsilon_i f_i(x,y) \text{ a.e. } x \in \Omega\},$$

and

$$\Phi(y,u) = \int_\Omega g(x,y(x))dx + \frac{b}{2}\|y\|^2_{L^2(\Omega)} + \frac{1}{2}\|u\|^2_{L^2(\Omega)}$$

for all $(y,u) \in L^2(\Omega) \times L^2(\Omega)$.

Clearly, the mappings A and C as defined above are closed linear operators with dense domains, and from condition (26.15) it follows that the functional $\Phi : L^2(\Omega) \times L^2(\Omega) \to \mathbb{R}$ is Lipschitz continuous on bounded subsets. Lebourg's mean value theorem (see, e.g., [7], p. 41) ensures that for a.a. $x \in \Omega$ and any $t \in \mathbb{R}$ there is $\zeta \in \partial g(x,t)$ such that $g(x,t) = g(x,0) + \zeta t$. This, in conjunction with (26.15) and the fact that $g(\cdot,0) \in L^\infty(\Omega)$, yields that there is a constant $\tilde{c}_1 > 0$ such that

$$|g(x,t)| \le \tilde{c}_1(1+t^2) \text{ on } \Omega \times \mathbb{R}. \tag{26.16}$$

We first discuss the existence of optimal pairs for problem $(\tilde{P})$.

Proposition 26.5. *If b is sufficiently large, namely $b > 2\tilde{c}_1$ with the constant $\tilde{c}_1$ in (26.16), there exists an optimal solution $(\overline{y}, \overline{u})$ for problem $(\tilde{P})$.*

Proof. The functional Φ is coercive on $L^2(\Omega) \times L^2(\Omega)$ as seen from (26.16) and the estimate

$$\Phi(y,u) \ge (\frac{b}{2} - \tilde{c}_1)\|y\|^2_{L^2(\Omega)} + \frac{1}{2}\|u\|^2_{L^2(\Omega)} - \tilde{c}_1|\Omega| \quad \forall (y,u) \in L^2(\Omega) \times L^2(\Omega).$$

So assumption (i) of Proposition 26.3 holds true.

Let a sequence $\{(y_n,u_n)\} \subset (H_0^1(\Omega) \cap H^2(\Omega)) \times L^2(\Omega)$ satisfy $y_n \to y$ weakly in $L^2(\Omega)$, $u_n \to u$ weakly in $L^2(\Omega)$ and

$$-\Delta y_n(x) + u_n(x) = \sum_{i=1}^{m} \varepsilon_i^n f_i(x, y_n(x)) \quad \text{a.e. } x \in \Omega, \tag{26.17}$$

with numbers $\varepsilon_i^n \in [0,a]$, $i = 1, ..., m$. We deduce from (26.14) that

$$|f_i(x,t)| \le |f_i(x,0)| + c_0|t| \quad \text{for a.e. } x \in \Omega, \ \forall t \in \mathbb{R}.$$

Then, for each $i = 1, ..., m$, the sequence $f_i(\cdot, y_n)$ is bounded in $L^2(\Omega)$. The compactness of $(-\Delta)^{-1} : L^2(\Omega) \to L^2(\Omega)$ and equality (26.17) ensure, up to a subsequence, that $y_n \to y$ strongly in $L^2(\Omega)$. Due to (26.14) we have $f_i(\cdot, y_n) \to f_i(\cdot, y)$ strongly in $L^2(\Omega)$ (see, e.g., [4], p. 16) for $i = 1, ..., m$. We may suppose that $\varepsilon_i^n \to \varepsilon_i$ as $n \to \infty$, with some $\varepsilon_i \in [0,a]$ for $i = 1, ..., m$. We get

$$-u_n + \sum_{i=1}^{m} \varepsilon_i^n f_i(\cdot, y_n) \to -u + \sum_{i=1}^{m} \varepsilon_i f_i(x,y) \quad \text{weakly in } L^2(\Omega).$$

Since $-\Delta$ is a closed linear operator, from (26.17) we infer that condition (*ii*) of Proposition 26.3 is satisfied.

It remains to justify (*iii*) of Proposition 26.3. Let a sequence $\{(y_n, u_n)\}$ satisfy $y_n \to y$ weakly in $L^2(\Omega)$, $u_n \to u$ weakly in $L^2(\Omega)$ and equality (26.17). We have already shown that along a relabeled subsequence one has $y_n \to y$ strongly in $L^2(\Omega)$, with $y \in H_0^1(\Omega) \cap H^2(\Omega)$, and that there exist numbers $\varepsilon_i \in [0,a]$, $i = 1, ..., m$, such that

$$-\Delta y(x) + u(x) = \sum_{i=1}^{m} \varepsilon_i f_i(x, y(x)) \quad \text{for a.e. } x \in \Omega.$$

This means that (y,u) is an admissible pair for problem $(\tilde{P})$, which means that $(y,u) \in M$ with M introduced in (26.1). The growth relation obtained in (26.16) and the convergence $y_n \to y$ strongly in $L^2(\Omega)$ imply that $g(\cdot, y_n) \to g(\cdot, y)$ strongly in $L^1(\Omega)$. We are led to $\Phi(y,u) \le \liminf_{n\to\infty} \Phi(y_n, u_n)$, which represents just assumption (*iii*) of Proposition 26.3. The application of Proposition 26.3 completes the proof.

We now present a necessary condition of optimality for problem $(\tilde{P})$.

Theorem 26.6. *If* $(\overline{y}, \overline{u}) \in L^2(\Omega) \times L^2(\Omega)$ *is an optimal solution for problem* $(\tilde{P})$, *then* $\overline{y}, \overline{u} \in H_0^1(\Omega) \cap H^2(\Omega)$ *and there are numbers* $\overline{\varepsilon}_1, ..., \overline{\varepsilon}_m \in [0,a]$ *such that*

$$-\Delta \overline{y} + \overline{u} = \sum_{i=1}^{m} \overline{\varepsilon}_i f_i(x, \overline{y}(x)) \quad \textit{for a.e. } x \in \Omega, \tag{26.18}$$

and

$$-\Delta \overline{u} - \sum_{i=1}^{m} \overline{\varepsilon}_i \frac{\partial}{\partial t} f_i(\cdot, \overline{y}) \overline{u} \in \partial g(\cdot, \overline{y}) + b\overline{y} \quad \textit{for a.e. } x \in \Omega. \tag{26.19}$$

Proof. Since $(\overline{y}, \overline{u}) \in L^2(\Omega) \times L^2(\Omega)$ is an admissible pair, there exist numbers $\overline{\varepsilon}_1, ..., \overline{\varepsilon}_m \in [0,a]$ such that (26.18) holds. Under hypothesis (26.14), the mapping from $L^2(\Omega)$ to $L^2(\Omega)$ given by $y \mapsto f_i(\cdot, y)$ $(i = 1, ..., m)$ is Gâteaux differentiable and its differential at y is the map $v \in L^2(\Omega) \mapsto \frac{\partial}{\partial t} f_i(\cdot, y) v \in L^2(\Omega)$ (see, e.g., Ambrosetti and Prodi [4], p. 19). Given $(y,u) \in (H_0^1(\Omega) \cap H^2(\Omega)) \times L^2(\Omega)$ such that there exist numbers $\varepsilon_i \in [0,a]$, $i = 1, ..., m$, with

$$-\Delta y(x) + u(x) = \sum_{i=1}^{m} \varepsilon_i f_i(x, y(x)) \quad \text{for a.e. } x \in \Omega, \tag{26.20}$$

we define the linear bounded operators $A_{y,u}, C_{y,u} \in L(L^2(\Omega), L^2(\Omega))$ by

$$A_{y,u} z = \sum_{i=1}^{m} \varepsilon_i \frac{\partial}{\partial t} f_i(\cdot, y) z \quad \forall z \in L^2(\Omega) \tag{26.21}$$

(so, in (26.21), $A_{y,u}$ is the Gâteaux differential of the mapping $y \in L^2(\Omega) \mapsto \sum_{i=1}^{m} \varepsilon_i f_i(\cdot, y) \in L^2(\Omega)$ at $y \in L^2(\Omega)$) and $C_{y,u} = 0$.

Let $(z,w) \in (H_0^1(\Omega) \cap H^2(\Omega)) \times L^2(\Omega)$ fulfill

$$-\Delta z + w = \sum_{i=1}^{m} \varepsilon_i \frac{\partial}{\partial t} f_i(\cdot, y) z \quad \text{for a.e. } x \in \Omega.$$

For every $s > 0$ and a.e. $x \in \Omega$ we have

$$-\Delta(y+sz)(x) + (u(x) + sw(x)) = \sum_{i=1}^{m} \varepsilon_i \left[f_i(x, y(x)) + s \frac{\partial}{\partial t} f_i(x, y(x)) z(x) \right].$$

This can be equivalently expressed as follows:

$$-\Delta(y+sz) + (u + s(w + q(s))) = \sum_{i=1}^{m} \varepsilon_i f_i(\cdot, y+sz),$$

where

$$q(s) = \sum_{i=1}^{m} \varepsilon_i \left[\frac{1}{s}(f_i(\cdot, y+sz) - f_i(\cdot, y)) - \frac{\partial}{\partial t} f_i(\cdot, y) z \right].$$

According to the Gâteaux differentiability of the Nemytskij operator $y \in L^2(\Omega) \mapsto f_i(\cdot, y) \in L^2(\Omega)$ for $i = 1, \ldots, m$, we have that $q(s) \to 0$ in $L^2(\Omega)$ as $s \to 0^+$. It turns out that assumption (H_1) holds with an arbitrary sequence $t_n \to 0^+$ and $p_n = 0$, $q_n = q(t_n)$.

Notice that assumption (H_2) is automatically verified since $R(C - C_{y,u}) = L^2(\Omega)$ for all $(y,u) \in L^2(\Omega) \times L^2(\Omega)$ satisfying (26.20). We are thus in a position to apply Theorem 26.1. Then there exist $\overline{p} \in D(A^*) = H_0^1(\Omega) \cap H^2(\Omega)$ (see (26.8) and because the linear operator A is self-adjoint) such that relation (26.9) is valid. Writing down (26.9) readily shows that $\overline{p} = \overline{u}$, thereby we obtain $\overline{u} \in H_0^1(\Omega) \cap H^2(\Omega)$. On the basis of (26.9) and (26.21) we find that

$$-\Delta \overline{u} - \sum_{i=1}^{m} \varepsilon_i \frac{\partial}{\partial t} f_i(\cdot, \overline{y}) \overline{u} \in \partial \left(\int_\Omega g(x, \cdot) dx \right)(\overline{y}) + b\overline{y}.$$

Now it suffices to apply the Aubin–Clarke theorem (see [7], p. 83) to conclude that (26.19) is true. This completes the proof. □

Remark 26.7. Denote by λ_1 the first eigenvalue of $-\Delta$ on $H_0^1(\Omega)$ and assume in addition that the constant c_0 in (26.14) satisfies

$$c_0 < \frac{\lambda_1}{ma}.$$

Then the Lax–Milgram theorem applied on $H_0^1(\Omega)$ shows that, for every $y \in L^2(\Omega)$, the linear operator on $L^2(\Omega)$ given by

$$w \mapsto -\Delta w - \sum_{i=1}^{m} \overline{\varepsilon}_i \frac{\partial}{\partial t} f_i(\cdot, y) w,$$

with domain $H_0^1(\Omega)\cap H^2(\Omega)$ has a continuous inverse on $L^2(\Omega)$. This enables us to solve the inclusion in (26.19) obtaining explicitly the optimal control

$$\overline{u}\in(-\Delta-\sum_{i=1}^{m}\overline{\varepsilon}_i\frac{\partial}{\partial t}f_i(\cdot,\overline{y}))^{-1}(\partial g(\cdot,\overline{y})+b\overline{y})\ \text{ a.e. in } \Omega.$$

References

1. S. Aizicovici, D. Motreanu and N. H. Pavel, Nonlinear programming problems associated with closed range operators, *Appl. Math. Optim.* **40** (1999), 211–228.
2. S. Aizicovici, D. Motreanu and N. H. Pavel, Fully nonlinear programming problems with closed range operators, in *Differential Equations and Control Theory* (S. Aizicovici and N. H. Pavel, eds.), *Lecture Notes Pure Appl. Math.*, Vol. 225, M. Dekker, New York, 2001, pp. 19–30.
3. S. Aizicovici, D. Motreanu and N. H. Pavel, Nonlinear mathematical programming and optimal control, *Dynamics of Continuous, Discrete and Impulsive Systems* **11** (2004), 503–524.
4. A. Ambrosetti and G. Prodi, *A Primer of Nonlinear Analysis*, Cambridge University Press, Cambridge, 1995.
5. J. Baier and J. Jahn, On subgradients of set-valued maps, *J. Optim. Theory Appl.* **100** (1999), 233–240.
6. H. Brézis, *Analyse fonctionnelle. Théorie et applications*, Masson, Parris, 1992.
7. F. H. Clarke, *Optimization and Nonsmooth Analysis*, John Wiley and Sons, New York, 1983.
8. S. C. Gao and N. H. Pavel, Optimal control of a functional equation associated with closed range self-adjoint operators, Proc. Amer. Math. Soc. **126** (1998), 2979–2986.
9. G. Isac, V. A. Bulavski and V. V. Kalashnikov, *Complementarity, Equilibrium, Efficiency and Economics*, Kluwer Academic Publishers, Dordrecht, 2002.
10. G. Isac and A. A. Khan, Dubovitskii-Milyutin approach in set-valued optimization, *SIAM J. Control Optim.* **126** (2008), 144–162.
11. V. K. Le and D. Motreanu, Some properties of general minimization problems with constraints, *Set-Valued Anal.* **14** (2006), 413–424.
12. D. Motreanu and N. H. Pavel, *Tangency, Flow-Invariance for Differential Equations and Optimization Problems*, Marcel Dekker, New York, 1999.
13. M. D. Voisei, First-order necessary optimality conditions for nonlinear optimal control problems, *PanAmer. Math. J.* **14** (2004), 1–44.

Chapter 27
On Variational Inequalities Involving Mappings of Type (S)

Dan Pascali

Dedicated to the memory of Professor George Isac

Abstract Variational inequalities can be converted into inclusions defined by a sum between a mapping of monotone type and a subdifferential. In our case, a topological approach of variational inequalities is based on a degree function for a *(S)*-operator F with maximal monotone perturbations T. The paper surveys some new advances on topological degree in the case $F+T$, removing the condition $0 \in T(0)$. In this way, the main difficulty is to determine the admissible homotopies. A graph homotopy for maximal monotone mappings is introduced. Finally, we mention some recent references regarding the related fixed point index.

27.1 Main Results

Let X be a real reflexive separable Banach space with dual space X^*. We denote by $\langle \cdot, \cdot \rangle : X^* \times X \mapsto \mathbb{R}$ their duality. Let K be a nonempty closed convex subset of X and $A : X \mapsto X^*$ a monotone-like (possibly, nonlinear and multivalued) mapping. For an element $f \in X^*$, the problem of finding an element $u \in K$ such that

$$\langle Au - f, v - u \rangle \geq 0, \ \forall v \in K, \tag{27.1}$$

is called a *variational inequality* $VI(A, f, K)$. More generally, if $\varphi : X \mapsto \mathbb{R} \cup \{+\infty\}$ is a convex lower semicontinuous function and $D(\varphi) = \{x \in X \mid \varphi(x) < \infty\}$ its effective domain, then finding an element $u \in D(\varphi)$ satisfying

$$\langle Au - f, v - u \rangle + \varphi(u) - \varphi(v) \geq 0, \ \forall v \in D(\varphi) \tag{27.2}$$

Dan Pascali
Courant Institute of Mathematical Sciences, New York University, New York, New York, USA, e-mail: dp39@nyu.edu

P.M. Pardalos et al. (eds.), *Nonlinear Analysis and Variational Problems*,
Springer Optimization and Its Applications 35, DOI 10.1007/978-1-4419-0158-3_27,

determines also a variational inequality $VI(A,f,\varphi)$. We note that (27.2) reduces to (27.1) when φ is replaced by the indicator function of K, i.e.,

$$I_K(x) = \begin{cases} 0 & \text{for } x \in X, \\ +\infty & \text{otherwise.} \end{cases}$$

Moreover, $D(\varphi) = K$ and $K \neq \emptyset \Leftrightarrow \varphi \neq +\infty$, *K closed* $\Leftrightarrow$ *φ l.s.c.* and *K convex* $\Leftrightarrow$ *φ convex*.

Recall that for a proper l.s.c. function $\varphi : X \to \mathbb{R} \cup \{+\infty\}$, *the subdifferential* $\partial\varphi : X \to 2^{X^*}$ was introduced by

$$\partial\varphi(x) = \{h \in X^* \mid \langle h, y-x\rangle \leq \varphi(y) - \varphi(x),\ \forall y \in X\}, \tag{27.3}$$

which is a pattern of the maximal monotone (multivalued) mapping. In particular, if I_K is the indicator function of a convex subset K of X, then

$$N_K(x) = \partial I_K(x) = \{g \in X^* \mid \langle g, y-x\rangle \leq 0,\ \forall y \in K\}$$

is called *normal cone* of K at x. We mention also that $\overline{D(\partial\varphi)} = \overline{D(\varphi)}$ holds.

According to the subgradient inequality (27.3), the variational inequality (27.2) is equivalent with the inclusion

$$f \in Au + \partial\varphi(u) \tag{27.4}$$

and, in particular, the variational inequality (27.1) is equivalent with

$$f \in Au + N_K(u). \tag{27.5}$$

Based on these equivalences, we approach the construction of a topological degree for variational inequalities as a topological degree for *(S)*-perturbations of multivalued maximal monotone mappings.

We denote by "$\to$" and "$\rightharpoonup$" the strong and weak convergence, respectively. We introduce the maps of monotone type only in the case of single-valued operators $A : D(A) \subseteq X \to X^*$, which is enough in what follows. Thus, A is called *monotone* if $\langle Ax - Ay, x - y\rangle \geq 0$ for all $x, y \in D(A)$, while the operator A is of *type (S)*, if each sequence $\{x_n\} \subset D(A)$ with $x_n \rightharpoonup x_0$ for which

$$limsup\langle Ax_n, x_n - x_0\rangle \leq 0 \tag{27.6}$$

is in fact strongly convergent to $x_0 \in D(A)$. In turn, A is called *pseudomonotone* if for any sequence $\{x_n\} \subset D(A)$ with $x_n \rightharpoonup x_0$, such the inequality (27.6) holds, it follows that $\langle Ax_n, x_n\rangle \to \langle Ax_0, x_0\rangle$ and $Ax_n \rightharpoonup Ax_0$ in X^*. Finally, A is *quasimonotone* if $\limsup\langle Ax_n, x_n - x_0 \geq 0$ for any sequence $\{x_n\} \subset D(A)$ with $x_n \to x_0$. When all mappings are demicontinuous and bounded, we have with obvious notations the inclusions $(S) \subset (PM) \subset (QM)$. Moreover, the class *(S)* is stable under quasimonotone perturbations [20], i.e.,

$$(S) + (QM) = (S).$$

The simplest example of a topological degree for variational inequalities is referred to an hemicontinuous operator $A : X \mapsto X^*$, strongly monotone, i.e., there is $\alpha > 0$ so that

$$\langle Ax - Ay, x - y \rangle \geq \alpha \|x - y\|^2, \quad \forall x, y \in X,$$

and a proper, convex, l.s.c. function $\varphi : X \mapsto \mathbb{R} \cup \{+\infty\}$, [9]. Clearly, A so defined is of *type (S)*. Then the mapping $\Pi_{A,\varphi} : X^* \mapsto D(\varphi)$ which associated to $f \in X^*$ the unique solutions of the variational inequality (27.2) is well-defined and satisfies

$$\left\| \Pi_{A,\varphi}(f) - \Pi_{A,\varphi}(f') \right\| \leq \alpha^{-1} \|f - f'\|_*, \quad \forall f, f' \in X^*.$$

Moreover, let $F : X \mapsto X^*$ be a compact operator [22]. Then the finding $u \in D(\varphi)$ satisfying the variational inequality

$$\langle Au - F(u), v - u \rangle + \varphi(v) - \varphi(u) \geq 0, \quad \forall v \in D(\varphi) \tag{27.7}$$

is equivalent to the fixed point problem of finding $u \in D(\varphi)$ such that

$$u = \Pi_{A,\varphi}(F(u)).$$

Provided that the variational inequality (27.7) does not admit solutions such that $\|u\| = r$, for some $r > 0$, the integer

$$deg\left(I_X - \Pi_{A,\varphi}(F(\cdot)), B(0,r), 0\right)$$

is well-defined. Indeed, the mapping $\Pi_{A,\varphi}(F(\cdot))$ is compact as the composite of a compact operator $F(\cdot)$ and a continuous one $\Pi_{A,\varphi}(\cdot)$. Therefore, the topological degree is assigned in the Leray–Schauder sense.

Without loss of generality, due to the Lindenstrauss–Asplund–Troyanskii results (see [21], pp. 4), we assume in the sequel that both X and X^* are locally uniform convex. Then, the (normalized) duality map $J : X \mapsto X^*$ given by

$$Jx = \left\{ f \in X^* \mid \langle f, x \rangle = \|x\|^2 = \|f\|^2 \right\}$$

is single-valued, bijective, bicontinuous, maximal and strictly monotone operator of type *(S)*.

Let $\mathscr{O}$ be the class of bounded open subsets of X. For a given $\Omega \in \mathscr{O}$, we define the *(S)-admissible class*

$$\mathscr{A}_\Omega(S) = \left\{ A : \overline{\Omega} \to X^* \mid A \in (S), \text{ bounded and demicontinuous} \right\}$$

and the *(S)-admissible homotopies*

$$\mathscr{H}_\Omega(S) = \left\{H_t : \overline{\Omega} \to X^*, t \in [0,1] \mid H_t \text{ bounded homotopy of type } (S)\right\},$$

where H_t is said to be a *bounded homotopy of type (S)* if for any sequences

$$\begin{cases} \{x_n\} \subset \Omega,\ x_n \rightharpoonup x \text{ in } X, \\ \{t_n\} \subset [0,1],\ t_n \to t \text{ with} \\ \limsup \langle H_{t_n}(x_n),\ x_n - x\rangle \leq 0, \end{cases}$$

it follows that $x_n \to x$ in X and $H_{t_n}(x_n) \rightharpoonup H_t(x)$ in X^*.

Using an elliptic super-regularization method, J. Berkovits and V. Mustonen [3], [6], extended the degree to the *(S)*-admissible class $\mathscr{A}_\Omega(S)$. This new topological degree deg_S verifies the classical axioms: existence of solution, additivity with respect to the domain, invariance under homotopies $\mathscr{H}_\Omega(S)$ and normalization map $J : X \mapsto X^*$. The *(S)*-degree was recently enlarged to mappings in nonreflexive Banach spaces [24].

More general, let 2^{X^*} be the collection of all nonempty subsets of X^*. For a multimap $T : X \mapsto 2^{X^*}$ we denote the domain $D(T) = \{x \in X \mid Tx \neq \oslash\}$, the range $R(T) = \{x^* \in X^* \mid x^* \in Tx, x \in D(T)\}$ and the graph $G(T) = D(T) \times R(T)$.

We say that $T : D(T) \subset X \mapsto 2^{X^*}$ is *monotone* if

$$\langle f - g, x - y\rangle \geq 0, \quad \forall x, y \in D(T),\ f \in Tu, g \in Ty,$$

and *maximal monotone* if it does not admit a monotone extension in $X \times X^*$. Clearly, the inverse multimap $T^{-1} : R(T) \mapsto 2^X$ is maximal monotone if and only if T is so.

If we identify X with X^{**} by reflexivity, the inverse $J^{-1} : X^* \mapsto X$ is the dual map of the dual space X^*. We use the following maximality criterion:

Lemma 27.1. *([21], p. 123). Let $T : D(T) \subseteq X \mapsto 2^{X^*}$ be monotone. Then T is maximal monotone if and only if $T + \lambda J$ is surjective for all $\lambda > 0$.*

Now, to any maximal monotone multimap $T : D(T) \subseteq X \mapsto 2^{X^*}$ we associated the family of *Yosida transformations*

$$T_\lambda = \left(T^{-1} + \lambda J^{-1}\right)^{-1}, \quad \lambda > 0.$$

By Lemma 27.1, $D(T_\lambda) = X$, and we easily see that $T_\lambda : X \mapsto X^*$ is single-valued. More precisely, we have

Lemma 27.2. *([7]). Let $T : D(T) \subseteq X \mapsto 2^{X^*}$ be a maximal monotone, with $0 \in T(0)$, and $A : D(A) \subset X \mapsto X^*$ a (single-valued) bounded, demicontinuous operator of type (S).*

a) Then $T_\lambda = \left(T^{-1} + \lambda J^{-1}\right)^{-1} : X \mapsto X^*$ is bounded, continuous, maximal monotone (in particular, pseudomonotone) for all $\lambda > 0$.

b) Assume that $g \notin (T + A)(\omega \cap D(T))$, where $\omega \subset \overline{D(A)}$ is a closed subset. Then there exists $\lambda_0 > 0$ such that $g \notin (T_\lambda + A)(\omega)$ for all $0 < \lambda < \lambda_0$.

The properties of continuity and pseudomonotonicity type of Yosida transformation T_λ are recently specified in [6].

Lemma 27.3. *Let $\lambda_n \to \lambda > 0$ and $x_n \to x$ in X. Then $T_{\lambda_n} x_n \to T_\lambda x$ in X^*.*

Lemma 27.4. *Assume that $\lambda_n \to \lambda > 0$ and $x_n \to x$ in X and*

$$\limsup \left\langle T_{\lambda_n} x_n, x_n - x \right\rangle \leq 0.$$

Then $\left\langle T_{\lambda_n} x_n, x_n \right\rangle \to \langle T_\lambda x, x\rangle$ and $T_{\lambda_n} x_n \rightharpoonup T_\lambda x$ in X^*.

With these prerequisites, we can state the precise meaning of a new extension of topological degree of type (S). Let $T : D(T) \subseteq X \mapsto 2^{X^*}$ be a maximal monotone, with $0 \in T(0)$, $A \in \mathscr{A}_\Omega(S)$ a single-valued perturbation, where Ω is a given open bounded subset of X, and $g \notin (T+A)(\partial\Omega \cap D(T))$. According to b) in Lemma 27.2 for $\omega = \partial\Omega$, there exists $\lambda_0 > 0$ such that $g \notin (T_\lambda + A)(\partial\Omega)$ for all $0 < \lambda < \lambda_0$. If $0 < \lambda_1 < \lambda_2 < \lambda_0$ are fixed, then $\Theta(\lambda, x) = T_\lambda x + Ax$, $x \in \overline{\Omega}$, $\lambda_1 \leq \lambda \leq \lambda_2$ defines a bounded homotopy of *type (S)*. Consequently, the *S-degree* [20], $d_S(\Theta(\lambda,\cdot),\Omega,g)$ makes sense and remains constant for all $\lambda \in [\lambda_1, \lambda_2]$, that is,

$$d_S(\Theta(\lambda_1,\cdot),\Omega,g) = d_S(\Theta(\lambda_2,\cdot),\Omega,g).$$

Thus, it is relevant to consider this common value of $d_S(T_\lambda + A, \Omega, g)$ to define another topological degree for the sum $A+T$ as

$$d(T+A,\Omega,g) = \lim_{\lambda \to 0^+} d_S(T_\lambda + A, \Omega, g). \tag{27.8}$$

We discuss now how the integer valued function defined by (27.8) satisfies the properties of a degree function.

(A) *Existence of solution:* If $f \notin (A+T)\left(\overline{\Omega}\right)$, it follows from Lemma 27.2, b), that $f \notin (T_\lambda + A)\left(\overline{\Omega}\right)$ for all $0 < \lambda < \lambda_0$. But $degs(T_\lambda + A, \Omega, f) = 0$ for all $0 < \lambda < \lambda_0$ yields $d(T+A,\Omega,f) = 0$. Therefore, $d(T+A,\Omega,f) \neq 0$ implies $f \in (A+T)\left(\overline{\Omega}\right)$.

(B) *Additivity with respect to the domain:* If Ω_1 and Ω_2 are open disjoint subsets of Ω such that $0 \notin (T+A)\left(\overline{\Omega} \setminus (\Omega_1 \cup \Omega_2)\right)$, then

$$d(T+A,\Omega,0) = d(T+A,\Omega_1,0) + d(T+A,\Omega_2,0).$$

We use Lemma 27.2, b), with $\omega = \overline{\Omega} \setminus (\Omega_1 \cup \Omega_2)$ and the additivity property of d_S.

(C) *Normalization:* $d(J,\Omega,g) = 1$ for all $g \in J(\Omega)$.

The properties (A)–(C) follow from the same relationships for the *(S)*-degree. Because the new degree function acts on a sum of two different kinds of mappings, a detailed discussion of the meaning of admissible homotopy is required.

For simplicity, we can identify further a multimap $T : X \mapsto 2^{X^*}$ with its graph and consider $\{T_t \,|\, t \in [0,1]\}$ a family maximal monotone multimaps $T_t : X \mapsto 2^{X^*}$ whose domains are nonempty. Owing to the invariance of monotonicity under translations, in the Browder's construction [7] of a degree for the sum $T+A$, the restrictions $0 \in T(0)$ and $0 \in T_t(0)$ were required. It was remarked ([16], Example 1.10) that these conditions are not fulfilled in the case of variational inequalities in the form (27.1), where K is nonsymmetric with respect to origin.

This is why we consider more general homotopies and relax the Browder restrictions. The family of maximal monotone multimaps $\{T_t \mid t \in [0,1]\}$ is called a *maximal monotone homotopy*, if for any $(x,f) \in G(T_t)$ and a sequence $t_n \to t$ in $[0,1]$, there exists a sequence $(x_n, f_n) \in G(T_{t_n})$ such that $x_n \to x$ and $f_n \to f$. The collection of such homotopies for various Ω in $\mathscr{O}$ will be denoted by $\mathscr{H}_X(MM)$. For all $\lambda > 0$ we define $T_{t,\lambda} = \left(T_t^{-1} + \lambda J^{-1}\right)^{-1}$.

Theorem 27.5. *[16] Let X be a real reflexive Banach space, a family $H_t : \overline{\Omega} \mapsto X^*$ of bounded homotopies of type (S), a family $T_t : X \mapsto 2^{X^*}$ of maximal monotone homotopies and a continuous curve $\{g(t) \mid t \in [0,1]\}$ in X^* such that $g(t) \notin (H_t + T_t)(\partial\Omega)$, for all $t \in [0,1]$. Then there exists $\overline{\lambda} > 0$ such that*

$$g(t) \notin \left(H_t + T_{t,\lambda}\right)(\partial\Omega) \quad \text{for all} \quad \lambda \in \left(0, \overline{\lambda}\right], \, t \in [0,1],$$

and $d_S\left(H_t + T_{t,\lambda}, \Omega, g(t)\right)$ is independent of t and λ.

We are in position to complete the list of properties of the new topological degree defined by (27.8).

(D) *Homotopy invariance:* Let $\{H_t \mid t \in [0,1]\}$ be a family of homotopies in $\mathscr{H}_\Omega(S)$, $\{T_t \mid 0 \le t \le 1\}$ a family of homotopies in $\mathscr{H}_X(MM)$, and $\{g(t) \mid t \in [0,1]\}$ a continuous curve in X^* such that $g(t) \notin (H_t + T_t)(\partial\Omega)$ for all $t \in [0,1]$. Then $d(H_t + T_t, \Omega, g(t))$ is constant for all $t \in [0,1]$.

We have obtained an extension of the Browder degree given in [16] which includes the case $T = N_K$, where N_K is a normal cone of a closed convex subset K of X. This degree for the sum $A + N_K$ corresponds to variational inequalities converted into inclusions (27.5). Moreover, it was proved [5], [16], [10] that the above degree is uniquely defined.

In a discussion about the homotopy invariance, the class of admissible homotopies must includes an affine homotopy between A_1 and A_2, that is,

$$A_t = (1-t)A_1 + tA_2, \quad A_1, A_2 \in \mathscr{A}_\Omega(S), \quad 0 \le t \le 1,$$

which are simple examples and important in applications. In order that an homotopy of the form $(1-t)(A_1 + T) + tA_2$ be included in the frame of invariance (D), it is necessary that $\overline{D(T)} = X$, see [17].

With regard to the the existence of a solution, we can consider a more general fixed point index. Denote, for simplicity, $F = T + A : X \to X^*$ and say that $a \in D(F)$ is an *isolated zero* of F if there is an open ball $B(a, r_a)$ in X of center a and radius r_a so that $F^{-1}(0) \cap B(a, r_a) = \{a\}$. Using the excision property of the degree, $d(F, B(a,r), 0)$ does not depend on $r \in (0, r_a)$. This justifies the definition of the *fixed point index of F at a* as integer

$$ind(F,a) = d(F, B(a,r), 0) = d(T+A, B(a,r), 0) = ind(T+A, a),$$

for r sufficiently small, where $d(T+A, B(a,r), 0)$ is the topological degree constructed above. Roughly speaking, for a bounded open subset $\Omega \subset X$ and

$$F^{-1}(0) = \{a_1, a_2, \ldots, a_m\}, \quad a_j \in D(F) \cap \Omega, \quad j = 1, 2, \ldots, m,$$

we have

$$deg(F, \Omega, 0) = \sum_{j=1}^{m} ind(F, a_j).$$

The properties of the fixed point index as the solution condition, additivity, invariance under admissible homotopies and normalization are inherited from those of origin topological degree. In addition, the index is related to the spectral theory of a linearized mapping and its computation provides information about solvability, estimates for the number of solutions and bifurcation of solutions [19]. A complete survey of classical theory of the fixed point index can be found in [25].

In the past decade, important progress is due to A.G. Kartsatos and I.V. Skrypnik [14]. In the above designated framework, they considered a class of operators $A : D(A) \subset X \mapsto X^*$ of *type (S)* with respect to a dense subspace L of X such that $\overline{L} = X$. We denote this class $(S)_L$. Using finite-dimensional Galerkin procedures, a degree is defined for the sum $A + T$, where $T : X \mapsto 2^{X^*}$ is a maximal monotone multimap, with $0 \in T(0)$. In [4], [1], [11], an alternative construction combines the Leray–Schauder degree with the Browder–Ton method of elliptic super-regularization. In addition, using a new notion of linearization, appropriate to the perturbations of *type* $(S)_L$. A.G. Kartsatos and I.V. Skrypnik [15] calculated the index of an isolated point of such a map.

In the same context, I. Benedetti and V. Obukhovskii [2] studied recently the index of solvability as a similar topological characteristic, whose difference from zero provides the existence of a solution for variational inequalities.

Related to the degree-theoretic approach of A. Szulkin [22], it is worth noting a recent advance in the study of existence of non-zero solutions to variational inequalities has been made, by using the fixed point index theory in the case of the contractive mappings [18], [23], [26].

Chapter 9 of G. Isac's book [12] contains a competent survey of the mappings of type *(S)*, a good substitute of compactness in proving the existence results for nonlinear problems. In addition, the operators of type $(S)^1$ in the case of unbounded maps are studied. The class of $(S)^1$-operators was introduced by G. Isac and M.S. Gowda [13] in the study of complementary problems. An extension of notion of $(S)^1$-operator to multivalued mappings has been used by P. Cubiotti and J.C. Yao [8] to investigate generalized variational inequalities.

References

1. D.R. Adhikari and A. Kartsatos, *Strongly quasibounded maximal monotone perturbations for the Berkovits-Mustonen topological degree theory*, J. Math. Anal. Appl. 348 (2008), 122–136; MR 2009k:47213.
2. I. Benedetti and V. Obukhovskii, *On the index of solvability for variational inequalities*, Set-Valued Analysis **16** (2008), 67–92.
3. J. Berkovits, *Some extensions of topological degree theory with applications to nonlinear problems*, Pitman Res. Notes Math. 365, 1–29, Longman, Harlow, 1997, MR 98m:47101.
4. J. Berkovits, *On the degree theory for densely defined mappings of class* $(S_+)_L$, Abstr. Appl. Anal., 4 (1999), 141–152; MR 2001m:47125.
5. J. Berkovits and M. Miettunen, *On the uniqueness of the Browder degree*, Proc. Amer. Math. Soc. 136 (2008), 3467–3476; MR 2009d:47072.
6. J. Berkovits and V. Mustonen, *An extension of Leray-Schauder degree and applications to nonlinear wave equations*, Differential Integral Equations 3 (1990), 945–963; MR 91j:35179.
7. F.E. Browder, *Fixed point theory and nonlinear problems*, Bull. Amer. Math. Soc. 9 (1983), 1–39; MR 84h:58027.
8. P. Cubiotti and J. C. Yao, *Multivalued* $(S)^1_+$ *operators and generalized variational inequalities*, Comput. Math. Appl. 29(1995), 49–56; MR 96d:49013.
9. D. Goeloven, D. Motreanu, Y. Dumont and M. Rhoedi, *Variational and hemivariational inequalities: Theory, methods and applications, Vol. I: Unilateral analysis and unilateral machanics*, Kluwer, 2003, Boston; MR 2004g:49004.
10. S. Hu and N.S. Papageorgiou, *Handbook of multivalued analysis*, Vol. I, Kluwer Acad. Publ., Dordrecht, 1997; MR 98k:47001.
11. B. Ibrahimou and A. Karsatos, *The Leray-Schauder approach the degree for* (S_+) *perturbations of maximal monotone operators in separable Banach spaces*, Nonlinear Anal. 70 (2009), 4350–4568.
12. G. Isac, *Topological methods in complementarity theory*, Kluwer Acad. Publ., Dordrecht, 2000; MR 2001c:90001.
13. G. Isac and M.S. Gowda, *Operators of class* $(S)^1_+$*, Altman's condition and the complementarity problems*, Journ. Fac. Sci., Univ. Tokyo, Sct. 1A, Math. 40, 1993, 1–16; MR 94d:49001.
14. A.G. Kartsatos and I.V. Skrypnik, *Topological degree for densely defined mappings involving operators of type* (S_+), Adv. Differential Equations 4 (1999), 413–456; MR 2000a:47127.
15. A.G. Kartsatos and I.V. Skrypnik, *The index of a critical point for densely defined operators of type* $(S_+)_L$ *in Banach spaces*, Trans. Amer. Math. Soc. 354 (2002), 1601–1640; MR 2002j:47087.
16. B.T. Kien, M.-M. Wong and N.-C. Wong, *On the degree theory for general mappings of monotone type*, J. Math. Anal. Appl. 340 (2008), 707–720; MR 2009c:47107.
17. J. Kobayashi and M. Ôtani, *Topological degree for* $(S)_+$ *mappings with maximal monotone perturbations and its applications to variational inequalities*, Nonlinear Anal. 59 (2004), 147–172; MR 2005f:49028.
18. Y.S. Lai, Y.G. Zhu and Y.B. Deng, *Fixed point index approach for solution of variational inequalities*, Int. J. Math. Mech. Sci. 2005, 12, 1879–1887; MR 2006e:49013.
19. V.K. Le and K. Schmitt, *Global bifurcation in variational inequalities, Applications to obstacle and unilateral problems*, Springer, New York, 1997; MR 98c:47101.
20. D. Pascali, *Topological methods in nonlinear analysis, Topological degree for monotone mappings*, Ovidius Univ. Constanţa, 2001.
21. D. Pascali and S. Sburlan, *Nonlinear mappings of monotone type*, Edit. Academiei, Bucureşti, or Sijthoff & Noordhoff Intern. Publ., Alphen aan den Rijn, 1978; MR 80g:47056.
22. A. Szulkin, *Positive solutions of variational inequalities: A degree-theoretic approach*, J. Diff. Equations **57** (1985), 90–111; MR 86i:47072.
23. B. Wang, *Nonzero solutions for generalized variational inequalities by index*, J. Math. Anal. Appl. 334 (2007), 621–626; MR 2008g: 49012.

24. F. Wang, Y.-Q. Chan and D. O'Regan, *Degree theory for* (S_+) *mappings in nonreflexive Banach spaces*, Appl. Math. Comput. 202 (2008), 229–232; MR 2009g:47164.
25. E. Zeidler, *Functional analysis and its applications*, I, *Fixed point theorems*, Springer, New York, 1986; MR 87f:47083.
26. Y. Zhu and G. Yuan, *The existence of nontrivial solutions for variational inequalities by index approach*, Fixed point theory and applications, Vol. 6, 197–203. Nova Sci. Publ. Inc., New York, 2007; MR 2008k:49009.

Chapter 28
Completely Generalized Co-complementarity Problems Involving p-Relaxed Accretive Operators with Fuzzy Mappings

Abul Hasan Siddiqi and Syed Shakaib Irfan

Dedicated to the memory of Professor George Isac

Abstract In the current work, we introduce and study completely generalized co-complementarity problems for fuzzy mappings (for short, CGCCPFM). By using the definitions of p-relaxed accretive and p-strongly accretive mappings, we propose an iterative algorithm for computing the approximate solutions of CGCCPFM. We prove that approximate solutions obtained by the proposed algorithm converge to the exact solutions of CGCCPFM.

28.1 Introduction

Due to the applications in areas such as optimization theory, structural engineering, mechanics, elasticity, lubrication theory, economics, equilibrium theory on networks, stochastic optimal control, etc., the complementarity problem is one of the interesting and important problems and was introduced by G. E. Lemke in 1964, but Cottle [5] and Cottle and Dantzig [6] formally defined the linear complementarity problem and called it the fundamental problem. In recent years, it has been generalized and extended to many different directions, see, for example, [1, 3, 7, 8, 9, 10, 12, 15, 16, 17] and references therein.

Inspired and motivated by the recent research work going on in this field, we consider in this paper the completely generalized co-complementarity problems for

Abul Hasan Siddiqi
B.M.A.S. Engineering College, Agra 282007, U.P., India, e-mail: siddiqi.abulhasan@gmail.com

Syed Shakaib Irfan
College of Engineering, Qassim University, P. O. Box 6677, Buraidah 51452, Al-Qassim, Kingdom of Saudi Arabia, e-mail: shakaib11@rediffmail.com

P.M. Pardalos et al. (eds.), *Nonlinear Analysis and Variational Problems*, Springer Optimization and Its Applications 35, DOI 10.1007/978-1-4419-0158-3_28,

fuzzy mappings (for short, CGCCPFM). By using the definitions of p-relaxed accretive and p-strongly accretive mappings, we propose an iterative algorithm for computing the approximate solutions of CGCCPFM. We prove that approximate solutions obtained by the proposed algorithm converge to the exact solutions of CGCCPFM. Our results are new and represent a significant improvement of previously known results.

28.2 Background of Problem Formulation

Let E be real Banach space equipped with norm $\|.\|, E^*$ the topological dual space of $E, \langle .,. \rangle$ the dual pairing between E and E^*, and $CB(E)$ the family of all nonempty closed and bounded subset of E; $\tilde{\mathscr{H}}(.,.)$ is the Hausdorff metric on $CB(E)$ defined by

$$\tilde{\mathscr{H}}(A,B) = \max\left\{ \sup_{x\in A} d(x,B), \sup_{y\in B} d(A,y) \right\}, \quad A,B \in CB(E)$$

where $d(x,B) = \inf_{y\in B} d(x,y)$ and $d(A,y) = \inf_{x\in A} d(x,y)$ and $J_p : E \to 2^{E^*}$ is the *generalized duality mapping* defined by

$$J_p(x)\{f^* \in E^* : \langle x, f^* \rangle = \|f^*\|.\|x\| \text{ and } \|f^*\| = \|x\|^{p-1}\} \;\forall\, x \in E,$$

where $1 < p < \infty$ is a constant. In particular, J_2 is the usual normalized duality mapping. It is known that, in general, $J_p(x) = \|x\|^{p-1} J_2(x)$, for all $x \neq 0$ and J_p is single-valued if E^* is strictly convex. If $E = H$ is a Hilbert space, then J_2 becomes the identity mapping on H.

Definition 28.1. Let E be a real Banach space and K be a nonempty subset of E. Then a multivalued mapping $T : K \to 2^E$ is said to be

(i) *accretive* if for any $x, y \in K, u \in T(x)$ and $v \in T(y)$ there exists $j_2(x-y) \in J_2(x-y)$ such that
$$\langle u - v, j_2(x-y) \rangle \geq 0,$$
or equivalently, there exists $j_p(x-y) \in J_p(x-y),\; 1 < p < \infty$ such that
$$\langle u - v, j_p(x-y) \rangle \geq 0;$$

(ii) *strongly accretive* if for any $x, y \in K, u \in T(x), v \in T(y)$ there exist $j_2(x-y) \in J_2(x-y)$ and a constant $k > 0$ such that
$$\langle u - v, j_2(x-y) \rangle \geq k\|x-y\|^2,$$
or equivalently, there exists $j_p(x-y) \in J_p(x-y),\; 1 < p < \infty$ such that
$$\langle u - v, j_p \rangle \geq k\|x-y\|^p.$$

In 1967, Browder [4] and Kato [13] introduced independently the concept of single-valued accretive mappings. An early fundamental result in the theory of accretive mappings which is due to Browder states that the following initial value problem,

$$\frac{du(t)}{dt}+Tu(t)=0,u(0)=u_0,$$

is solvable if T is locally Lipschitzian and accretive on E.

Now let $\mathscr{F}(E)$ be a collection of all fuzzy sets over E. A mapping $F:E\to\mathscr{F}(E)$ is said to be fuzzy mapping. For each $x\in E,F(x)$ (denote it by F_x, in the sequel) is a fuzzy set on E and $F_x(y)$ is the membership function of y in F_x.

A fuzzy mapping $F:E\to\mathscr{F}(E)$ is said to be closed if for each $x\in E$, the function $y\to F_x(y)$ is upper semi-continuous, i.e., for any given net $\{y_\alpha\}\subset E$ satisfying $y_\alpha\to y_0\in E$, $\lim\sup_\alpha F_x(y_\alpha)\leq F_x(y_0)$.

For $A\in\mathscr{F}(E)$ and $\lambda\in[0,1]$, the set $(A)_\lambda=\{x\in E:A_x\geq\lambda\}$ is called a λ-cut set of A.

A closed fuzzy mapping $A:E\to\mathscr{F}(E)$ is said to satisfy the condition $(*)$ if there exists a function $a:E\to[0,1]$ such that for each $x\in E$, $(A_x)_{a(x)}$ is a nonempty bounded subset of E. It is clear that if A is a closed fuzzy mapping satisfying the condition $(*)$, then for each $x\in E$, the set $(A_x)_{a(x)}\in CB(E)$.

In fact, let $\{y_\alpha\}_{\alpha\in\Gamma}\subset(A_x)_{a(x)}$ be a net and $y_\alpha\to y_0\in E$. Then $(A_x)y_\alpha\geq a(x)$ for each $\alpha\in\Gamma$. Since A is closed, we have

$$(A_x)(y_0)\geq\ \lim\sup_{\alpha\in\Gamma}A_x(y_\alpha)\geq a(x).$$

This implies that $y_0\in(A_x)_{a(x)}$ and so $(A_x)_{a(x)}\in CB(x)$.

Let $F,G,H,T:E\to\mathscr{F}(E)$ be closed fuzzy mappings satisfying condition $(*)$. Then there exist functions $a,b,c,d:E\to[0,1]$ such that for each $x\in E$, we have $(F_x)_{a(x)},(G_x)_{b(x)},(H_x)_{c(x)},(T_x)_{d(x)}\in CB(E)$. Therefore we can define multi-valued mappings $\tilde{F},\tilde{G},\tilde{H},\tilde{T}:E\to CB(E)$ by $\tilde{F}(x)=(F_x)_{a(x)},\tilde{G}(x)=(G_x)_{b(x)},\tilde{H}(x)=(H_x)_{c(x)},\tilde{T}(x)=(T_x)_{d(x)}$ for each $x\in E$. In the sequel, $\tilde{F},\tilde{G},\tilde{H}$ and $\tilde{T}$ are called the multi-valued mappings induced by the fuzzy mappings F,G,H, and T, respectively.

Let $N:E\times E\times E\to E$ and $f,g,h,t,m:E\to E$ be single-valued mappings. Let $F.G,H,T:E\to\mathscr{F}(E)$ be fuzzy mappings. Let $a,b,c,d:E\to[0,1]$ be given functions and X be a fixed closed convex cone of E. Define $K:E\to 2^E$ by

$$K(z)=m(z)+X\ \forall\, x\in E,\ z\in T(x).$$

We consider the following *completely generalized co-complementarity problem with fuzzy mappings (CGCCPFM)*:

$$\text{(CGCCPFM)}\quad\begin{cases}\text{Find } x,u,v,w,z\in E \text{ such that}\\ F_x(u)\geq a(x),G_x(v)\geq b(x),\\ H_x(w)\geq c(x),T_x(z)\geq d(x),g(x)\in K(z)\text{ and}\\ N(f(u),h(v),t(w))\in J(K(z)-g(x))^*\end{cases}$$

where $J(K(z)-g(x))^*$ is the dual cone of the set $J(K(z)-g(x))$.

We remark that for suitable choices of $F,G,H,T,f,g,h,t,m,$ and N, (CGC-CPFM) reduces to various new as well as known classes of complementarity problems and variational inequalities (e.g., [7, 11, 12, 15, 16] and the references therein).

28.3 The Characterization of Problem and Solutions

In this section, we consider some basic concepts and results, which will be used throughout the paper. The modulus of smoothness of real Banach space E is the function $\rho_E : [0,\infty) \to [0,\infty)$ defined by

$$\rho_E(\tau) = \sup\left\{\frac{\|x+y\|+\|x-y\|}{2} - 1 : \|x\| = 1,\ \|y\| = \tau\right\}.$$

A Banach space E is called uniformly smooth if

$$\lim_{t\to 0}\frac{\rho_E(\tau)}{\tau} = 0.$$

E is called p-uniformly smooth if there exists a constant $c > 0$ such that

$$\rho_E(\tau) \le c\tau^p, \text{ for } p > 1.$$

Note that J_p is single-valued if E is uniformly smooth.

Remark 28.2. It is known that all Hilbert spaces and Banach spaces $L_p; l_p$ and $W_m^p (1 < p < \infty)$ are uniformly smooth and

$$\rho_E(\tau) < \begin{cases} \frac{1}{p}\tau^p. \ 1 < p < 2 \\ \frac{p-1}{2}\tau^2,\ p \le 2. \end{cases} \qquad (E = L_p, l_p \text{ or } W_m^p)$$

Therefore E is a p-uniformly smooth Banach space with modulus of smoothness of power type $p < 1$ and J_p will always represent the single-valued mapping.

Definition 28.3. [2] Let E be a p-uniformly smooth Banach space and let Ω be a nonempty closed convex subset of E. A mapping $Q_\Omega : E \to \Omega$ is said to be

(i) *retraction* on Ω if $Q_\Omega^p = Q_\Omega$;
(ii) *nonexpansive retraction* if

$$\|Q_\Omega(x) - Q_\Omega(y)\| \le \|x-y\|; \qquad \forall\, x,y \in E,$$

(iii) *sunny retraction*

$$Q_\Omega(Q_\Omega(x) + t(x - Q_\Omega(x))) = Q_\Omega(x) \qquad \forall\, x,y \in E,\ t \in (-\infty,\infty).$$

Lemma 28.4. *[2] Q_Ω is sunny nonexpansive retraction if and only if*

$$\langle x - Q_\Omega^*, J(Q_\Omega(x-y))\rangle \geq 0, \quad \forall\, x,y \in E.$$

Lemma 28.5. *Let E be a real Banach space and $J_p : E \to 2^{E^*}$, $1 < p < \infty$ be the duality mapping. Then for any $x,y \in E$,*

$$\|x+y\|^p \leq \|x\|^p + p\langle y, j_p(x+y)\rangle, \quad \forall\, j_p(x+y) \in J_p(x+y).$$

Theorem 28.6. *[2] Let E be a real Banach space, Ω a nonempty closed convex subset of E, and $m : E \to E$ be a single-valued mapping. Then we have*

$$Q_{\Omega|m(z)}(x) = m(z) + Q_\Omega(x - m(z)), \quad \forall\, x,y \in E.$$

Theorem 28.7. *(x,u,v,w,z) is a solution of (CGCCPFM) if and only if it satisfies the relation*

$$x = x - g(x) + m(z) + Q_X[g(x) - \tau N(f(u),h(v),t(w)) - m(z)]$$

where $u \in \tilde{F}(x), v \in \tilde{G}(x), w \in \tilde{H}(x), z \in \tilde{T}(x)$ and $\tau > 0$ is a constant.

The following lemma will be used in our main results.

Lemma 28.8. *Let E be a real Banach space and $j_p : E \to 2^{E^*}$, $1 < p < \infty$ a duality mapping. Then, for any $x,y \in E$, we have*

$$\langle x-y, j_p(x) - j_p(y)\rangle \leq 2d^p \rho_E\left(\frac{4\|x-y\|}{d}\right), \text{ where } d^p = \frac{\|x\|^2 + \|y\|^2}{2}.$$

Proof. For the proof see [7]. □

28.4 Iterative Algorithm and Pertinent Concepts

Using Theorem 28.7 and Nadler's theorem [14], we establish an iterative algorithm for finding the approximate solutions of (CGCCPFM) as follows:

Algorithm 28.9. *Let $F,G,H,T : E \to \mathscr{F}(E)$ be closed fuzzy mappings satisfying the condition $(*)$ and let $\tilde{F},\tilde{G},\tilde{H},\tilde{T} : E \to CB(E)$ be the multi-valued mappings induced by the fuzzy mappings $F,G,H,$ and T, respectively. Let $N : E \times E \times E \to E$ and $f,g,h,t,m : E \to E$ be the single-valued mappings.*

For given $x_0 \in E, u_0 \in \tilde{F}(x_0), v_0 \in \tilde{G}(x_0), w_0 \in \tilde{H}(x_0)$ and $z_0 \in \tilde{T}(x_0)$ and let

$$x_1 = x_0 - g(x_0) + m(z_0) + Q_X[g(x_0) - \tau N(f(u_0),h(v_0),t(w_0)) - m(z_0)],$$

where $\tau > 0$ is a constant.

Since $\tilde{F}(x_0), \tilde{G}(x_0), \tilde{H}(x_0), \tilde{T}(x_0) \in CB(E)$ by Nadler's theorem [14], there exist $u_1 \in \tilde{F}(x_1), v_1 \in \tilde{G}(x_1), w_1 \in \tilde{H}(x_1)$ and $z_1 \in \tilde{T}(x_1)$ such that

$$\|u_1 - u_0\| \leq (1+(1+0)^{-1})\mathscr{H}(\tilde{F}(x_1)\tilde{F}(x_0));$$

$$\|v_1 - v_0\| \leq (1+(1+0)^{-1})\mathscr{H}(\tilde{G}(x_1)\tilde{G}(x_0));$$

$$\|w_1 - w_0\| \leq (1+(1+0)^{-1})\mathscr{H}(\tilde{H}(x_1)\tilde{H}(x_0));$$

$$\|z_1 - z_0\| \leq (1+(1+0)^{-1})\mathscr{H}(\tilde{T}(x_1)\tilde{T}(x_0)).$$

Let

$$x_2 = x_1 - g(x_1) + m(z_1) + Q_X[g(x_1) - \tau N(f(u_1), h(v_1), t(w_1)) - m(z_1)].$$

Since $u_1 \in \tilde{F}(x_1) \in CB(E), v_1 \in \tilde{G}(x_1) \in CB(E), w_1 \in \tilde{H}(x_1) \in CB(E)$ and $z_1 \in \tilde{T}(x_1) \in CB(E)$, there exists $u_2 \in \tilde{F}(x_2), v_2 \in \tilde{G}(x_2), w_2 \in \tilde{H}(x_2)$ and $z_2 \in \tilde{T}(x_2)$ such that

$$\|u_2 - u_1\| \leq (1+(1+1)^{-1})\mathscr{H}(\tilde{F}(x_2)\tilde{F}(x_1));$$

$$\|v_2 - v_1\| \leq (1+(1+1)^{-1})\mathscr{H}(\tilde{G}(x_2)\tilde{G}(x_1));$$

$$\|w_2 - w_1\| \leq (1+(1+1)^{-1})\mathscr{H}(\tilde{H}(x_2)\tilde{H}(x_1));$$

$$\|z_2 - z_1\| \leq (1+(1+1)^{-1})\mathscr{H}(\tilde{T}(x_2)\tilde{T}(x_1));$$

continuing the above process inductively, we obtain the sequences $\{x_n\}, \{u_n\}, \{v_n\}, \{w_n\}$ and $\{z_n\}$ satisfying

$$x_{n+1} = x_n - g(x_n) + m(z_n) + Q_X[g(x_n) - \tau N(f(u_n), h(v_n), t(w_n)) - m(z_n)],$$

$$u_n \in \tilde{F}(x_n), \|u_{n+1} - u_n\| \leq (1+(1+n)^{-1})\mathscr{H}(\tilde{F}(x_{n+1})\tilde{F}(x_n));$$

$$v_n \in \tilde{G}(x_n), \|v_{n+1} - v_n\| \leq (1+(1+n)^{-1})\mathscr{H}(\tilde{G}(x_{n+1})\tilde{G}(x_n));$$

$$w_n \in \tilde{H}(x_n), \|w_{n+1} - w_n\| \leq (1+(1+n)^{-1})\mathscr{H}(\tilde{H}(x_{n+1})\tilde{H}(x_n));$$

$$z_n \in \tilde{T}(x_n), \|z_{n+1} - z_n\| \leq (1+(1+n)^{-1})\mathscr{H}(\tilde{T}(x_{n+1})\tilde{T}(x_n));$$

$n = 0, 1, 2, 3, \cdots$, where $\tau > 0$ is a constant.

We remark that Iterative Algorithm 28.9 includes as special cases many known iterative algorithms (e.g., [7, 11, 12, 15, 16] and the references therein).

Definition 28.10. A single-valued mapping $g : E \to E$ is said to be

(i) *p-strongly accretive* if there exists $j_p(x-y) \in J_p(x-y)$ and $k > 0$ such that

$$\langle g(x) - g(y), j_p(x-y)\rangle \geq k\|x-y\|^p, \quad \forall\, x, y \in E;$$

(ii) *Lipschitz continuous* if there exists a constant $\lambda_g > 0$ such that

$$\|g(x) - g(y)\| \leq \lambda_g \|x-y\|, \quad \forall\, x, y \in E.$$

Definition 28.11. A multi-valued mapping $F : E \to CB(E)$ is said to be $\mathscr{H}$-Lipschitz continuous if there exists a constant $\lambda_F > 0$ such that

$$\mathscr{H}(F(x), F(y)) \le \lambda_F \|x - y\|, \qquad \forall\, x, y \in E,$$

where $\mathscr{H}(.,.)$ is Hausdorff metric defined on $CB(E)$.

Definition 28.12. A mapping $N : E \times E \times E \to E$ is said to be

(i) *p-relaxed accretive with respect to the first argument* if there exists $j_p(x-y) \in J_p(x-y)$ and $\alpha > 0$ such that

$$\langle N(u_n, v_n, w_n) - N(u_{n-1}, v_n, w_n), j_p(x_n, x_{n-1})\rangle \ge -\alpha \|x_n - x_{n-1}\|^p,$$

$$\forall\, u_n \in F(x_n),\ u_{n-1} \in F(x_{n-1}) \text{ and } x_n, x_{n-1} \in E.$$

(ii) *p-relaxed accretive with respect to the second argument* if there exists $j_p(x-y) \in J_p(x-y)$ and $\beta > 0$ such that

$$\langle N(u_n, v_n, w_{n-1}) - N(u_n, v_{n-1}, w_n), j_p(x_n, x_{n-1})\rangle \ge -\beta \|x_n - x_{n-1}\|^p,$$

$$\forall\, v_n \in G(x_n),\ v_{n-1} \in G(x_{n-1}) \text{ and } x_n, x_{n-1} \in E.$$

(iii) *p-relaxed accretive with respect to the third argument* if there exists $j_p(x-y) \in J_p(x-y)$ and $\gamma > 0$ such that

$$\langle N(u_n, v_n, w_n) - N(u_n, v_n, w_{n-1}), j_p(x_n, x_{n-1})\rangle \ge -\gamma \|x_n - x_{n-1}\|^p,$$

$$\forall\, w_n \in H(x_n),\ w_{n-1} \in H(x_{n-1}) \text{ and } x_n, x_{n-1} \in E.$$

(iv) *Lipschitz continuous with respect to the first argument* if there exists a constant λ_{N_1} such that

$$\|N(u_n, v_n, w_n) - N(u_{n-1}, v_n, w_n)\| \le \lambda_{N_1} \|u_n - u_{n-1}\|,$$

$$\forall\, u_n \in F(x_n),\ u_{n-1} \in F(x_{n-1}) \text{ and } x_n, x_{n-1} \in E.$$

(v) *Lipschitz continuous with respect to the second argument* if there exists a constant λ_{N_2} such that

$$\|N(u_n, v_n, w_n) - N(u_n, v_{n-1}, w_n)\| \le \lambda_{N_2} \|v_n - v_{n-1}\|,$$

$$\forall\, v_n \in G(x_n),\ v_{n-1} \in G(x_{n-1}) \text{ and } x_n, x_{n-1} \in E.$$

(vi) *Lipschitz continuous with respect to the third argument* if there exists a constant λ_{N_3} such that

$$\|N(u_n, v_n, w_n) - N(u_n, v_n, w_{n-1})\| \le \lambda_{N_3} \|w_n - w_{n-1}\|,$$

$$\forall\, w_n \in H(x_n),\ w_{n-1} \in H(x_{n-1}) \text{ and } x_n, x_{n-1} \in E.$$

28.5 Existence and Convergence Result for CGCCPFM

Theorem 28.13. *Let E be a p-uniformly smooth Banach space with $\rho_E(t) \leq c\tau^p$ for some $c > 0$, $0 < \tau < \infty$ and $1 < p < \infty$. Let X be a closed convex cone of E. Let $f,g,h,t,m : E \to E$ be Lipschitz continuous mappings with constants $\lambda_f, \lambda_g, \lambda_h, \lambda_t$, and λ_m, respectively, g is p-strongly accretive with constant k, and N is p-relaxed accretive with respect to first, second, and third argument with constants α, β, and γ, respectively, and Lipschitz continuous with respect to first, second, and third argument with constants $\lambda_{N_1}, \lambda_{N_2}$, and λ_{N_3}, respectively. Let $F,G,H,T : E \to \mathscr{F}(E)$ be closed fuzzy mappings satisfying the condition $(*)$ and let $\tilde{F}, \tilde{G}, \tilde{H}, \tilde{T} : E \to CB(E)$ be the multi-valued mappings induced by the fuzzy mappings F,G,H, and T, respectively. Let $\tilde{F}, \tilde{G}, \tilde{H}$, and $\tilde{T}$ be $\mathscr{H}$-Lipschitz continuous mappings with constants $\lambda_F, \lambda_G, \lambda_H$, and λ_T, respectively. Let $K : E \to 2^E$ be such that $K(z) = m(z) + X$, for all $x \in E$, $z \in \tilde{T}(x)$ and suppose the following conditions are satisfied*

$$q + \phi + \psi + \xi < 1 \tag{28.1}$$

where

$$q = 2(1 - pk + 2^{2p+1}cp\lambda_g^p)^{1/p} + 2\lambda_m\lambda_T,$$

$$\phi = (1 + p\tau\gamma + 2^{2p+1}p\tau^p c\lambda_{N_3}^p \lambda_t^p)^{1/p},$$

$$\psi = (1 + p\tau\alpha + 2^{2p+1}p\tau^p c\lambda_{N_1}^p \lambda_f^p)^{1/p}$$

$$\text{and } \xi = (1 + p\tau\beta + 2^{2p+1}p\tau^p c\lambda_{N_2}^p \lambda_h^p)^{1/p}.$$

Then the iterative sequences $\{x_n\}, \{u_n\}, \{v_n\}, \{w_n\}$, and $\{z_n\}$ generated by Algorithm 28.9 converge strongly to x,u,v,w, and z, respectively, and (x,u,v,w,z) is a solution of (CGCCPFM).

Proof. From Algorithm 28.9 and Definition 28.3, we have

$$\begin{aligned}
\|x_{n+1} - x_n\| &= \|x_n - g(x_n) + m(z_n) + Q_X[g(x_n) - \tau N(f(u_n), h(v_n), t(w_n)) \\
&\quad - m(z_n)] - x_{n-1} + g(x_{n-1}) - m(z_{n-1}) - Q_X[g(x_{n-1}) \\
&\quad - \tau N(f(u_{n-1}), h(v_{n-1}), t(w_{n-1})) - m(z_{n-1})]\| \\
&\leq \|x_n - x_{n-1} - (g(x_n) - g(x_{n-1}))\| + \|m(z_n) - m(z_{n-1})\| \\
&\quad + \|Q_X[g(x_n) - \tau N(f(u_n), h(v_n), t(w_n)) - m(z_n)] \\
&\quad - Q_X[g(x_{n-1}) - \tau N(f(u_{n-1}), h(v_{n-1}), t(w_{n-1})) - m(z_{n-1})]\| \\
&\leq 2\|x_n - x_{n-1} - (g(x_n) - g(x_{n-1}))\| + 2\|m(z_n) - m(z_{n-1})\| \\
&\quad + \|x_n - x_{n-1} - \tau(N(f(u_n), h(v_n), t(w_n)) \\
&\quad - N(f(u_{n-1}), h(v_{n-1}), t(w_{n-1})))\|
\end{aligned}$$

$$\begin{aligned}
&\leq 2\|x_n - x_{n-1} - (g(x_n) - g(x_{n-1}))\| + 2\|m(z_n) - m(z_{n-1})\| \\
&\quad + \|x_n - x_{n-1} - \tau(N(f(u_n),h(v_n),t(w_n)) - N(f(u_n),h(v_n),t(w_{n-1})))\| \\
&\quad + N(f(u_n),h(v_n),t(w_{n-1})) - N(f(u_{n-1}),h(v_n),t(w_{n-1})) \\
&\quad + N(f(u_{n-1}),h(v_n),t(w_{n-1})) - N(f(u_{n-1}),h(v_{n-1}),t(w_{n-1}))\| \\
&\leq 2\|x_n - x_{n-1} - (g(x_n) - g(x_{n-1}))\| + 2\|m(z_n) - m(z_{n-1})\| \\
&\quad + \|x_n - x_{n-1} - \tau(N(f(u_n),h(v_n),t(w_n)) - N(f(u_n),h(v_n),t(w_{n-1})))\| \\
&\quad + \|x_n - x_{n-1} - \tau(N(f(u_n),h(v_n),t(w_{n-1})) - N(f(u_{n-1}),h(v_n),t(w_{n-1})))\| \\
&\quad + \|x_n - x_{n-1} - \tau(N(f(u_{n-1}),h(v_n),t(w_{n-1})) \\
&\quad - N(f(u_{n-1}),h(v_{n-1}),t(w_{n-1})))\|. \qquad (28.2)
\end{aligned}$$

Since g is p-strongly accretive and Lipschitz continuous with constants k and λ_g respectively and using Lemma 28.5, and Lemma 28.8 we obtain

$$\|x_n - x_{n-1} - (g(x_n) - g(x_{n-1}))\|^p$$

$$\begin{aligned}
&\leq \|x_n - x_{n-1}\|^p - p\langle g(x_n) - g(x_{n-1}), j_p(x_n - x_{n-1} - (g(x_n) - g(x_{n-1})))\rangle \\
&\leq \|x_n - x_{n-1}\|^p - p\langle g(x_n) - g(x_{n-1}), j_p(x_n - x_{n-1})\rangle - p\langle g(x_n) - g(x_{n-1}), \\
&\quad j_p(x_n - x_{n-1} - (g(x_n) - g(x_{n-1}))) - j_p(x_n - x_{n-1})\rangle \\
&\leq \|x_n - x_{n-1}\|^p - pk\|x_n - x_{n-1}\|^p + 2pd^p\rho_E\left(\frac{4\|g(x_n) - g(x_{n-1})\|}{d}\right) \\
&\leq \|x_n - x_{n-1}\|^p - pk\|x_n - x_{n-1}\|^p + \frac{2pd^p c4^p\|g(x_n) - g(x_{n-1})\|^p}{d^p} \\
&\leq \|x_n - x_{n-1}\|^p - pk\|x_n - x_{n-1}\|^p + 2^{p+1}cp\lambda_g^p\|x_n - x_{n-1}\|^p \\
&\leq (1 - pk + c2^{2p+1}p\lambda_g^p)\|x_n - x_{n-1}\|^p. \qquad (28.3)
\end{aligned}$$

By the Lipschitz continuity of m and $\mathscr{H}$-Lipschitz continuity of $\tilde{T}$, we have

$$\begin{aligned}
\|m(z_n) - m(z_{n-1})\| &\leq \lambda_m\|z_n - z_{n-1}\| \\
&\leq \lambda_m(1 + n^{-1})\mathscr{H}(\tilde{T}(x_n), \tilde{T}(x_{n-1})) \\
&\leq \lambda_m(1 + n^{-1})\lambda_T\|x_n - x_{n-1}\|. \qquad (28.4)
\end{aligned}$$

Since N is p-relaxed accretive and Lipschitz continuous with respect to first, second, and third argument with constants $\alpha, \beta, \gamma, \lambda_{N_1}, \lambda_{N_2}$, and λ_{N_3}, respectively, and f, h, and t are Lipschitz continuous with constants λ_f, λ_h, and λ_t, respectively, we have

$$\|x_n - x_{n-1} - \tau(N(f(u_n),h(v_n),t(w_n)) - N(f(u_n),h(v_n),t(w_{n-1})))\|^p$$

$$
\begin{aligned}
&\leq \|x_n - x_{n-1}\|^p - p\tau\langle N(f(u_n),h(v_n),t(w_n)) - N(f(u_n),h(v_n),t(w_{n-1})),\\
&\quad j_p(x_n - x_{n-1} - \tau(N(f(u_n),h(v_n),t(w_n)) - N(f(u_n),h(v_n),t(w_{n-1}))))\rangle\\
&\leq \|x_n - x_{n-1}\|^p\\
&\quad - p\tau\langle N(f(u_n),h(v_n),t(w_n)) - N(f(u_n),h(v_n),t(w_{n-1})), j_p(x_n - x_{n-1})\rangle\\
&\quad - p\langle \tau(N(f(u_n),h(v_n),t(w_n)) - N(f(u_n),h(v_n),t(w_{n-1}))), j_p(x_n - x_{n-1}\\
&\quad - \tau(N(f(u_n),h(v_n),t(w_n)) - N(f(u_n),h(v_n),t(w_{n-1})))) - j_p(x_n - x_{n-1})\rangle\\
&\leq \|x_n - x_{n-1}\|^p + p\tau\gamma\|x_n - x_{n-1}\|^p\\
&\quad + 2pd^p\rho_E\left(\frac{4\tau\|N(f(u_n),h(v_n),t(w_n)) - N(f(u_n),h(v_n),t(w_{n-1}))\|}{d}\right)\\
&\leq \|x_n - x_{n-1}\|^p + p\tau\gamma\|x_n - x_{n-1}\|^p\\
&\quad + 2p\tau^p c4^p\|N(f(u_n),h(v_n),t(w_n)) - N(f(u_n),h(v_n),t(w_{n-1}))\|^p\\
&\leq \|x_n - x_{n-1}\|^p + p\tau\gamma\|x_n - x_{n-1}\|^p + 2^{2p+1}p\tau^p c\lambda_{N_3}^p\|t(w_n) - t(w_{n-1})\|^p\\
&\leq \|x_n - x_{n-1}\|^p + p\tau\gamma\|x_n - x_{n-1}\|^p + 2^{2p+1}p\tau^p c\lambda_{N_3}^p\lambda_t^p\|w_n - w_{n-1}\|^p\\
&\leq \|x_n - x_{n-1}\|^p + p\tau\gamma\|x_n - x_{n-1}\|^p + 2^{2p+1}p\tau^p c\lambda_{N_3}^p\lambda_t^p(1+n^{-1})\|x_n - x_{n-1}\|^p\\
&\leq (1 + p\tau\gamma + 2^{2p+1}p\tau^p c\lambda_{N_3}^p\lambda_t^p(1+n^{-1}))\|x_n - x_{n-1}\|^p. \qquad (28.5)
\end{aligned}
$$

Similarly we have

$$
\begin{aligned}
&\|x_n - x_{n-1} - \tau(N(f(u_n),h(v_n),t(w_{n-1})) - N(f(u_{n-1}),h(v_n),t(w_{n-1})))\|^p\\
&\qquad \leq (1 + p\tau\alpha + 2^{2p+1}p\tau^p c\lambda_{N_1}^p\lambda_f^p(1+n^{-1}))\|x_n - x_{n-1}\|^p, \qquad (28.6)
\end{aligned}
$$

and

$$
\begin{aligned}
&\|x_n - x_{n-1} - \tau(N(f(u_{n-1}),h(v_n),t(w_{n-1})) - N(f(u_{n-1}),h(v_{n-1}),t(w_{n-1})))\|^p\\
&\qquad \leq (1 + p\tau\beta + 2^{2p+1}p\tau^p c\lambda_{N_2}^p\lambda_h^p(1+n^{-1}))\|x_n - x_{n-1}\|^p. \qquad (28.7)
\end{aligned}
$$

from (28.5)–(28.7), we obtain

$$
\begin{aligned}
\|x_{n+1} - x_n\| &\leq 2(1 - pk + 2^{2p+1}cp\lambda_g^p)^{1/p}\|x_n - x_{n-1}\| + 2\lambda_m\lambda_T(1+n^{-1})\|x_n - x_{n-1}\|\\
&\quad + (1 + p\tau\gamma + 2^{2p+1}p\tau^p c\lambda_{N_3}^p\lambda_t^p(1+n^{-1}))^{1/p}\|x_n - x_{n-1}\|
\end{aligned}
$$

$$
\begin{aligned}
&+(1+p\tau\alpha+2^{2p+1}p\tau^p c\lambda_{N_1}^p\lambda_f^p(1+n^{-1}))^{1/p}\|x_n-x_{n-1}\| \\
&+(1+p\tau\beta+2^{2p+1}p\tau^p c\lambda_{N_2}^p\lambda_h^p(1+n^{-1}))^{1/p}\|x_n-x_{n-1}\| \\
\le\ &[2(1-pk+2^{2p+1}cp\lambda_g^p)^{1/p}+2\lambda_m\lambda_T(1+n^{-1}) \\
&+(1+p\tau\gamma+2^{2p+1}p\tau^p c\lambda_{N_3}^p\lambda_t^p(1+n^{-1}))^{1/p} \\
&+(1+p\tau\alpha+2^{2p+1}p\tau^p c\lambda_{N_1}^p\lambda_f^p(1+n^{-1}))^{1/p} \\
&+(1+p\tau\beta+2^{2p+1}p\tau^p c\lambda_{N_2}^p\lambda_h^p(1+n^{-1}))^{1/p}]\|x_n-x_{n-1}\| \\
\le\ &[q_n+\phi_n+\psi_n+\xi_n]\|x_n-x_{n-1}\| \\
\le\ &\theta_n\|x_n-x_{n-1}\|, \qquad (28.8)
\end{aligned}
$$

where

$$
\begin{aligned}
\theta_n &= q_n+\phi_n+\psi_n+\xi_n, \\
q_n &= 2(1-pk+2^{2p+1}cp\lambda_g^p)^{1/p}+2\lambda_m\lambda_T(1+n^{-1}) \\
\phi_n &= (1+p\tau\gamma+2^{2p+1}p\tau^p c\lambda_{N_3}^p\lambda_t^p(1+n^{-1}))^{1/p} \\
\psi_n &= (1+p\tau\alpha+2^{2p+1}p\tau^p c\lambda_{N_1}^p\lambda_f^p(1+n^{-1}))^{1/p} \\
\text{and } \xi_n &= (1+p\tau\beta+2^{2p+1}p\tau^p c\lambda_{N_2}^p\lambda_h^p(1+n^{-1}))^{1/p}.
\end{aligned}
$$

Let

$$\theta = q+\phi+\psi+\xi$$

and

$$
\begin{aligned}
q &= 2(1-pk+2^{2p+1}cp\lambda_g^p)^{1/p}+2\lambda_m\lambda_T \\
\phi &= (1+p\tau\gamma+2^{2p+1}p\tau^p c\lambda_{N_3}^p\lambda_t^p)^{1/p} \\
\psi &= (1+p\tau\alpha+2^{2p+1}p\tau^p c\lambda_{N_1}^p\lambda_f^p)^{1/p} \\
\xi &= (1+p\tau\beta+2^{2p+1}p\tau^p c\lambda_{N_2}^p\lambda_h^p)^{1/p}.
\end{aligned}
$$

Letting $n\to\infty$, we see that $\theta_n\to\theta$. Since $\theta<1$ by condition (28.1), $\theta_n<1$ for n sufficiently large. Therefore (28.1) implies that $\{x_n\}$ is a Cauchy sequence in E, and hence there exists $x\in E$ such that $x_n\to x$. By $\mathscr{H}$-Lipschitz continuity of $\tilde{F},\tilde{G},\tilde{H}$, and $\tilde{T}$, we have

$$
\begin{aligned}
\|u_n-u_{n-1}\| &\le (1+n^{-1})\mathscr{H}(\tilde{F}(x_n),\tilde{F}(x_{n-1})) \\
&\le (1+n^{-1})\lambda_F\|x_n-x_{n-1}\|; \\
\|v_n-v_{n-1}\| &\le (1+n^{-1})\mathscr{H}(\tilde{G}(x_n),\tilde{G}(x_{n-1})) \\
&\le (1+n^{-1})\lambda_G\|x_n-x_{n-1}\|; \\
\|w_n-w_{n-1}\| &\le (1+n^{-1})\mathscr{H}(\tilde{H}(x_n),\tilde{H}(x_{n-1})) \\
&\le (1+n^{-1})\lambda_H\|x_n-x_{n-1}\|;
\end{aligned}
$$

$$\begin{aligned}\|z_n - z_{n-1}\| &\leq (1+n^{-1})\mathscr{H}(\tilde{T}(x_n),\tilde{T}(x_{n-1}))\\ &\leq (1+n^{-1})\lambda_T\|x_n - x_{n-1}\|.\end{aligned}$$

It follows that $\{u_n\},\{v_n\},\{w_n\}$, and $\{z_n\}$ are the cauchy sequences in E. Hence there exist u,v,w, and $z \in E$ such that $u_n \to u, v_n \to v, w_n \to w$, and $z_n \to z$ as $n \to \infty$. Further, since $f,g,h,t,m,N,\tilde{F},\tilde{G},\tilde{H},\tilde{T}$, and Q_X are all continuous, we have

$$x = x - g(x) + Q_X[g(x) - \tau N(f(u),h(v),t(w)) - m(z)].$$

Since $u_n \in \tilde{F}(x_n)$, we have

$$\begin{aligned}d(u,\tilde{F}(x_n)) &\leq \|u-u_n\| + d(u_n,\tilde{F}(x))\\ &\leq \|u-u_n\| + \mathscr{H}(\tilde{F}(x_n),\tilde{F}(x))\\ &\leq \|u-u_n\| + \lambda_F\|x_n - x\| \to 0 \text{ as } n \to \infty,\end{aligned}$$

and hence $u \in \tilde{F}(x)$. Similarly $v \in \tilde{G}(x), w \in \tilde{H}(x)$ and $z \in \tilde{T}(x)$. By Theorem 28.7, it follows that (x,u,v,w,z) is a solution of (CGCCPFM). □

References

1. R. Ahmad, S. S. Irfan, Generalized quasi-complementarity problems with fuzzy set-valued mappings. International J. Fuzzy System. **5(3)**, 194–199 (2003)
2. Ya Alber, Metric and Generalized Projection Operators in Banach Spaces, Properties and Applications, in *Theory and Applications of Nonlinear Operators of Monotone and Accretive Type*, ed. by A. Kartsatos (Marcel Dekker, New York, 1996), p. 15–50
3. Q. H. Ansari, A. P. Farajzadeh, S. Schaible, Existence of solutions of vector variational inequalities and vector complementarity Problems. J. Glob. Optim. **45(2)**, 297–307 (2009)
4. F. E. Browder, Nonlinear mappings of nonexpansive and accretive type in Banach spaces. Bull. Amer. Math. Soc. **73**, 875–885 (1967)
5. R. W. Cottle, Nonlinear programs with positively bounded Jacobians. SIAM J. Appl. Math. **14**, 125–147 (1966)
6. R. W. Cottle, G. B. Dantzing, Complementarity pivot theory of mathematical programming. Linear Algebra Appl. **1**, 163–185 (1968)
7. M. F. Khan, Salahuddin, Generalized co-complementarity problems in p-uniformly smooth Banach spaces. J. Inequal. Pure and Appl. Math. **7(2) 66**, 1–11 (2006)
8. S. J. Habetler, A. L. Price, Existence theory for generalized nonlinear complementarity problems. J. Optim. Theo. Appl. **7**, 223–239 (1971)
9. G. Isac, On the implicit complementarity problem in Hilbert spaces. Bull. Austral. Math. Soc. **32**, 251–260 (1985)
10. G. Isac, Fixed point theory and complementarity problem in Hilbert spaces. Bull. Austral. Math. Soc. **36**, 295–310 (1987)
11. C. R. Jou, J. C. Yao, Algorithm for generalized multivalued variational inequalities in Hilbert spaces. Comput. Math. Appl. **25(9)**, 7–13 (1993)
12. S. Karmardian, Generalized complementarity problems. J. Optim. Theo. Appl. **8**, 161–168 (1971)
13. T. Kato, Nonlinear semigroups and evolution equations. J. Math. Soc. Japan. **18/19**, 508–520 (1967)
14. S. B. Nadler, Jr., Multivalued contraction mappings. Pacific J. Math. **30**, 475–488 (1969)

15. M. A. Noor, Nonlinear quasi-complementarity problems. Appl. Math. Lett. **2(3)**, 251–254 (1980)
16. M. A. Noor, On nonlinear complementarity problems. J. Math. Anal. Appl. **123**, 455–460 (1987)
17. A. H. Siddiqi, Q. H. Ansari, On the nonlinear implicit complementarity problem. Int. J. Math. and Math. Sci. **16(4)**, 783–790 (1993)

Chapter 29
Generating Eigenvalue Bounds Using Optimization

Henry Wolkowicz

Dedicated to the memory of Professor George Isac

Abstract This paper illustrates how optimization can be used to derive known and new theoretical results about perturbations of matrices and sensitivity of eigenvalues. More specifically, the Karush–Kuhn–Tucker conditions, the shadow prices, and the parametric solution of a fractional program are used to derive explicit formulae for bounds for functions of matrix eigenvalues.

29.1 Introduction

Many classical and new inequalities can be derived using optimization techniques. One first formulates the desired inequality as the maximum (minimum) of a function subject to appropriate constraints. The inequality, along with conditions for equality to hold, can then be derived and proved, provided that the optimization problem can be explicitly solved.

For example, consider the *Rayleigh principle*

$$\lambda_{\max} = \max\{\langle x, Ax\rangle : \ x \in \mathbb{R}^n, \|x\| = 1\}, \tag{29.1}$$

where A is an $n \times n$ Hermitian matrix, $\lambda_{\max}$ is the largest eigenvalue of A, $\langle\cdot,\cdot\rangle$ is the Euclidean inner product, and $\|\cdot\|$ is the associated norm. Typically, this principle is proved by maximizing the quadratic function $\langle x, Ax\rangle$ subject to the equality constraint, $\|x\|^2 = 1$. An explicit solution can be found using the classical and well known, Euler–Lagrange multiplier rule of calculus (see Example 29.2 below). It is

Henry Wolkowicz
Department of Combinatorics and Optimization, University of Waterloo, Waterloo, Ontario N2L 3G1, Canada.

P.M. Pardalos et al. (eds.), *Nonlinear Analysis and Variational Problems*, Springer Optimization and Its Applications 35, DOI 10.1007/978-1-4419-0158-3_29,

an interesting coincidence that λ is the standard symbol used in the literature for both eigenvalues and Lagrange multipliers; and the eigenvalue and Lagrange multiplier coincide in the above derivation. Not so well known are the multiplier rules for inequality constrained programs. The *Holder inequality*

$$\langle x,y\rangle = \sum_{i=1}^{n} x_i y_i \leq \left(\sum_{i=1}^{n} x_i^p\right)^{1/p} \left(\sum_{i=1}^{n} y_i^q\right)^{1/q},$$

where $x,y \in \mathbb{R}^n_+, p > 1, q = p/(p-1)$, can be proved by solving the optimization problem $h(x) := \max_y\{\sum_i x_i y_i : \sum_i y_i^q - 1 \leq 0, y_i \geq 0, \forall i\}$. The John multiplier rule yields the explicit solution (see [8] and Example 29.5 below). The classical *arithmetic–geometric mean inequality* $(\alpha_1 \ldots \alpha_n)^{1/n} \leq \frac{1}{n}(\alpha_1 + \ldots + \alpha_n)$, where $\alpha_i > 0, i = 1,\ldots,n$, can be derived by solving the geometric programming problem

$$\max\{\Pi_i \alpha_1 : \sum_{i=1}^{n} \alpha_i = 1, \alpha_i \geq 0, \forall i\}.$$

Convexity properties of the functions, which arise when reformulating the inequalities as programming problems, can prove very helpful. For example, convexity can guarantee that sufficiency, rather than only necessity, holds in optimality conditions. The *quasi-convexity* of the function

$$\phi(f) = \int f d\nu \int (1/f) d\mu, \tag{29.2}$$

where μ and ν are two nontrivial positive measures on a measurable space X, can be used to derive the Kantorovich inequality, [1]. (We prove the Kantorovich inequality using optimization in Example 29.3.) We rely heavily on the convexity and pseudo-convexity of the functions.

Optimality conditions, such as the Lagrange and Karush–Kuhn–Tucker multiplier rules, are needed to numerically solve mathematical programming problems. The purpose of this paper is to show how to use optimization techniques to generate known, as well as new, explicit eigenvalue inequalities. Rather than include all possible results, we concentrate on just a few, which allow us to illustrate several useful techniques. For example, suppose that A is an $n \times n$ complex matrix with real eigenvalues $\lambda_1 \geq \ldots \geq \lambda_n$. A lower bound for λ_k can be found if we can explicitly solve the problem

$$\begin{array}{lll} \min & \lambda_k & \\ \text{subject to} & \sum_{i=1}^n \lambda_i = \operatorname{trace} A & \\ & \sum_{i=1}^n \lambda_i^2 \leq \operatorname{trace} A^2 & \\ & \lambda_k - \lambda_i \leq 0, & i = 1,\ldots,k-1 \\ & \lambda_i - \lambda_k \leq 0, & i = k+1,\ldots,n. \end{array} \tag{29.3}$$

We can use the *Karush–Kuhn–Tucker* necessary conditions for optimality to find the explicit solution (see Theorem 29.9.) Sufficiency guarantees that we actually have the solution. This yields the best lower bound for λ_k based on the known data. (Further results along these lines can be found in [4].)

In addition, the Lagrange multipliers, obtained when solving the program (29.3), provide *shadow prices*. These shadow prices are sensitivity coefficients with respect to perturbations in the right-hand sides of the constraints. We use these shadow prices to improve the lower bound in the case that we have additional information about the eigenvalues (see, e.g., Corollaries 29.11 and 29.12).

29.1.1 Outline

In Section 29.2, we introduce the optimality conditions and use them to prove the well known: (i) Rayleigh principle; (ii) Holder inequality; and (iii) Kantorovich inequality. In Section 29.3, we show how to use the convex multiplier rule (or the Karush–Kuhn–Tucker conditions) to generate bounds for functions of the eigenvalues of an $n \times n$ matrix A with real eigenvalues. Some of these results have appeared in [7, 12, 4]. Included are bounds for $\lambda_k, \lambda_k + \lambda_\ell$ and $\lambda_k - \lambda_\ell$. We also show how to use the Lagrange multipliers (shadow prices) to strengthen the bounds. Section 29.4 uses fractional programming techniques to generate bounds for the ratios $(\lambda_k - \lambda_\ell)/(\lambda_k + \lambda_\ell)$. Some of the inequalities obtained here are given in [7, 12] but with proofs using elementary calculus techniques rather than optimization.

29.2 Optimality Conditions

29.2.1 Equality Constraints

First, consider the program

$$\min\{f(x) : h_k(x) = 0, k = 1, \ldots, q,\ x \in U\}, \tag{29.4}$$

where U is an open subset of $\mathbb{R}^n$ and the functions f, h_k, $k = 1, \ldots, q$, are continuously differentiable. The function f is called the *objective function* of the program. The *feasible set*, denoted by $\mathscr{F}$, is the set of points in $\mathbb{R}^n$ which satisfy the constraints. Then, the classical Euler–Lagrange multiplier rule states, e.g. [8],

Theorem 29.1. *Suppose that* $a \in \mathbb{R}^n$ *solves* (29.4) *and that the gradients* $\nabla h_1(a), \ldots, \nabla h_q(a)$ *are linearly independent. Then,*

$$\nabla f(a) + \sum_{k=1}^{q} \lambda_k \nabla h_k(a) = 0, \tag{29.5}$$

for some (Lagrange multipliers) $\lambda_k \in \mathbb{R}, k = 1, \ldots, q$.

Example 29.2. Suppose that A is an $n \times n$ Hermitian matrix with eigenvalues $\lambda_1 \geq \ldots \geq \lambda_n$. To prove the Rayleigh principle (29.1), consider the equivalent program

$$\text{minimize} \left\{ -\langle x, Ax \rangle : 1 - \sum_{i=1}^{n} x_i^2 = 0, x \in \mathbb{R}^n \right\}. \tag{29.6}$$

Since the objective function is continuous while the feasible set is compact, the minimum is attained at some $a \in \mathscr{F} \subset \mathbb{R}^n$. If we apply Theorem 29.1, we see that there exists a Lagrange multiplier $\lambda \in \mathbb{R}$ such that

$$2Aa - 2\lambda a = 0,$$

i.e., a is an eigenvector corresponding to the eigenvalue equal to the Lagrange multiplier λ. Since the objective function

$$\langle a, Aa \rangle = \lambda \langle a, a \rangle = \lambda,$$

we conclude that λ must be the largest eigenvalue and we get the desired result. If we now add the constraint that x be restricted to the $n-1$ dimensional subspace orthogonal to a, then we recover the second largest eigenvalue. Continuing in this manner, we get all the eigenvalues. More precisely, if $a_1, a_2, \ldots, a_k$ are k mutually orthonormal eigenvectors corresponding to the k largest eigenvalues of A, $\lambda_1 \geq \ldots \geq \lambda_k$, then we solve (29.6) with the added constraints

$$\langle x, a_i \rangle = 0, \quad i = 1, \ldots, k.$$

The gradients of the constraints are necessarily linearly independent since the vectors x, and a_i, $i = 1, \ldots, k$, are (mutually) orthonormal. Now if a is a solution, then (29.5) yields

$$2Aa - 2\lambda a + \sum_{i=1}^{k} \alpha_i a_i = 0,$$

for some Lagrange multipliers $\lambda, \alpha_i, i = 1, \ldots, k$. However, taking the inner product with fixed a_i, and using the fact that

$$\langle Aa, a_i \rangle = \langle a, Aa_i \rangle = \lambda_i \langle a, a_i \rangle = 0,$$

we see that $\alpha_i = 0$, $i = 1, \ldots, k$, and so $Aa = \lambda a$, i.e., a is the eigenvector corresponding to the $(k+1)-st$ largest eigenvalue. This argument also shows that A necessarily has n (real) mutually orthonormal eigenvectors.

Example 29.3. Consider the Kantorovich inequality, e.g., [1, 3],

$$1 \leq \langle x, Ax \rangle \langle x, A^{-1}x \rangle \leq \frac{1}{4} \left(\sqrt{\frac{\lambda_1}{\lambda_n}} + \sqrt{\frac{\lambda_n}{\lambda_1}} \right)^2, \tag{29.7}$$

where A is an $n \times n$ positive definite Hermitian matrix with eigenvalues $\lambda_1 \geq \ldots \geq \lambda_n > 0, x \in \mathbb{R}^n$, and $\|x\| = 1$. This inequality is useful in obtaining bounds for the rate of convergence of the method of steepest descent, e.g., [5]. To prove the inequality we consider the following (two) optimization problems

$$\begin{array}{ll} \min(\max)\ f_1(a) := \left(\sum_{i=1}^n a_i^2 \lambda_i\right) \left(\sum_{i=1}^n a_i^2 \lambda_i^{-1}\right) \\ \text{subject to } \ g(a) := 1 - \sum_{i=1}^n a_i^2 = 0, \end{array} \tag{29.8}$$

where $a = (a_i) \in \mathbb{R}^n, a_i = <x, u_i>$ and $u_i, i = 1, \ldots, n$, is an orthonormal set of eigenvectors of A corresponding to the eigenvalues $\lambda_i, \ \ i = 1, \ldots, n$, respectively. Thus, $f_1(a)$ is the middle expression in (29.7). Suppose that the vector $a = (a_i)$ solves (29.8). Then, the necessary conditions of optimality state that (μ is the Lagrange multiplier)

$$a_i \lambda_i \left(\sum_j a_j^2 \lambda_j^{-1} \right) + a_i \lambda_i^{-1} \left(\sum_j a_j^2 \lambda_j \right) - \mu a_i = 0, i = 1, \ldots, n; \sum_i a_i^2 = 1. \tag{29.9}$$

Thus,

$$f_2(a) := \lambda_i \left(\sum_j a_j^2 \lambda_j^{-1} \right) + \lambda_i^{-1} \left(\sum_j a_j^2 \lambda_j \right) = \mu, \text{ if } a_i \neq 0. \tag{29.10}$$

On the other hand, if we multiply (29.9) by a_i and sum over i, we get

$$\mu = 2 \left(\sum_j a_j^2 \lambda_j \right) \left(\sum_j a_j^2 \lambda_j^{-1} \right) = 2 f_1(a). \tag{29.11}$$

By (29.10) and (29.11), we can replace $f_1(a)$ in (29.8) by the middle expression in (29.10), i.e., by $f_2(a)$. The new necessary conditions for optimality (with μ playing the role of the Lagrange multiplier again and $a_i \neq 0$) are

$$a_j \frac{\lambda_i}{\lambda_j} + a_j \frac{\lambda_j}{\lambda_i} - a_j \mu = 0, \quad j = 1, \ldots, n.$$

Now, if both $a_j \neq 0, a_i \neq 0$, we get

$$f_3(a) := \frac{\lambda_i}{\lambda_j} + \frac{\lambda_j}{\lambda_i} = \mu. \tag{29.12}$$

And, multiplying (29.12) by a_j and summing over j yields

$$\mu = \sum_j a_j^2 \left(\frac{\lambda_i}{\lambda_j} + \frac{\lambda_j}{\lambda_i} \right) = f_2(a).$$

Thus, we can now replace $f_2(a)$ (and so $f_1(a)$) in (29.8) by $f_3(a)$. Note that a does not appear explicitly in $f_3(a)$. However, the i and j must correspond to $a_i \neq 0$ and $a_j \neq 0$. Consider the function

$$h(x,y) = \frac{x}{y} + \frac{y}{x}, \tag{29.13}$$

where $0 < \alpha \leq x \leq y \leq \beta$. Since

$$\frac{d}{dx}h(x,y) = \frac{y(x^2 - y^2)}{(xy)^2} \leq 0 \quad (< 0 \text{ if } x \neq y),$$

and, similarly $\frac{d}{dy}h(x,y) \geq 0$ (> 0 if $x \neq y$)), we see that h attains its maximum at $x = \alpha$ and $y = \beta$, and it attains its minimum at $x = y$. This shows that $2 \leq f_3(a)$ and that f_3 attains its maximum at $\frac{\lambda_1}{\lambda_n} + \frac{\lambda_n}{\lambda_1}$, i.e., at $a_1 \neq 0$ and $a_n \neq 0$. The left-hand side of (29.7) now follows from $2f_1(a) = f_2(a) = f_3(a)$. Now, to have $f_3(a) = f_2(a)$, we must choose $a_1 = a_n = \frac{1}{2}$, and $a_i = 0, \forall 1 < i < n$. Substituting this choice of a in $f_1(a)$ yields the right-hand side of (29.7).

29.2.2 Equality and Inequality Constraints

Now suppose that program (29.4) has, in addition, the inequality constraints (continuously differentiable)

$$g_i(x) \leq 0, \quad i = 1, \ldots, m. \tag{29.14}$$

Then, we obtain the John necessary conditions of optimality (see, e.g., [8]).

Theorem 29.4. *Suppose that $a \in \mathbb{R}^n$ solves (29.4) with the additional constraints* (29.14). *Then, there exist Lagrange multiplier vectors $\lambda \in \mathbb{R}^{m+1}_+, \alpha \in \mathbb{R}^q$, not both zero, such that*

$$\begin{array}{c} \lambda_0 \bigtriangledown f(a) + \sum_{i=1}^m \lambda_i \bigtriangledown g_i(a) + \sum_{j=1}^q \alpha_j \bigtriangledown h_j(a) = 0, \\ \lambda_i g_i(a) = 0, \quad i = 1, \ldots, m. \end{array} \tag{29.15}$$

The first condition in (29.15) is *dual feasibility*. The second condition in (29.15) is called *complementary slackness*. It shows that either the multiplier $\lambda_i = 0$ or the constraint is *binding*, i.e., $g_i(a) = 0$. The Karush–Kuhn–Tucker conditions (e.g., [8]) assume a *constraint qualification* and have $\lambda_0 = 1$.

Example 29.5. Holder's inequality states that if $x, y \in \mathbb{R}^n_{++}$, are (positive) vectors, $p > 1$, and $q = p/(p-1)$, then

$$\langle x, y \rangle = \sum_{i=1}^n x_i y_i \leq \left(\sum_i x_i^p \right)^{1/p} \left(\sum_i y_i^q \right)^{1/q} = \|x\|_p \|y\|_q.$$

We now include a proof of this inequality using the John multiplier rule. (This proof corrects the one given in [8].)

Fix $y = (y_i) \in \mathbb{R}^n_{++}$ and consider the program

$$\begin{array}{rll} \min & f(x) := -\sum_{i=1}^n x_i y_i & \\ \text{subject to} & g(x) := \sum_{i=1}^n x_i^p - 1 & \leq 0 \\ & h_i(x) := -x_i & \leq 0,\ i = 1,\dots,n. \end{array}$$

Holder's inequality follows if the optimal value is $-\|y\|_q$. Since the feasible set is compact, the minimum is attained at say $a = (a_i) \in \mathbb{R}^n_+$. Then, there exist constants (Lagrange multipliers) $\lambda_0 \geq 0, \lambda_1 \geq 0, \gamma_i \geq 0$, not all zero, such that

$$\begin{array}{l} -\lambda_0 y_i + \lambda_1 p a_i^{p-1} - \gamma_i = 0,\ \gamma_i \geq 0, \forall i \\ \lambda_1 g(a) = 0, \quad \gamma_i a_i = 0, \forall i. \end{array}$$

This implies that, for each i we have

$$\begin{array}{rl} -\lambda_0 y_i + \lambda_1 p a_i^{p-1} = \gamma_i = 0, & \text{if } a_i > 0, \\ -\lambda_0 y_i = \gamma_i \geq 0, & \text{if } a_i = 0. \end{array}$$

Since $y_i \geq 0$ and $\lambda_0 \geq 0$, we conclude that $\lambda_0 y_i = \gamma_i = 0$, if $a_i = 0$. Therefore, we get

$$-\lambda_0 y_i + \lambda_1 p a_i^{p-1} = \gamma_i = 0, \forall i. \tag{29.16}$$

The remainder of the proof now follows as in [8]. More precisely, since not all the multipliers are 0, if $\lambda_0 = 0$, then $\lambda_1 > 0$. This implies that

$$g(a) = 0, \tag{29.17}$$

and, by (29.16) that $a = 0$, contradiction. On the other hand, if $\lambda_1 = 0$, then $\lambda_0 > 0$ which implies $y = 0$, contradiction. Thus, both λ_0 and λ_1 are positive and we can assume, without loss of generality, that $\lambda_0 = 1$. Moreover, we conclude that (29.17) holds. From (29.16) and (29.17) we get

$$\begin{aligned} -f(a) &= \textstyle\sum_{i=1}^n a_i y_i \\ &= \lambda_1 p \textstyle\sum_{i=1}^n a_i^p \\ &= \lambda_1 p. \end{aligned}$$

Since $q = p/(p-1)$, (29.16) and (29.17) now imply that

$$\sum_{i=1}^n y_i^q = \sum_{i=1}^n (\lambda_1 p)^q a_i^q = (\lambda_1 p)^q = -f(a)^q.$$

29.2.3 Sensitivity Analysis

Consider now the *convex (perturbed) program*

$$(P_\varepsilon) \qquad \begin{array}{rl} \mu(\varepsilon) = \min & f(x) \\ & \text{subject to } g_i(x) \leq \varepsilon_i, \ i = 1, \ldots, m, \\ & h_j(x) = \varepsilon_j, \ j = m+1, \ldots, q, \\ & x \in U, \end{array} \tag{29.18}$$

where U is an open subset of $\mathbb{R}^n$, and the functions f and g_i, $i = 1, \ldots, m$, are convex and h_j, $j = m+1, \ldots, q$, are affine. The *generalized Slater constraint qualification (CQ)* for (P_ε) states that

$$\begin{array}{c} \text{there exists } \hat{x} \in \operatorname{int} U \text{ such that} \\ g_i(\hat{x}) < \varepsilon_i, i = 1, \ldots, m, \text{ and } h_j(\hat{x}) = \varepsilon_j, j = m+1, \ldots, q. \end{array} \tag{29.19}$$

We can now state the convex multiplier rule and the corresponding shadow price interpretation of the multipliers (see, e.g., [8, 9]).

Theorem 29.6. *Suppose that the CQ in* (29.19) *holds for* (P_0) *in* (29.18). *Then,*

$$\mu(0) = \min\{f(x) + \sum_{i=1}^{m} \lambda_i g_i(x) + \sum_{j=m+1}^{q} \lambda_j h_j(x) : \ x \in U\}, \tag{29.20}$$

for some $\lambda_j \in R$, $j = m+1, \ldots q$, *and* $\lambda_i \geq 0$, $i = 1, \ldots, m$. *If* $a \in \mathscr{F}$ *solves* (P_0), *then in addition*

$$\lambda_i g_i(a) = 0, \quad i = 1, \ldots, m. \tag{29.21}$$

Theorem 29.7. *Suppose that* $a \in \mathscr{F}$. *Then, (29.20) and (29.21) imply that a solves* (P_0).

Theorem 29.8. *Suppose that* a^1 *and* a^2 *are solutions to* (P_{ε^1}) *and* (P_{ε^2}), *respectively, with corresponding multiplier vectors* λ^1 *and* λ^2. *Then,*

$$(\varepsilon^2 - \varepsilon^1, \lambda^2) \leq f(a^1) - f(a^2) \leq (\varepsilon^2 - \varepsilon^1, \lambda^1). \tag{29.22}$$

Note that since the functions are convex and the problem (29.20) is an unconstrained minimization problem, we see that if $a \in \mathscr{F}$ solves (P_0), then (29.20) and (29.21) are equivalent to the system

$$\begin{array}{l} \nabla f(a) + \sum_{i=1}^{m} \lambda_i \nabla g_i(a) + \sum_{j=m+1}^{q} \lambda_j \nabla h_j(a) = 0 \\ \lambda_i \geq 0, \ \ \lambda_i g_i(a) = 0, \ \ i = 1, \ldots, m. \end{array} \tag{29.23}$$

Moreover, since $f(a^i) = \mu(\varepsilon^i)$, when a^i solves (P_{ε^i}), (29.22) implies that $-\lambda^i \in \partial\mu(\varepsilon^i)$, i.e., the negative of the multiplier λ^i is in the subdifferential of the perturbation function $\mu(\varepsilon)$ at ε^i. In fact (see [9])

$$\partial\mu(0) = \{-\lambda : \lambda \text{ is a multiplier vector for } (P_0)\}.$$

If λ is unique, this implies that μ is differentiable at 0 and $\nabla\mu(0) = -\lambda$. Note that

$$\partial\mu(a) = \{\phi \in \mathbb{R}^n : \ (\phi, \eta - a) \le \mu(\eta) - \mu(a)\}.$$

We will apply the convex multiplier rule in the sequel. Note that the necessity of (29.23) requires a constraint qualification, such as Slater's condition, while sufficiency does not. Thus, in our applications we do not have to worry about any constraint qualification. For, as soon as we can solve (29.23), the sufficiency guarantees optimality. Note that necessity is used in numerical algorithms.

29.3 Generating Eigenvalue Bounds

We consider the $n \times n$ matrix A which has real eigenvalues $\lambda_1 \ge \ldots \ge \lambda_n$. We have seen how to apply optimization techniques in order to prove several known inequalities. Now suppose that we are given several facts about the matrix A, e.g., n, trace$A$ and/or det$A$ etc... In order to find upper (lower) bounds for $f(\lambda)$, a function of the eigenvalues, we could then maximize (minimize) $f(\lambda)$ subject to the constraints corresponding to the given facts about A. An explicit solution to the optimization problem would then provide the, previously unknown, best upper (lower) bounds to $f(\lambda)$ given these facts. To be able to obtain an explicit solution, we must choose simple enough constraints and/or have a lot of patience.

Suppose we wish to obtain a lower bound for λ_k, the k-th largest eigenvalue, given the facts that

$$K := \operatorname{trace} A, \quad m := \frac{K}{n}, \quad L := \operatorname{trace} A^2, \quad s^2 := \frac{L}{n} - m^2.$$

Then we can try and solve the program

$$\begin{array}{ll} \min & \lambda_k \\ \text{subject to} & (a)\ \sum_{i=1}^n \lambda_i = K, \\ & (b)\ \sum_{i=1}^n \lambda_i^2 \le L, \\ & (c)\ \lambda_k - \lambda_i \le 0, \ \ i = 1, \ldots, k-1, \\ & (d)\ \lambda_i - \lambda_k \le 0, \ \ i = k+1, \ldots, n. \end{array} \tag{29.24}$$

This is a program in the variables λ_i with n, k, K, and L fixed. We have replaced the constraint $\sum \lambda_i^2 = L$ with $\sum \lambda_i^2 \le L$. This increases the feasible set of vectors $\lambda = (\lambda_i)$ and so the solution of (29.24) still provides a lower bound for λ_k. However, the program now becomes a convex program. Note that $(\operatorname{trace} A)^2 = (\sum \lambda_i)^2 \le n \sum \lambda_i^2 = n\operatorname{trace} A^2$, by the Cauchy–Schwartz inequality, with equality if and only if $\lambda_1 = \lambda_2 = \ldots = \lambda_n$. Thus, if $(\operatorname{trace} A)^2 = n\operatorname{trace} A^2$, then we can immediately conclude that $\lambda_i = \operatorname{trace} A/n, \ \ i = 1, \ldots, n$. Moreover, if $nL \ne K^2$, then $nL > K^2$, and we can always find a feasible solution to the constraints which strictly satisfies $\sum \lambda_i^2 < L$, and hence we can always satisfy the generalized Slater CQ.

Theorem 29.9. *Suppose that $K^2 < nL$. If $1 < k \le n$, then the (unique) explicit solution to* (29.24) *is*

$$\begin{aligned} \lambda_1 = \ldots = \lambda_{k-1} = m + s\left(\tfrac{n-k+1}{k-1}\right)^{\frac{1}{2}}, \\ \lambda_k = \ldots = \lambda_n = m - s\left(\tfrac{k-1}{n-k+1}\right)^{\frac{1}{2}}, \end{aligned} \tag{29.25}$$

with Lagrange multipliers for the constraints (a) to (d) in (29.24) *being*

$$\begin{aligned} \alpha &= \tfrac{-m}{ns}\left(\tfrac{k-1}{n-k+1}\right)^{\frac{1}{2}} - \tfrac{1}{n}, \\ \beta &= \left(\tfrac{k-1}{n-k+1}\right)^{\frac{1}{2}} \tfrac{1}{ns}, \\ \gamma_i &= 0, \quad i = 1, \ldots, k-1, \\ \gamma_i &= \tfrac{1}{n-k+1}, \quad i = k+1, \ldots, n, \end{aligned} \tag{29.26}$$

respectively.

Proof. Since (29.24) is a convex program, the Karush–Kuhn–Tucker conditions are sufficient for optimality. Thus, we need only verify that the above solution satisfies both the constraints and (29.23). However, let us suppose that the solution is unknown beforehand, and show that we can use the necessity of (29.23) to find it. We get

$$\alpha + \beta\lambda_i - \gamma_i = 0, \quad i = 1, \ldots, k-1 \tag{29.27a}$$

$$1 + \alpha + \beta\lambda_k + \sum_{i=1}^{k-1} \gamma_i - \sum_{i=k+1}^{n} \gamma_i = 0 \tag{29.27b}$$

$$\alpha + \beta\lambda_i + \gamma_i = 0, i = k+1, \ldots, n, \tag{29.27c}$$

$$a \in R, \beta, \gamma_i > 0, \beta\left(\sum_{i=1}^{n} \lambda_i^2 - L\right) = 0, \gamma_i(\lambda_i - \lambda_k) = 0, i = 1, \ldots, n. \tag{29.27d}$$

Now, if $\beta = 0$, then

$$\alpha = \gamma_i = -\gamma_j, i = 1, \ldots, k-1, j = k+1, \ldots, n.$$

This implies that they are all 0, (or all > 0 if $k = n$) which contradicts (29.27b). Thus, $\beta > 0$ and, by (29.27d),

$$\sum_i \lambda_i^2 = L. \tag{29.28}$$

From (29.27d), we now have

$$\lambda_i = \frac{-\alpha}{\beta} + \frac{\gamma_i}{\beta}, i = 1, \ldots, k-1,$$

$$\lambda_i = \frac{-\alpha}{\beta} - \frac{\gamma_i}{\beta}, i = k+1, \ldots, n,$$

$$\lambda_k = \frac{-\alpha}{\beta} - \frac{1}{\beta} - \sum_{i=1}^{k-1} \frac{\gamma_i}{\beta} + \sum_{j=k+1}^{n} \frac{\gamma_j}{\beta}.$$

Suppose $\gamma_{i_0} > 0$, where $1 < i_0 < k-1$. Then (29.27a) and (29.27d) imply that $\lambda_k = \lambda_{i_0} = \frac{-\alpha}{\beta} + \frac{\gamma_{i_0}}{\beta}$. On the other hand, since we need $\lambda_i = \frac{-\alpha}{\beta} + \frac{\gamma_i}{\beta} \geq \lambda_k, i = 1, \ldots, k-1$, we must have $\gamma_i > 0, i = 1, \ldots, k-1$ and, by complementary slackness,

$$\lambda_i = \lambda_k = \frac{-\alpha}{\beta} + \gamma_i, i = 1, \ldots, k-1. \tag{29.29}$$

But then

$$\lambda_k = \frac{-\alpha}{\beta} - \frac{1}{\beta} - \sum_{j=1}^{k-1} \frac{\gamma_j}{\beta} + \sum_{j=k+1}^{n} \frac{\gamma_j}{\beta} = \frac{-\alpha}{\beta} + \frac{\gamma_i}{\beta}, i = 1, \ldots, k-1,$$

which implies $\sum_{j=k+1}^{n} \gamma_j > 0$. But $\gamma_j > 0$ implies $\lambda_j = \lambda_k$. This yields $\lambda_1 = \ldots = \lambda_k = \ldots = \lambda_n$, a contradiction since we assumed $nL > K^2$. We thus conclude that

$$\gamma_i = 0, \quad i = 1, \ldots, k-1.$$

Now if $\gamma_{j_0} > 0$, for some $k < j_0 < n$, then $\lambda_{j_0} = \frac{-\alpha}{\beta} - \frac{1}{\beta} + \sum_{j=k+1}^{n} \frac{\gamma_j}{\beta}$. Since $\lambda_j = \frac{-\alpha}{\beta} - \frac{\lambda_j}{\beta} \leq \lambda_k$, we must have $\gamma_j > 0$ for all $j = k+1, \ldots, n$. Note that $\gamma_j = 0$ for all $j = k+1, \ldots, n$, leads to a contradiction since then $\lambda_j = \frac{-\alpha}{\beta} > \lambda_k = \frac{-\alpha}{\beta} - \frac{1}{\beta}$. Thus we have shown that the λ_i's split into two parts,

$$\lambda_1 = \ldots = \lambda_{k-1}, \lambda_k = \ldots = \lambda_n. \tag{29.30}$$

The Lagrange multipliers also split into two parts,

$$\gamma_1 = \ldots = \gamma_{k-1} = 0, \gamma_{k+1} = \ldots = \gamma_n = \gamma.$$

We now explicitly solve for $\lambda_1, \lambda_k, \alpha, \beta$, and γ. From the first two constraints and (29.28) we get

$$\begin{aligned} (k-1)\lambda_1 + (n-k+1)\lambda_k &= K, \\ (k-1)\lambda_1^2 + (n-k+1)\lambda_k^2 &= L. \end{aligned} \tag{29.31}$$

Eliminating one of the variables in (29.31) and solving the resulting quadratic yields (29.25). Uniqueness of (29.25) follows from the necessity of the optimality conditions. It also follows from the strict convexity of the quadratic constraint in the program (29.24). Using the partition in (29.30), we can substitute (29.27c) in (29.27b) to get

$$1 + \alpha + \beta\left(\frac{-\alpha - \gamma}{\beta}\right) - (n-k)\gamma = 0,$$

i.e.,

$$\gamma = \frac{1}{n-k+1}. \tag{29.32}$$

In addition, $\lambda_1 - \lambda_k = \frac{-\alpha}{\beta} - \left(\frac{-\alpha}{\beta} - \frac{\gamma}{\beta}\right)$ implies

$$\beta = \frac{\gamma}{\lambda_1 - \lambda_k} = \left(\frac{k-1}{n-k+1}\right)^{\frac{1}{2}} \frac{1}{ns}, \tag{29.33}$$

while

$$\alpha = -\lambda_1 \beta = \frac{-m}{ns}\left(\frac{k-1}{n-k+1}\right)^{\frac{1}{2}} - \frac{1}{n}. \tag{29.34}$$

□

In the above, we have made use of the necessity of the Karush–Kuhn–Tucker (**KKT**) conditions to eliminate non-optimal feasible solutions. Sufficiency of the **KKT** conditions in the convex case then guarantees that we have actually found the optimal solution and so we need not worry about any constraint qualification. We can verify our solution by substituting into (29.27).

The explicit optimal solution yields the lower bound as well as conditions for it to be attained.

Corollary 29.10. *Let* $1 < k \leq n$. *Then*

$$\lambda_k \geq m - \left(\frac{k-1}{n-k+1}\right)^{\frac{1}{2}} s, \tag{29.35}$$

with equality if and only if $\lambda_1 = \ldots = \lambda_{k-1}, \lambda_k = \ldots = \lambda_n$. □

The above corollary is given in [12] but with a different proof. From the proof of Theorem 29.9, we see that $\beta = 0$ if $k = 1$ and so the quadratic constraint $\sum \lambda_i^2 \leq L$ may not be binding at the optimum. Thus the solution may violate the fact that $\sum \lambda_i^2 = \operatorname{trace} A^2$. This suggests that we can do better if we replace the inequality constraint by the equality constraint $\sum \lambda_i^2 = L$. We, however, lose the nice convexity properties of the problem. However, applying the John conditions, Theorem 29.4, and using a similar argument to the proof of Theorem 29.9, yields the explicit solution

$$\begin{aligned} \lambda_1 = \ldots = \lambda_{n-1} &= m + s/(n-1)^{\frac{1}{2}} \\ \lambda_n &= m - (n-1)^{\frac{1}{2}} s, \end{aligned}$$

i.e., we get the lower bound

$$\lambda_1 \geq m + s/(n-1)^{\frac{1}{2}},$$

with equality if and only if $\lambda_1 = \ldots = \lambda_{n-1}$. (This result is also given in [12] but with a different proof.)

The Lagrange multipliers obtained in Theorem 29.9 also provide the sensitivity coefficients for program (29.24). (In fact, the multipliers are unique and so the perturbation function is differentiable.) This helps in obtaining further bounds for the eigenvalues when we have some additional information, e.g., from Gerŝgorin discs. We can now improve our lower bound and also obtain lower bounds for other eigenvalues.

Corollary 29.11. *Let* $1 < k < n$. *Suppose that we know*

$$\lambda_{k+i} - \lambda_k \leq -\varepsilon_i, \tag{29.36}$$

where $\varepsilon_i \geq 0, i = 1, \dots, n-k$. *Then*

$$\lambda_k \geq m - \left(\frac{k-1}{n-k+1}\right)^{\frac{1}{2}} s + \frac{1}{n-k+1} \sum_{i=1}^{n-k} \varepsilon_i. \tag{29.37}$$

Proof. The result follows immediately from the left-hand side of (29.22), if we perturb the constraints in program (29.24) as given in (29.36) and use the multipliers $\gamma_i = \frac{1}{n=k+1}$. Note that λ_k remains the k-th largest eigenvalue. □

Corollary 29.12. *Let* $1 < k < n$. *Suppose that we know*

$$\lambda_{k+i-1} - \lambda_{k+t} \leq \varepsilon_i,$$

for some $\varepsilon_i \geq 0, \;\; i = 1, \dots, t$. *Then*

$$\lambda_{k+t} \geq m - \left(\frac{k-1}{n-k+1}\right)^{\frac{1}{2}} s - \frac{1}{n-k+1} \sum_{i=1}^{t} \varepsilon_i.$$

Proof. Suppose that we perturb the constraints in program (29.24) to obtain

$$\lambda_{k+i} - \lambda_k \leq \varepsilon_i, i = 1, \dots, t. \tag{29.38}$$

Since $\varepsilon_i \geq 0$, this allows a change in the ordering of the λ_i, for then we can have $\lambda_{k+i} = \lambda_k + \varepsilon_i > \lambda_k$. Thus the perturbation in the hypothesis is equivalent to (29.38). From (29.22), the result follows, since the k-th ordered λ_i has become the $(k+t)$-th order λ_i. □

The results obtained using perturbations in the above two corollaries can be approached in a different way. Since the perturbation function μ is convex (see, e.g., [9]) we are obtaining a lower estimate of the perturbed value $\mu(\varepsilon)$ by using the multiplier whose negative is an element of the sub-differential $\partial\mu(\varepsilon)$. We can however obtain better estimates by solving program (29.24) with the new perturbed constraints.

Theorem 29.13. *Under the hypotheses of Corollary 29.11, we get*

$$\lambda_k \geq m - \left(\frac{k-1}{n-k+1}\right)^{\frac{1}{2}} s_\varepsilon + \frac{1}{n-k+1} \sum_{j=1}^{t} \varepsilon_j, \tag{29.39}$$

where

$$s_\varepsilon^2 = s^2 - \frac{(n-k+1)\sum_{j=1}^t \varepsilon_j^2 - \left(\sum_{j=1}^t \varepsilon_j\right)^2}{n(n-k+1)}.$$

Equality holds if and only if

$$\lambda_1 = \ldots = \lambda_{k-1}; \lambda_{k+i} - \lambda_k = -\varepsilon_i, i = 1, \ldots, t.$$

Proof. We replace the last set of constraints in program (29.24) by the perturbed constraints (29.36), for $i = k+1, \ldots, k+t$. The arguments in the proof of Theorem 29.9 show that the solution must satisfy (29.28) and

$$\lambda_1 = \ldots = \lambda_{k-1}; \lambda_{k+j} - \lambda_k = -\varepsilon_j, j = 1, \ldots, t.$$

We can assume that $k+t = n$, since we must have $\lambda_{k+t+j} \leq \lambda_{k+t}$ and so we can add the constraints

$$\lambda_{k+t+j} - \lambda_k \leq -\varepsilon_t, j > 1,$$

if required. This leads to the system

$$\begin{array}{l}(k-1)\lambda_1 + \sum_{j=1}^t (\lambda_k - \varepsilon_j) = K, \\ (k-1)\lambda_1^2 + \sum_{j=1}^t (\lambda_k - \varepsilon_j)^2 = L.\end{array} \tag{29.40}$$

Let $\varepsilon := \sum_{j=1}^t \varepsilon_j$ and $\bar{\varepsilon} := \sum_{j=1}^t \varepsilon_j^2$. Then (29.40) reduces to

$$\begin{array}{l}(k-1)\lambda_1 + (n-k+1)\lambda_k = K_\varepsilon := K + \varepsilon, \\ (k-1)\lambda_1^2 + (n-k+1)\lambda_k^2 - 2\varepsilon\lambda_k = L_\varepsilon := L - \bar{\varepsilon}.\end{array}$$

Then

$$\lambda_k = (K_\varepsilon - (k-1)\lambda_1)/(n-k+1).$$

Substituting for λ_k yields the quadratic

$$n(k-1)\lambda_1^2 - 2(k-1)(K_\varepsilon - \varepsilon)\lambda_1 + K_\varepsilon^2 - 2\varepsilon K_\varepsilon - (n-k+1)L_\varepsilon = 0,$$

which implies

$$\lambda_1 = \frac{K}{n} + \left(\frac{n-k+1}{k-1}\right)^{\frac{1}{2}} \left\{ \frac{(n-k+1)L_\varepsilon + \varepsilon^2}{n(n-k+1)} - \left(\frac{K}{n}\right)^2 \right\}^{\frac{1}{2}} \tag{29.41}$$

and

$$\lambda_k = \frac{K}{n} + \frac{\varepsilon}{n-k+1} - \left(\frac{k-1}{n-k+1}\right)^{\frac{1}{2}} \left\{ \frac{L}{n} - \frac{(n-k+1)\bar{\varepsilon} + \varepsilon^2}{n(n-k+1)} - \left(\frac{K}{n}\right)^2 \right\}^{\frac{1}{2}} \tag{29.42}$$

□

Note that the partial derivative with respect to $-\varepsilon_j$, at $\varepsilon_j = 0$, of the lower bound for λ_k in (29.39) is $-1/(n-k+1)$. This agrees with the fact that the corresponding multiplier is $\gamma_j = 1/(n-k+1)$.

Corollary 29.12 can be improved in the same way that Theorem 29.13 improves Corollary 29.11. We need to consider the program (29.24) with the new constraints

$$\lambda_{k-i} - \lambda_k \leq \varepsilon_i, i = 1, \ldots, t,$$

where $\varepsilon_i \geq 0$ and k has replaced $k+t$. Further improvements can be obtained if more information is known. For example, we might know that

$$\lambda_{t+i} - \lambda_t \leq -\varepsilon_i, \quad i = 1, \ldots, s,$$

where $l+s < t+s < k$ or $k+s < t+s < n$. In these cases we would obtain a result as in Theorem 29.13.

In the remainder of this section we consider bounds for $\lambda_k + \lambda_\ell$ and $\lambda_k - \lambda_\ell$. To obtain a lower bound for $\lambda_k + \lambda_\ell$ we consider the program

$$\begin{array}{lll} \text{minimize} & \lambda_k + \lambda_\ell & \\ \text{subject to} & (a) \sum \lambda_i = K, & \\ & (b) \sum \lambda_i^2 \leq L, & \\ & (c) \lambda_i - \lambda_k \leq 0, i = k+1, \ldots, \ell & \\ & (d) \lambda_j - \lambda_\ell \leq 0, j = \ell+1, \ldots, n. & \end{array} \tag{29.43}$$

Note that we have ignored the constraints $\lambda_i - \lambda_k \geq 0, i = 1, \ldots, k-1$. From our previous work in the proof of Theorem 29.9, we see that the Lagrange multipliers for these constraints should all be 0, i.e., we can safely ignore these constraints without weakening the bound.

Theorem 29.14. *Suppose that $K^2 < nL$ and $1 \leq k < \ell \leq n$. Then the explicit solution to* (29.43) *is*

1. If $n - \ell > \ell - k - 1$, then

$$\begin{aligned} \lambda_1 &= \ldots = \lambda_{k-1} = m + s\left(\tfrac{n-k+1}{k-1}\right)^{\frac{1}{2}}, \\ \lambda_k &= \ldots = \lambda_n = m - s\left(\tfrac{k-1}{n-k+1}\right)^{\frac{1}{2}}, \end{aligned} \tag{29.44}$$

with Lagrange multipliers for the constraints

$$\begin{aligned} &\alpha = -2\beta\lambda_k - \tfrac{2}{(n-k+1)}, \\ &\beta = \tfrac{1}{ns}\left(\tfrac{k-1}{n-k+1}\right)^{\frac{1}{2}}, \\ &\gamma_i = \delta_j = 2/(n-k+1), i = k+1, \ldots, \ell-1, j = \ell+1, \ldots, n, \\ &\gamma_\ell = \tfrac{2(n-\ell+1)}{n-k+1} - 1. \end{aligned} \tag{29.45}$$

2. If $n - \ell \leq \ell - k - 1$, then λ_1 is the solution of the quadratic (29.51) *and*

$$
\begin{array}{l}
\lambda_1 = \ldots = \lambda_{k-1}, \\
\lambda_k = \ldots = \lambda_{\ell-1} = \lambda_1 + \frac{K-n\lambda_1}{2(\ell-k)} \\
\lambda_\ell = \ldots = \lambda_n = \lambda_1 + \frac{K-n\lambda_1}{2(n-\ell+1)}
\end{array} \tag{29.46}
$$

with Lagrange multipliers for the constraints being

$$
\begin{array}{l}
\alpha = -\beta\lambda_1, \\
\beta = \frac{2}{n\lambda_1 - K}, \\
\gamma_i = 1/(\ell-k), i = k+1,\ldots,\ell-1, \gamma_\ell = 0, \\
\delta_j = 1/(n-\ell+1), j = \ell+1,\ldots,n.
\end{array}
$$

Proof. Let $\beta \leftarrow 2\beta$. The Karush–Kuhn–Tucker conditions for (29.43) yield

$$
\begin{array}{ll}
(a) & \alpha + \beta\lambda_i = 0,\ i = 1,\ldots,k-1, \\
(b) & 1 + \alpha + \beta\lambda_k - \sum_{i=k+1}^{\ell} \gamma_i = 0, \\
(c) & \alpha + \beta\lambda_i + \gamma_i = 0,\ i = k+1,\ldots,\ell-1, \\
(d) & 1 + \alpha + \beta\lambda_\ell - \sum_{j=\ell+1}^{n} \delta_j + \gamma_\ell = 0, \\
(e) & \alpha + \beta\lambda_j + \delta_j = 0,\ j = \ell+1,\ldots,n, \\
(f) & \beta, \gamma_i, \delta_j \geq 0, \beta\left(\sum_1^n \lambda_i^2 - L\right) = 0,\ \forall i,j, \\
(g) & \gamma_i(\lambda_i - \lambda_k) = 0, \delta_j(\lambda_j - \lambda_\ell) = 0,\ \forall i,j.
\end{array} \tag{29.47}
$$

First suppose that $\beta = 0$. If $k > 1$, we get that $\alpha = 0$ and so $\gamma_i = \delta_j = 0$, for all i, j. This contradicts (29.47)(b). If $k = 1$, we get $\alpha = -\delta_j = -\gamma_i$, for all i, j. So if $\alpha \neq 0$, we must have $\lambda_1 = \ldots = \lambda_n = m$. So we can let $k > 1$ and assume that $\beta > 0$. Then we get

$$
\begin{array}{lll}
\lambda_i &= \frac{-\alpha}{\beta}, & i = 1,\ldots,k-1, \\
\lambda_k &= \frac{-1}{\beta} - \frac{\alpha}{\beta} + \sum_{i=k+1}^{\ell} \frac{\gamma_i}{\beta}, & \\
\lambda_i &= \frac{-\alpha}{\beta} - \frac{\gamma_i}{\beta}, & i = k+1,\ldots,\ell-1 \\
\lambda_\ell &= \frac{-1}{\beta} - \frac{\alpha}{\beta} + \frac{\sum_{j=\ell+1}^{n} \delta_j}{\beta} - \frac{\gamma_\ell}{\beta}, & \\
\lambda_j &= \frac{-\alpha}{\beta} - \frac{\delta_j}{\beta}, & j = \ell+1,\ldots,n.
\end{array}
$$

To simplify notation, the index i will now refer to $i = k+1,\ldots,\ell-1$ while the index j will refer to $j = \ell+1,\ldots,n$. Since $\lambda_{i_0} \leq \lambda_k, i_0 = k+1,\ldots,\ell-1$, we get

$$
\sum \gamma_i \geq 1 - \gamma_{i_0}. \tag{29.48}
$$

Therefore, there exists at least one $\gamma_{i_0} > 0$. This implies that $\gamma_{i_0} = \lambda_k$, and

$$
\gamma_{i_0} = 1 - \sum_{i=k+1}^{\ell} \gamma_i.
$$

Now if $\gamma_{i_1} = 0$, then

$$\lambda_{i_1} = \frac{-\alpha}{\beta} > \frac{-\alpha}{\beta} - \frac{\gamma_{i_0}}{\beta} = \lambda_{i_0} = \lambda_k,$$

which is a contradiction. We conclude

$$\lambda_{i_0} = \lambda_k, \ \gamma_{i_0} = 1 - \sum_{i=k+1}^{\ell} \gamma_i, \ i_0 = k+1, \dots, \ell - 1.$$

Note that if $\gamma_\ell = 0$, we get

$$\gamma_i = 1/(\ell - k), i = k+1, \dots, \ell - 1.$$

Similarly, since $\lambda_{j_0} \leq \lambda_\ell, j_0 = \ell + 1, \dots, n$, we get

$$\sum \delta_j - \gamma_\ell \geq 1 - \delta_j,$$

i.e., at least one $\delta_{j_0} > 0$ and so $\lambda_{j_0} = \lambda_\ell$. But if $\delta_{j_1} = 0$, then

$$\lambda_{j_1} = \frac{-\alpha}{\beta} > \frac{-\alpha}{\beta} - \frac{\delta_{j_0}}{\beta} = \lambda_{j_0} = \lambda_\ell,$$

a contradiction. We conclude

$$\lambda_{j_0} = \lambda_\ell, \ \delta_{j_0} = 1 - \sum_{l+1}^{n} \delta_j + \gamma_\ell, \ j_0 = \ell + 1, \dots, n.$$

So that if $\gamma_\ell = 0$, we also have

$$\delta_j = 1/(n - \ell + 1), j = \ell + 1, \dots, n.$$

There now remains two cases to consider:

$$\gamma_\ell = 0 \text{ and } \gamma_\ell > 0.$$

Since $\lambda_k \geq \lambda_\ell$, we must have

$$\sum_{i=k+1}^{\ell-1} \gamma_i + 2\gamma_\ell \geq \sum_{j=\ell+1}^{n} \delta_j.$$

Moreover

$$\lambda_j \leq \lambda_\ell \leq \lambda_k = \lambda_i, \text{ for all } i, j,$$

which implies that

$$\delta_j \geq \gamma_i, \text{ for all } i, j.$$

So that if $\gamma_\ell = 0$, we must have

$$\ell - k - 1 > n - \ell.$$

From the expressions for γ_i, δ_j, we get

$$\begin{aligned} \lambda_k &= \tfrac{-1}{\beta} - \tfrac{\alpha}{\beta} + \tfrac{\ell-k-1}{(\ell-k)\beta}, \\ &= \lambda_i = \tfrac{-\alpha}{\beta} - \tfrac{1}{(\ell-k)\beta}, \qquad i = k+1,\ldots,\ell-1, \\ \lambda_\ell &= \tfrac{-1}{\beta} - \tfrac{\alpha}{\beta} + \tfrac{n-\ell}{(n-\ell+1)\beta}, \\ &= \lambda_j = \tfrac{-\alpha}{\beta} - \tfrac{1}{(n-\ell+1)\beta}, \; j = \ell+1,\ldots,n. \end{aligned}$$

Thus

$$\begin{aligned} \lambda_k &= \lambda_1 - \tfrac{1}{\beta(\ell-k)}, \\ \lambda_\ell &= \lambda_1 - \tfrac{1}{\beta(n-\ell+1)}. \end{aligned} \tag{29.49}$$

After substitution, this yields

$$\frac{-1}{\beta} = \frac{K - n\lambda_1}{2}. \tag{29.50}$$

Since $\beta > 0$, we can apply complementary slackness and substitute for λ_k and λ_ℓ. We get the quadratic

$$(k-1)\lambda_1^2 + (\ell-k)\left(\lambda_1 + \frac{K-n\lambda_1}{2(\ell-k)}\right)^2 + (n-\ell+1)\left(\lambda_1 + \frac{K-n\lambda_1}{2(n-\ell+1)}\right)^2 = L, \tag{29.51}$$

or equivalently

$$\begin{aligned} &\left\{4(\ell-k)(n-\ell+1)(k-1) + (n-l+1)(2(\ell-k)-n)^2 + (l-k)(2(n-\ell+1)-n^2\right\}\lambda_1^2 \\ &\quad +2k\left\{(n-l+1)(2(\ell-k)-n) + (l-k)(2(n-\ell+1)-n)\right\}\lambda_1 \\ &\quad\quad +\left\{(n-l+1)K^2 + (l-k)K^2 - 4(\ell-k)(n-\ell+1)L\right\} = 0 \end{aligned}$$

Note that the above implies

$$\lambda_k + \lambda_\ell = 2\lambda_1 + \frac{K-n\lambda_1}{2}\left(\frac{1}{\ell-k} + \frac{1}{n-\ell+1}\right). \tag{29.52}$$

In the case that $\ell-k-1 < n-\ell$, we get $\gamma_\ell > 0$. Thus, $\lambda_\ell = \lambda_k$ and

$$\lambda_i = \lambda_k, i = k+1,\ldots,n. \tag{29.53}$$

Substitution yields the desired optimal values for λ. Moreover,

$$\gamma_i = \delta_j = 1 - \sum_{t=k+1}^{\ell} \gamma_t = 1 - \sum_{s=\ell+1}^{n} \delta_s + \gamma_\ell.$$

Let $\gamma = \gamma_i$ and $\delta = \delta_j$, then we get

$$\gamma = \delta = 1 - (\ell-k-1)\gamma - \gamma_\ell = 1 - (n-\ell)\delta + \gamma_\ell.$$

This implies

$$\begin{aligned} \gamma &= \delta = 2/(n-k+1) \\ \gamma_\ell &= \tfrac{2(n-\ell+1)}{n-k+1} - 1. \end{aligned} \tag{29.54}$$

Now if $k > 1$, we see that

$$\beta = \frac{\gamma_i}{\lambda_1 - \lambda_i} = 2\left(\frac{k-1}{n-k+1}\right)^{\frac{1}{2}} /(ns).$$

Then

$$\alpha = -\beta\lambda_k - \gamma_i.$$

□

To obtain an upper bound for $\lambda_k - \lambda_\ell$, we consider the program

$$\begin{aligned} &\text{minimize } -\lambda_k + \lambda_\ell \\ &\text{subject to } \sum \lambda_i = K, \\ &\qquad \sum \lambda_i^2 \le L, \\ &\qquad \lambda_k - \lambda_i \le 0, i = 1,\ldots,k-1, \\ &\qquad \lambda_j - \lambda_\ell \le 0, j = \ell+1,\ldots,n. \end{aligned} \tag{29.55}$$

Theorem 29.15. *Suppose that $K^2 < nL$ and $1 < k < \ell < n$. Then the explicit solution to program* (29.55) *is*

$$\begin{aligned} \lambda_1 &= \ldots = \lambda_k = m + \tfrac{1}{2k\beta} \\ \lambda_{k+1} &= \ldots = \lambda_{\ell-1} = m, \\ \lambda_\ell &= \ldots = \lambda_n = m - \tfrac{1}{2(n-\ell+1)\beta}, \end{aligned} \tag{29.56}$$

with Lagrange multipliers for the four sets of constraints being

$$\begin{aligned} \alpha &= -2m\beta, \\ \beta &= \frac{\sqrt{\frac{1}{k} + \frac{1}{n-\ell+1}}}{2\sqrt{ns}}, \\ \gamma_i &= 1/k, i = 1,\ldots,k-1, \\ \delta_j &= 1/(n-\ell+1), j = \ell+1,\ldots,n, \end{aligned} \tag{29.57}$$

respectively.

Proof. The proof is similar to that in Theorem 29.14. Alternatively, sufficiency of the KKT can be used. □

The theorem yields the upper bound

$$\lambda_k - \lambda_\ell \le n^{\frac{1}{2}} s \left(\frac{1}{k} + \frac{1}{n-\ell+1}\right)^{\frac{1}{2}}.$$

29.4 Fractional Programming

We now apply techniques from the theory of *fractional programming* to derive bounds for the *Kantorovich ratio*

$$\frac{\lambda_k - \lambda_\ell}{\lambda_k + \lambda_\ell}. \tag{29.58}$$

This ratio is useful in deriving rates of convergence for the accelerated steepest descent method, e.g., [6].

Consider the *fractional program* (e.g., [10, 11])

$$\max\left\{\frac{f(x)}{g(x)} : x \in \mathscr{F}\right\}. \tag{29.59}$$

If f is concave and g is convex and positive, then $h = \frac{f}{g}$ is a *pseudo-concave* function, i.e., $h : \mathbb{R}^n \to R$ satisfies $(y-x)^t \bigtriangledown h(x) \le 0$ implies $h(y) \le h(x)$. The convex multiplier rules still hold if the objective function is pseudo-convex. We could therefore generate bounds for the ratio (29.58) as was done for λ_k in Section 29.3. However, it is simpler to use the following parametric technique. Let

$$h(q) := \max\{f(x) - qg(x) : x \in \mathscr{F}\}. \tag{29.60}$$

Lemma 29.16. *[2] Suppose that $g(x) > 0$, for all $x \in \mathscr{F}$, and that q is a zero of $h(q)$ with corresponding solution $\bar{x} \in \mathscr{F}$. Then $\bar{x}$ solves* (29.59).

Proof. Suppose not. Then there exists $x \in \mathscr{F}$ such that

$$q = \frac{f(\bar{x})}{g(\bar{x})} < \frac{f(x)}{g(x)},$$

which yields $0 < f(x) - qg(x)$. This contradicts the definition of q. □

We also need the following

Lemma 29.17. *[12] Let $w, \lambda \in \mathbb{R}^n$ be real, nonzero vectors, and let*

$$m = \lambda^T e/n \text{ and } s^2 = \lambda^T C\lambda/n,$$

where e is the $n \times 1$ vector of ones, and the centering matrix $C = I - ee^T/n$. Then

$$-s(nw^T Cw)^{\frac{1}{2}} \le w^T\lambda - mw^T e = w^T C\lambda \le s(nw^T Cw)^{\frac{1}{2}}.$$

Equality holds on the left (resp. right) if and only if

$$\lambda = aw + be$$

for some scalars a and b, where $a \le 0$ (resp. $a \ge 0$).

We now use the above techniques to derive an upper bound for the *Kantorovich ratio* in (29.58). Consider the program

$$\begin{array}{ll} \max & \gamma_{k\ell} = \frac{\lambda_k - \lambda_\ell}{\lambda_k + \lambda_\ell} \\ \text{subject to} & \sum \lambda_i = K \\ & \sum \lambda_i^2 \leq L \\ & \lambda_k - \lambda_i \leq 0, i = 1, \ldots, k-1 \\ & \lambda_i - \lambda_\ell \leq 0, i = \ell+1, \ldots, n. \end{array} \tag{29.61}$$

Theorem 29.18. *Suppose that* $1 < k < \ell < n$, $K^2 < nL$, *and Theorem 29.14 guarantees* $\lambda_k + \lambda_\ell > 0$. *Then the explicit solution to* (29.61) *is*

$$\begin{aligned} \lambda_1 = \ldots = \lambda_k &= \bar{p}\frac{(n-\ell+1+k)-(n-\ell+1)(1-\hat{p}^{\frac{1}{2}})}{k(n-\ell+1+k)} \\ \lambda_{k+1} = \ldots = \lambda_{\ell-1} &= \frac{\operatorname{trace} A^2}{\operatorname{trace} A} \\ \lambda_\ell = \ldots = \lambda_n &= \bar{p}\frac{1-\hat{p}^{\frac{1}{2}}}{n-\ell+1+k}, \end{aligned} \tag{29.62}$$

$$\gamma_{k\ell} = \frac{(p+k)(n-\ell+1-p)^{\frac{1}{2}}(n-\ell+1+k)}{2(p+k)(k(n-\ell+1))^{\frac{1}{2}} + \{(p+k)(n-\ell+1-p)\}^{\frac{1}{2}}(n-\ell+1+k)},$$

where

$$\begin{aligned} p &:= \tfrac{K^2}{L} - (\ell-1) \\ \bar{p} &:= K - (\ell-k-1)\tfrac{L}{K} \\ \hat{p} &:= 1 - \tfrac{k}{n-\ell+1}(n-\ell+1+k)\left(\tfrac{1}{k} + \tfrac{\ell-k-1}{\bar{p}^2}\left(\tfrac{L}{K}\right)^2 - \tfrac{L}{\bar{p}^2}\right). \end{aligned}$$

Proof. Let $\mathscr{F}$ denote the feasible set of (29.61), i.e., the set of $\lambda = (\lambda_i) \in \mathbb{R}^n$ satisfying the constraints. We consider the following parametric program

$$(P_q) \qquad h(q) := \max\{(\lambda_k - \lambda_\ell) - q(\lambda_k + \lambda_\ell) : \ \lambda \in \mathscr{F}\}.$$

Then $h(q)$ is a strictly decreasing function of q and, if λ^* solves (P_q) with $h(q) = 0$, then, by the above Lemma 29.16, λ^* solves the initial program (29.61) also.

The objective function of (P_q) can be rewritten as $\min -(1-q)\lambda_k + (1+q)\lambda_\ell$. The Karush–Kuhn–Tucker conditions for (P_q) now yield:

$$\begin{array}{c} \\ \\ \\ k-th \\ \\ \\ l-th \\ \\ \\ \end{array} \begin{pmatrix} \cdots \\ \cdots \\ \cdots \\ -(1-q) \\ \cdots \\ \cdots \\ 1+q \\ \cdots \\ \cdots \end{pmatrix} + \alpha \begin{pmatrix} 1 \\ \cdots \\ \cdots \\ \cdots \\ \cdots \\ \cdots \\ \cdots \\ \cdots \\ 1 \end{pmatrix} + \beta \begin{pmatrix} \lambda_1 \\ \cdots \\ \cdots \\ \cdots \\ \cdots \\ \cdots \\ \cdots \\ \cdots \\ \lambda_n \end{pmatrix} + \ldots + \begin{pmatrix} \cdots \\ -\delta_i \\ \cdots \\ \\ \cdots \\ \cdots \\ \cdots \\ \cdots \\ \cdots \end{pmatrix} + \begin{pmatrix} \cdots \\ \cdots \\ \cdots \\ \sum_{i=\ell+1}^n \delta_i \\ \cdots \\ \cdots \\ -1 \\ 1 \\ \cdots \end{pmatrix}$$

$$
+\begin{pmatrix} \cdots \\ \cdots \\ \cdots \\ \cdots \\ \cdots \\ \cdots \\ \sum_{i=\ell+1}^{n}\gamma_i \\ \cdots \\ \cdots \end{pmatrix} + \ldots + \ldots \begin{pmatrix} \cdots \\ \cdots \\ \cdots \\ \cdots \\ \cdots \\ \cdots \\ \cdots \\ \gamma_j \\ \cdots \end{pmatrix} = 0
$$

$$
\beta \geq 0; \quad \delta_i \geq 0, \forall i = 1,\ldots,k-1; \quad \gamma_j \geq 0, \forall j = \ell+1,\ldots,n;
$$
$$
\lambda \in \mathscr{F},
$$

$$
\begin{array}{c} \beta\left(\sum \lambda_i^2 - K^2\right) = 0; \quad \delta_i(\lambda_k - \lambda_i) = 0, \forall i = 1,\ldots,k-1; \\ \gamma_j(\lambda_j - \lambda_\ell) = 0, \forall j = \ell+1,\ldots,n. \end{array}
$$

Since $\lambda_k \geq \lambda_\ell$ and we seek q such that $h(q) = 0$, we need only consider $q > 0$. Further, if $\beta = 0$, then we get the following cases:

$$
\begin{array}{rl} k < i < \ell: & 0 = \alpha + \beta\lambda_i \text{ implies } \alpha = 0 \\ i < k: & 0 = \alpha + \beta\lambda_i - \delta_i \text{ implies } \alpha = \delta_i = 0 \\ \ell < i: & 0 = \alpha = \beta\lambda_i + \gamma_i \text{ implies } \alpha = -\gamma_i = 0 \\ i = k: & 0 = -(1-q) + \alpha + \beta\lambda_k + \sum\delta_i \text{ implies } \alpha = -\sum\delta_i + 1 - q \\ \ell = i: & 0 = +(1+q) + \alpha + \beta\lambda_\ell - \sum\delta_i \text{ implies } \alpha = \sum\gamma_i - (1+q). \end{array} \tag{29.63}
$$

These equations are inconsistent. Therefore, we can assume $\beta > 0$, which implies that $\sum \lambda_i^2 = L$.

Now, for $i < k$, either $\lambda_k = \lambda_i$ or $\delta_i = 0$ which implies that $\lambda_i = -\alpha/\beta$. Similarly, for $\ell < i$, $\lambda_\ell = \lambda_i$ or $\lambda_i = -\alpha/\beta$. And, for $k < i < \ell$, $\lambda_i = -\alpha/\beta$. We can therefore see that our solution must satisfy

$$
\begin{array}{l} \lambda_i = \lambda_k, \; i = 1,\ldots,k \\ \lambda_i = \lambda, \;\; i = k+1,\ldots,\ell-1 \\ \lambda_i = \lambda_\ell, \; i = \ell,\ldots,n. \end{array}
$$

Now rather than continuing in this way, we can apply Lemma 29.17. Let $w = (w_i)$, with

$$
\begin{array}{ll} w_i = \frac{1-q}{k}, & i = 1,\ldots,k \\ w_i = 0, & i = k+1,\ldots,\ell-1 \\ w_i = \frac{-(1+q)}{n-\ell+1}, & i = \ell,\ldots,n. \end{array}
$$

Then

$$
\begin{aligned} (1-q)\lambda_k - (1+q)\lambda_\ell &= \tfrac{(1-q)}{k}\textstyle\sum_{i=1}^{k}\lambda_i - \tfrac{(1+q)}{n-\ell+1}\sum_{i=\ell}^{n}\lambda_i = w^T\lambda; \\ m w^T e &= m(1-q-1-q) = -2mq; \\ w^T C w &= w^T I w - \tfrac{1}{n} w^T e e^T w \\ &= \tfrac{(1-q)^2}{k} + \tfrac{(1+q)^2}{n-\ell+1} - \tfrac{1}{n}(4q^2) \end{aligned}
$$

$$nw^T Cw = \tfrac{n(1-q)^2}{k} + \tfrac{n(1+q)^2}{n-\ell+1} - 4q^2.$$

Therefore, Lemma 29.17 yields

$$(1-q)\lambda_k - (1+q)\lambda_\ell \le -2mq + s\left\{\frac{n(1-q)^2}{k} + \frac{n(1+q)^2}{n-\ell+1} - 4q^2\right\}^{\frac{1}{2}} \quad (29.64)$$

with equality if and only if

$$\lambda = aw + be$$

for some scalars a and b with $a \ge 0$.

Therefore the right-hand side of (29.64) equals $h(q)$, the maximum value of (P_q). We now need to find q such that $h(q) = 0$, i.e.,

$$4m^2q^2 = s^2\left\{\frac{n(1-q)^2}{k} + \frac{n(1+q)^2}{n-\ell+1} - 4q^2\right\};$$

$$k(n-\ell+1)4m^2q^2 = (n-\ell+1)s^2n(1-q)^2 + ks^2n(1+q)^2 - k(n-\ell+1)s^2 4q^2;$$

$$\begin{aligned}(-k(n-\ell+1)4m^2 + s^2n(n-\ell+1+k) - k(n-\ell+1)s^2 4)q^2 \\ +2s^2n(-(n-\ell+1)+k)q \\ +s^2n((n-\ell+1)+k) = 0\end{aligned}$$

$$\begin{aligned}[(ns^2 - 4km^2 - 4s^2k)(n-\ell+1) + ns^2k]q^2 + 2ns^2(k-(n-\ell+1))q \\ +ns^2(n-\ell+1+k) = 0\end{aligned}$$

$$q = \frac{-ns^2(k-(n-\ell+1)) - n^2s^4(k-(n-\ell+1))^2 - [\text{as above}]ns^2(n-\ell+1+k)^{\frac{1}{2}}}{[\text{as above}]}.$$

We have chosen the negative radical for the root, since the quantity in [ ] is negative and we need $q > 0$. The conditions for equality in (29.64) yield:

$$\begin{aligned}\lambda_i &= a\tfrac{(1-q)}{k} + b, i = 1,\dots,k \\ \lambda_i &= b, i = k+1,\dots,\ell-1 \\ \lambda_i &= \tfrac{-a(1+q)}{n-\ell+1} + b, i = \ell,\dots,n\end{aligned}$$

or

$$\begin{aligned}\frac{\lambda_k - \lambda_\ell}{\lambda_k + \lambda_\ell} &= \frac{a\left(\frac{1-q}{k}\right) + b + \frac{a(1-q)}{n-\ell+1} - b}{\frac{a(1-q)}{k} - \frac{a(1+q)}{n-\ell+1} + 2b} \\ &= \frac{a(1-q)(n-\ell+1) + ak(1+q)}{a(n-\ell+1)(1-q) - a(1+q)k + 2bk(n-\ell+1)}.\end{aligned}$$

We now solve for a and b by substituting for λ_i in $\sum \lambda_i = K$ and $\sum \lambda_i^2 = L$:

$$k\left(\frac{a(1-q)}{k}\right)+b+(\ell-k-1)b+(n-\ell+1)\left(\frac{-a(1+q)}{n-\ell+1}\right)+b=K$$

or

$$(k+\ell-k-1+n-\ell+1)b=K-a(1-q)+a(1+q)$$
$$b=\frac{2aq+K}{n}.$$

And

$$k\left(\frac{a(1-q)}{k}\right)+b^2+(\ell-k-1)b^2+(n-\ell-1)\left(\frac{-a(1+q)}{n-\ell+1}+b^2\right)=K^2$$

or

$$[k+\ell-k-1+n-\ell+1]b^2+[2a(1-q)-2a(1+q)]b+\frac{a^2(1-q)^2}{k}+$$
$$\frac{a^2(1+q)^2}{n-\ell+1}-K=0$$

$$k\left(a\frac{1-q}{k}+\frac{2aq+K}{n}\right)^2+(\ell-k-1)\left(\frac{2aq+K}{n}\right)^2+(n-\ell+1)$$
$$\left(\frac{-a(1+q)^2}{n-\ell+1}+\frac{2aq+K}{n}\right)-K=0$$

$$k\left(a\left(\frac{1-q}{k}+\frac{2q}{n}\right)+\frac{K}{n}\right)^2+(\ell-k-1)\left(\frac{2aq}{n}+\frac{K}{n}\right)^2$$
$$+(n-\ell+1)\left(+a\left(\frac{-(1-q)}{n-\ell+1}+\frac{2q}{n}\right)+\frac{K}{n}\right)^2-L=0;$$

$$[k\left(\frac{1-q}{k}+\frac{2q}{n}\right)^2+(\ell-k-1)\frac{4q^2}{n^2}+(n-\ell+1)\left(\frac{-(1-q)}{n-\ell+1}+\frac{2q}{n}\right)^2]a^2$$
$$+a2[k\left(\frac{1-q}{k}+\frac{2q}{n}\right)\frac{K}{n}+(\ell-k-1)\frac{2q}{n}\frac{K}{n}+(n-\ell+1)\left(\frac{2q}{n}-\frac{(1+q)^2}{n-\ell+1}\right)\frac{K}{n}]$$
$$+k\left(\frac{K}{n}\right)^2+(\ell-k-1)\left(\frac{K}{n}\right)^2+(n-\ell+1)\left(\frac{K}{n}\right)^2-L=0$$
$$\left[\frac{-4q^2}{n}+\frac{(1-q)^2}{k}+\frac{(1-q)^2}{n-\ell+1}\right]a^2+nm^2-L=0$$
$$a=\frac{-nm^2+L}{\frac{-4q^2}{n}+\frac{(1-q)^2}{k}+\frac{(1+q)^2}{(n-\ell+1)}}^{\frac{1}{2}}.$$

Substitution for the λ_i yields the desired results. □

Let $\gamma = \lambda_k/\lambda_\ell$. Then

$$\gamma_{k\ell} = \frac{\gamma - 1}{\gamma + 1}$$

and

$$\frac{d_{\gamma_{k\ell}}}{d_\gamma} = \frac{2}{(\gamma+1)^2} > 0.$$

Thus γ is isotonic to $\gamma_{k\ell}$. This yields an upper bound to $\gamma_{k\ell}$, see [13]: If (to guarantee $\lambda_\ell > 0$) we have $(\ell - 1)L < L$, then

$$\frac{\lambda_k}{\lambda_\ell} \leq \frac{c + k + \left\{\frac{n-\ell+1}{k}(c+k)(n-\ell+1-c)\right\}^{\frac{1}{2}}}{c + k - \left\{\frac{k}{n-\ell+1}(c+k)(n-\ell+1-c)\right\}^{\frac{1}{2}}},$$

where

$$c = \frac{(K)^2}{L} - (\ell - 1).$$

(These inequalities are also given in [7].) Note that

$$\frac{\gamma+1}{\gamma-1} = \frac{\lambda_k + \lambda_\ell}{\lambda_k - \lambda_\ell},$$

is reverse isotonic to γ. Thus we can derive a lower bound for this ratio.

29.5 Conclusion

We have used optimization techniques to derive bounds for functions of the eigenvalues of an $n \times n$ matrix A with real eigenvalues. By varying both the function to be minimized (maximized) and the constraints of a properly formulated program, we have been able to derive bounds for the k-th largest eigenvalue, as well as for sums, differences and ratios of eigenvalues. Additional information about the eigenvalues was introduced to improve the bounds using the shadow prices of the program. Many more different variations remain to be tried.

The results obtained are actually about ordered sets of numbers $\lambda_1 \geq \ldots \geq \lambda_n$ and do not depend on the fact that these numbers are the eigenvalues of a matrix. We can use this to extend the bounds to complex eigenvalues. The constraints on the traces can be replaced by

$$\sum v_i = \operatorname{trace} T, \sum (v_i)^2 \leq \operatorname{trace} T^*T,$$

where v_i can take on the real, imaginary, and modulus of the eigenvalues λ_i, and the matrix T can become $(A + A^*)/2$, $(A - A^*)/2i$. Further improvements can be made by using improvements of the Schur inequality $\sum(v_i)^2 \leq \operatorname{trace} T^*T$. This approach is presented in [12] and [13].

Acknowledgment Research supported by The Natural Sciences and Engineering Research Council of Canada. The author thanks Wai Lung Yeung for his help in correcting many statements in the paper.

References

1. A. Clausing. Kantorovich-type inequalities. *Amer. Math. Monthly*, 89(5):314, 327–330, 1982.
2. W. Dinkelbach. On nonlinear fractional programming. *Management Sci.*, 13:492–498, 1967.
3. L.V. Kantorovich and G.P. Akilov. *Functional analysis*. Pergamon Press, Oxford, second edition, 1982. Translated from the Russian by Howard L. Silcock.
4. R. Kumar. Bounds for eigenvalues. Master's thesis, University of Alberta, 1984.
5. D.G. Luenberger. *Optimization by Vector Space Methods*. John Wiley, 1969.
6. D.G. Luenberger. Algorithmic analysis in constrained optimization. In *Nonlinear programming (Proc. Sympos., New York, 1975)*, pages 39–51. SIAM–AMS Proc., Vol. IX, Providence, R. I., 1976. Amer. Math. Soc.
7. J. Merikoski, G.P.H. Styan, and H. Wolkowicz. Bounds for ratios of eigenvalues using traces. *Linear Algebra Appl.*, 55:105–124, 1983.
8. B.H. Pourciau. Modern multiplier methods. *American Mathematical Monthly*, 87(6):433–451, 1980.
9. R.T. Rockafellar. *Convex analysis*. Princeton Landmarks in Mathematics. Princeton University Press, Princeton, NJ, 1997. Reprint of the 1970 original, Princeton Paperbacks.
10. S. Schaible. Fractional programming—state of the art. In *Operational research '81 (Hamburg, 1981)*, pages 479–493. North-Holland, Amsterdam, 1981.
11. S. Schaible. Fractional programming. In *Handbook of global optimization*, volume 2 of *Nonconvex Optim. Appl.*, pages 495–608. Kluwer Acad. Publ., Dordrecht, 1995.
12. H. Wolkowicz and G.P.H. Styan. Bounds for eigenvalues using traces. *Linear Algebra Appl.*, 29:471–506, 1980.
13. H. Wolkowicz and G.P.H Styan. More bounds for eigenvalues using traces. *Linear Algebra Appl.*, 31:1–17, 1980.

MIX
Papier aus verantwortungsvollen Quellen
Paper from responsible sources
FSC® C105338

If you have any concerns about our products,
you can contact us on
ProductSafety@springernature.com

In case Publisher is established outside the EU,
the EU authorized representative is:
Springer Nature Customer Service Center GmbH
Europaplatz 3, 69115 Heidelberg, Germany

Printed by Libri Plureos GmbH
in Hamburg, Germany